T0236145

Lecture Notes in Computer Science 8596

Commenced Publication in 1973
Founding and Former Series Editors:
Gerhard Goos, Juris Hartmanis, and Jan van Leeuwen

Editorial Board

More information about this series at http://www.springer.com/series/7407

Pierre Fouilhoux · Luis Eduardo Neves Gouveia
A. Ridha Mahjoub · Vangelis T. Paschos (Eds.)

Combinatorial Optimization

Third International Symposium, ISCO 2014
Lisbon, Portugal, March 5–7, 2014
Revised Selected Papers

Springer

Editors
Pierre Fouilhoux
LIP6
Université Pierre et Marie Curie
Paris
France

Luis Eduardo Neves Gouveia
Universidade de Lisboa
Lisbon
Portugal

A. Ridha Mahjoub
Vangelis T. Paschos
LAMSADE
Université Paris-Dauphine
Paris Cedex 16
France

ISSN 0302-9743 ISSN 1611-3349 (electronic)
ISBN 978-3-319-09173-0 ISBN 978-3-319-09174-7 (eBook)
DOI 10.1007/978-3-319-09174-7

Library of Congress Control Number: 2014945216

LNCS Sublibrary SL1 – Theoretical Computer Science and General Issues

Springer Cham Heidelberg New York Dordrecht London

Springer is part of Springer Science+Business Media (www.springer.com)

Preface

This volume contains the regular papers presented at ISCO 2014, the Third International Symposium on Combinatorial Optimization, held in Lisbon during March 5–7, 2014. ISCO 2014 was preceded by the Spring School on "Stochastic Programs with Integer Variables: Theory, Algorithms and Applications" given by Rüdiger Schultz and Jonas Schweiger. ISCO is a new biennial symposium. The first edition was held in Hammamet, Tunisia, in March 2010, and the second in Athens, Greece, in April 2012. The symposium aims to bring together researchers from all the communities related to combinatorial optimization, including algorithms and complexity, mathematical programming, operations research, stochastic optimization, graphs, and combinatorics. It is intended to be a forum for presenting original research on all aspects of combinatorial optimization, ranging from mathematical foundations and theory of algorithms to computational studies and practical applications, and especially their intersections.

In response to the call for papers, ISCO 2014 received 97 regular submissions. Each submission was reviewed by at least two Program Committee (PC) members with the assistance of external reviewers. The submissions were judged on their originality and technical quality and the PC had to discuss in length the reviews and make tough decisions. As a result, the PC selected 37 regular papers to be presented in the symposium giving an acceptance rate of 38 % (64 short papers were also selected from both regular and short submissions). Three eminent invited speakers, Michel Balinski (CNRS and Ecole Polytechnique, Paris), Martin Grötshel (Zuse Institute Berlin), and Matteo Fischetti (Badova University), gave talks at the symposium. The revised versions of the accepted regular papers and extended abstracts of the invited talks are included in this volume.

We would like to thank all the authors who submitted their work to ISCO 2014, and the PC members and external reviewers for their excellent work. We would also like to thank our invited speakers as well as the speakers of the Spring School for their exciting lectures. They all greatly contributed to the quality of the symposium.

Finally, we would like to thank the Organizing Committee members for their dedicated work in preparing this conference, and we gratefully acknowledge our sponsoring institutions for their assistance and support.

April 2014

Pierre Fouilhoux
Luis Eduardo Neves Gouveia
A. Ridha Mahjoub
Vangelis T. Paschos

Organization

ISCO 2014 was organized by the "Centro de Investigao Operacional" (Operations Research Center) of the Faculty of Sciences of the University of Lisbon, in cooperation with LAMSADE Laboratory of Université Paris-Dauphine.

Program Committee Co-chairs

Luis Eduardo Neves Gouveia   CIO, University of Lisbon, Portugal
A. Ridha Mahjoub   Paris Dauphine University, France

Program Committee

Agostinho Agra	University of Aveiro, Portugal
Mourad Baiou	LIMOS, CNRS, University Blaise-Pascal, Clermont-Ferrand, France
Francisco Barahona	IBM T.J. Watson Research Center, New York, USA
Domingos Cardoso	University of Aveiro, Portugal
Miguel Fragoso Constantino	CIO, University of Lisbon, Portugal
Ángel Corberán	University of Valencia, Spain
José Correa	University of Chile, Chile
Bernard Fortz	Université Libre de Bruxelles, Belgium
Pierre Fouilhoux	Université Pierre et Marie Curie, France
Satoru Fujishige	Kyoto University, Japan
Bernard Gendron	CIRRELT, University of Montréal, Canada
Oktay Gunluk	IBM, New York, USA
Mohamed Haouari	Qatar University, Quatar
Brahim Hnich	Izmir University of Economics, Turkey
Martine Labbé	Université Libre de Bruxelles, Belgium
Leo Liberti	IBM, New York, USA
Abdel Lisser	University of Paris-Sud, France
Ivana Ljubic	University of Vienna, Austria
Andrea Lodi	University of Bologna, Italy
Abílio Lucena	Federal University of Rio de Janeiro, Brazil
Nelson Maculan	Federal University of Rio de Janeiro, Brazil
Isabel Méndez-Diaz	University of Buenos Aires, Argentina
Ioannis Milis	Athens University of Economics and Business, Greece
Jerome Monnot	Paris Dauphine University, France
Adam Ouorou	Orange Labs, Paris, France
Vangelis Paschos	Paris Dauphine University, France

Pierre Pesneau	University of Bordeaux, France
Giovanni Rinaldi	IASI, Rome, Italy
Juan Jose Salazar Gonzalez	University of La Laguna, Spain
Rüdiger Schultz	University of Duisburg-Essen, German
Maria Grazia Scutella	University of Pisa, Italy
Cid de Sousa	University of Campinas, Brazil
Eduardo Uchoa	UFF Rio de Janeiro, Brazil
Francois Vanderbeck	University of Bordeaux, France
Hande Yaman	Bikent University, Turkey

Additional Reviewers

Marek Adamczyk	Augusto Eusbio	Silvano Martello
Pablo Adasme	Lionel Eyraud-Dubois	Alfredo Marn
Amirali Ahmadi	Javier Faulin	Marta Mesquita
Laurent Alfandari	Rosa Figueiredo	Philippe Michelon
Aysegul Altin	Samuel Fiorini	Juan Jose Miranda Bront
Eduardo Alvarez-Miranda	Alexandre Freire	Stefano Moretti
Milica Andelic	Yuri Frota	Pedro Moura
Miguel Anjos	Takuro Fukunaga	Cëcile Murat
Antonios Antoniadis	Fabio Furini	Fernando Ordonez
Cesar Beltran-Royo	Aristotelis Giannakos	Aris Pagourtzis
Andreas Bley	Kristiaan Glorie	Ana Paias
Christian Bliek	Laurent Gourves	Aline Parreau
Edouard Bonnet	Vincent Guigues	Fanny Pascual
Laurent Bulteau	Magnus M. Halldorsson	Arnaud Pecher
Paola Cappanera	Jinil Han	Dilson Lucas Pereira
Jean Cardinal	Said Hanafi	Raffaele Pesenti
Paula Carvalho	Hiplito Hernndez-Prez	Ulrich Pferschy
Jordi Castro	Olivier Hudry	Daniel Porumbel
Daniele Catanzaro	Andrew Hughes	Michael Poss
Andre Cire	Philipp Hungerlaender	Lionel Pournin
Franois Clautiaux	Imed Kacem	Pablo Pérez-Lantero
Andrea Clementi	Enver Kayaaslan	Maurice Queyranne
Joo Clmaco	Sandi Klavzar	Steffen Rebennack
Johanne Cohen	Alexander Kononov	Cristina Requejo
Denis Cornaz	Carlile Lavor	Mauricio Resende
Pedro Cruz	Markus Leitner	Jorge Riera-Ledesma
Alexandre Cunha	Dimitrios Letsios	Bernard Ries
Amaro de Sousa	Janny Leung	Inmaculada
Paolo Detti	Roel Leus	Rodriguez-Martin
Ibrahima Diarrassouba	Carlos Luz	Mario Ruthmair
Thang Dinh	Jens Lysgaard	Ignaz Rutter
Charles Dominic	Enrico Malaguti	Ruslan Sadykov
Emrah Edis	Javier Marenco	Saket Saurabh

Marco Senatore
Luidi Simonetti
Markus Sinnl
Georgios Stamoulis
Leen Stougie
Anand Subramanian
Kenjiro Takazawa

Raouia Taktak
Orestis Telelis
Claudio Telha
Lydia Tlilane
Artur Tomaszewski
Sebastián Urrutia
Paolo Ventura

Gustavo Vulcano
Annegret Wagler
Bang Ye Wu
Eduardo Xavier
Georgios Zois

Organizing Committee

Maria Conceição Fonseca   CIO, University of Lisbon, Portugal
Rodrigo Marques   CIO, University of Lisbon, Portugal
Pedro Moura   CIO, University of Lisbon, Portugal
Ana Paias   CIO, University of Lisbon, Portugal

Sponsoring Institutions

Associação Portuguesa de Investigação Operacional
Centro de Investigação Operacional, Faculdade de Ciências da Universidade de
    Lisboa, Portugal
Fundaçao pàra a Ciência e a Tecnologia, Portugal
Instituto Nacional de Estatística, Portugal
LAMSADE, Université Paris-Dauphine, France

Invited Talks

Judge: Don't Vote!

Michel Balinski

CNRS and Ecole Polytechnique Paris, France

This talk argues that the traditional methods of voting and judging contestants (e.g., figure skaters, movies, wines, beauty queens, political candidates) fail in both theory and practice and should be replaced by a new method, "majority judgment." Majority judgment best meets five essential properties: (1) It avoids the Condorcet and Arrow paradoxes, (2) it elicits honest voting, (3) it is meaningful (in the sense of measurement theory), (4) it resists manipulation, and (5) it heeds the majority's will.

BRANCHstorming
(Brainstorming About Tree Search)

Matteo Fischetti

Padova University, Italy

Quoting from Wikipedia (http://en.wikipedia.org/wiki/System_dynamics): "System Dynamics is an aspect of systems theory as a method for understanding the dynamic behavior of complex systems. The basis of the method is the recognition that the structure of any system the many circular, interlocking, sometimes time-delayed relationships among its components is often just as important in determining its behavior as the individual components themselves. Examples are chaos theory and social dynamics. It is also claimed that because there are often properties-of-the-whole which cannot be found among the properties-of-the-elements, in some cases the behavior of the whole cannot be explained in terms of the behavior of the parts."

No doubts that tree search is a very complex process with its own dynamics, that sometimes behaves as a chaotic system due to its high dependency on the initial conditions.

However, tree search is seldom studied as a whole by the Mathematical Programming community, perhaps because it is often perceived as a shame—we should be able to solve our problems at the root node, don't we? As a matter of fact, its main ingredients (e.g., cut generation and selection) are often studied in vitro—i.e., evaluated "at the root node"—and then just transplanted in the enumeration body with significant organ-rejection rates.

The study of the properties-of-the-whole of tree search is of course a long-term project. In this talk we will make a mandatory preliminary step by addressing a number of preconceptions about it.

We will start reviewing recent work on the role of erraticism in the design and validation of tree-search algorithms, thus addressing the prejudice that enumeration is a stable mechanism whose performance only depends on how clever we are in designing its single elements.

We will then address a main source of erraticism in a branch-and-bound scheme, namely, the existence of multiple relaxation solutions. In particular, we will comment about the risk of overfitting due to the common practice of uncritically picking one such solution (or just few) for guiding the search.

Routing Problems: Standard and Unusual Cases

Martin Grötschel

Zuse Institute Berlin, Germany

Shortest path, Chinese postman and symmetric travelling salesman problems are combinatorial optimization problems with a rich theory. They have undergone extensive computational studies and can be viewed as "solved" for the majority of their practical applications. However, most routing problems are not so nicely structured. They often come with various combinations of side constraints such as capacity, depot and ordering constraints as well as time windows, with online or real-time requirements and possibly multiple objective functions. Such routing problems are notoriously difficult and a typical playground for heuristics.

In the last 30 years my research group has covered a large variety of routing problems in public transport, logistics, general transportation, machine and emergency scheduling, etc.

I plan to give a broad survey on these problems as well as on successful solution approaches, and I will concentrate on particular cases that we are currently working on. These include train scheduling (high-speed trains ICE in Germany with uncommon "regularity requirements") and a quite unusual routing problem where vehicles have to be routed "optimally" to catch trucks on the German Autobahn that try to avoid the payment of road tolls. Needless to say, that these inspection vehicles have to satisfy several nonstandard legal requirements.

Contents

Maximum Throughput Network Routing Subject to Fair Flow Allocation . . . 1
Edoardo Amaldi, Stefano Coniglio, and Leonardo Taccari

Study of Identifying Code Polyhedra for Some Families of Split Graphs. . . . 13
Gabriela Argiroffo, Silvia Bianchi, and Annegret Wagler

Parametric Multiroute Flow and Its Application to Robust Network
with k Edge Failures. 26
*Jean-François Baffier, Vorapong Suppakitpaisarn, Hidefumi Hiraishi,
and Hiroshi Imai*

The Dominating Set Polytope via Facility Location 38
Mourad Baïou and Francisco Barahona

Solving Graph Partitioning Problems Arising in Tagless Cache Management . . . 50
Sandro Bartolini, Iacopo Casini, and Paolo Detti

Lagrangean Decomposition for Mean-Variance Combinatorial Optimization . . . 62
Frank Baumann, Christoph Buchheim, and Anna Ilyina

Maximum Generalized Assignment with Convex Costs. 75
Marco Bender and Stephan Westphal

An Integer Programming Formulation for the Maximum k-Subset
Intersection Problem 87
*Eduardo T. Bogue, Cid C. de Souza, Eduardo C. Xavier,
and Alexandre S. Freire*

b-Coloring is NP-Hard on Co-Bipartite Graphs
and Polytime Solvable on Tree-Cographs 100
Flavia Bonomo, Oliver Schaudt, Maya Stein, and Mario Valencia-Pabon

Proactive Reactive Scheduling in Resource Constrained Projects
with Flexibility and Quality Robustness Requirements 112
Mario Brčić, Damir Kalpić, and Marija Katić

Active Set Methods with Reoptimization for Convex Quadratic Integer
Programming 125
Christoph Buchheim and Long Trieu

Fixed-Parameter Algorithms for Scaffold Filling 137
Laurent Bulteau, Anna Paola Carrieri, and Riccardo Dondi

Finding Totally Independent Spanning Trees with Linear
Integer Programming.. 149
   Alexandre Salles da Cunha and Fernanda Sumika Hojo de Souza

Coupled-Tasks in Presence of Bipartite Compatibilities Graphs 161
   *Benoit Darties, Gilles Simonin, Rodolphe Giroudeau,
and Jean-Claude König*

The Computational Complexity of Stochastic Optimization 173
   Cassio Polpo de Campos, Georgios Stamoulis, and Dennis Weyland

A Hybrid Heuristic Approach Based on a Quadratic Knapsack Formulation
for the Max-Mean Dispersion Problem 186
   Federico Della Croce, Michele Garraffa, and Fabio Salassa

A Constraint Generation Approach for the Two-Machine Flow Shop Problem
with Jobs Selection 198
   Federico Della Croce, Christos Koulamas, and Vincent T'kindt

Rectilinear Shortest Path and Rectilinear Minimum Spanning
Tree with Neighborhoods.. 208
   Yann Disser, Matúš Mihalák, Sandro Montanari, and Peter Widmayer

Lovász and Schrijver N_+-Relaxation on Web Graphs 221
   Mariana Escalante and Graciela Nasini

The Envy-Free Pricing Problem and Unit-Demand Markets 230
   *Cristina G. Fernandes, Carlos E. Ferreira, Álvaro J.P. Franco,
and Rafael C.S. Schouery*

Mathematical Programming Models for Traffic Engineering in Ethernet
Networks Implementing the Multiple Spanning Tree Protocol 242
   Bernard Fortz, Luís Gouveia, and Martim Moniz

Graph Compact Orthogonal Layout Algorithm...................... 255
   Kārlis Freivalds and Jans Glagoļevs

State Space Reduced Dynamic Programming for the Aircraft Sequencing
Problem with Constrained Position Shifting........................ 267
   Fabio Furini, Martin Philip Kidd, Carlo Alfredo Persiani, and Paolo Toth

Decomposition Algorithm for the Single Machine Scheduling Polytope 280
   Ruben Hoeksma, Bodo Manthey, and Marc Uetz

Subexponential Fixed-Parameter Algorithms for Partial Vector Domination ... 292
   Toshimasa Ishii, Hirotaka Ono, and Yushi Uno

Efficient Approximation Schemes for the Maximum Lateness Minimization
on a Single Machine with a Fixed Operator or Machine
Non-Availability Interval.. 305
  Imed Kacem, Hans Kellerer, and Maryam Seifaddini

A Multi-period Bi-level Stochastic Programming with Decision
Dependent Uncertainty in Supply Chains.......................... 315
  Yohanes Kristianto

{k}-Packing Functions of Graphs................................... 325
  Valeria Alejandra Leoni and Erica G. Hinrichsen

Robust Shift Scheduling in Call Centers 336
  Sara Mattia, Fabrizio Rossi, Mara Servilio, and Stefano Smriglio

A Tabu Search Heuristic for the Equitable Coloring Problem 347
  Isabel Méndez Díaz, Graciela Nasini, and Daniel Severín

Linear Arrangement Problems and Interval Graphs.................. 359
  Alain Quilliot and Djamal Rebaine

On the Asymmetric Connected Facility Location Polytope 371
  Markus Leitner, Ivana Ljubić, Juan-José Salazar-González,
  and Markus Sinnl

Heuristic Approaches for the Robust Vehicle Routing Problem 384
  Elyn L. Solano-Charris, Christian Prins, and Andréa Cynthia Santos

A Fast Large Neighborhood Search for Disjunctively Constrained
Knapsack Problems.. 396
  Mhand Hifi, Sagvan Saleh, and Lei Wu

Approximating the k-Set Packing Problem by Local Improvements 408
  Martin Fürer and Huiwen Yu

Multi-Objective Cuckoo Search with Leader Selection Strategies 421
  Kamel Zeltni and Souham Meshoul

Vulnerability Assessment of Spatial Networks: Models and Solutions...... 433
  Eduardo Álvarez-Miranda, Alfredo Candia-Véjar, Emilio Carrizosa,
  and Francisco Pérez-Galarce

Author Index 445

Maximum Throughput Network Routing Subject to Fair Flow Allocation

Edoardo Amaldi[1], Stefano Coniglio[2], and Leonardo Taccari[1]($\boxtimes$)

[1] Dipartimento di Elettronica, Informazione e Bioingegneria,
Politecnico di Milano, Milan, Italy
{edoardo.amaldi,leonardo.taccari}@polimi.it
[2] Lehrstuhl II Für Mathematik, RWTH Aachen University, Aachen, Germany
coniglio@math2.rwth-aachen.de

Abstract. We investigate a bilevel network routing problem where, given a directed graph with a capacity for each arc and a set of elastic traffic demands specified by the corresponding origin-destination pairs, the network operator has to select a single path for each pair so as to maximize the total throughput while assuming that the flows are allocated over the chosen paths according to a *fairness principle*. We consider max-min fair flow allocation as well as maximum bottleneck flow allocation. After presenting a complexity result, we discuss MILP formulations for the two problem versions, describe a Branch-and-Price algorithm and report some computational results.

Keywords: Networks · Routing · Fairness · Computational complexity · Integer programming

1 Introduction

Network routing problems with elastic traffic demands (specified by an origin-destination pair without a prescribed flow value) and fair allocation of flows have been attracting a growing attention. Our original motivation arises from best-effort service in Internet Protocol (IP) networks [2] where, for instance, several users simultaneously download data between different hosts with no guaranteed rate, but wish to do so as fast as possible. Although the IP network operator, which aims at maximizing a utility function such as the total throughput, can select the routing paths, it has no direct control over the transport protocol (TCP). In this setting, the flow of each origin-destination pair is adapted by TCP based on the available capacity (which depends on the current traffic load) and the distributed congestion control mechanism is expected to allocate the flows in a fair way, that is, without privileging any user.

Consider a capacitated network defined by a directed graph $G = (V, A)$ with a capacity c_{ij} for each arc $(i, j) \in A$ and a set K of k origin-destination pairs (s, t). Let $\phi \in \mathbb{R}^k$ denote a flow vector whose i-th component ϕ_i corresponds to the flow allocated to the i-th origin-destination pair.

© Springer International Publishing Switzerland 2014
P. Fouilhoux et al. (Eds.): ISCO 2014, LNCS 8596, pp. 1–12, 2014.
DOI: 10.1007/978-3-319-09174-7_1

A widely used notion of fairness in networks is that of Max-Min Fairness (MMF). Indeed, in IP networks common congestion avoidance mechanisms aim at realizing a max-min fair allocation of the flows over the routing paths provided by the IP layer, see e.g. [11] and the references therein.

Definition 1. *Let σ be the sorting operator permuting the components of $\underline{\phi}$ in nondecreasing order, i.e., such that $\sigma(\underline{\phi})_i \leq \sigma(\underline{\phi})_j$ whenever $i < j$. A flow vector $\underline{\phi} \in \mathbb{R}^k$ is* Max-Min Fair *(MMF) if, for any other flow vector $\underline{\phi}' \in \mathbb{R}^k$, $\sigma(\underline{\phi})$ lexicographically dominates $\sigma(\underline{\phi}')$, i.e., either $\sigma(\underline{\phi}) = \sigma(\underline{\phi}')$ or there exists an integer ℓ, with $1 \leq \ell \leq k$, such that $\sigma(\underline{\phi})_q = \sigma(\underline{\phi}')_q$ for all $q < \ell$ and $\sigma(\underline{\phi})_\ell > \sigma(\underline{\phi}')_\ell$.*

If the routing paths are given, the set of feasible flows is convex and equivalently a flow vector is MMF if and only if there is no way to increase the flow of any origin-destination pair without decreasing the flow of a pair with an equal or smaller flow.

When restricting the attention to the first element $\sigma(\underline{\phi})_1$ of the lexicographically sorted allocation vector $\sigma(\underline{\phi})$, we obtain the simpler and relaxed fairness criterion where $\min_{(s,t) \in K} \phi_{st}$ is maximized.

Definition 2. *A flow vector $\underline{\phi} \in \mathbb{R}^k$ is* Max-Bottleneck *(MB) if, for any other flow vector $\underline{\phi}' \in \mathbb{R}^k$, $\sigma(\underline{\phi})_1 \geq \sigma(\underline{\phi}')_1$.*

Along the lines of [2,3], in this work we consider the Max-Throughput Single-Path Network Routing problem subject to fair flow allocation.

MT-SPNR-Fair: *Given a directed graph $G = (V, A)$ with capacities c_{ij} and a set of k origin-destination pairs, select a single routing path for each pair so as to maximize the total throughput $\sum_{i=1}^k \phi_i$, subject to the constraint that the amount of flow allocated to the origin-destination pairs is fair w.r.t. the chosen paths.*

Depending on the adopted notion of fairness, the problem version is referred to as MT-SPNR-MMF or MT-SPNR-MB.

MT-SPNR-fair is a bilevel problem where, at the upper level, the leader (e.g., the network operator) selects a routing path for each origin-destination pair and, at the lower level, the follower (e.g., the TCP protocol) allocates the flows to the chosen paths according to the MMF or MB principle. For each origin-destination pair $(s, t) \in K$, let P^{st} be the set of all the simple paths connecting s to t. Let the binary variable λ_p^{st} be 1 if the path of index p is selected for the pair (s, t) and 0 otherwise, and let $\underline{\phi} \in \mathbb{R}_+^k$ be the vector of flow allocations. Denote by $\Lambda = \{\lambda_p^{st} \in \{0, 1\}, (s, t) \in K, p \in P^{st} : \sum_{p \in P^{st}} \lambda_p^{st} = 1\}$ the feasible region of the leader, namely, the set of all vectors $\underline{\lambda}$ with exactly one path for each origin-destination pair (s, t). For any given choice of paths prescribed by the vector $\underline{\lambda}$, the feasible region $\Phi(\underline{\lambda})$ for the follower amounts to the set of all vectors $\underline{\phi}$ of flows that can be allocated (according to the considered fairness principle) along those paths without exceeding the arc capacities.

A high-level formulation for MT-SPNR-MMF is:

$$\max_{\underline{\phi}, \underline{\lambda}} \ \underline{1}^T \underline{\phi} \tag{1}$$

$$\text{s.t.} \ \underline{\lambda} \in \Lambda \tag{2}$$

$$\underline{\phi} \in \operatorname*{arglexmax}_{\underline{\phi} \in \Phi(\underline{\lambda})} \left\{ \sigma(\underline{\phi}) \right\}, \tag{3}$$

where $\underline{1}$ is the all-one vector and arglexmax is used as in Definition 1. The formulation for the case of a Max-Bottleneck flow allocation, MT-SPNR-MB, is obtained by substituting Constraint (3) with:

$$\underline{\phi} \in \operatorname*{argmax}_{\underline{\phi} \in \Phi(\underline{\lambda})} \left\{ \min_{(s,t) \in K} \left\{ \phi_{st} \right\} \right\}. \tag{4}$$

2   Related Work

A thorough treatment on fairness in network routing and design can be found in [15]. For a good introduction to Max-Min Fairness, see the tutorial [13].

So far, most of the attention has been devoted to the problem of finding a solution which is "as fair as possible", where fairness is the problem objective. If the routing paths are given, a simple polynomial algorithm, known as *water (or progressive) filling*, allows to allocate the flows in an MMF way, as explained in Chap. 6 of [5]. If the routing paths are not known *a priori*, the problem consists in determining a network routing such that the flow allocation is as (max-min) fair as possible. An important distinction is between the cases with unsplittable or splittable routing, that is where the flow between each origin-destination pair can be routed over one or many (not necessarily disjoint) paths. In the splittable case, the problem can be solved via a sequence of at most k Linear Programming (LP) problems, one per origin-destination pair [14,18,20]. In the unsplittable case, it can be solved via a sequence of at most k Mixed Integer Linear Programming (MILP) problems with binary variables [17].

Following the work of [12], where the MMF (splittable) single-source routing problem is originally introduced, the authors of [10] provide hardness results and a 2-approximation scheme for the extension of the problem where the unsplittable fairest routing is sought. Note that the results of [10,12] are, in general, only valid for the case of a single-source or a single-sink. A game theoretic approach to MMF routing can be found in [9], where routing games that converge to a fair equilibrium are considered.

Different definitions of fairness have also be investigated. In [19], an approach similar to that of [14,18,20] is adopted to derive fair flow allocations balancing fairness and efficiency (utility). In [1], the splittable case with a weighted MMF criterion is considered. The authors of [6] propose a relaxation of the notion of MMF with the definition of Upward Max-Min Fairness, whereas in [7] a practical algorithm to balance throughput and fairness for splittable routing is presented.

To the best of our knowledge, the work in [2] is the first to consider the fair flow allocation as a constraint of a more general network routing problem, rather than the optimization objective. Since the TCP flow allocation can be approximated by MMF, [2] proposes the bilevel problem MT-SPNR-MMF and an arc-based MILP formulation for it. As shown in [3], the problem is $\mathcal{NP}$-hard and the gap (in terms of either maximum throughput or smallest flow) between optimal solutions of MT-SPNR-MMF and of Max-Min fairest routing can be arbitrarily large. In this work, we pursue the study of MT-SPNR-MMF and extend it to the relaxed case of maximum bottleneck fairness.

3   Problem Versions and Complexity

Since, for any path selection vector $\underline{\lambda}$, the set of feasible flow vectors $\Phi(\underline{\lambda})$ for MMF is a subset of $\Phi(\underline{\lambda})$ for MB, MT-SPNR-MB is a relaxation of MT-SPNR-MMF and, hence, the optimal objective function value of the former is an upper bound on that of the latter.

The following simple example shows that, in general, the optimal solution values of the unconstrained throughput maximization problem (referred to as MT-SPNR or MT), the MT-SPNR-MMF, and the MT-SPNR-MB differ.

Example. Consider the graph with the arc capacities and the $k = 6$ origin-destination pairs reported in the figure on the right, where $c_{ae} = \varepsilon$ is a positive value smaller than 1. Note that for $i = 2, \ldots, 5$ there is a unique path to route (s_i, t_i), while there are two paths for (s_1, t_1) and three paths for (s_6, t_6). It is easy to verify that the optimal value of MT can be obtained by allocating a flow $\phi_1 = 0$ to the pair (s_1, t_1), and by routing a flow $\phi_6 = \varepsilon$ over the arc (a, e). The resulting flow allocation vector is $\underline{\phi} = (0, 3, 3, 2, 2, \varepsilon)$, with a total throughput $\tau = 10 + \varepsilon$. If the flow allocation is subject to the Max-Bottleneck

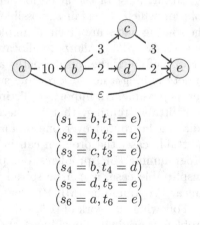

$(s_1 = b, t_1 = e)$
$(s_2 = b, t_2 = c)$
$(s_3 = c, t_3 = e)$
$(s_4 = b, t_4 = d)$
$(s_5 = d, t_5 = e)$
$(s_6 = a, t_6 = e)$

Constraint (4), it is easy to see that the maximum throughput is obtained by routing again (s_6, t_6) over the arc (a, e) and by assigning a flow $\phi_6 = \varepsilon$. If (s_1, t_1) is routed over the path through node c, the resulting allocation is $\underline{\phi} = (\varepsilon, 3, 3, 2 - \varepsilon, 2 - \varepsilon, \varepsilon)$, with $\tau = 10$. If that through node d is used, we obtain $\underline{\phi} = (\varepsilon, 3 - \varepsilon, 3 - \varepsilon, 2, 2, \varepsilon)$, with the same $\tau = 10$. Finally, if the flow allocation is required to be Max-Min Fair, the reader can verify that, in the optimal solution, (s_6, t_6) is routed over the arc (a, e), and (s_1, t_1) is routed through d. The allocation vector is then $\underline{\phi} = (1, 3, 3, 1, 1, \varepsilon)$, with $\tau = 9 + \varepsilon$. Notice that, in this case, routing (s_1, t_1) through c gives a smaller, suboptimal total throughput.

The connection between our two problems and the fundamental and extensively studied problem of finding edge-disjoint paths in directed and undirected graphs with or without congestion leads to the following inapproximability results.

Proposition 1. *MT-SPNR-MMF and MT-SPNR-MB are $\mathcal{NP}$-hard to approximate within any factor smaller than 2.*

Proof. By polynomial-time reduction from the following $\mathcal{NP}$-hard problem [8]:

> 2-DIR-PATH: Given a directed graph $G = (V, A)$ and distinct vertices s_1, s_2, t_1, t_2, decide whether there exist two edge-disjoint paths, one from s_1 to t_2 and the other from s_2 to t_2.

For any instance of 2-DIR-PATH, consider the special instance of MT-SPNR-MMF with the same graph G, the same two origin-destination pairs and $c_{ij} = 1$ for all $(i, j) \in A$. Clearly, Yes instances of 2-DIR-PATH are mapped onto MT-SPNR-MMF instances whose optimal flow allocation has total throughput of value 2, and No instances onto those with total throughput of at most 1. Thus, a ρ-approximation algorithm for MT-SPNR-MMF with $1 \leq \rho < 2$ would solve 2-DIR-PATH in polynomial time.

The same argument can be applied to MT-SPNR-MB. $\square$

A stronger inapproximability result for MT-SPNR-MMF and MT-SPNR-MB can be proved by building on that presented in [4] for the well-known problem of finding edge-disjoint paths with low congestion in undirected graphs.

4 Single-Level MILP Path Formulations

As shown in [2], the bilevel formulation (1)–(3) can be cast as a single-level MILP formulation by exploiting a simple characterization of the unique optimal MMF flow allocation for a given set of paths, which is based on the notion of bottleneck arc.

Definition 3. *An arc $(i, j) \in A$ is* bottleneck *for the pair (s, t) if*

1. *the arc capacity c_{ij} is saturated,*
2. *the flow allocated to (s, t) is greater than or equal to the value of the flow allocated to any other pair that shares the arc (i, j).*

The characterization is as follows:

Proposition 2 (Bertsekas, Gallager [5]). *Given a directed graph $G = (V, A)$, a set K of origin-destination pairs and a simple path for each $(s, t) \in K$, a feasible flow allocation vector ϕ is MMF if and only if there is at least a bottleneck arc for each pair $(s, t) \in K$.*

According to this proposition, the second level problem in Formulation (1)–(3) can be replaced with a set of linear constraints and binary variables that ensure the existence of a bottleneck arc for each (s, t) pair, see [2, 20]. Let P^{st} be the set of all the simple paths connecting the origin-destination pair $(s, t) \in K$. Let the parameter σ_{ij}^{pst} be 1 if the path $p \in P^{st}$ contains the arc (i, j) and 0 otherwise. Let the binary variable λ_p^{st} be 1 if the path of index

p is selected for the pair (s,t) and 0 otherwise. For each pair $(s,t) \in K$, let $\phi^{st} \in \mathbb{R}_+$ be the flow allocated to it and let f_{ij}^{st} the amount of flow on the arc $(i,j) \in A$. Notice that f_{ij}^{st} is either ϕ^{st} or 0. Let u_{ij} be an upper bound on the flows over the arc $(i,j) \in A$. The binary variables y_{ij}^{st} indicate whether an arc (i,j) is a bottleneck for (s,t). We obtain the following MILP formulation, originally introduced in [3]:

$$\max \quad \sum_{(s,t) \in K} \phi^{st} \tag{5}$$

$$\text{s.t.} \quad \sum_{(i,j) \in \delta^+(i)} f_{ij}^{st} - \sum_{(j,i) \in \delta^-(i)} f_{ji}^{st} = \begin{cases} \phi^{st} & \text{if } i = s \\ -\phi^{st} & \text{if } i = t \\ 0 & \text{else} \end{cases} \quad i \in V, (s,t) \in K \tag{6}$$

$$\sum_{(s,t) \in K} f_{ij}^{st} \le c_{ij} \qquad\qquad (i,j) \in A \tag{7}$$

$$\sum_{p \in P^{st}} \lambda_{st}^p = 1 \qquad\qquad (s,t) \in K \tag{8}$$

$$f_{ij}^{st} \le c_{ij} \sum_{p \in P^{st}} (\sigma_{ij}^{pst} \lambda_{st}^p) \qquad\qquad (i,j) \in A, (s,t) \in K \tag{9}$$

$$\sum_{(i,j) \in A} y_{ij}^{st} \ge 1 \qquad\qquad (s,t) \in K \tag{10}$$

$$\sum_{(o,d) \in K} f_{ij}^{od} \ge c_{ij} y_{ij}^{st} \qquad\qquad (i,j) \in A, (s,t) \in K \tag{11}$$

$$u_{ij} \ge f_{ij}^{st} \qquad\qquad (i,j) \in A, (s,t) \in K \tag{12}$$

$$u_{ij} \le f_{ij}^{st} + c_{ij}(1 - y_{ij}^{st}) \qquad\qquad (i,j) \in A, (s,t) \in K \tag{13}$$

$$\phi^{st} \ge \frac{\min_{(i,j) \in A} c_{ij}}{|K|} \qquad\qquad (s,t) \in K \tag{14}$$

$$\phi^{st}, f_{ij}^{st}, u_{ij} \ge 0 \qquad\qquad (i,j) \in A, (s,t) \in K \tag{15}$$

$$\lambda_{st}^p \in \{0,1\} \qquad\qquad (s,t) \in K, p \in P^{st} \tag{16}$$

$$y_{ij}^{st} \in \{0,1\} \qquad\qquad (i,j) \in A, (s,t) \in K. \tag{17}$$

Constraints (6)–(7) are standard flow conservation and capacity constraints. Constraints (8) guarantee that only one path is chosen for each $(s,t) \in K$. Constraints (9) ensure that f_{ij}^{st} is 0 if the selected path $p \in P^{st}$ does not contain the arc $(i,j) \in A$. Constraints (10)–(13) impose that the flow vector is MMF for the selected paths, according to Proposition 2. More specifically, Constraints (10) guarantee that at least an arc is bottleneck for each (s,t). Constraints (11) ensure that arc (i,j) is saturated if it is bottleneck for some pair (s,t). Constraints (12) make sure that u_{ij} is equivalent to the largest flow allocated over arc (i,j). Constraints (13) impose that the flow of a pair (s,t) through its bottleneck arc (i,j) is as large as the largest flow through (i,j) for all the other pairs. Finally, Constraints (14) introduce a valid lower bound on the value of the flow allocations, which is tight in the case where all the flows are routed over the same arc with minimum capacity. For an equivalent arc formulation, see [2].

We now derive a characterization of the optimal solution for the Max-Bottle-neck version, similar to that of Proposition 2.

Proposition 3. *Given a directed graph $G = (V, A)$, a set K of origin-destination pairs and a simple path for each $(s, t) \in K$, a feasible flow allocation vector $\underline{\phi}$ is optimal for the problem of maximizing the minimum flow allocated to any pair if and only if there is at least an arc $(i, j) \in A$ (referred to as* global bottleneck*), satisfying the following properties:*

1. *the arc capacity is saturated,*
2. *the arc capacity is equally divided among all the origin-destination pairs that share the arc,*
3. *the flow allocated to the pairs that share the arc is the smallest among the flow values allocated to the pairs in K.*

Proof. Suppose that arc (i, j) is a global bottleneck and let $\eta := \min_{i=1,\dots,k}\{\phi_i\}$. Due to (3), all the flows sharing the global bottleneck have a value of η. Since, due to (1) and (2), the capacity c_{ij} is equally divided among the pairs, it is not possible to improve one of the allocations of value η without decreasing another one (thus, decreasing η). Since η is independent of the flow value for pairs not using (i, j), it follows that the solution is optimal.

Conversely, suppose that $\underline{\phi}$ is optimal. Consider again the smallest flow allocation η in $\underline{\phi}$. Assume that there is no global bottleneck. Then, for all (s, t) pairs with flow value η, either all the arcs in their path are nonsaturated, or the capacity is not equally shared. In both cases, the flow can be increased (in the latter case, by decreasing a larger flow). □

To obtain a formulation for MT-SPNR-MB, it suffices to remove the y_{ij}^{st} variables while introducing the binary variables b_{ij}, each of which equals 1 if the corresponding arc (i, j) is a global bottleneck. Then, we can replace the MMF constraints (10)–(13) and the variables y_{ij}^{st} with the following ones:

$$\sum_{(i,j)\in A} b_{ij} \geq 1 \tag{18}$$

$$\sum_{(s,t)\in K} f_{ij}^{st} \geq c_{ij}b_{ij} \qquad (i, j) \in A \tag{19}$$

$$\eta \leq \phi^{st} \qquad (s, t) \in K \tag{20}$$

$$\eta \geq f_{ij}^{st} - c_{ij}(2 - b_{ij} - \sum_{p \in P^{st}} \sigma_{ij}^{pst}\lambda_{st}^p) \qquad (i, h) \in A, (s, t) \in K \tag{21}$$

$$\eta \geq 0, \ b_{ij} \in \{0, 1\} \qquad (i, j) \in A. \tag{22}$$

5   Branch-and-Price

Clearly, Formulation (5)–(17) has an exponential number of variables w.r.t. the size of the graph. A natural idea is to use a Branch-and-Price algorithm where the paths, and their associated λ_p^{st} variables, are dynamically generated. Since the same approach can be used for both MT-SPNR-MMF and MT-SPNR-MB, what follows is valid for both variants, unless differently specified.

The restricted master problem is obtained by restricting each set P^{st} to the set $\bar{P}^{st}$ of the paths that have been generated so far. In order to facilitate branching, we adopt an extended formulation where we also include the arc variables $x_{ij}^{st} \in \{0,1\}$, whose value is 1 if arc (i,j) is used by $(s,t) \in K$ and 0 otherwise. Path variables λ_p^{st} and arc variables x_{ij}^{st} are linked as follows:

$$\sum_{p \in P^{st}} \lambda_p^{st} \sigma_{ij}^{pst} = x_{ij}^{st} \qquad\qquad (i,j) \in A, \ (s,t) \in K. \qquad (23)$$

Notice that, assuming that the paths in $\bar{P}^{st}$ are distinct, Constraints (23) imply the integrality of the variables λ_p^{st}. Constraints (16) can hence be dropped, thus yielding a MILP model where branching only involves the x_{ij}^{st} and y_{ij}^{st} variables. **Pricing.** Let $\omega^{st} \in \mathbb{R}$, $\pi_{ij}^{st} \geq 0$ and $\nu_{ij}^{st} \in \mathbb{R}$ be the dual variables associated to, respectively, Constraints (8), (9), and (23). The dual constraint associated to a variable λ_p^{st} is $\omega^{st} + \sum_{(i,j) \in A} \sigma_{ij}^{pst}(\nu_{ij}^{st} - \pi_{ij}^{st} c_{ij}) \leq 0$.

In order to generate an improving column for a given $(s,t) \in K$, we need to generate a simple path $p \in P^{st}$ whose associated λ_p^{st} variable has a positive reduced cost, i.e., such that $\sum_{(i,j) \in A} \sigma_{ij}^{pst}(\nu_{ij}^{st} - \pi_{ij}^{st} c_{ij}) > -\omega^{st}$. This is equivalent to finding, for each $(s,t) \in K$, a longest simple path $p \in P^{st}$ over the original graph G where the weight of each arc is given by $\pi_{ij}^{st} c_{ij} - \nu_{ij}^{st}$ if the arc $(i,j) \in A$ belongs to the path and equals 0 otherwise.

The longest simple path problem is well known to be hard even to approximate. Here, we cast it as a MILP which is based on the standard LP formulation of the shortest path problem. Therefore we add continuous variables and constraints taken from the extended formulation proposed by Wong in [21] for the TSP, which prevents subtours. A desirable property of the MILP approach is its flexibility, which allows us to easily incorporate the branching information in the pricing subproblems.

Let $V_s := V \setminus \{s\}$ and $V_{st} := V \setminus \{s,t\}$, for a given $(s,t) \in K$. Let the variable $\sigma_{ij} \in \{0,1\}$ be 1 if the path that we are looking for contains arc (i,j) and 0 otherwise. Let the variable $z_h \in \{0,1\}$, for $h \in V_s$, be 1 if the path contains node h. Let $q_{ij}^h \leq 0$ be the value of an auxiliary flow from node s to node $h \in V_s$, and 0 otherwise. The pricing subproblem reads as follows:

$$\max \quad \sum_{(i,j) \in A} (c_{ij}\pi_{ij} - \nu_{ij})\sigma_{ij} \qquad\qquad\qquad\qquad (24)$$

$$\text{s.t.} \quad \sum_{(i,j) \in \delta^+(i)} \sigma_{ij} - \sum_{(j,i) \in \delta^-(i)} \sigma_{ji} = \begin{cases} 1 & \text{if } i = s \\ -1 & \text{if } i = t \\ 0 & \text{else} \end{cases} \quad i \in V \qquad (25)$$

$$\sum_{(i,j) \in \delta^+(i)} \sigma_{ij} \leq 1 \qquad\qquad\qquad\qquad i \in V \qquad (26)$$

$$q_{ij}^h \leq \sigma_{ij} \qquad\qquad\qquad (i,j) \in A, h \in V_s \quad (27)$$

$$\sum_{(i,j)\in\delta^+(i)} q_{ij}^h - \sum_{(j,i)\in\delta^-(i)} q_{ji}^h = \begin{cases} z_h & \text{if } i = s \\ -z_h & \text{if } i = h \\ 0 & \text{else} \end{cases} \qquad h \in V_s \quad (28)$$

$$\sum_{(i,h)\in\delta^-(h)} \sigma_{ih} = z_h \qquad\qquad\qquad h \in V_s \quad (29)$$

$$q_{ij}^h \geq 0 \qquad\qquad\qquad (i,j) \in A, h \in V_s \quad (30)$$

$$\sigma_{ij}, z_h \in \{0,1\} \qquad\qquad\qquad (i,j) \in A, h \in V_s \quad (31)$$

Constraints (25) are standard flow balance constraints. Constraints (26) limit the outgoing degree to 1, hence guaranteeing that the flow be unsplittable. Constraints (27) ensure that the auxiliary flow q_{ij}^h be 0 for all $h \in V_s$ if the arc (i,j) is not contained in the path. Constraints (28) are flow balance constraints for the auxiliary flow, guaranteeing that each node $h \in V_s$ be the sink node of the corresponding auxiliary flow. Constraints (29) impose that z_h be 1 if the path contains an arc entering the node $h \in V_s$ and 0 otherwise.

Pool Initialization. In order to initialize the Branch-and-Price algorithm, we populate the initial pool of columns by generating a set of initial paths for each $(s,t) \in K$. This is achieved by repeatedly finding a shortest path from s to t in the graph, using unit arc weights. Then, each time a new path is generated, the cost of the arcs therein contained is increased, thus promoting diversity among the paths. For each (s,t) pair in K, we generate a number of initial paths that is proportional to the minimum (s,t)-cut in the graph with unit arc capacities.

Primal Heuristics. Very often, finding even a feasible solution with state-of-the-art MILP solvers is extremely hard for both MT-SPNR-MMF and MT-SPNR-MB. Thus, the introduction of *ad hoc* primal heuristics is crucial to the effectiveness of any algorithm based on Branch-and-Bound.

We propose three such heuristics. The first two are based on rounding:

- *Standard rounding*: starting from a feasible solution of the continuous relaxation, for each (s,t) select the path $p \in \bar{P}^{st}$ with the largest λ_p^{st}; alternatively (randomized variant) pick a path $p \in \bar{P}^{st}$ with a probability equal to λ_p^{st}.
- *Shortest-path rounding*: starting from a feasible solution of the continuous relaxation, for each (s,t) find a shortest path in G with weights $1 - x_{ij}^{st}$ for each arc $(i,j) \in A$. A variant is obtained by sampling the weight of each arc from a uniform distribution in $(0, 1 - x_{ij}^{st})$. If the path that is found is not already in $\bar{P}^{st}$, a new column is added to the pool.

For both rounding heuristics, once a path $p \in P^{st}$ has been selected or generated for each (s,t) pair, a complete feasible solution is constructed by running, for MT-SPNR-MMF, the water filling algorithm or, for MT-SPNR-MB, a simple 2-stage algorithm. The latter consists in solving a bottleneck maximization problem followed by an LP that maximizes the total throughput over the residual graph.

The third algorithm is a MILP-based heuristic:

– *Restricted MILP*: the restricted master problem is solved by imposing the integrality constraints on λ_p^{st}, using only the columns that have been generated so far. Since solving this restricted MILP is still hard, in order to use it within the Branch-and-Price framework, we adopt a short time limit to provide good solutions quickly.

6   Some Computational Results

We have carried out computational experiments on a set of network topologies taken from the SND library [16]. We report results for MT-SPNR-MMF and MT-SPNR-MB on three network topologies: atlanta ($|V| = 15, |A| = 44$), france ($|V| = 25, |A| = 90$), and nobel-us ($|V| = 14, |A| = 42$). The arc capacities are randomly assigned to the arcs from a predefined set of values, see [2].

We compare the results of the Branch-and-Price algorithm to those obtained with an arc formulation of the problem, similar to that presented in [2]. We solve it with SCIP, using CPLEX as the underlying LP solver. The Branch-and-Price method is implemented in C++ within the SCIP framework. The machine used for the experiments is equipped with Intel Xeon E5645 CPUs and 16 GB of RAM.

The following table summarizes the computational results. Column k reports the number of origin-destination pairs. Column *opt* represents the optimal value of the unconstrained Max-Throughput problem (MT). The "-arc" suffix denotes the results for the arc formulation of the two problem variants, while "-B&P" represents the Branch-and-Price algorithm. Columns *UB*, *best*, and *gap* report, respectively, the best upper bound found within the time limit, the value of the best feasible solution, and the relative gap percentage. Column *time* represents the solution time in seconds. An asterisk is used when the time limit is reached (3600 s). When no feasible solution is found, the gap is reported as "inf".

For MT-SPNR-MB, the arc formulation is tractable on the two smaller topologies (atlanta and nobel-us), where small gaps are obtained for all instances. The Branch-and-Price algorithm has a similar behaviour, with solutions that are only marginally different. On france, however, the problem quickly becomes very hard for the arc formulation as k grows. On the contrary, the Branch-and-Price algorithm is capable of finding the optimal solution on more than half of the instances, with gaps exceeding 4 % only twice.

MT-SPNR-MMF proves to be very challenging to solve on all topologies. With the arc formulation, SCIP is able to find a feasible solution in only 10 of the 36 instances, with optimal solutions only for smaller instances. On the other hand, the Branch-and-Price algorithm finds good solutions for all the instances, achieving a gap of 2.6 % on average, with optimal solutions for 12 instances. The gap is never larger than 10 %.

It is interesting to point out that the optimal values for MT-SPNR-MB and MT-SPNR-MMF are very close (often identical) to those for the simple unsplittable Max Throughput problem (MT). In particular, for all instances solved to optimality except one, we obtain the same optimal values. Notice that there are only few cases in which the upper bound for MT-SPNR-MB and MT-SPNR-MMF is strictly smaller than the optimal value of MT, for example `nobel-9-1`. This small difference in optimal objective function values is good news from the point of view of the leader (the IP network operator), since it suggests that, for the instances under consideration, the throughput that can be achieved when the flows are allocated fairly by the follower (the TCP protocol) is almost the same as when fairness is completely neglected. Note however that the paths in the optimal solutions of our bilevel network routing problem subject to fairness constraints can substantially differ from those of MT, where the leader has full control on the flow allocation as well as on the paths.

	k	MT opt	MT-MB-arc UB	best	gap	time	MT-MB-B&P UB	best	gap	time	MT-MMF-arc UB	best	gap	time	MT-MMF-B&P UB	best	gap	time
at11	12	48.5	48.5	48.5	0	19	48.5	48.5	0	8	48.5	48.5	0	23	48.5	48.5	0	3
at12	12	56	56	56	0	10	56	56	0	0	56	56	0	30	56	56	0	1
at21	20	62.5	62.5	62.5	0	38	62.5	62.5	0	75	62.5	62.5	0	338	62.5	62.5	0	56
at22	20	63	62.9	62.5	0.6	*	63	62.5	0.7	*	62.5	62.5	0	1717	63	62.5	0.7	*
at31	30	98.5	98.3	97.1	1.3	*	98.3	96.8	1.6	*	98.6	–	inf	*	98.3	95	3.5	*
at32	30	83.5	83.5	83.3	0.2	*	83.5	83.3	0.2	*	83.5	–	inf	*	83.5	82	1.8	*
at41	42	76	75.9	75.6	0.4	*	75.9	75.6	0.5	*	76	–	inf	*	75.9	73.7	3	*
at42	42	119.5	119.5	119.5	0	228	119.5	119.5	0	999	119.5	–	inf	*	119.5	117.9	1.3	*
at51	56	87.5	87.3	86	1.5	*	87.3	86.1	1.3	*	87.3	–	inf	*	87.3	80.6	8.3	*
at52	56	141	140.7	139.7	0.7	*	140.7	139.4	1	*	140.8	–	inf	*	141.4	132.3	6.9	*
fr21	10	52.5	52.5	52.5	0	1358	52.5	52.5	0	23	52.5	52.5	0	1701	52.5	52.5	0	32
fr22	10	48.5	48.5	48.5	0	2028	48.5	48.5	0	713	49.5	48.5	2	*	48.5	48.5	0	1132
fr31	15	56.5	56.5	50.5	11.9	*	56.5	56.5	0	30	56.5	–	inf	*	56.5	56.5	0	65
fr32	15	67	67	58.5	14.5	*	67	67	0	49	67	59	13.6	*	67	67	0	195
fr41	21	77	77.3	75.2	2.8	*	77	77	0	1763	77.5	–	inf	*	77.7	76.5	1.6	*
fr42	21	72	72.7	53	37.2	*	72	72	0	208	72.5	–	inf	*	72	72	0	726
fr51	28	113.5	113.5	–	inf	*	113.5	112.5	0.9	*	113.5	–	inf	*	113.5	113.5	0	51
fr52	28	101	102	–	inf	*	101	101	0	1159	102.1	–	inf	*	102.2	101	1.2	*
fr61	36	116	116	–	inf	*	116.3	114.8	1.3	*	116.2	–	inf	*	116.2	114.5	1.5	*
fr62	36	95.5	95.5	–	inf	*	95.5	95.3	0.3	*	95.5	–	inf	*	95.5	93.3	2.4	*
fr71	45	131.5	131.7	124.2	6.1	*	131.4	125.7	4.5	*	–	–	inf	*	132.2	120.5	9.7	*
fr72	45	110	111	–	inf	*	111	106	4.7	*	–	–	inf	*	111.2	100.3	11	*
nb31	15	63.5	63.5	63.5	0	31	63.5	63.5	0	1	63.5	63.5	0	1266	63.5	63.5	0	4
nb32	15	49.5	49.5	49.5	0	45	49.5	49.5	0	29	49.5	44.5	11.2	*	49.5	49.5	0	22
nb41	21	76	75.9	75.2	1	*	76	75.2	1.1	*	76	–	inf	*	76	75	1.3	*
nb42	21	85.5	85.5	85.5	0	112	85.5	85.5	0	20	85.5	85.5	0	966	85.5	85.5	0	118
nb51	28	106	105.9	105.5	0.4	*	106	105.5	0.4	*	106.4	–	inf	*	106	105.2	0.7	*
nb52	28	100.5	100.5	100.2	0.3	*	100.5	100.2	0.3	*	100.5	–	inf	*	100.5	99.1	1.4	*
nb61	36	92	91.9	91	1	*	91.9	91.5	0.5	*	91.9	–	inf	*	91.9	89.7	2.5	*
nb62	36	117	116.9	115.8	0.9	*	116.9	116	0.8	*	116.9	–	inf	*	116.9	115	1.6	*
nb71	42	104	103.9	102.3	1.6	*	103.9	103.5	0.4	*	103.9	–	inf	*	103.9	98.7	5.3	*
nb72	42	127	126.9	125.8	0.9	*	126.9	125.8	0.9	*	126.9	–	inf	*	126.9	125.6	1	*
nb91	50	84.5	84.3	82.8	1.8	*	84.3	83.3	1.3	*	84.3	–	inf	*	84.3	78.5	7.4	*
nb92	50	118	118	115.9	1.8	*	118	117.6	0.3	*	118	–	inf	*	118	111.6	5.7	*
nb81	56	89.5	89.3	88.3	1.2	*	89.3	87.8	1.8	*	89.3	–	inf	*	89.3	82.3	8.5	*
nb82	56	129	128.8	128	0.7	*	128.8	127.8	0.9	*	128.8	–	inf	*	128.9	121.7	6	*

References

1. Allalouf, M., Shavitt, Y.: Maximum flow routing with weighted max-min fairness. In: Solé-Pareta, J., Smirnov, M., Van Mieghem, P., Domingo-Pascual, J., Monteiro, E., Reichl, P., Stiller, B., Gibbens, R.J. (eds.) QofIS 2004. LNCS, vol. 3266, pp. 278–287. Springer, Heidelberg (2004)

2. Amaldi, E., Capone, A., Coniglio, S., Gianoli, L.G.: Network optimization problems subject to max-min fair flow allocation. IEEE Commun. Lett. **17**(7), 1463–1466 (2013)
3. Amaldi, E., Coniglio, S., Gianoli, L.G., Ileri, C.U.: On single-path network routing subject to max-min fair flow allocation. Electron. Notes Discrete Math. **41**, 543–550 (2013)
4. Andrews, M., Chuzhoy, J., Guruswami, V., Khanna, S., Talwar, K., Zhang, L.: Inapproximability of edge-disjoint paths and low congestion routing on undirected graphs. Combinatorica **30**(5), 485–520 (2010)
5. Bertsekas, D., Gallager, R.: Data Networks. Prentice-Hall, Upper Saddle River (1992)
6. Danna, E., Hassidim, A., Kaplan, H., Kumar, A., Mansour, Y., Raz, D., Segalov, M.: Upward max min fairness. In: Proceedings IEEE INFOCOM 2012, pp. 837–845, March 2012
7. Danna, E., Mandal, S., Singh, A.: A practical algorithm for balancing the max-min fairness and throughput objectives in traffic engineering. In: Proceedings IEEE INFOCOM 2012, pp. 846–854, March 2012
8. Fortune, S., Hopcroft, J., Wyllie, J.: The directed subgraph homeomorphism problem. Theor. Comput. Sci. **10**(2), 111–121 (1980)
9. Harks, T., Hoefer, M., Schewior, K., Skopalik, A.: Routing games with progressive filling. CoRR abs/1308.3161, abs/1308.3161 (2013)
10. Kleinberg, J., Rabani, Y., Tardos, É.: Fairness in routing and load balancing. In: 40th Annual Symposium on Foundations of Computer Science (FOCS), pp. 568–578. IEEE (1999)
11. Massoulié, L., Roberts, J.: Bandwidth sharing: objectives and algorithms. IEEE/ACM Trans. Netw. **10**(3), 320–328 (2002)
12. Megiddo, N.: Optimal flows in networks with multiple sources and sinks. Math. Program. **7**(1), 97–107 (1974)
13. Nace, D., Pióro, M.: Max-min fairness and its applications to routing and load-balancing in communication networks: a tutorial. Commun. Surv. Tutorials **10**(4), 5–17 (2008)
14. Nace, D., Doan, N.L., Klopfenstein, O., Bashllari, A.: Max-min fairness in multi-commodity flows. Comput. Oper. Res. **35**(2), 557–573 (2008)
15. Nilsson, P.: Fairness in communication and computer network design. Ph.D. thesis, Lund University, Sweden (2006)
16. Orlowski, S., Wessäly, R., Pióro, M., Tomaszewski, A.: SNDlib 1.0 - survivable network design library. Networks **55**(3), 276–286 (2010)
17. Pioro, M.: Fair routing and related optimization problems. In: International Conference on Advanced Computing and Communications (ADCOM), pp. 229–235. IEEE (2007)
18. Radunovic, B., Boudec, J.Y.L.: A unified framework for max-min and min-max fairness with applications. IEEE/ACM Trans. Netw. **15**(5), 1073–1083 (2007)
19. Salles, R.M., Barria, J.A.: Lexicographic maximin optimisation for fair bandwidth allocation in computer networks. Eur. J. Oper. Res. **185**(2), 778–794 (2008)
20. Tomaszewski, A.: A polynomial algorithm for solving a general max-min fairness problem. Eur. Trans. Telecommun. **16**(3), 233–240 (2005)
21. Wong, R.: Integer programming formulations of the travelling salesman problem. In: Proceedings IEEE Conference on Circuits and Computers, pp. 149–152 (1980)

Study of Identifying Code Polyhedra for Some Families of Split Graphs

Gabriela Argiroffo[1], Silvia Bianchi[1], and Annegret Wagler[2]([⊠])

[1] Facultad de Ciencias Exactas, Ingeniería y Agrimensura,
Universidad Nacional de Rosario, Rosario, Argentina
{garua,sbianchi}@fceia.unr.edu.ar
[2] University Blaise Pascal (LIMOS, UMR 6158 CNRS),
Clermont-Ferrand, France
wagler@isima.fr

Abstract. The identifying code problem is a newly emerging search problem, challenging both from a theoretical and a computational point of view, even for special graphs like bipartite graphs and split graphs. Hence, a typical line of attack for this problem is to determine minimum identifying codes of special graphs or to provide bounds for their size.

In this work we study the associated polyhedra for some families of split graphs: headless spiders and complete suns. We provide the according linear relaxations, discuss their combinatorial structure, and demonstrate how the associated polyhedra can be entirely described or polyhedral arguments can be applied to find minimum identifying codes for special split graphs. We discuss further lines of research in order to apply similar techniques to obtain strong lower bounds stemming from linear relaxations of the identifying code polyhedron, enhanced by suitable cutting planes to be used in a B&C framework.

Keywords: Identifying code problem · Polyhedral approach · Split graphs

1 Introduction

Many practical applications can be stated as set covering problems, among them newly emerging search problems for identifying codes [12]. Consider a graph $G = (V, E)$ and denote by $N[i] = \{i\} \cup N(i)$ the closed neighborhood of i. A subset $C \subseteq V$ is *dominating* (resp. *identifying*) if $N[i] \cap C$ are non-empty (resp. distinct) sets for all $i \in V$. An *identifying code* of G is a node subset which is dominating and identifying, and the *identifying code number* $\gamma^{ID}(G)$ of a graph G is the minimum cardinality of an identifying code of G.

Determining a minimum identifying code in a graph $G = (V, E)$ can be formulated as set covering problem $\min \mathbf{1}^T x, M_{ID}(G) \geq 1, x \in \{0,1\}^{|V|}$ by:

This work was supported by an ECOS-MINCyT cooperation France-Argentina, A12E01.

© Springer International Publishing Switzerland 2014
P. Fouilhoux et al. (Eds.): ISCO 2014, LNCS 8596, pp. 13–25, 2014.
DOI: 10.1007/978-3-319-09174-7_2

$$\min \mathbf{1}^T x$$
$$x(N[j]) = \textstyle\sum_{i \in N[j]} x_i \geq 1 \quad \forall j \in V \qquad \text{(domination)}$$
$$x(N[j] \bigtriangleup N[k]) = \textstyle\sum_{i \in N[j] \triangle N[k]} x_i \geq 1 \quad \forall j,k \in V, j \neq k \text{ (identification)}$$
$$x \in \{0,1\}^{|V|}$$

We call

$$M_{ID}(G) = \begin{pmatrix} N[G] \\ \bigtriangleup[G] \end{pmatrix},$$

the *identifying code matrix* of G, encoding the closed neighborhoods of the nodes of G ($N[G]$) and their symmetric differences ($\bigtriangleup[G]$), and define the *identifying code polyhedron* of G as $P_{ID}(G) = \mathrm{conv}\{x \in \mathbb{Z}_+^{|V|} : M_{ID}(G)\, x \geq \mathbf{1}\}$. It is clear by construction that $\gamma^{ID}(G)$ equals the covering number $\tau(M_{ID}(G)) := \min\{\mathbf{1}^T x : x \in P_{ID}(G)\}$. In addition, a graph G has an identifying code or is *identifiable* if and only if $M_{ID}(G)$ has no zero-row. As $N[G]$ has clearly no zero-row, G is identifiable if and only if $\bigtriangleup[G]$ has no zero-row which is equivalent to the known condition that G is identifiable if and only if it has no true twins, i.e., nodes i, j with $N[i] = N[j]$, see [12].

As $M_{ID}(G)$ may contain rows which are equal to or dominated by other rows in $M_{ID}(G)$, we define the corresponding clutter matrix, the *identifying code clutter* $C_{ID}(G)$ of G, obtained by removing repeated or dominated rows from $M_{ID}(G)$. We clearly have that

$$P_{ID}(G) = \mathrm{conv}\{x \in \mathbb{Z}_+^{|V|} : C_{ID}(G)\, x \geq \mathbf{1}\},$$

and obtain as a linear relaxation the *fractional identifying code polyhedron*

$$Q_{ID}(G) = \{x \in \mathbb{R}_+^{|V|} : C_{ID}(G)\, x \geq \mathbf{1}\}.$$

In [2,3] we characterized when $P_{ID}(G)$ is full-dimensional and which constraints of $Q_{ID}(G)$ define facets of $P_{ID}(G)$:

Lemma 1 [2,3]. *Let G be a graph without isolated nodes and let $V_1(G)$ be the set of nodes $k \in V(G)$ such that $\{k\} = N[i] \bigtriangleup N[j]$ for two different nodes i and j in $V(G)$. Then,*

- *$P_{ID}(G)$ is full-dimensional if and only if $V_1(G) = \emptyset$.*
- *The constraint $x_i \geq 0$ defines a facet of $P_{ID}(G)$ if and only if $i \notin V_1(G)$.*
- *All constraints from $C_{ID}(G)\, x \geq \mathbf{1}$ define facets of $P_{ID}(G)$.*

Due to the possible formulation as set covering problem, it is immediate that the identifying code problem is hard in general. It even remains hard for several graph classes where many other in general hard problems are easy to solve, including bipartite graphs [6], split graphs [8] and, therefore, chordal graphs (see Sect. 2 for details).

Our aim is to study identifying codes in split graphs from a polyhedral point of view. In this work we study the associated polyhedra for some families of split graphs: headless spiders and complete suns. We provide the according linear relaxations, discuss their combinatorial structure, and demonstrate how the associated polyhedra can be entirely described or polyhedral arguments can be applied to find minimum identifying codes for special split graphs, see Sect. 2. We discuss further lines of research in order to apply similar techniques to obtain strong lower bounds stemming from linear relaxations of the identifying code polyhedron, enhanced by suitable cutting planes to be used in a B&C framework, see Sect. 3.

1.1 Preliminary Definitions

Given a set F of vectors in $\{0,1\}^n$, we say $y \in F$ is a dominating vector (of F) if there exits $x \in F$ with $x \leq y$. It can be also said that x is dominated by y.

From now on, every matrix has $0, 1$-entries, no zero columns and no dominating rows.

As there is a one-to-one correspondence between a vector $x \in \{0,1\}^n$ and the subset $S_x \subset \{1, \ldots, n\}$ having x as characteristic vector, we write x instead of S_x. Remind that a *cover* of a matrix M is a vector $x \in \{0,1\}^n$ such that $Mx \geq 1$. According to the previous convention, a cover of M is a subset of columns ($\{1, \ldots, n\}$) that intersects all the rows of M.

In addition, the *cardinality* of a cover x is denoted by $|x|$ and equals $1x$. A cover x is *minimum* if it has the minimum cardinality and in this case $|x|$ is called the *covering number* of the matrix M, denoted by $\tau(M)$. Recall that the set covering polyhedron of M, denoted by $Q^*(M)$, is defined as the convex hull of its covers. The polytope $Q(M) = \{x \in [0,1]^n : Mx \geq 1\}$ is known as the *linear relaxation* of $Q^*(A)$. When $Q^*(A) = Q(A)$ the matrix A is *ideal* and the set covering problem can be solved in polynomial time (in the size of M).

A cover of M is *minimal* if it does not dominate any other cover of M. The *blocker* of M, denoted by $b(M)$, is the matrix whose rows are the minimal covers of M. It is known that $b(b(M)) = M$ and also that a matrix M is ideal if and only if its blocker is (see [13]). In addition, since $b(b(M)) = M$ we can refer to $Q^*(M)$ and $Q(b(M))$ as a blocking pair of polyhedra. Moreover, **a** is an extreme point of $Q(b(M))$ if and only if $\mathbf{a}^T x \geq 1$ is a facet defining inequality of $Q^*(M)$ (see [10]). In the sequel we will refer to this property as *blocking duality*.

Given a matrix M and $j \in \{1, \ldots, n\}$, we introduce two matrix operations: the contraction of j, denoted by M/j, means that column j is removed from M as well as the resulting dominating rows and hence, corresponds to setting $x_j = 0$ in the constraints $Mx \geq 1$. The deletion of j, denoted by $M \backslash j$ means that column j is removed from M as well as all the rows with a 1 in column j and this corresponds to setting $x_j = 1$ in the constraints $Mx \geq 1$. Then, given M and $V_1, V_2 \subset \{1, \ldots, n\}$ disjoint, we will say that $M/V_1 \backslash V_2$ is a *minor* of M and this minor does not depend on the order of operations or elements in $\{1, \ldots, n\}$. It is clear that M is always a minor of itself and we will say that a minor $M/V_1 \backslash V_2$ is *proper* if $V_1 \cup V_2 \neq \emptyset$. It is not hard to see that $b(M/j) = b(M) \backslash j$ and $b(M \backslash j) = b(M)/j$ for every $j \in \{1, \ldots, n\}$. In addition, if a matrix is ideal then so are all its minors (see [7] for further details).

A *rank* inequality is

$$\sum_{i \in M'} x_i \geq \tau(M') \tag{1}$$

associated with a minor $M' = M \setminus U$. If (1) is a facet of $Q^*(M')$, then it is also a facet of $Q^*(M)$ (see [14]).

In addition, if the rank associated with some minor induces a facet defining inequality of $Q^*(M)$ then this inequality is also induced by a minor obtained by deletion (see [1] for further details).

2   Identifying Code Polyhedra of Some Split Graphs

A graph $G = (C \cup S, E)$ is a *split graph* if its node set can be partitioned into a clique C and a stable set S. Hence, split graphs are closed under taking complements by definition. Moreover, they form the complementary core of chordal graphs (graphs without chordless cycles of length ≥ 4) since G is a split graph if and only if G and $\overline{G}$ are chordal [9]. This is also reflected in terms of forbidden subgraphs since a graph is a split graph if and only if it is $(C_4, \overline{C}_4, C_5)$-free [9] (note that C_5 is self-complementary and that $\overline{C}_4$ occurs as induced subgraph in any chordless cycle C_k with $k \geq 6$ such that all chordless cycles C_k with $k \geq 4$ are excluded in G as well as in $\overline{G}$). The relation between chordal and split graphs can also be interpreted in terms of intersection graphs: while chordal graphs are the intersection graphs of distinct subtrees of a tree, split graphs are the intersection graphs of distinct substars of a star, see e.g. [5].

Our aim is to study identifying codes in split graphs from a polyhedral point of view. First note that a split graph is identifiable if and only if no two nodes in C have the same neighbors in S. For instance, a complete split graph (i.e., a split graph where all edges between C and S are present) is not identifiable as soon as C contains 2 nodes (as any two nodes in C are true twins).

Next, recall that finding a minimum identifying code in split graphs is NP-hard [8]. So far, $\gamma^{ID}(G)$ is only known for two families: on the one hand, stars (the complete split graphs $G = (C \cup S, E)$ with $|C| = 1$) are the only identifiable complete split graphs and have $\gamma^{ID}(G) = |S|$; on the other hand, split graphs $G = (C \cup S, E)$ where every node in S is connected to a distinct 2-node subset of C have $\gamma^{ID}(G)$ of order $\log(|S| + |C|)$, see [8]. The two families show the wide range of the possible size of minimum identifying codes in split graphs: while the lowest possible lower bound of $\log n$ is attained for the latter, stars achieve almost the highest possible value n.

Moreover, a split graph is connected if and only if no node in S is isolated. Every non-connected split graph G contains a connected split graph G' and a non-empty subset $S' \subset S$ of isolated nodes, and clearly $\gamma^{ID}(G) = \gamma^{ID}(G') + |S'|$.

This motivates the study of identifying codes in non-complete, connected split graphs G. We concentrate on three families of split graphs with a regular structure. This allows us to benefit from a certain combinatorial structure of the identifying code clutter $C_{ID}(G)$ of G and to draw conclusions for the polyhedra $P_{ID}(G)$ and the identifying code number $\gamma^{ID}(G)$ in a similar way as discussed

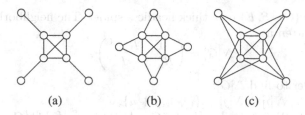

(a)   (b)   (c)

Fig. 1. (a) thin headless spider, (b) complete sun, (c) thick headless spider.

for families of bipartite graphs in [2,3]. In particular, note that stars $K_{1,n}$ are bipartite graphs as well as split graphs. Their identifying code clutter is related to q-roses R_n^q, $0,1$-matrices with n columns whose rows encode the incidence vectors of all the q-element subsets of $\{1,\ldots,n\}$. We have:

Theorem 1 [2,3]. *For a star $K_{1,n} = (V,E)$ with $n \geq 3$, we have*

- $C_{ID}(K_{1,n}) = R_{n+1}^2$;
- $P_{ID}(K_{1,n})$ *is entirely described by the inequalities $x(V') \geq |V| - 1$ for all nonempty subsets $V' \subset V$;*
- $\gamma^{ID}(K_{1,n}) = n$.

In this paper, we study three families of split graphs with $|S| = |C| \geq 2$ having a regular structure. A *headless spider* is a split graph $G = (C \cup S, E)$ with $S = \{s_1,\ldots,s_n\}$, $C = \{c_1,\ldots,c_n\}$, and $n \geq 2$. In a *thin headless spider*, s_i is adjacent to c_j if and only if $i = j$, and in a *thick headless spider*, s_i is adjacent to c_j if and only if $i \neq j$. It is straightforward to check that the complement of a thin spider is a thick spider, and vice-versa. Moreover, headless spiders where s_i is adjacent to exactly c_i and c_{i+1} for all $1 \leq i \leq n$ are called *complete suns*.

It is easy to see that for $n = 2$, the path P_4 equals the thin and thick headless spider, whereas the complete sun is not identifiable. For $n = 3$, the thin headless spider equals the net, and thick headless spider and complete sun its complement, the 3-sun. We consider headless spiders with $n \geq 4$; Fig. 1 illustrates all studied three families for $n = 4$. The partition (C,S) is called the *spider partition* and can be found in linear time [11].

2.1   Thick Headless Spiders

For simplicity, we will denote thick headless spiders by their partition and we will consider that $C = \{1,\ldots,n\}$ and $S = \{n+1,\ldots,2n\}$. Also, $\mathbf{E}$ denotes a matrix with all entries at value one.

Lemma 2. *For a thick headless spider $G = (C \cup S, E)$ with $n \geq 4$, we have*

$$C_{ID}(G) = \begin{pmatrix} R_n^{n-1} & I \\ 0 & R_n^2 \end{pmatrix}.$$

Proof. Let $G = (C \cup S, E)$ be a thick headless spider. The neighborhood matrix of G can be written as

$$N[G] = \begin{pmatrix} \mathbf{E} & R_n^{n-1} \\ R_n^{n-1} & I \end{pmatrix}.$$

Now, in order to find $\triangle[G]$:

(1) If $i, j \in C$, $N[i] \triangle N[j] = \{i + n, j + n\}$.

(2) If $i, j \in S$, $N[i] \triangle N[j] = [\{i\} \cup (C - \{i - n\})] \triangle [\{j\} \cup (C - \{j - n\})] = \{i, j, i - n, j - n\}$ and is dominated by a row of the case (1).

(3.a) If $i \in C$ and $j \in S$, $j \neq i + n$, $N[i] \triangle N[j] = [C \cup (S - \{i + n\})] \triangle [\{j\} \cup (C - \{j - n\}) = \{j - n\} \cup (S - \{j, i + n\})$ and is dominated by a row of the case (1) as $n \geq 4$.

(3.b) If $i \in C$ and $j = i + n \in S$, $N[i] \triangle N[j] = [C \cup (S - \{i + n\})] \triangle [\{j\} \cup (C - \{j - n\}) = [C \cup (S - \{i + n\})] \triangle [\{i + n\} \cup (C - \{i\}) = \{i\} \cup S$ and are dominated by a row of the case (1).

As the first n rows of the matrix $N[G]$ above are also dominated, we have that the clutter matrix $C_{ID}(G)$ can be written as

$$C_{ID}(G) = \begin{pmatrix} R_n^{n-1} & I \\ 0 & R_n^2 \end{pmatrix}.$$

As an immediate consequence, we obtain:

Corollary 1. *Let $G = (C \cup S, E)$ be a thick headless spider. Then,*

- $P_{ID}(G)$ *is full-dimensional.*
- *The constraint $x_i \geq 0$ defines a facet of $P_{ID}(G)$ for each $i \in C \cup S$.*
- *All constraints from $C_{ID}(G) \, x \geq 1$ define facets of $P_{ID}(G)$.*

Observe that $\tau(I, I) = n$. Then if $G = (C \cup S, E)$ is a thick headless spider, $\gamma^{ID}(G) \geq n$. In fact, we have:

Corollary 2. *Let $G = (C \cup S, E)$ be a thick headless spider. Then S is a minimum identifying code and, thus, $\gamma^{ID}(G) = n$.*

In [4], the set covering polyhedron $Q^*(R_n^q) = \text{conv}\{x \in \mathbb{Z}_+^n : R_n^q x \geq 1\}$ of complete q-roses was studied.

Theorem 2 [4]. *Let $n \geq q \geq 2$. A non-Boolean inequality $ax \geq 1$ is a facet defining inequality for $Q^*(R_n^q)$ if and only if $ax \geq 1$ can be written as $x(A_s) \geq |A_s| - q + 1$ for some $A_s \subset \{1, \ldots, n\}$ where $s \in \{0, \ldots, n - q - 1\}$ and $|A_s| = n - s$.*

As R_n^2 is a minor of $C_{ID}(G)$ obtained from the deletion of the nodes of C, in the underlying graph $G = (C \cup S, E)$, we have:

Corollary 3. *Let $G = (C \cup S, E)$ be a thick headless spider. Then, for all nonempty subsets $A \subset S$, the inequalities $x(A) \geq |A| - 1$ are facets of $P_{ID}(G)$.*

In order to study the remaining facets we need a description of the blocker of $C_{ID}(G)$.

From now on we consider vectors in $\{0,1\}^{l+k}$ of the form $e_i \oplus f_j$ where e_i for $i = 1,\dots,l$ is the unit vector in $\{0,1\}^l$ and f_j for $j = 1,\dots,k$ is a vector in $\{0,1\}^k$ such that $(f_j)_t = 0$ if $j = t$ and $(f_j)_t = 1$ if $j \neq t$.

Theorem 3. *Let $C_{ID}(G)$ be the clutter matrix of a thick headless spider $G = (C \cup S, E)$. Every minimal cover x of $C_{ID}(G)$ is minimum. Moreover, either $x = \mathbf{0} \oplus \mathbf{1}$ where $\mathbf{0}, \mathbf{1} \in \{0,1\}^n$ or $x = e_i \oplus f_j$ where $e_i, f_j \in \{0,1\}^n$ with $i \neq j$.*

Proof. From Corollary 2, S is a minimum cover of $C_{ID}(G)$.

Now, let x be a minimal cover of $C_{ID}(G)$ such that $x_j = 0$ for some $j \in \{n+1,\dots,2n\}$. It is known that every row of R_n^{n-1} covers $(e_i + e_j)$ for every $i, j = 1,\dots,n$ [4], then any cover x with $x_{n+j} = 0$ for some $j \in \{n+1,\dots,2n\}$ must be of the form $x = y \oplus f_j$ with some $y \in \{0,1\}^n$. In order to cover the rows of submatrix (R_n^{n-1}, I) it is enough to consider $x^{i,j} = e_i \oplus f_j$ for $i, j = 1,\dots,n$ and $i \neq j$. Then $|x| = |x^{i,j}| = n$ for every $i, j = 1,\dots,n$ and $i \neq j$ and they are all minimum covers.

Now, let v be a cover of $C_{ID}(G)$, with $v_{n+j} = 0$. Then $v = y \oplus f_j$ and $y \in \{0,1\}^n$. But y must be a cover of R_n^{n-1}, i.e., $y = e_i + h$ for some $i \neq j$ and some $h \in \{0,1\}^n$. Then y is not minimal.

We can further prove the following:

Corollary 4. *If $\bar{x} \in \mathbb{R}^{2n}$ is an extreme point of $Q(b(C_{ID}(G)))$ such that $x_i \neq 0$ then $\bar{x} = \frac{1}{n}\mathbf{1} \in \mathbb{R}^{2n}$.*

Proof. From Theorem 3 it follows that every row of $b(C_{ID}(G))$ is either $\mathbf{0} \oplus \mathbf{1}$ where $\mathbf{0}, \mathbf{1} \in \{0,1\}^n$ or $e_i \oplus f_j$ for $i, j = 1,\dots,n$ and $i \neq j$. Then they all have n ones per row. One can show that there are $2n$ linearly independent rows. It follows that if $\bar{x} = \frac{1}{n}\mathbf{1} \in \mathbb{R}^n$ then it satisfies $b(C_{ID}(G))\bar{x} = 1$. Hence $\bar{x}$ is a fractional extreme point of $Q(b(C_{ID}(G)))$. Now, if $\bar{y}$ is an extreme point of $Q(b(C_{ID}(G)))$ with all nonzero components then it must satisfy $2n$ linearly independent inequalities of $Q(b(C_{ID}(G)))$ at equality. It follows that $\bar{y} = \bar{x}$.

Using blocking duality it can be seen that Corollary 4 gives an alternative proof of $\gamma^{ID}(G) = n$ and states that the only facet of $P_{ID}(G)$ with full support is the rank inequality associated with $C_{ID}(G)$.

With the help of some technical lemmas, we can further show:

Theorem 4. *Let $b(C_{ID}(G))$ be the blocker of the identifying clutter matrix of a thick headless spider $G = (C \cup S, E)$. Let $\bar{x} \in \mathbb{R}^{2n}$ be a fractional extreme point of $Q(b(C_{ID}(G)))$ such that the set $A = \{i : \bar{x}_i = 0\}$ is nonempty. Then either*

1. *$A \subsetneq S$ and $\bar{x}_i = \frac{1}{n-|A|}$ when $i \notin A$ or*
2. *$C \subsetneq A$ and $|A| \leq 2(n-1)$ and $\bar{x}_i = \frac{1}{|A-C|-1}$ for all $i \notin A$.*

As a consequence of Theorem 4 and blocking duality, we conclude:

Corollary 5. *Let $G = (C \cup S, E)$ be a thick headless spider and $S' \subset S$ non-empty. Then, the inequalities $x(C) + x(S') \geq n - |S - S'|$ when $2 \leq |S'| \leq n - 1$ and $x(S') \geq |S'| - 1$ when $2 \leq |S'| \leq n$ are facets of $P_{ID}(G)$.*

As a consequence of Corollary 5 and Theorem 4, we obtain the main result of this section:

Corollary 6. *Let $G = (C \cup S, E)$ be a thick headless spider. Then, the facets of $P_{ID}(G)$ are:*

- *the constraint $x_i \geq 0$ for all $i \in C \cup S$;*
- *the constraints $C_{ID}(G)\, x \geq 1$;*
- *the constraints $x(C) + x(S') \geq n - |S - S'|$ and $x(S') \geq |S'| - 1$ for every $S' \subseteq S$ with $2 \leq |S'|$.*

2.2   Thin Headless Spiders

Lemma 3. *For a thin headless spider $G = (C \cup S, E)$ with $n \geq 4$, we have*

$$C_{ID}(G) = \begin{pmatrix} I & I \\ 0 & R_n^2 \\ R_n^{n-1} & 0 \end{pmatrix}.$$

Proof. Let $G = (C \cup S, E)$ be a thin headless spider having $C = \{1, \ldots, n\}$ and $S = \{n + 1, \ldots, 2n\}$. The neighborhood matrix of (C, S) can be written as

$$N[G] = \begin{pmatrix} \mathbf{E}\, I \\ I\, I \end{pmatrix}.$$

Now, in order to find $\triangle[G]$:

(1) If $i, j \in C$, $N[i] \triangle N[j] = \{i + n, j + n\}$.

(2) If $i, j \in S$, $N[i] \triangle N[j] = \{i, i - n\} \triangle \{j, j - n\}$ and are dominated by $N[i]$.

(3.a) If $i \in C$ and $j \in S$, $j \neq i + n$, $N[i] \triangle N[j] = [C \cup \{i + n\})] \triangle \{j, j - n\} = \{j, i + n\} \cup (C - \{j - n\})$ is dominated by N[i].

(3.b) If $i \in C$ and $j = i + n \in S$, $N[i] \triangle N[j] = [C \cup \{i + n\})] \triangle [\{i, i + n\} = C - \{i\}$.

As the first n rows of the matrix N above are also dominated, we have that the clutter matrix $C_{ID}(G)$ can be written as

$$C_{ID}(G) = \begin{pmatrix} I & I \\ 0 & R_n^2 \\ R_n^{n-1} & 0 \end{pmatrix}.$$

As an immediate consequence, we obtain:

Corollary 7. *Let $G = (C \cup S, E)$ be a thin headless spider. Then,*

- $P_{ID}(G)$ is full-dimensional.
- The constraint $x_i \geq 0$ defines a facet of $P_{ID}(G)$ for all $i \in C \cup S$.
- All constraints from $C_{ID}(G)$ $x \geq 1$ define facets of $P_{ID}(G)$.

Observe that $\tau(I, I) = n$. Then if $G = (C \cup S, E)$ is a thin headless spider, $\gamma^{ID}(G) \geq n$. In fact, we have:

Corollary 8. *Let $G = (C \cup S, E)$ be a thin headless spider. Then, $\gamma^{ID}(G) = n + 1$.*

Moreover, we obtain:

Corollary 9. *Let $G = (C \cup S, E)$ be a thin headless spider. Then,*

1. *the inequalities $x(A) \geq |A| - 1$ for all nonempty subsets $A \subset S$ are facets of $P_{ID}(G)$,*
2. *the inequality $x(C) \geq 2$ is a facet of $P_{ID}(G)$.*

Proof. As R_n^2 is a minor of $C_{ID}(G)$ obtained after deletion of the nodes in C, as a consequence of Theorem 2 we have that the inequalities $x(A) \geq |A| - 1$ for all nonempty subsets $A \subset S$ are facets of $P_{ID}(G)$.

Also, R_n^{n-1} is a minor of $C_{ID}(G)$ obtained after deletion of the nodes in S, and again using Theorem 2 we obtain $x(C) \geq 2$ as a facet of $P_{ID}(G)$.

As an immediate observation the rank inequality $x(C, S) \geq n + 1$ is not a facet of $P_{ID}(G)$ since it can be obtained as the sum of the facets $x(S) \geq n - 1$ and $x(C) \geq 2$.

Based on our computational experience, we conjecture that the identifying code polyhedra $P_{ID}(G)$ of thin headless spiders have rank facets of a special structure only:

Conjecture 1. Let $G = (C, S)$ be a thin headless spider. Then, the facets of $P_{ID}(G)$ are:

- the constraint $x_i \geq 0$ for all $i \in C \cup S$;
- all constraints from $C_{ID}(G)$ $x \geq 1$;
- the constraint $x(C) \geq 2$;
- the constraints $x(S') \geq |S'| - 1$ for all nonempty subsets $S' \subset S$.

2.3 Complete Suns

As third family of headless spiders $G = (C \cup S, E)$ having a regular structure, we consider complete suns, where $S = \{s_1, \ldots, s_n\}$, $C = \{c_1, \ldots, c_n\}$ and s_i is adjacent to exactly c_i and c_{i+1} for all $1 \leq i \leq n$ (indices are taken modulo n).

In contrary to thin and thick headless spiders whose identifying code clutters are composed by few q-roses, the identifying code clutters of complete suns have a more complex structure, involving different combinations of submatrices with a circular structure, where some submatrices occur for all $n \geq 4$, others not (depending on the parity of n and the size of the graph).

A circulant matrix is a square matrix where each row vector is rotated one element to the right relative to the preceding row vector. We denote by C_n^k a matrix in $\{0,1\}^{n \times n}$ having as first row the vector starting with k 1-entries and having 0-entries otherwise. Moreover, we denote by C_n^{k+k} a matrix in $\{0,1\}^{n \times n}$ with $n \geq 2k+2$ having as first row the vector starting with k 1-entries, then having 0-entries, again k 1-entries, and 0-entries otherwise.

Lemma 4. *For a complete sun $G = (C \cup S, E)$ with $n \geq 4$, the identifying code clutter $C_{ID}(G)$ is composed by the following submatrices*

$$
\begin{array}{ll}
(\ C_n^2 \ | \quad I \quad) & \forall n \geq 4 \\
(\ \ 0 \ \ | \ C_n^{1+1} \ ) & \forall n \geq 5 \\
(\ \ 0 \ \ | \ C_{\frac{n}{2}}^1, C_{\frac{n}{2}}^1 \ ) & for \ n = 4 \\
(\ \ 0 \ \ | \ C_n^{2+2} \ ) & \forall n \geq 9 \\
(\ \ 0 \ \ | \ C_{\frac{n}{2}}^2, C_{\frac{n}{2}}^2 \ ) & \forall n \geq 8, n \ even \\
(\ C_n^{n-2} \ | \quad I \quad) & \forall n \geq 4 \\
(\ C_n^{1+1} \ | \quad C_n^2 \quad) & \forall n \geq 4
\end{array}
$$

where the first part refers to C, the second part of the matrices to S.

Proof. Let $G = (C \cup S, E)$ be a complete sun. The neighborhood matrix $N[G]$ of (C, S) is composed from

$$
\begin{array}{ll}
(\ \mathbf{E} \ | \ C_n^2 \ ) & for \ N[C], \\
(\ C_n^2 \ | \ I \ ) & for \ N[S].
\end{array}
$$

Thus, only $N[S]$ is in $C_{ID}(G)$. In order to find $\triangle[G]$, we distinguish three cases.
Case 1: the symmetric differences between two nodes in C have the form

- $N[c_i] \triangle N[c_{i+1}] = \{s_{i-1}, s_{i+1}\}$;
- $N[c_i] \triangle N[c_{i+j}] = \{s_{i-1}, s_i, s_{i+j-1}, s_{i+j}\}$ for $1 < j \leq \frac{n}{2}$.

For all $n \geq 4$, the former symmetric differences remain in $C_{ID}(G)$ as submatrix

$$
(\ 0 \ | \ C_n^{1+1} \ ) \qquad \forall n \geq 5
$$

(but yield for $n = 4$ not the whole circulant matrix). The latter symmetric differences are dominated by the former if $j = 2, 3$. Thus, for each $4 \leq j \leq \frac{n}{2}$, the symmetric differences $N[c_i] \triangle N[c_{i+j}]$ remain in $C_{ID}(G)$ as submatrix

$$
(\ 0 \ | \ C_n^{2+2} \ ) \qquad \forall n \geq 9
$$

(but yield for $j = \frac{n}{2}$ not the whole circulant matrix).
Case 2: the symmetric differences between nodes in C and S have the form

- $N[c_i] \triangle N[s_i] = (C - \{c_i, c_{i+1}\}) \cup \{s_{i-1}\}$;
- $N[c_i] \triangle N[s_{i-1}] = (C - \{c_{i-2}, c_{i-1}\}) \cup \{s_i\}$;
- $N[c_i] \triangle N[s_j] = (C - \{c_{j-1}, c_j\}) \cup \{s_{i-1}, s_i\}$ for $j \neq i, i - 1$.

Thus, $N[c_i] \triangle N[s_j]$ is dominated by $N[s_i]$ if $j \neq i, i-1$, and remains in $C_{ID}(G)$ for $j = i, i-1$, forming two submatrices of the form

$$(\ C_n^{n-2} \ | \ I \ ) \qquad \forall n \geq 4.$$

Case 3: the symmetric differences between two nodes in S have the form

- $N[s_i] \triangle N[s_{i+1}] = \{c_i, c_{i+2}, \} \cup \{s_i, s_{i+1}\}$;
- $N[s_i] \triangle N[s_j] = \{c_i, c_{i+1}, c_j, c_{j+1}\} \cup \{s_i, s_j\}$ for $j \neq i$.

Thus, $N[s_i] \triangle N[s_j]$ is dominated by $N[s_i]$ if $j \neq i+1$, and remains in $C_{ID}(G)$ for $j = i+1$, forming a submatrix of the form

$$(\ C_n^{1+1} \ | \ C_n^2 \ ) \qquad \forall n \geq 4.$$

This together completely describes the identifying code clutter $C_{ID}(G)$.

As an immediate consequence, we obtain:

Corollary 10. *Let $G = (C \cup S, E)$ be a complete sun with $n \geq 4$.*

- *$P_{ID}(G)$ is full-dimensional.*
- *The constraint $x_v \geq 0$ defines a facet of $P_{ID}(G)$ for each $v \in C \cup S$.*
- *All constraints from $C_{ID}(G) \ x \geq 1$ define facets of $P_{ID}(G)$.*

Unfortunately, the whole system of facet-defining inequalities for the identifying code polyhedra $P_{ID}(G)$ of complete suns is not easy to describe since non-rank facets are required for all cases $n \geq 4$ (in fact, most facets of $P_{ID}(G)$ are non-rank and involve large coefficients). However, from a careful analysis of the constraints involved in the identifying code clutter $C_{ID}(G)$ of complete suns, we derive at the following conjecture:

Conjecture 2. For a complete sun $G = (C \cup S, E)$ with $n \geq 4$, the stable set S is a minimum identifying code.

Note that it is easy to see that S is always an identifying code for a complete sun $G = (C \cup S, E)$, since all rows of $C_{ID}(G)$ have at least one 1-entry in S. Hence, $\gamma^{ID}(G) \leq |S| = n$ follows. On the other hand, for some cases, it has been already verified that S is a minimum identifying code, by generating the full rank constraint $x(C) + x(S) \geq |S| = n$ by means of the Chátal-Gomory procedure. This implies $\gamma^{ID}(G) \geq |S| = n$, and together equality follows for these cases. Our goal is to find a general construction of this type for all $n \geq 4$.

3   Concluding Remarks

The identifying code problem is hard in general and challenging both from a theoretical and a computational point of view, even for special graphs like split graphs [8]. In this paper, we studied three families of split graphs with $|S| = |C| \geq 2$ having a regular structure: thin headless spiders, thick headless spiders, and complete suns. For all three families, we determined the identifying code clutter and discussed according consequences. In the case of thin and thick spiders G, $C_{ID}(G)$ is composed from certain q-roses. Based on related results from [4,14], we could give the complete description of $P_{ID}(G)$ for thick spiders, and arrived at a profound conjecture for thin spiders. For both classes, we found the exact value for $\gamma^{ID}(G)$: $|S|$ for thick spiders and $|S| + 1$ for thin spiders. It turned out that the identifying code clutters of complete suns have a more complex structure involving different circulant matrices and, accordingly, more involved facets are required to describe $P_{ID}(G)$. For this class, we showed $\gamma^{ID}(G) \leq |S|$ and conjecture that $\gamma^{ID}(G) = |S|$ holds. So, all three families seem to have small minimum identifying codes close to the lower bound of order $\log(|S| + |C|)$.

This demonstrates how the polyhedral approach can be applied to find identifying codes of minimum size for special graphs G, just by determining and analyzing the identifying code clutter $C_{ID}(G)$, even in cases where no complete description of $P_{ID}(G)$ is known yet.

As future lines of research, we plan to apply similar and more advanced techniques to obtain either the identifying code of minimum size or strong lower bounds stemming from linear relaxations of the identifying code polyhedron, enhanced by suitable cutting planes. For that, note that facets associated with deletion minors of $C_{ID}(G)$ remain facets in $P_{ID}(G)$, so according facets identified for special graphs are relevant for every graph having such subgraphs.

References

1. Argiroffo, G., Bianchi, S.: On the set covering polyhedron of circulant matrices. Discrete Optim. **6**(2), 162–173 (2009)
2. Argiroffo, G., Bianchi, S., Wagler, A.: Polyhedra associated to identifying codes (extended abstract), In: Proceedings of the VII Latin-American Algorithms, Graphs and Optimization Symposium (LAGOS 2013), Electronic Notes in Discrete Mathematics, vol. 44, pp. 175–180 (2013)
3. Argiroffo, G., Bianchi, S., Wagler, A.: Polyhedra associated with identifying codes, submitted to Discrete Applied Mathematics
4. Argiroffo, G., Carr, M.: On the set covering polyhedron of q-roses. In: Proceedings of the VI ALIO/EURO Workshop on Applied Combinatorial Optimization 2008, Buenos Aires, Argentina (2008)
5. Brandstdt, A., Le, V.B., Spinrad, J.: Graph Classes: A Survey. SIAM Monographs on Discrete Mathematics and Applications. SIAM, Philadelphia (1999)
6. Charon, I., Hudry, O., Lobstein, A.: Minimizing the size of an identifying or locating-dominating code in a graph is NP-hard. Theoret. Comput. Sci. **290**, 2109–2120 (2003)
7. Cornuéjols, G.: Combinatorial optimization: packing and covering. SIAM, CBMS, vol. 74 (2001)
8. Foucaud, F.: The complexity of the identifying code problem in restricted graph classes. In: Lecroq, T., Mouchard, L. (eds.) IWOCA 2013. LNCS, vol. 8288, pp. 150–163. Springer, Heidelberg (2013)
9. Földes, S., Hammer, P.: Split graphs, In: Proceedings of the VIII Southeastern Conference on Combinatorics, Graph Theory and Computing (Baton Rouge, La.), Congressus Numerantium XIX, Winnipeg: Utilitas Math., pp. 311–315 (1977)

10. Fulkerson, D.: Blocking polyhedra. In: Haris, B. (ed.) Graph Theory and its Applications, pp. 93–112. Academic Press, New York (1970)
11. Jamison, B., Olariu, S.: Recognizing P_4-tidy graphs in linear time. SIAM J. Comput. **21**, 381–406 (1992)
12. Karpovsky, M.G., Chakrabarty, K., Levitin, L.B.: On a new class of codes for identifying vertices in graphs. IEEE Trans. Inf. Theory **44**, 599–611 (1998)
13. Lehman, A.: On the width-length inequality. Math. Program. **17**, 403–417 (1979)
14. Sassano, A.: On the facial structure of the set covering polytope. Math. Program. **44**, 181–202 (1989)

Parametric Multiroute Flow and Its Application to Robust Network with k Edge Failures

Jean-François Baffier[1,2], Vorapong Suppakitpaisarn[3,4(✉)],
Hidefumi Hiraishi[1,4], and Hiroshi Imai[1]

[1] The University of Tokyo, Tokyo, Japan
[2] JFLI, CNRS, Université Paris-Sud, Orsay, France
[3] National Institute of Informatics, Tokyo, Japan
vorapong@nii.ac.jp
[4] JST, ERATO, Kawarabayashi Large Graph Project, Tokyo, Japan

Abstract. In this work, we investigate properties of the function taking the real value h to the max h-route flow value, and apply the result to solve robust network flow problems. We show that the function is piecewise hyperbolic, and modify a parametric optimization technique, the ES algorithm, to find this function. The running time of the algorithm is $O(\lambda mn)$, when λ is a source-sink edge connectivity of our network, m is the number of links, and n is the number of nodes. We can use the result from that algorithm to solve two max-flow problems against k edge failures, referred to as max-MLA-robust flow and max-MLA-reliable flow. When h is optimally chosen from the function, we show that the max-h-route flow is an exact solution of both problems for graphs in a specific class. Our numerical experiments show that 98 % of random graphs generated in the experiment are in that specific class. Given a parametric edge e, we also show that the function taking the capacity of e to the max-h-route flow value is linear piecewise. Hence we can apply our modified ES algorithm to find that function in $O(h^2 mn)$.

1 Introduction

Since its introduction by Ford and Fulkerson [1], the maximum flow problem (max-flow) has been widely studied due to its many theoretical and practical applications. There are many polynomial-time algorithms with which to solve this problem, including the recent results by Orlin [2] where the max-flow of a network with n nodes and m links is solved in $O(mn)$.

An h-route flow is a nonnegative linear combination of h edge-disjoint paths. The notion was introduced by Kishimoto and Takeuchi in [3], where they extend the max-flow/min-cut duality property to the multiroute flow context and provide an algorithm to compute a max-h-route flow based on h iterations of a classical max-flow algorithm. The duality proof is simplified by Bagchi et al. [4], and the improved algorithm, in which the number of max-flow iterations is less than h is some networks, is proposed by Aggarwal and Orlin [5].

It is shown in [6] that the max-$(k + 1)$-route flow is a $(k + 1)$-approximation of two natural variants of the max-flow problem against k edge failures, referred

© Springer International Publishing Switzerland 2014
P. Fouilhoux et al. (Eds.): ISCO 2014, LNCS 8596, pp. 26–37, 2014.
DOI: 10.1007/978-3-319-09174-7_3

to, in this work, as the *maximum multilink attack robust flow (max-MLA-robust flow)* and *maximum multilink attack reliable flow (max-MLA-reliable flow)* problems. The max-MLA-robust flow problem is to find the minimum max-flow value among $\binom{m}{k}$ networks obtained by deleting each set of k edges. The problem can be considered as a special case of R-MAX-FLOW-KCU, which is proposed and shown to be NP-hard in [7]. The max-MLA-reliable flow is to find a max-flow of the network such that the flow value is maximum against any set of k edge failures, when deleting the corresponding flow to those k edges in the original flow. The problem can be considered as a special case of the minimax problem to combat link attacks, which is proposed and solved heuristically in [8]. For the case when $k = 1$, we can solve this problem by the method for δ-reliable flow proposed in [9]. Throughout this paper, we will also refer to the edge failures as edge attacks, since the best attacks are equivalent to the worst failures.

The parametric optimization scheme is introduced to several problems in network design. In this scheme, our goal is to find an algorithm that outputs a function taking some network parameters to an optimization result. For instance, the parametric max-flow algorithm proposed in [10] outputs a function taking a capacity of a parametric edge e to a max-flow value. Several works have introduced the variants of that algorithm to solving problems in database [11] and computer vision [12]. For those variants, the output function is shown to be linear piecewise. They introduce a method called ES algorithm to find that linear piecewise function in $O(pT)$ when p is the number of line segments in that function and T is the running time to get an optimization result for a specific parameter value.

While the most common parameter considered in this scheme is the edge capacity, we consider the route number h as a parameter in this work. In Sect. 3, we consider the definition of max-h-route flow for the case when the route number h can be non-integer proposed in [13]. We show that a function taking the value h to the flow value is piecewise hyperbolic. We call the problem as *max-route-parametric h-route flow*, and propose an algorithm to find that function. The bottleneck part of the algorithm is the calculation of a linear piecewise function with at most λ line segments. Because we compute that function by ES algorithm, our computation time is $O(\lambda T)$ where λ is a source-sink edge connectivity of our network and T is the computation time of max-flow algorithm. We note that we can find max h-route flow for any h in $O(\log \lambda)$ time from our function. Those include the case that $h = \lambda$ where the state-of-the-art algorithm also takes $O(\lambda T)$ only to compute this case.

Although the max-route-parametric h-route flow itself can be applied to solving problems in network design, we show that it can also be applied to exactly solve max-MLA-robust flow and max-MLA-reliable flow in Sect. 4. When the route number h is optimally chosen from the function obtained from Sect. 3, we show that the max-h-route flow is an exact solution for both problems in a specific class of network. Our numerical experiments show that 98 % of random graphs generated in the experiment are in that specific class.

Shown in Sect. 5, we also have a contribution when the parameter is edge capacities. Given a parametric edge e, we show that the function taking the capacity of e to the max-h-route flow value is linear piecewise with at most $h+1$ line segments. Using ES algorithm, we can find that function in $O(hT)$, when T is the computation time of max-h-route flow algorithm. We will refer to this problem as *max-edge-parametric h − route flow* in this paper.

2   Preliminaries

In this section, we provide the notation that we will use throughout this article. The definition and properties of the multiroute flow are provided in Subsect. 2.1, while max-MLA-robust flow and max-MLA-reliable flow will be discussed in Subsect. 2.2.

Let $G = (V, E, c)$ be a network, where V is a set of nodes, E is a set of links, and $c : E \to \mathbb{R}^+$ is a capacity function. Let $s, t \in V$ be a source node and a sink node, respectively. Throughout this paper, we will consider single-commodity flows from s to t. All terminologies are based on that setting unless otherwise specified. The set $\mathscr{C}$ (resp., $\mathscr{F}$) refers to the set of all s-t cuts (resp., the set all possible s-t flows) of G. λ refers to the s-t edge connectivity of G, and k refers to the number of edges that the attacker can remove.

Definition 2.1 (MLA-robust capacity [14]). *Given a cut $X \in \mathscr{C}$, let $\{e_0, e_1, \dots, e_p\}$ be the cut-set of X, where $c(e_i) \geqslant c(e_{i+1})$ for any $0 \leqslant i < p$. For $0 \leqslant k \leqslant p$, we define the MLA-robust capacity of X, $\alpha_k(X)$ as $\alpha_k(X) = \sum_{i=k}^{p} c(e_i)$. For $k > p$, we define $\alpha_k(X) = 0$.*

2.1   Multiroute Flow

In this subsection, we will briefly describe the h-route flow introduced in [3], which, for $h \geqslant 2$, is also called a *multiroute flow*.

Definition 2.2 (h-route flow). *A h-route flow is a nonnegative linear combination of h edge-disjoint s-t paths with unitary flow, in which the value on each edge does not exceed the edge capacity.*

Definition 2.3 (max-h-route flow). *A max-h-route flow is a h-route flow such that its value is at least as large as the value of any other h-route flow.*

Next, we explain an algorithm that can efficiently calculate the max-h-route flow proposed by Kishimoto and Takeuchi [3]. The running time of that algorithm is $O(hT)$, where T is the computation time of the max-flow problem. Let $G^p = (V, E, c_p)$, where $c_p(e) = \min(c(e), p)$. If p^* is the value such that the max-flow value of G^{p^*} is equal to hp^*, then the max-flow of G^{p^*} is shown to be the max-h-route flow of G. The paper proposes an effective method to search for that p^* based on the max-flow value of G^p for at most h distinct values of p. In the same paper, Kishimoto and Takeuchi also extend the max-flow/min-cut duality property to h-route flow. That duality is as follows.

Definition 2.4 (h-capacity [14]). *The h-capacity of a cut X is given by*
$$\beta_h(X) = \min_{0 \leqslant i \leqslant h-1} \left(\frac{h}{h-i} \cdot \alpha_i(X) \right).$$

A min-h-route cut is a cut minimizing the h-capacities over all the cuts in the network. Now, we can state the h-route duality theorem.

Theorem 2.1 (h-route duality [3,4]). *The value of a max-h-route flow is equal to the h-capacity of a min-h-route cut, $\min_{X \in \mathscr{C}} \beta_h(X)$.*

The definition of h-route flow is extended to the case when h can be non-integer by Aneja et al. [13]. While we omit a precise definition due to the page limitation, we give an equivalent notation of the flow in Definition 2.5. By the definition, the non-integer case share several properties with the integer case. Those properties include the result shown in Lemma 2.1.

Definition 2.5 [13]. *For $h \in \mathbb{R}^+$, a flow F with value v is an h-route flow if $F(e) \leqslant \frac{v}{h}$ for all $e \in E$.*

Lemma 2.1 [13]. *For real number $h < \lambda$, a max-h-route flow value is equal to v^*, if $v^* > 0$ and the max-flow value of $G^{\frac{v^*}{h}}$ is equal to v^*.*

2.2 Max-MLA-Robust Flow and Max-MLA-Reliable Flow [6]

The maximum multilink attack robust flow problem (max-MLA-robust flow) is to find the minimum max-flow value among $\binom{m}{k}$ networks obtained by deleting each set of k edges. Define ϕ_S as the value of a max-flow of the network $(V, E \backslash S)$. The formal definition of max-MLA-robust flow is as follows.

Definition 2.6 (max-MLA-robust flow). *Let $S^* = \operatorname*{argmin}_{S \subseteq E : |S| = k} \phi_S$. A max-MLA-robust flow with k edge failures is a max-flow of the network $(V, E \backslash S^*)$. We denote its value as P_k.*

The maximum multilink attack reliable flow problem (max-MLA-reliable flow) is to find a max-flow of the network such that the flow value is maximum against any set of k edge failures, when deleting the corresponding flow to those k edges in the original flow. In this setting, we will choose a flow $F \in \mathscr{F}$ before the attacker selects the edges to attack. Assume that our choice is F. We will call the value of the flow that remains after the attack the k-effectiveness of F, and we define it as follows.

Definition 2.7 (Effectiveness of a Flow). *Let F be a valid flow, and let ϕ_S^F be a max-flow of a network $G' = (V, E \backslash S, f)$, where $f(e)$ is a value of the flow F on edge e. We define the k-effectiveness of F as $C_F = \min_{S \subseteq E : |S| = k} \phi_S^F$.*

We note that ϕ_S^F, defined in Definition 2.7, is actually the amount of flow F that remains after all the edges in S have been removed. The formal definition of the max-MLA-reliable flow is as follow.

Definition 2.8 (max-MLA-reliable flow). *The max-MLA-reliable flow can be defined as a solution of $F^* = \underset{F \in \mathscr{F}}{\operatorname{argmax}} C_F$.*

For any h and k, we know that the k-effectiveness of max-h-route flow value cannot be larger than the max-MLA-reliable flow value by their definition. Also, it is shown that the max-MLA-reliable flow value cannot be larger than the max-MLA-robust flow value.

Theorem 2.2. *The algorithm for the max-$(k + 1)$-route flow problem in [3] is also a $(k + 1)$-approximation algorithm for the max-MLA-robust flow and the max-MLA-reliable flow problems with k edge failures.*

3 Route-Parametric Multiroute Flow Problem

In this section, we will apply the concept of parametric optimization to the max-h-route flow. The parameter we will consider in this section is the route number h. In other words, we will propose an algorithm to find a function taking a value $h \in \mathbb{R}_+$ to max-h-route flow value.

Define a network G^x as (V, E, c_x) when $c_x(e) := \min(c(e), x)$. One of the most important elements of our algorithm is a parametric function $\mathcal{F}$ that takes a value x to the max-flow value of G^x. We will discuss its properties, and propose an algorithm to find this function in Subsect. 3.1. In Subsect. 3.2, we will extend the notation of the max-h-route flow to the case when h can be non-integer, and use the result of Subsect. 3.1 as a part to solve our problem.

3.1 Parametric Function $\mathcal{F}$

The parametric function $\mathcal{F}$ is first studied in [5], where the following property is shown.

Proposition 3.1 (Piecewise-Linear Property of $\mathcal{F}$ [5]). *Consider a network $G(V, E, c)$ such that $c(e) \in \{1, \ldots, U\}$ for all $e \in U$. The parametric function $\mathcal{F}$ is linear piecewise with at most $|E|^2 U$ line segments.*

Recall P_k the max-MLA-robust flow value with k edge failures defined in Subsect. 2.2. In Theorem 3.1, we give the relationship between $\mathcal{F}$ and those P_k.

Theorem 3.1. $\mathcal{F}(x) = \underset{0 \leqslant i \leqslant \lambda}{\min} (ix + P_i)$.

Proof. Consider a cut $X \in \mathscr{C}$ with a cutset $E = \{e_0, \ldots, e_p\}$ such that $c(e_0) \geqslant c(e_1) \geqslant \cdots \geqslant c(e_p)$. Denote $\mathcal{C}_X(x)$ a capacity of this cut in the graph G^x. We know that $\mathcal{C}_X(x) = \underset{i:c(e_i) < x}{\sum} c(e_i) + \underset{i:c(e_i) \geqslant x}{\sum} x.$

Recall Definition 2.1 where we denote $\sum_{i=\ell}^{p} c(e_i)$ as $\alpha_\ell(X)$. Then, we get $\mathcal{C}_X(x) = (p + 1)x = \alpha_{p+1}(X) + (p + 1)x$ if $x \leqslant c(e_p)$, $\mathcal{C}_X(x) = \alpha_\ell(X) + \ell x$ if $c(e_\ell) < x \leqslant c(e_{\ell-1})$, and $\mathcal{C}_X(x) = \alpha_0(X)$ otherwise.

Next, we show that the step function shown above can be simplified to $C_X(x) = \min_{0 \leqslant j \leqslant p+1} (\alpha_j(X) + jx)$. Let $c_\ell(X) < x \leqslant c_{\ell-1}(X)$. We know from the previous paragraph that $C_X(x) = \alpha_\ell(X) + \ell x$. To prove that $\alpha_\ell(X) + \ell x = \min_{0 \leqslant j \leqslant p+1} (\alpha_j(X) + jx)$, we will show that $\alpha_\ell(X) + \ell x \leqslant \alpha_j(X) + jx$ for any $j \neq \ell$.

$$\text{For } j < \ell, \quad \alpha_j(X) + jx = \sum_{i=j}^{p} c(e_i) + jx = \sum_{i=\ell}^{p} c(e_i) + \sum_{i=j}^{\ell-1} c(e_i) + jx$$

$$\geqslant \sum_{i=\ell}^{p} c(e_i) + (\ell - j)x + jx = \alpha_\ell(X) + \ell x.$$

$$\text{For } j > \ell, \quad \alpha_j(X) + \ell x = \sum_{i=j}^{p} c(e_i) + jx = \sum_{i=j}^{p} c(e_i) + (j - \ell)x + \ell x$$

$$\geqslant \sum_{i=j}^{p} c(e_i) + \sum_{i=\ell}^{j-1} c(e_i) + jx = \alpha_\ell(X) + \ell x.$$

We can also use the similar argument to show the case when $x > c(e_0)$ and $x \leqslant c(e_p)$. Since $\alpha_i(X) = 0$ for $i \geqslant p$, we can further simplify the formula to $C_X(x) = \min_{j \geqslant 0} (\alpha_j(X) + jx)$.

It is easy to see that $\mathcal{F}(x) = \min_{X \in \mathscr{C}} C_X(x)$. Hence,

$$\mathcal{F}(x) = \min_{X \in \mathscr{C}} \min_{i \geqslant 0} (ix + \alpha_i(X)) = \min_{i \geqslant 0} \left(ix + \min_{X \in \mathscr{C}} \alpha_i(X) \right) = \min_{i \geqslant 0} (ix + P_i).$$

Since $P_i = 0$ for $i \geqslant \lambda$, we get $\mathcal{F}(x) = \min_{0 \leqslant i \leqslant \lambda} (ix + P_i)$. $\qquad \square$

By Theorem 3.1, we know that the parametric function $\mathcal{F}$ is a minimum of $\lambda + 1$ linear functions. All of them have a non-negative slope, and one of them is constant. We get the following corollary from that.

Corollary 3.1. *The parametric function $\mathcal{F}$ is linear piecewise, continuous, and derivative non-increasing. The function contains at most $\lambda + 1$ line segments, and there exists x^* such that $\mathcal{F}$ is constant on $[x^*, \infty)$.*

Corollary 3.1 improve Proposition 3.1 in two aspects. We improve the upper bound for the number of line segments from $|E|^2 U$ to $\lambda + 1$, and our result does not need the restriction that the edge capacities have to be an integer less than U.

It is shown in [11, 15] that a function with such properties can be computed efficiently using a method called ES algorithm. The computation time is equal to $O(pT)$ when p is the number of line segments and T is the computation time of a function $\mathcal{F}$ at a specific point x. In our setting, T is the computation time of max-flow, and p is $\lambda + 1$. Hence, we get the following corollary.

Corollary 3.2. *Using ES algorithm, we can find the function $\mathcal{F}$ in $O(\lambda mn)$.*

For the remaining part of this paper, we will assume that the output of ES algorithm is $\mathcal{S} = \{(\eta_0, \mathcal{F}(\eta_0)), \ldots, (\eta_q, \mathcal{F}(\eta_q))\}$, when $0 = \eta_0 < \eta_1 < \cdots < \eta_q$, and the derivation of $\mathcal{F}$ at η_i^- is not equal to the derivation at η_i^+ for $1 \leqslant i \leqslant q$. We call the pair $(\eta_i, \mathcal{F}(\eta_i))$ a breaking point.

Those breaking points can fully described our function $\mathcal{F}$, because we know that

$$\mathcal{F}(x) = \begin{cases} \mathcal{F}(\eta_i) + \mathcal{F}'(\eta_i^+)(x - \eta_i) & \text{for } \eta_i \leqslant x \leqslant \eta_{i+1}, \\ \mathcal{F}(\eta_q) & \text{for } x \geqslant \eta_q, \end{cases}$$

when $\mathcal{F}'(\eta_i^+) := \frac{\mathcal{F}(\eta_{i+1}) - \mathcal{F}(\eta_i)}{\eta_{i+1} - \eta_i}$. We can store the description in $O(\lambda)$ memory as we know from Corollary 3.1 that $q \leqslant \lambda + 1$.

Consider the example of the network in Fig. 1a. First, we will use a costly method [6] based on direct computation of max-MLA-robust flow. By Definition 2.6, we know that $P_0 = 7$, $P_1 = 3$, $P_2 = 1$, and $P_3 = 0$. From Theorem 3.1, we get $\mathcal{F}(x) = \min(7, 3 + x, 1 + 2x, 3x)$ as shown in Fig. 1b. Second, we will obtain an equivalent result by using our ES algorithm. At the initial phase, we calculate a line segment at the points $x = 0^+$ and $x \to \infty$. As a result, we get the lines $y = 3x$ and $y = 7$. Then, we find the intersection of those two functions. The point is $(\frac{7}{3}, 7)$. We test if it is a breaking point of $\mathcal{F}$ by calculating $\mathcal{F}(\frac{7}{3})$. If $\mathcal{F}(\frac{7}{3}) = 7$, the point $(\frac{7}{3}, 7)$ is the breaking point of $\mathcal{F}$. Unfortunately, $\mathcal{F}(\frac{7}{3}) = \frac{16}{3}$. We compute the line corresponding to that point, and get the line $y = 3 + x$. Next, we find the points where the line $y = 3 + x$ intersects $y = 3x$ and $y = 7$, and get $(1.5, 4.5)$ and $(4, 7)$. By calculating $\mathcal{F}(1.5)$ and $\mathcal{F}(4)$, we know that $(1.5, 4.5)$ is not a breaking point, while $(4, 7)$ is. We add the line corresponding to $\mathcal{F}(1.5)$, that is $y = 1 + 2x$. Now, all the cut points are the breaking points of $\mathcal{F}$. The result that we get from ES algorithm is $\{(0, 0), (1, 3), (2, 5), (4, 7)\}$.

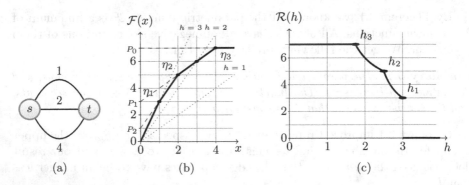

Fig. 1. (a) The network that we consider in this figure (b) the parametric function $\mathcal{F}$ of the network (c) The function $\mathcal{R}$ taking the value h to the max-h-route flow value

3.2   Non-integer Parametric Multiroute Flow

Recall Definition 2.5 and Lemma 2.1. By the definition of function $\mathcal{F}$, we can imply from the definition that a max-h-route flow value is equal to v^* if $v^* = \mathcal{F}(\frac{v^*}{h})$ and $v^* > 0$ for $h < \lambda$. When $h \geqslant \lambda$, we can obtain the max-h-route flow value following the similar idea. The value is $\mathcal{F}(\eta_1)$ when $h = \lambda$, and it is 0 for any $h > \lambda$. Define $\mathcal{R}$ as our desired output function, i.e. $\mathcal{R}$ is a function taking $h \in \mathbb{R}_+$ to the max-h-route flow value. By Lemma 2.1 and Theorem 3.1, we get the following result.

Theorem 3.2. *For* $1 \leqslant i \leqslant q$, *let* $h_i := \frac{\mathcal{F}(\eta_i)}{\eta_i}$. *Also, for* $0 \leqslant i \leqslant q$, *let* $\mu_i := \mathcal{F}'(\eta_i^+) := \frac{\mathcal{F}(\eta_{i+1}) - \mathcal{F}(\eta_i)}{\eta_{i+1} - \eta_i}$ *be a derivation of* $\mathcal{F}$ *at* η_i^+, *and* $\gamma_i := \mathcal{F}(\eta_i) - \mathcal{F}'(\eta_i^+)\eta_i$.

$$\mathcal{R}(h) = \begin{cases} \frac{h\gamma_i}{h - \mu_i} & \text{for } h_{i+1} < h \leqslant h_i, \\ \gamma_q & \text{for } 0 < h \leqslant h_q, \\ 0 & \text{for } h > h_1. \end{cases}$$

Proof. Recall from Theorem 3.1 that $\mathcal{F}(x) = \min\limits_{0 \leqslant i \leqslant \lambda} (ix + P_i)$. Since $P_i > 0$ for $i < \lambda$ and $P_\lambda = 0$, there exists $x^* > 0$ such that $\mathcal{F}(x) = \lambda x + P_\lambda = \lambda x$ for $x \leqslant x^*$. Thus, $(\eta_1, \mathcal{F}(\eta_1)) = (\eta_1, \lambda\eta_1)$, and $h_1 = \lambda$. We know that $\mathcal{F}(h) = 0$ for $h > h_1$ by the argument done previously in this subsection.

Recall from Lemma 2.1 that the max-h-route flow value is equal to hx if $hx = \mathcal{J}(x)$. Our task is to find the point (x, hx) where the line hx cut the function $\mathcal{F}(x)$. That is equivalent to the task of finding a value x such that $\mathcal{L}(x) := \mathcal{F}(x) - hx$ is equal to 0. When $h < \lambda$, the function is increasing for a small x and decreasing after the derivation of $\mathcal{F}$ becomes less than h. When h becomes larger, the point x such that $\mathcal{F}$ begin to decrease comes faster, and the decrement are faster. Thus, the value x such that $\mathcal{L}(x) = 0$ becomes smaller in that case.

Consider the case when $h = h_i$ for some $1 \leqslant i \leqslant q$. Since $h_i := \frac{\mathcal{F}(\eta_i)}{\eta_i}$, we know that $h_i\eta_i = \mathcal{F}(\eta_i)$. Hence, the max-$h$-route flow is $h_i\eta_i$ and the cut point is $(\eta_i, h_i\eta_i)$.

When $h = h_q$, the value x such that $\mathcal{L}(x) = 0$ is η_q. We know from the previous paragraph that x will become larger for $h < h_q$. By Corollary 3.2 and the definition of breaking points, we know that $\mathcal{F}(x) = \mathcal{F}(\eta_q) = \gamma_q$ for all $x \geqslant \eta_q$. Hence, if $h < h_q$, the point that the line hx cut $\mathcal{F}(x)$ is (x^*, γ_q) for some x^*. The max-h-route flow is γ_q.

When $h_{i+1} < h \leqslant h_i$, we know that the value x such that $\mathcal{L}(x) = 0$ is in between η_i and η_{i+1}. Because of that, the line hx cut the function $\mathcal{F}(x)$ at the line segment linking $(\eta_i, \mathcal{F}(\eta_i))$ and $(\eta_{i+1}, \mathcal{F}(\eta_{i+1}))$. By some algebra, we know that the line segment is $\mu_i x + \gamma_i$, and the cut point is $(\frac{\gamma_i}{h - \mu_i}, \frac{h\gamma_i}{h - \mu_i})$. Hence the max-$h$-route flow is $\frac{h\gamma_i}{h - \mu_i}$. $\qquad\square$

By Theorem 3.2, we also get the following corollary.

Corollary 3.3. *The function $\mathcal{R}$ taking the value h to max-h-route flow is hyperbolic piecewise with at most $\lambda + 1$ hyperbolic segments. The function can be computed in $O(\lambda mn)$.*

Proof. From Corollary 3.2, we can get the set of breaking points of $\mathcal{R}$ in $O(\lambda mn)$. Then, we can compute $\mathcal{R}$ from those breaking points in $O(\lambda)$ as shown in Theorem 3.2.   □

Using the result function we get from Corollary 3.3, we can compute the max h-route flow value for any h in $O(\log \lambda)$. Those include the case when $h = \lambda$. In [5], the max-h-route flow algorithm has been improved, but the algorithm still need $O(\lambda mn)$ only to compute the max-λ-route flow in the worst case. Because of this, we can say that our algorithm outputs more general results without increasing the time complexity.

Consider the function $\mathcal{F}$ that we get in Fig. 1b. By Theorem 3.2, we get $h_1 = 3$, $h_2 = \frac{5}{2}$, and $h_3 = \frac{7}{4}$. Also, $\mu_0 = 3$, $\mu_1 = 2$, $\mu_2 = 1$, $\mu_3 = 0$, and $\gamma_0 = 0$, $\gamma_1 = 1$, $\gamma_2 = 3$, $\gamma_3 = 7$. Hence, as shown in Fig. 1c, $\mathcal{R}(h) = 0$ when $h > 3$, $\mathcal{R}(h) = \frac{h}{h-2}$ when $\frac{5}{2} < h \leqslant 3$, $\mathcal{R}(h) = \frac{3h}{h-1}$ when $\frac{7}{4} < h \leqslant \frac{5}{2}$, and $\mathcal{R}(h) = 7$ when $0 < h \leqslant \frac{7}{4}$.

4   Applications to Max-Flow with k Edge Failures

In this section, we apply the result of Sect. 3 to two variants of the max-flow problem, max-MLA-robust flow and max-MLA-reliable flow. Recall the parametric function $\mathcal{F}$ defined in the previous section, we get the following results.

Corollary 4.1. *If there exists a real number x such that $kx + P_k = \mathcal{F}(x)$, then the max-$\left(\frac{\mathcal{F}(x)}{x}\right)$-route flow value is equal to the max-MLA-robust flow and max-MLA-reliable flow value with k edge failures.*

Proof. Let $h := \frac{\mathcal{F}(x)}{x}$. The cut point of lines $y = hx$ and $y = \mathcal{F}(x)$ is $(x, kx + P_k)$. From Theorem 2.1, we know that the max-h-route flow value is $kx + P_k$. By our definition of h-route flow, the amount of flow remaining after k edge attacks is at least $\frac{h-k}{h}(kx + P_k)$. We get

$$\frac{h-k}{h}(kx + P_k) = \frac{\frac{\mathcal{F}(x)}{x} - k}{\frac{\mathcal{F}(x)}{x}}(kx + P_k) = \frac{\frac{kx+P_k}{x} - k}{\frac{kx+P_k}{x}}(kx + P_k) = P_k.$$

Hence, the max-h-route value is equal to max-MLA-robust flow value. Since the max-MLA-reliable flow is in between max-MLA-robust flow and max-h-route flow value, the max-MLA-reliable flow is also equal to that max-h-route flow value.   □

Corollary 4.1 shows that the maximum multiroute flow is not only an approximate result of two robust network problems as in [6], but it provides an exact solutions for the problems if we can find an appropriate route number h. The only weak point of the result is the requirement that there must be some real x such that $kx + P_k = \mathcal{F}(x)$.

We perform a set of experiments to show that such x does exists in most random network using the same setting as in [6]. We generate 100 networks with $|V| = 20$ and $|E| = 80$. To have a network with higher source-sink connectivity, each node is chosen with probability $\frac{1}{|V|} = 0.05$ to be the tail endpoint of an edge. Exceptions to this are the source node and sink node, where the probabilities are $\frac{2}{|V|} = 0.1$, and 0, respectively. Similarly, the probabilities that the source node, the sink node, and the other nodes are chosen to be a head endpoint are 0, $\frac{2}{|V|} = 0.1$, and $\frac{1}{|V|} = 0.05$, respectively. The capacity of each edge was picked uniformly at random between 0 and 20.

In 98 out of those 100 random networks, the requirement in Corollary 4.1 is satisfied for all $0 \leqslant k \leqslant \lambda$. For a larger graph, we also perform an experiment on the case when $|V| = 40$, $|E| = 160$. We also get the similar result for that case, as 98 out of 100 networks are satisfying the equation for all $0 \leqslant k \leqslant \lambda$. Since the value of max-MLA-reliable flow and max-MLA-robust flow are equal when the requirement is satisfied, we can also imply from the experimental results that those two values are equal in most of the random networks.

We also performed experiments on the graph where a small number of edges had capacities that were larger than those of others. Let $G = (V, E, c)$ be a network used for the experiment mentioned in the previous paragraph. We modified the network to $G' = (V, E, c')$, where $c'(e) = \frac{c(e)^h}{20^{h-1}}$ for $2 \leqslant h \leqslant 3$. By doing this, the capacity will follow a power-law distribution with h as the exponent [16]. As a result, we found that all those random networks with a power-law capacity satisfy the equation for all $0 \leqslant k \leqslant \lambda$.

5   Edge-Parametric Multiroute Flow Problem

In this section, we will consider a network such that the capacity of a given edge e can be any real number, while the capacities of the other edges are fixed. From here, h is a fixed integer. We call the edge e as *parametric edge*, and denote the capacity of edge c as a variable z. We refer to $\alpha_i^z(X)$, $\beta_i^z(X)$ as a value $\alpha_i(X)$ and $\beta_i(X)$ when the capacity of that parametric edge is equal to z. Also, we denote $\mathscr{C}_e$ as the set of cuts containing the parametric edge e, and $\overline{\mathscr{C}_e} := \mathscr{C} \backslash \mathscr{C}_e$.

Lemma 5.1. *For any cut $X \in \mathscr{C}_e$, $\alpha_i^z(X) = \min(\alpha_{i-1}^0(X), \alpha_i^0(X) + z)$, for any $i \geqslant 1$ and $\alpha_0^z(X) = \alpha_0^0(X) + z$.*

Proof. The first statement is obvious. To prove the second statement, let the cut-set of X be $\{e_0, e_1, \ldots, e_p\} \cup \{e\}$ where $c(e_0) \geqslant c(e_1) \geqslant \cdots \geqslant c(e_p)$ and $e_i \neq e$ for all i. We note that $\alpha_i^0(X) = \sum_{j=i}^{p} c(e_j)$.

If $z \geqslant c(e_{i-1})$, then $\{e_0, \ldots, e_{i-2}, e\}$ will be the set of edges with the i largest capacities. Hence, $\alpha_i^z(X) = \sum_{j=i-1}^{p} c(e_j) = \alpha_{i-1}^0(X)$.

If $z < c(e_{i-1})$, then $\{e_0, \ldots, e_{i-1}\}$ will be the set of edges of size i with the i largest capacities. Hence, $\alpha_i^z(X) = \sum_{j=i}^{p} c(e_j) + z = \alpha_i^0(X) + z$. $\square$

Theorem 5.1. *Let $C_M(z)$ be a value of a max-h-flow when the capacity of the parametric edge is z. The function C_M is $(h+1)$-piecewise linear, continuous, and derivative non-increasing. Also, there exists some z^* such that the derivative of C_M is 0 for all $z \geqslant z^*$.*

Proof. Let $P_i^z = \min\limits_{X \in \mathscr{C}} \alpha_i^z(X)$. We know from Lemma 5.1 that

$$P_i^z = \min\left(\min\limits_{X \in \mathscr{C}_e} (\alpha_{i-1}^0(X), \alpha_i^0(X) + z), \min\limits_{X \in \mathscr{C}} \alpha_i(X) \right).$$

Let $J_i := \min\left(\min\limits_{X \in \mathscr{C}_e} \alpha_{i-1}^0(X), \min\limits_{X \in \mathscr{C}} \alpha_i(X) \right)$, and $H_i := \min\limits_{X \in \mathscr{C}_e} \alpha_i^0(X)$. We get $P_i^z = \min(J_i, H_i + z)$. By Definition 2.4, we know that $C_M(z) = \min\limits_{0 \leqslant i < h} \left(\frac{h P_i^z}{h-i} \right)$.

Thus, $C_M(z) = \min\left(J, \min\limits_{0 \leqslant i < h} \left[\frac{h(H_i+z)}{h-i} \right] \right)$, when $J := \min\limits_{0 \leqslant i < h} \left(\frac{h J_i}{h-i} \right)$. It is easy to see that $\mu_i(z) := \frac{h(H_i+z)}{h-i}$ is a linear function of z. Since the function C_M is the minimum of those linear functions, we know that the function is $(h+1)$-piecewise linear, continuous, and derivative non-increasing. As we know that the functions μ_i is increasing, there exists some z^* such that $\mu_i(z) \geqslant J$ for all i and $z \geqslant z^*$. We get $C_M(z) = J$ for those z. □

The theorem leads us to the following corollary.

Corollary 5.1. *Using ES algorithm, the breaking points of function C_M can be found in $O(h^2mn)$.*

Proof. As shown in Theorem 5.1, the function C_M at most $h+1$ line segments. The calculation of $C_M(z)$ at a specific z is the computation of one max-h-route flow value. By the algorithm in [5], we can compute the flow value in $O(hmn)$. Hence, it takes $O(h^2mn)$ to find all breaking points of C_M. □

Similar to Sect. 3, we can calculate $C_M(z)$ for a specific z from the set of breaking points obtained from ES algorithm. Assume that the set is $\{(\eta_0, C_M(\eta_0)), \ldots, (\eta_q, C_M(\eta_q))\}$. The calculation can be done as follows.

$$C_M(z) = \begin{cases} (C_M(\eta_{i+1}) - C_M(\eta_i))\frac{z - \eta_i}{\eta_{i+1} - \eta_i}, & \text{for } \eta_i \leqslant z \leqslant \eta_{i+1} \\ C_M(\eta_q), & \text{for } z \geqslant \eta_q \end{cases}$$

6   Conclusion and Future Works

When we tune parameters in our network, the conventional way to evaluate each parameter value is the amount of max-flow. However, the amount of max-h-route flow can also be another design criteria, as it is an approximate value of the max-flow when edges are attacked. We propose algorithms that help that tuning in this work. The algorithms output a trade-off function between edge capacity, route number h, and the max-h-route flow value.

In a specific class of network, we can prove that the max-h-route flow can be an exact solution of max-flow with k edge attacks, if h is optimally chosen. Our experimental results show that most of the random networks are in that class, and we are currently finding their theoretical properties.

Acknowledgement. We would like to thank Prof. Abdel Lisser and anonymous reviewers for giving us valuable comments during the course of this research.

References

1. Ford, L.R., Fulkerson, D.R.: Flows in Networks. Princeton University Press, Princeton (1955)
2. Orlin, J.B.: Max flows in $O(nm)$ time, or better. In: Proceedings of the STOC 2013, pp. 765–774 (2013)
3. Kishimoto, W.: A method for obtaining maximum multi-route flows in a network. Networks **27**(4), 279–291 (1996)
4. Bagchi, A., Chaudhary, A., Kolman, P., Sgall, J.: A simple combinatorial proof of duality of multiroute flows and cuts. Technical report, Charles Univ. (2004)
5. Aggarwal, C., Orlin, J.B.: On multi-route maximums flows in networks. Networks **39**(1), 43–52 (2002)
6. Baffier, J.-F., Suppakitpaisarn, V.: A $(k + 1)$-approximation robust network flow algorithm and a tighter heuristic method using iterative multiroute flow. In: Pal, S.P., Sadakane, K. (eds.) WALCOM 2014. LNCS, vol. 8344, pp. 68–79. Springer, Heidelberg (2014)
7. Minoux, M.: On robust maximum flow with polyhedral uncertainty sets. Optim. Lett. **3**(3), 367–376 (2009)
8. Lee, P.P., Misra, V., Rubenstein, D.: Distributed algorithms for secure multipath routing. In: Proceedings of the INFOCOM 2005, vol. 3, pp. 1952–1963. IEEE (2005)
9. Kishimoto, W., Takeuchi, M.: A method for obtaining the maximum δ-reliable flow in a network. IEICE Trans. Fundam. **81**(5), 776–783 (1998)
10. Diallo, M., Gueye, S., Berthomé, P.: Impact of a varying capacity on the all pairs 2-route network flows. Electron. Notes Discrete Math. **35**(3), 59–64 (2009)
11. Eisner, M.J., Severance, D.G.: Mathematical techniques for efficient record segmentation in large shared databases. JACM **23**(4), 619–635 (1976)
12. Kolmogorov, V., Boykov, Y., Rother, C.: Applications of parametric maxflow in computer vision. In: Proceedings of the ICCV 2007, pp. 1–8 (2007)
13. Aneja, Y.P., Chandrasekaran, R., Kabadi, S.N., Nair, K.: Flows over edge-disjoint mixed multipaths and applications. Discrete Appl. Math. **155**(15), 1979–2000 (2007)
14. Chandrasekaran, R., Nair, K., Anejac, Y., Kabadib, S.: Multi-terminal multipath flows: synthesis. Discrete Appl. Math. **143**, 182–193 (2004)
15. Hulgeri, A., Sudarshan, S.: Parametric query optimization for linear and piecewise linear cost functions. In: Proceedings of the VLDB 2002, pp. 167–178 (2002)
16. Faloutsos, M., Faloutsos, P., Faloutsos, C.: On power-law relationships of the internet topology. ACM SIGCOMM Comput. Commun. Rev. **29**, 251–262 (1999)

The Dominating Set Polytope
via Facility Location

Mourad Baïou[1]($\boxtimes$) and Francisco Barahona[2]

[1] CNRS and Université Clermont II, Campus des Cézeaux, BP 125,
63173 Aubière Cedex, France
baiou@isima.fr
[2] IBM T. J. Watson Research Center, Yorktown Heights, NY 10589, USA

Abstract. In this paper we present an extended formulation for the dominating set polytope via facility location. We show that with this formulation we can describe the dominating set polytope for cacti graphs, though its description in the natural node variables dimension has been only partially obtained. Moreover, the inequalities describing this polytope have coefficients in $\{-1, 0, 1\}$. This is not the case for the dominating set polytope in the node-variables dimension. It is known from [1] that for any integer p, there exists a facet defining inequality having coefficients in $\{1, \ldots, p\}$. We also show a decomposition theorem by means of 1-sums. Again this decomposition is much simpler with the extended formulation than with the node-variables formulation given in [2].

1 Introduction

Let $G = (V, A)$ be a directed graph, not necessarily connected, where each arc and each node has a cost (or a profit) associated with it. Consider the following version of the *uncapacitated facility location problem* (UFLP), where each location $v \in V$ has a weight $w(v)$ that corresponds to the revenue obtained by opening a facility at that location, minus the cost of building this facility. Each arc $(u, v) \in A$ has a weight $w(u, v)$ that represents the revenue obtained by assigning the customer u to the opened facility at location v, minus the cost originated by this assignment. The goal is to select some nodes where facilities are opened and assign to them the non selected node in such a way that the overall profit is maximized. This version of the $UFLP$ is called the *prize-collecting uncapacitated facility location* (pc-UFLP). The following is a natural linear relaxation of the pc-UFLP.

$$\max \quad \sum_{(u,v)\in A} w(u,v)x(u,v) + \sum_{v\in V} w(v)y(v) \tag{1}$$

$$\sum_{(u,v)\in A} x(u,v) + y(u) \leq 1 \quad \forall u \in V, \tag{2}$$

$$x(u,v) \leq y(v) \quad \forall (u,v) \in A, \tag{3}$$

This work has been supported by project PICS05891, CNRS-IBM.

$$x(u,v) \geq 0 \quad \forall (u,v) \in A, \tag{4}$$

$$y(v) \geq 0 \quad \forall v \in V. \tag{5}$$

Let $P(G)$ be the polytope defined by (2)–(4), and let $UFLP'(G)$ be the convex hull of $P(G) \cap \{0,1\}^{|V|+|A|}$. Clearly $UFLP'(G) \subseteq P(G)$.

Given a directed graph $G = (V, A)$, a subgraph induced by the nodes $v_1, \ldots, v_r$ of G is called a *bidirected cycle* if the only arcs in this induced subgraph are (v_i, v_{i+1}) and (v_{i+1}, v_i), for $i = 1, \ldots, r$, with $v_{r+1} = v_1$. We denote it by BIC_r. The first part of this paper is devoted to the study of $UFLP'(G)$, when G is a bidirected cycle. At first sight, the description of $UFLP'(BIC_n)$ seems easy because of the simple structure of BIC_n. We will show that we need to add the so-called *lifted g-odd cycle inequalities*, to complete its description. These inequalities define facets of $UFLP'(BIC_n)$, and are valid for $UFLP'(G)$ for any graph G. We also give a linear time algorithm to separate these inequalities.

To complete the description of $UFLP'(G)$ in a more general class of graphs, we consider the graphs $G = (V, A)$ that decompose by means of 1-sum. As a consequence we obtain a complete description of $UFLP'(G)$ when G can be decomposed as 1-sums of bidirected cycles.

In the second part of this paper we discuss the consequences of these results when applied to the dominating set problem. More precisely, let $G = (V, E)$ be an undirected graph. A subset $D \subseteq V$ is called a *dominating set* if every node of $V \setminus D$ is adjacent to a node of D. The *minimum weight dominating set problem* (MWDSP) is to find a dominating set D that minimizes $\sum_{v \in V} w(v)$, where $w(v)$ is a weight associated with each node $v \in V$. A natural linear relaxation of the MWDSP is defined by the linear program below

$$\min \sum_{v \in V} w(v) x(v) \tag{6}$$

$$x(N[v]) \geq 1 \quad \forall v \in V, \tag{7}$$

$$x(v) \geq 0 \quad \forall v \in V, \tag{8}$$

$$x(v) \leq 1 \quad \forall v \in V, \tag{9}$$

where $N[v]$ denotes the set of neighbors of v including it. Define $DSP(G)$ to be the convex hull of the integer vectors satisfying (7)–(9).

The MWDSP is a special case of the set covering problem. It is NP-hard even when all the weights are equal to 1, this can be shown using a simple reduction from the vertex cover problem. A large literature is devoted to this case and many of its variants, for a deep understanding of the subject we refer to [3,4]. It has been shown that when the weights are all equal to 1, the MWDSP is solvable in many classes of graphs, a non-exhaustive list is cactus graphs, series-parallel graphs, permutation graphs, cocomparability graphs (see Chap. 2 in [4] for more classes). For the weighted case of the MWDSP we have a short list of graphs where this problem can be solved in polynomial time, for threshold graphs [5], for cycles [6] and for strongly chordal graphs [7]. Little is known from the point of view of polyhedral approach and particularly few complete characterizations of the polytope associated with the MDWSP are known. For the case of strongly chordal graph Farber [7] gives a primal-dual algorithm to solve the MWDSP this shows that $DSP(G)$ is defined by (7)–(9). $DSP(G)$ has been described for threshold graphs [5]. And it has been, first, characterized for cycle graphs in [6] and later published in [1]. This result has also been established in [8] using a different approach. One can also use the results related to the set covering polytope [9–13], to cite a few, to establish new results for the MWDSP. The set covering polytope is the convex hull of $\{x \in \mathbb{R}^n : Ax \geq 1, x \in \{0,1\}^n\}$, where A is an $m \times n$ matrix with $0,1$ entries. For example, the polytope $DSP(G)$ when G is a cycle with n nodes coincide with the set covering polytope when A is the C_n^3 circulant matrix. Recently in [14] a complete description of the set covering polytope is established when A is the circulant matrix C_{2k}^k or C_{3k}^k, $k \geq 3$.

We give an extended formulation via facility location to completely characterize the $DSP(G)$ when G is a cactus. This description has been studied in the original dimension that is $\mathbb{R}^{|V|}$ in [1,6]. They developed several facet defining inequalities for this case, and showed that this polytope has a more complicated structure than the case when G is a cycle. Even with the 1-sum composition developed in [2], the complete characterization of $DSP(G)$ in cactus graphs has not been found. The main difficulty reported in [1,6] is the description of the polytope when restricted to the auxiliary graphs obtained after the decomposition. In our work we show that with the extended formulation this task is easy and allows us to completely describe this polytope in a higher dimension. Moreover in [1,6], it has been shown that for any fixed integer p, there exist a cactus G such that $DSP(G)$ has a facet defining inequality with coefficients $1, \ldots, p$. In our description all the facets defining inequalities have coefficients in $\{0, -1, +1\}$.

This paper is organized as follows. In Sect. 2, we give some useful definitions and notations. Section 3 is devoted to the characterization of $UFLP'(G)$ when G is a bidirected cycle. In Sect. 4, we show how the results of the previous sections apply to the dominating set polytope using a composition theorem. Finally, in Sect. 5 we present the algorithmic consequences of our approach. In particular, we devise the first polynomial time algorithm to solve the MWDSP in cacti. This is done via a linear time separation algorithm of the inequalities we introduced.

2   Definitions and Notations

Recall that a bidirected cycle BIC_r of a directed graph $G = (V, A)$ is a sequence of nodes $v_1, \ldots, v_n$ in V and arcs (v_i, v_{i+1}), (v_{i+1}, v_i) in A, for $i = 1, \ldots, n$, where $v_{n+1} = v_1$. The arcs of BIC_n are denoted by $A(BIC_n)$. To simplify the notation, we will denote the nodes of BIC_n by $1, \ldots, n$, and the arcs by $(i, i+1)$ and $(i+1, i)$ for $i = 1, \ldots, n$. When we use numbers $i + j$ or $i - j$, $i, j \in \{1, n\}$, the positive numbers are taken modulo n and the negative ones are taken modulo $-n$. The number zero represents the node n. A *bidirected path* P of the graph BIC_n is an ordered sequence of consecutive nodes of BIC_n, where the arcs $(i, i+1)$ and $(i+1, i)$ of any two consecutive nodes i and $i+1$ of P, are both considered in the path. Here $i + 1$ is taken modulo n. The size of P

is the number of its nodes minus one. Given a directed graph $G = (V, A)$ its intersection graph denoted by $I(G)$ is obtained by associating a node for each arc of A. Two nodes are adjacent if the tail of one of the corresponding arcs coincides with the tail or the head of the other corresponding arc. It is easy to see that $I(BIC_n)$ consists of the following *circulant graph* $G_{2n} = (A(BIC_n), E)$, where $A = \{a_1, \ldots, a_{2n}\}$ and the set of edges E consists of the edges $\{a_i, a_{i+1}\}$ and $\{a_i, a_{i+2}\}$, for $i = 1, \ldots, 2n$; the indices are taken modulo $2n$.

For a directed graph $D = (V, A)$, and $S \subseteq V$, we denote by $\delta^+(S)$ the set of arcs $(u, v) \in A$ with $u \in S$ and $v \in V \setminus S$. For a node $v \in V$ we write $\delta^+(v)$ instead of $\delta^+(\{v\})$. If there is a risk of confusion we use δ_G^+.

Given an undirected graph $G = (V, E)$, a subset $S \subseteq V$ is called *stable* if there is no edge between any pair of nodes of S. The convex hull of the incidence vectors of the stable sets in G is called the *stable set polytope* and is denoted by $SSP(G)$. When each node $v \in V$ has an associated weight $w(v)$, the *maximum weight stable set problem* (MWSSP) is to find a stable set $S \subseteq V$ maximizing $\sum_{v \in S} w(v)$. A set $K \subseteq V$ is called a *clique* if there is an edge between every pair of nodes in K.

For a ground set U and a function f from U to $\mathbb{R}$, we use $f(S)$ to denote $f(S) = \sum_{a \in S} f(a)$, whenever $S \subseteq U$.

3   The Characterization of $UFLP'(BIC_n)$

First we will give two families of valid inequalities for $UFLP'(G)$, when G is any directed graph.

Let $G = (V, A)$ be any directed graph. Let BIC_r a bidirected cycle included in G. The inequality below is called a *bidirected cycle inequality* and has been introduced in [15],

$$\sum_{a \in A(BIC_r)} x(a) \leq \left\lfloor \frac{2|r|}{3} \right\rfloor. \tag{10}$$

Now let us introduce the g-odd cycle inequalities. For any directed graph $G = (V, A)$, a simple cycle C is an ordered sequence $v_0, a_0, v_1, a_1, \ldots, a_{p-1}, v_p$, where $v_0 = v_p$ and for $i = 0, \ldots, p-1$, v_i and a_i are distinct nodes and arcs, respectively. For $i = 0, \ldots, p-1$, the nodes v_i and v_{i+1} are the endnodes of a_i.

By setting $a_p = a_0$, we associate with C three more sets as below.

- We denote by $\hat{C}$ the set of nodes v_i, such that v_i is the head of a_{i-1} and also the head of a_i, $1 \leq i \leq p$.
- We denote by $\dot{C}$ the set of nodes v_i, such that v_i is the tail of a_{i-1} and also the tail of a_i, $1 \leq i \leq p$.
- We denote by $\tilde{C}$ the set of nodes v_i, such that either v_i is the head of a_{i-1} and also the tail of a_i, or v_i is the tail of a_{i-1} and also the head of a_i, $1 \leq i \leq p$.

Notice that $|\hat{C}| = |\dot{C}|$. A cycle will be called *g-odd* (generalized odd) if $p + |\hat{C}|$ (or $|\dot{C}| + |\tilde{C}|$) is odd, otherwise it will be called *g-even*. A cycle C with $\dot{C} = \hat{C} = \emptyset$ is a *directed* cycle. The set of arcs in C is denoted by $A(C)$.

Let C be a g-odd cycle. Now we define a set of arcs $\tilde{A}(C)$ as follows. For each node $v_i \in \dot{C}$ we have two cases. Let v_{i-1} and v_{i+1} be the two neighbors of v_i in C.

- If v_{i-1} and v_{i+1} are in $\tilde{C}$, we pick arbitrarily one arc from $\{(v_{i-1}, v_i), (v_{i+1}, v_i)\}$ and add it to $\tilde{A}(C)$.
- If only one of the neighbors of v_i is in $\tilde{C}$, say the node $v_j \in \{v_{i-1}, v_{i+1}\}$. We add (v_j, v_i) to $\tilde{A}(C)$.

Once the *lifting set* $\tilde{A}(C)$ has been defined, a *lifted g-odd cycle inequality* has the form

$$\sum_{a \in A(C)} x(a) + \sum_{a \in \tilde{A}(C)} x(a) - \sum_{v \in \hat{C}} y(v) \leq \frac{|\tilde{C}| + |\hat{C}| - 1}{2}. \tag{11}$$

One can easily show that this is a Gomory-Chvátal cut of rank one. Notice that given a g-odd cycle C, we might have several lifting sets $\tilde{A}(C)$, therefore we might have several lifted g-odd cycle inequalities. Similar inequalities called lifted odd cycle inequalities have been studied in [16–19].

The main result of this section is the following theorem.

Theorem 1. *$UFLP'(BIC_n)$ is described by the constraints (2)–(5), the bidirected cycle inequality (10) with respect to BIC_n and the lifted g-odd cycle inequalities (11).*

The remainder of this section is devoted to prove this theorem. It is easy to see that $UFLP'(G)$ is full dimensional for any graph G. Now assume that

$$\alpha x + \beta y \leq \rho, \tag{12}$$

is a valid inequality defining a facet of $UFLP'(BIC_n)$. Let $F_{\alpha,\beta} = \{(x,y) \in UFLP(G) \cap \{0,1\}^{|V|+|A|} : \alpha x + \beta y = \rho\}$. We will show that (12) is one of the inequalities (2)–(5), (10) or (11). We assume in this section that (12) is different from (2)–(5) and (10). We will recall this when needed. In the proof we will implicitly use the following remark.

Remark 1. There exist always a feasible 0-1 solution in $F_{\alpha,\beta}$ that satisfies inequalities (2)–(5) as a strict inequalities (not necessarily at the same time). Otherwise (12) is one of the inequalities (2)–(5).

Now we give a series of technical lemmas that will be used in the discussion that complete the proof in Sect. 3.1. For a detailed proofs see [20].

Lemma 1. *We have $\alpha(u,v) \in \{0,1\}$ for each $(u,v) \in A(BIC_n)$ and $\beta(u) \in \{0,-1\}$ for each $u \in V(BIC_n)$.*

Proof. The main idea of the proof is a transformation to the stable set polytope. We add a slack variable to each inequality (2), then we eliminate the y's variables using the equations obtained from (2) after the additions of the slack variables. It is not difficult to see that the convex hull of the 0-1 solutions in this new system is exactly the stable set polytope of a graph $H = (U, E)$. Each column corresponds to a node in H, and two nodes are adjacent if there is some inequality so that the two respective columns appear with non zero coefficients. We can observe that this graph is quasi-line. Using the results in [21], we show that the inequalities defining the stable set polytope in this new graph can have coefficients in $\{0, 1, 2\}$. And we know that any valid inequality of $UFLP'(G)$ can be obtained from a valid inequality of that stable set polytope by eliminating the slack variables using the equations obtained from (2). This yield to a valid inequality with coefficients in $\{0, 1\}$ for the x's variables and with coefficients in $\{0, -1\}$ for the y's variables.   □

The following four lemmas are easy to prove, see [20].

Lemma 2. *We cannot have* $\alpha(u, v) = 1$ *for all* $(u, v) \in A(BIC_n)$ *and* $\beta(u) = -1$ *for all* $u \in V(BIC_n)$.

Lemma 3. *Let i be a node of BIC_n with $\beta(i) = -1$. Then $\alpha(i + 1, i) = \alpha(i - 1, i) = 1$.*

Lemma 4. *Let i be a node of BIC_n with $\beta(i) = -1$. If $\alpha(i, i - 1) = \alpha(i - 1, i) = 1$, then $\beta(i - 1) = -1$.*

Lemma 5. *Let i be a node of BIC_n with $\beta(i) = -1$. If $\alpha(i, i - 1) = 1$, then $\alpha(i, i + 1) = 1$.*

Lemma 6 below summarizes the implications of Lemmas 2, 3, 4 and 5.

Lemma 6. *Let i be a node of BIC_n with $\beta(i) = -1$. Then the following assumptions hold*

(a1) $\alpha(i + 1, i) = \alpha(i - 1, i) = 1$, and
(a2) $\alpha(i, i - 1) = \alpha(i, i + 1) = 0$.

Proof. (a1) is obtained from Lemma 3. Now if we suppose that (a2) is not true, then Lemma 4 and Lemma 5 imply that $\alpha(u, v) = 1$ for each $(u, v) \in A(BIC_n)$ and $\beta(u) = -1$ for each $u \in V(BIC_n)$. But this contradicts Lemma 2.   □

Lemma 7. *If $\alpha(i - 1, i) = 1$ and $\beta(i) = 0$, then $\alpha(i, i + 1) = 1$.*

Proof. There is a vector $x \in F_{\alpha,\beta}$ with $y(i - 1) + x(i - 1, i) + x(i - 1, i - 2) = 0$.

- If $y(i) = 1$, we set $x(i - 1, i) = 1$ and violate the inequality; so $y(i) = 0$.
- If $x(i, i + 1) = 0$, then we can set $y(i) = 1$ and proceed as before; so $x(i, i + 1) = 1$.
- If $\alpha(i, i+1) = 0$, we set $x(i, i+1) = 0$ and proceed as before; so $\alpha(i, i+1) = 1$.□

Lemma 8. *Suppose that we are not dealing with the bidirected cycle inequality. If $\alpha(i, i + 1) = \alpha(i + 1, i) = 1$ then $\alpha(i + 2, i + 1) = \alpha(i - 1, i) = 0$.*

Proof. Assume $i = 1$. The proof is based on the statements below.

– It follows from Lemma 6 that $\beta(1) = \beta(2) = 0$.
– It follows from Lemma 7 that $\alpha(2,3) = \alpha(1,n) = 1$.
– Since this is not a bidirected cycle inequality, we assume that there is an index $k \geq 2$ such that:
  - $\beta(j) = \beta(j+1) = 0$, $\alpha(j, j+1) = \alpha(j+1, j) = 1$, for $1 \leq j \leq k$.
  - $\alpha(n, 1) = \alpha(k+2, k+1) = 0$.
  - $\alpha(1, n) = \alpha(k+1, k+2) = 1$.
– There is a vector $x \in F_{\alpha,\beta}$ with $y(k-1) + x(\delta^+(k-1)) = 0$. We modify x as below to obtain a vector that violates the inequality.
  - If $y(k) = 1$ we just set $x(k-1, k) = 1$.
  - If $y(k) = 0$ and $x(k, k+1) = 0$, we set $y(k) = 1$ and proceed as above.
  - If $y(k) = 0$ and $x(k, k+1) = 1$, we set $y(k+1) = x(k, k+1) = x(k+2, k+1) = 0$, and $y(k) = x(k-1, k) = x(k+1, k) = 1$. □

Lemma 9. *If $\alpha(i-1, i) = \alpha(i+1, i) = 1$, then $\beta(i) = -1$.*

Proof. Suppose $\beta(i) = 0$. It follows from Lemma 7 that $\alpha(i, i-1) = \alpha(i, i+1) = 1$. This contradicts Lemma 8. □

Lemma 10. *We have at least one of the values $\alpha(i, i+1)$ or $\alpha(i+1, i)$ equal to 1, for each $i = 1, \ldots, n$.*

3.1 The Proof of Theorem 1

Let G_α be the graph induced by the arcs $(i, j) \in A(BIC_n)$ with $\alpha(i, j) = 1$, we call this graph the *support graph* of (12). Recall that a bidirected path P of a graph $G = (V, A)$ is a sequence of nodes $P = 1, 2, \ldots, k$ with $(i, i+1)$ and $(i+1, i)$ are both in A, for $i = 1, \ldots, k-1$. The size of P is $k-1$. We say that P is *maximal* if we cannot extend it to a bidirected path from one of its endnodes.

Notice that by definition the support graph of any g-odd lifted cycle inequality satisfy the following three properties

– it contains a cycle as a subgraph,
– each maximal bidirected path is of size 1. Moreover, if $P = i, i+1$ is such a path, then $(i-1, i)$ and $(i+2, i+1)$ do not appear, and
– if C is the lifted cycle and i a node in $\dot{C}$, then the support graph must contain exactly one of the arcs $(i-1, i)$ or $(i+1, i)$ when both nodes $i-1$ and $i+1$ are in $\tilde{C}$, it contains none of the arcs if both of these nodes are in $\hat{C}$ and finally if, say $i+1$ is in $\tilde{C}$, we must have the arc $(i+1, i)$.

Let us see that these properties are satisfied by G_α. Lemma 10 implies that G_α contains at least one cycle as a subgraph. Choose any such a cycle and call it C. Lemma 8 implies that each maximal bidirected path is of size one, and that for any such bidirected path $P = i, i+1$ the arcs $(i-1, i)$ and $(i+2, i+1)$ are not in G_α. Again Lemma 10 implies that $(i, i-1)$ and $(i+1, i+2)$ belong to G_α.

Let $i \in \dot{C}$, and let $i-1$ and $i+1$ be the neighbors of i in G_α. Notice that G_α must contain at most one of the arcs $(i-1, i)$ and $(i+1, i)$ since the size of maximal bidirected path is one.

If both $i - 1$ and $i + 1$ are in $\hat{C}$, then Lemma 9 implies that $\beta(i - 1) = -1 = \beta(i + 1)$, and using Lemma 6 we obtain that $\alpha(i - 1, i) = 0 = \alpha(i + 1, i)$. So in this case the arcs $(i - 1, i)$ and $(i + 1, i)$ are not in G_α.

Assume that $i + 1$ is in $\tilde{C}$ and that G_α contains none of the arcs $(i - 1, i)$ or $(i + 1, i)$, that is $\alpha(i - 1, i) = \alpha(i + 1, i) = 0$. By definition $\alpha(i, i - 1) = \alpha(i, i + 1) = 1$. Lemma 6 implies that $\beta(i) = 0$ and since $i + 1$ is in $\tilde{C}$, we must have $\alpha(i + 1, i + 2) = 1$ and then again Lemma 6 implies that $\beta(i + 1) = 0$. We can assume that there is a solution $(x, y) \in F_{\alpha, \beta}$ with $x(i + 1, i) = 1$, otherwise (12) is the trivial inequality $x(i + 1, i) \geq 0$. Now if we set $x(i, i + 1)$ and $y(i + 1)$ to 1; $x(i + 1, i)$ and $y(i)$ to 0 and possibly $x(i - 1, i)$ to 0, we obtain a feasible solution that violates (12). Therefore, we must have exactly one of the arcs $(i - 1, i)$ or $(i + 1, i)$ in G_α. Moreover, if the node $i - 1$ is in $\hat{C}$, Lemma 9 implies that $\beta(i - 1) = -1$, and Lemma 6 implies that $\alpha(i - 1, i) = 0$, so $(i - 1, i)$ is not an arc of G_α.

The above discussion shows that the support graph G_α coincides with the support graph of the lifted g-odd cycle inequality defined from C. Moreover, from Lemma 6, each node i with $\beta(i) = -1$ must be in $\hat{C}$. And from Lemma 9, for each node $i \in \hat{C}$ we have $\beta(i) = -1$. For a g-odd cycle inequality it is easy to find a 0-1 vector of $UFLP'(BIC_n)$ that satisfies it with equation. Then we have $\rho \geq (|\hat{C}| + |\tilde{C}| - 1)/2$. Now the proof of Theorem 1 is complete.

4 Application to the Dominating Set Polytope

Let $G = (V, E)$ an undirected connected graph. The graph G is a *cactus* if each edge of G is contained in at most one cycle of G. For example every tree is a cactus. The main result of this section is a complete description of the dominating set polytope $DSP(G)$ in $\mathbb{R}^{|V| + 2|E|}$ when G is a cactus. This description can be seen as an extended formulation of $DSP(G)$. We will show that with this extended formulation, to obtain the polytope associated with a cactus, it suffices to characterize the polytope associated with the maximal two-connected components. Given an undirected graph $G = (V, E)$. We say that G is a 1-sum of $G_1 = (V_1, E_1)$ and $G_1 = (V_1, E_1)$ if $|V_1 \cap V_2| = 1$, $V = V_1 \cup V_2$, $E = E_1 \cup E_2$. Consider the following equalities obtained from (2).

$$\sum_{(u,v) \in A} x(u, v) + y(u) = 1 \quad \forall u \in V. \tag{13}$$

Define $UFLP(G)$ to be the convex hull of the feasible 0-1 vectors satisfying (13) and (3)–(5). This is the classical uncapacitated facility location polytope. Now given an undirected graph $G = (V, E)$, define the directed graph $\overleftrightarrow{G} = (V, A)$ that have the same node-set as G, and its arc-set A is defined from E by replacing each edge $uv \in E$ by two arcs (u, v) and (v, u).

Lemma 11. *For any undirected graph $G = (V, E)$, the projection of $UFLP(\overleftrightarrow{G})$ onto the y's variables is exactly $DSP(G)$.*

Proof. We have to prove, $DSP(G) = \{y \mid$ there is a vector x such that $(x, y) \in UFLP(\overleftrightarrow{G})\}$. First consider $\bar{y} \in DSP(G)$. We have $\bar{y} = \sum \alpha_i y^i$, $\sum \alpha_i = 1$, $\alpha \geq 0$, where $\{y^i\}$ are extreme points of $DSP(G)$. Consider now a particular vector y^k. Let $D^k = \{u \mid y^k(u) = 1\}$. For each $v \in V \setminus D^k$, there is at least one of its neighbors in D^k, w_v say. We set $x^k(v, w_v) = 1$. We set $x^k(i, j) = 0$ for all other arcs (i, j) in $\overleftrightarrow{G}$. Each vector (x^k, y^k) is an extreme point of $UFLP(\overleftrightarrow{G})$. So $(\bar{x}, \bar{y}) = \sum \alpha_i(x^i, y^i)$ is a vector in $UFLP(\overleftrightarrow{G})$. Consider now $(\bar{x}, \bar{y}) \in UFLP(\overleftrightarrow{G})$. We have $(\bar{x}, \bar{y}) = \sum \alpha_i(x^i, y^i)$, $\sum \alpha_i = 1$, $\alpha \geq 0$, where each vector (x^i, y^i) is an extreme point of $UFLP(\overleftrightarrow{G})$. Then each vector y^i is the incidence vector of a dominating set D^i, therefore it is an extreme point of $DSP(G)$. Then $\bar{y} = \sum \alpha_i y^i$ is a vector in $DSP(G)$. □

Theorem 2 [22]. *Let D be a directed graph that is a 1-sum of $D_1 = (V_1, A_1)$ and $D_2 = (V_2, A_2)$, with $V_1 \cap V_2 = \{u\}$. Let D_1' be the graph obtained from D_1 by replacing u with u', and D_2' is obtained from D_2 by replacing u with u''. Suppose that the system*

$$Az' \leq b \tag{14}$$

$$z'\left(\delta_{D_1'}^+(u')\right) + z'(u') \leq 1 \tag{15}$$

describes $UFLP'(D_1')$. Suppose that (14) contains the inequalities (2)–(5) except for (15). Similarly suppose that

$$Cz'' \leq d \tag{16}$$

$$z''\left(\delta_{D_2'}^+(u'')\right) + z''(u'') \leq 1 \tag{17}$$

describes $UFLP'(D_2')$. Also (16) contains the inequalities (2)–(5) except for (17). Then the system below describes an integral polyhedron.

$$Az' \leq b \tag{18}$$

$$Cz'' \leq d \tag{19}$$

$$z'\left(\delta_{D_1'}^+(u')\right) + z''\left(\delta_{D_2'}^+(u'')\right) + z'(u') \leq 1 \tag{20}$$

$$z'(u') = z''(u''). \tag{21}$$

Thus the theorem below follows from Theorem 1 and Theorem 2.

Theorem 3. *If G is a cactus, then $UFLP'(\overleftrightarrow{G})$ is described by the constraints (2)–(5), the bidirected cycle inequalities (10), and the lifted g-odd cycle inequalities (11).*

$UFLP(\overleftrightarrow{G})$ is a face of $UFLP'(\overleftrightarrow{G})$. From Lemma 11, $DSP(G)$ is a projection of $UFLP(\overleftrightarrow{G})$. Therefore we have an extended formulation for $DSP(G)$.

5   Algorithmic Consequences

In [1] the authors give the first polynomial algorithm to solve the minimum weighted dominating set problem (MWDSP) in a cycle. They showed that the separation of the inequalities defining the dominating set polytope in a cycle can be done in $O(n^2)$. Below we will show that the separation of our inequalities can be done in linear time. From Theorem 3 it suffices to develop a polynomial time algorithm to solve the separation problem associated with inequalities (10) and (11). Recall that $\overleftrightarrow{G}$ can be decomposed by means of 1-sum into bidirected cycles and bidirected paths of size one. The number of bidirected cycles is at most $\dfrac{n}{3}$, where n is the number of nodes of G. It follows that one can easily introduce the bidirected cycle inequalities (10) in any linear program. Therefore we only need to solve the separation problem for the lifted g-odd inequalities (11) for each component of $\overleftrightarrow{G}$ that is a bidirected cycle.

Separating Lifted g-Odd Inequalities in a Bidirected Cycle. Given a vector (x, y) we want to verify if there is a lifted g-odd cycle inequality (11) violated by (x, y) if there is any.

Theorem 4. *The g-odd lifted cycle inequalities can be separated in linear time for bidirected cycles.*

Proof. A lifted g-odd cycle inequality (11) can also be written as

$$\sum_{a \in A(C)} (1 - 2x(a)) - \sum_{a \in \tilde{A}(C)} 2x(a) + \sum_{v \in \hat{C}} (2y(v) - 1) \geq 1. \qquad (22)$$

Thus we look for a cycle that violates (22). For that we create a directed graph $D' = (V', A')$ as follows. For every arc $(i, i+1)$ and $(i+1, i)$ we create a node in D'. The arcs in A' are as below.

- From $(i, i+1)$ to $(i+1, i+2)$ we create an arc with weight $1 - 2x(i+1, i+2)$ and label "odd."
- From $(i, i+1)$ to $(i+2, i+1)$ we create an arc with weight $2y(i+1) - 2x(i+2, i+1)$ and label "even."
- From $(i+1, i)$ to $(i+1, i+2)$ we create an arc with weight $1 - 2x(i+1, i+2)$ and label "odd."
- From $(i+1, i)$ to $(i+2, i+1)$ we create an arc with weight $1 - 2x(i+2, i+1)$ and label "odd."
- From $(i, i-1)$ to $(i+1, i+2)$ we create an arc with weight $2 - 2x(i, i+1) - 2x(i+1, i) - 2x(i+1, i+2)$, and label "even." This arc corresponds to the case when either $(i, i+1)$ or $(i+1, i)$ is in the lifting set $\tilde{A}(C)$.

Then we look for a minimum weight directed cycle with an odd number of odd arcs in D'. If the weight of such a cycle is less than one, we have found a violated inequality. Now we give the details of how to find a minimum weight directed cycle with an odd number of odd arcs. We pick and index i, and remove the arcs entering $(i, i+1)$ and $(i+1, i)$. We add an extra node s and connect it to $(i, i+1)$ and $(i+1, i)$ with even arcs of weight zero. For each node v in D' let $f_o(v)$ (resp. $f_e(v)$) be the weight of a shortest path from s to v having an odd (resp. even) number of odd arcs. We set $f_e(s) = 0$, $f_o(s) = f_o(v) = f_e(v) = \infty$ for every other node v in D'. We call the labels of s permanent and all others temporary. For each arc (u, v) we denote by $w(u, v)$ its weight. Then for a node v such that all its predecessors have permanent labels we update its labels as below.

$$f_o(v) = \min \Big\{ \; \min_u\{f_o(u) + w(u,v) : (u,v) \text{ is even}\},$$

$$\min_u\{f_e(u) + w(u,v) : (u,v) \text{ is odd}\}\Big\}$$

$$f_e(v) = \min \Big\{ \; \min_u\{f_o(u) + w(u,v) : (u,v) \text{ is odd}\},$$

$$\min_u\{f_e(u) + w(u,v) : (u,v) \text{ is even}\}\Big\}.$$

Then the labels of v are called permanent, and we continue. Once all labels are permanent, we use the arcs entering $(i, i+1)$ and $(i+1, i)$ to find a shortest directed cycle with an odd number of odd arcs and including either $(i, i+1)$ or $(i+1, i)$. Next we have to consider the case when neither $(i, i+1)$ nor $(i+1, i)$ is in the shortest cycle. This is when the arc from $(i, i-1)$ to $(i+1, i+2)$ is part of the shortest cycle. For that we repeat the same procedure with $i' = i+1$. Since the indegree of each node in D' is at most three, the labels are computed in constant time for each node. Therefore this is a linear time algorithm. □

References

1. Bouchakour, M., Contenza, T.M., Lee, C.W., Mahjoub, A.R.: On the dominating set polytope. Eur. J. Comb. **29**(3), 652–661 (2008)
2. Bouchakour, M., Mahjoub, A.R.: One-node cutsets and the dominating set polytope. Discrete Math. **165–166**(15), 101–123 (1997)
3. Haynes, T., Hedetniemi, S., Slater, P.: Fundamentals of Domination in Graphs. Monographs and Textbooks in Pure and Applied Mathematics. Taylor & Francis, Boca Raton (1998)
4. Haynes, T., Hedetniemi, S., Slater, P.: Domination in Graphs: Advanced Topics. Monographs and Textbooks in Pure and Applied Mathematics, vol. 2. Taylor & Francis, Boca Raton (1998)
5. Mahjoub, A.R.: Polytope des absorbants dans une classe de graphe a seuil. North-Holland Math. Stud. **75**, 443–452 (1983)
6. Bouchakour, M.: I: Composition de graphes et le polytope des absorbants, II: Un algorithme de coupes pour le problème du flot à coût fixes. Ph.D. thesis, Université de Rennes 1, Rennes, France, December 1996
7. Farber, M.: Domination, independent domination, and duality in strongly chordal graphs. Discrete Appl. Math. **7**(2), 115–130 (1984)
8. Saxena, A.: Some results on the dominating set polytope of a cycle. GSIA Working paper 2004-E28 (2004)

9. Cornuéjols, G., Sassano, A.: On the 0, 1 facets of the set covering polytope. Math. Program. **43**, 45–55 (1989)
10. Balas, E., Ng, S.M.: On the set covering polytope: I. All the facets with coefficients in 0, 1, 2. Math. Program. **43**, 57–69 (1989)
11. Balas, E., Ng, S.M.: On the set covering polytope: II. Lifting the facets with coefficients in 0, 1, 2. Math. Program. **45**, 1–20 (1989)
12. Cornuéjols, G., Novick, B.: Ideal 0, 1 matrices. J. Comb. Theory Ser. B **60**(1), 145–157 (1994)
13. Sánchez-García, M., Sobrón, M., Vitoriano, B.: On the set covering polytope: facets with coefficients in {0, 1, 2, 3}. Ann. Oper. Res. **81**, 343–356 (1998)
14. Bianchi, S., Nasini, G., Tolomei, P.: The minor inequalities in the description of the set covering polyhedron of circulant matrices. Preprint (2012)
15. Avella, P., Sassano, A., Vasilév, I.: Computational study of large-scale p-median problems. Math. Program. **109**, 89–114 (2007)
16. Padberg, M.W.: On the facial structure of set packing polyhedra. Math. Program. **5**, 199–215 (1973)
17. Cornuejols, G., Thizy, J.M.: Some facets of the simple plant location polytope. Math. Program. **23**(1), 50–74 (1982)
18. Cho, D.C., Johnson, E.L., Padberg, M., Rao, M.R.: On the uncapacitated plant location problem. I. Valid inequalities and facets. Math. Oper. Res. **8**(4), 579–589 (1983)
19. Cho, D.C., Padberg, M.W., Rao, M.R.: On the uncapacitated plant location problem. II. Facets and lifting theorems. Math. Oper. Res. **8**(4), 590–612 (1983)
20. Baïou, M., Barahona, F.: Simple extended formulation for the dominating set polytope via facility location. IBM Research Report RC25325t (2012)
21. Eisenbrand, F., Oriolo, G., Stauffer, G., Ventura, P.: The stable set polytope of quasi-line graphs. Combinatorica **28**(1), 45–67 (2008)
22. Baïou, M., Barahona, F.: On the integrality of some facility location polytopes. SIAM J. Discrete Math. **23**(2), 665–679 (2009)

Solving Graph Partitioning Problems Arising in Tagless Cache Management

Sandro Bartolini, Iacopo Casini, and Paolo Detti[✉]

Dipartimento di Ingegneria dell'Informazione Scienze Matematiche,
University of Siena, Siena, Italy
{bartolini,detti}@dii.unisi.it,
casiniaco@gmail.com

Abstract. The instruction cache is a critical component in any microprocessor. It must have high performance to enable fetching of instructions on every cycle. In this paper, we consider an optimization problem arising in the management of a new hybrid hardware and linker-assisted approach for cache memory management. A graph partitioning formulation is presented and different ILP formulations are proposed, obtained by strengthening and/or relaxing constraints and by reducing the number of integer variables. The formulations are tested on large benchmarks (with thousands of nodes and edges) arising from real applications.

Keywords: Cache memory · Graph partitioning · Integer programming

1 Introduction and Problem Description

One big technological constraint which since many years affects microprocessor and computer systems design is known as the *memory wall* [11]: CPUs are much faster than the main memory speed. Cache memories are adopted just to address this problem. They must have high performance to enable fetching of instructions on every cycle and, because of that, consume a big amount of energy and occupy significant space on chip, which in turn determines their building cost. In general, the more a given cache is able to capture the largest portion of the used memory space, the faster access latency the processor perceives or, from a design point of view, the smallest (and less energy hungry) cache module can be used to match a fixed performance/energy objective.

Therefore, numerous hardware and software solutions have been proposed to improve the cache ability to serve a higher fraction of the main memory space or to limit its power consumption.

As cache memories are much smaller than main memory, each of their elementary locations (blocks) can accommodate many memory blocks in a mutually exclusive fashion. For instance, if the cache is 16 kByte, organized in 32-byte blocks (512 blocks), and main memory is 512 MByte (about 16 millions of blocks), at a given time each cache block can host one of the 32768 (16 millions/512) memory blocks. For this reason, along with the data, the cache needs

© Springer International Publishing Switzerland 2014
P. Fouilhoux et al. (Eds.): ISCO 2014, LNCS 8596, pp. 50–61, 2014.
DOI: 10.1007/978-3-319-09174-7_5

to maintain also information about *which* memory block is stored, i.e.: "tag" information. In a cache, tags constitute auxiliary overhead chip area and energy consumption, needed for correct operation. Recently, a special cache has been proposed [8] in which, under some conditions and hypotheses on the software, tag accesses can be avoided at all, with a great benefit (more than 50 %) in cache power consumption. Such conditions and hypotheses dictate that, according to the proposed cache hardware features, the software (memory content) needs to be split into a few special *tagless* regions (e.g.: eight 4-kByte regions for a 32 kByte cache) where the most used instructions need to be stored. When accessing these tagless regions, tag checks can be typically omitted. To avoid some intrinsic overheads and get the most out of this scheme, we have to:

– fill tagless regions carefully, privileging frequently used instruction groups (named basic-blocks);
– limit as much as possible both successive access to different tagless regions and successive accesses to tagless and non-tagless memory regions.

The behavior of a running application determines the access pattern to basic blocks (BBs), and thus to regions. A BB is a sequence of consecutive instructions ending with a jump instruction, and not having other jumps inside. Hence, BBs are executed sequentially and can have various lengths. The exact access pattern of an application run is represented by a graph called the *Control Flow Graph (CFG)*, a weighted directed graph where nodes correspond to BBs and weights on edges between two BBs represent the number of times program execution jumps from the predecessor BB to the successor BB.

In the exposed scenario, the limitation of the overheads of the tagless cache can be achieved with a careful assignment of BBs to available tagless regions, taking into account the CFG and the following qualitative criteria:

($c1$) We want to spend as much time as possible executing instructions from a tagless region. Therefore, nodes in CFG, i.e., basic blocks, having incoming edges with high weights (the sum of the weights of the incoming edges in a given node is called the *execution count* of the node) should be preferred over those with edges with low weights.

($c2$) We want to minimize switching amongst tagless regions, and at a more limited extension to/from BBs outside tagless regions, as this induces overhead due to partial cache flushing. Hence, BBs connected by edges with high weights should be preferably kept in the same tagless region.

($c3$) We want to fill the tagless regions as much as possible to maximize the tagless accessing. This poses further challenges as, for instance, BBs have different sizes and tagless regions can be filled using leftover BBs that do not significantly impact the other two goals.

Another phenomenon, called *leeway management*, poses further challenges as the compiler increases the size of a basic block, according to the basic blocks it can possibly jump to that are allocated *far*, e.g., in another tagless region or outside any tagless region. Such information cannot be derived by the CFG, which represent only one sample run of the application, but by a static graph, provided by the compiler, which exposes a directed edge for any possible jump between BBs. More precisely, the size of a BB i decreases with the number of successors of i in the static graph, that are assigned to the same tagless region.

In this paper, a problem arising in the management of tagless cache memories is addressed. The problem can be formulated as a graph partitioning problem with side constraints and features. Different ILP formulations are proposed, obtained by strengthening and/or relaxing constraints and by reducing the number of integer variables. Some theoretical results are presented that allow to compare the different formulations. The formulations are tested on large benchmarks (with thousands of nodes and edges) arising from real applications. Computational results show that optimal or near optimal solutions can be achieved by using a state-of-the-art mathematical programming solver.

The paper is organized as follows. In Sect. 2, results from the literature are presented. In Sect. 3, the problem is formally defined and an ILP formulation is presented. In Sect. 4, other ILP formulations are presented and a theoretical comparative analysis is performed. Section 5 reports on the computational experiments performed on real world instances. Finally, conclusions follow.

2   Literature Review

The problem addressed in this paper generalizes a problem known in the literature as Graph Partitioning (GP) [6]. In GP the vertex set of a node-weighted and arc-weighted graph is to be partitioned into k disjoint clusters, such that the sum of the weights of the nodes belonging to the same cluster does not exceed a given value F, and the sum of the arc weights within the clusters is maximized (or equivalently the sum of the arc weights between different clusters is minimized). As it will be clear later the tagless cache management problem addressed in this paper can be formulated as a GP problem with additional constraints and features.

Graph partitioning is NP-complete [6] and, hence, the tagless cache management problem is NP-complete, too. GP has several applications arising in a wide area of applications, such as compiler design [7], VLSI (Very Large Scale Integration) design [2], qualitative data analysis [3], finite element computation [4] and manufacturing systems [1]. Branch-and-cut and branch-and-bound [10] algorithms have been presented in the literature for solving GP problems. In [7], a branch-and-cut algorithm has been proposed that includes cuts to reduce the symmetric nature of the suggested problem formulation. In [4], a study of classes of cutting planes for the GP problem is presented, and in [5] a computational study is proposed. In [9], a branch-and-price method is presented.

The tagless cache management problem addressed in this paper has been first introduced in [8]. In the same paper, a constructive four-phase heuristic algorithm is proposed, that, in our experiments, is used to provide a starting feasible solution to the branch-and-cut algorithm.

3   Problem Definition and Formulations

Let $G = (V, A)$ be the Control Flow Graph introduced in Sect. 1. For each BB, a node i exists in V. An edge (i, j) exists in A, if the execution of BB j passes through the execution of BB i. A positive weight w_{ij} is associated to each edge $(i, j) \in A$, corresponding to the number of times the execution passes from BB i and BB j. A weight v_i and a *gross* size s_i are associated to each node $i \in V$, where v_i correspond to the size of node i times its execution count, i.e.,
$v_i = s_i \sum_{(j,i) \in \delta^-(i)} w_{ji}$ (in which $\delta^-(i)$ denotes the set of edges in A entering into node (i), and s_i is the size of node i, when no size reduction occurs. In fact, as explained in Sect. 1, the size of a given node i can be reduced if its successors in the static graph provided by the compiler, are assigned to the same tagless region of i. Let $H = (V, E)$ denote the static graph. H is defined on the same node set V of G. An edge (i, j) exists in E, if it is possible to jump from BBs i and j, $i, j \in V$. A weight $a_{ij} \geq 0$ is associated to each edge $(i, j) \in E$, representing the size reduction occurring to node i when node j is assigned to the same cluster of i. Let $C = \{1, \ldots, K\}$ denote the set of tagless regions, called *clusters* in the following, and let D_k be the size of cluster $k \in C$. Let σ be an assignment of the nodes in V to the clusters in C, and let S_k be the set of nodes in V assigned to the cluster $k \in C$. Then, the *net size* of a node $i \in S_k$ is defined as

$$s_i - \sum_{j \in S_k : (i,j) \in E} a_{ij}. \tag{1}$$

Moreover, given an edge $(i, j) \in A$, we say that:

- (i, j) is an *inter-cluster* edge if nodes i and j are assigned to different clusters in σ;
- (i, j) is an *outbound* edge if node i is assigned to a cluster and j is not assigned to any cluster in σ.

A feasible solution of the tagless cache management problem is an assignment of nodes of V to the clusters in C, in a such way that the sum of the net sizes of the nodes assigned to each cluster k does not exceed the capacity D_k. According to the criteria $c1$, $c2$ and $c3$, the tagless cache management problem consists in finding a feasible assignment of nodes to the clusters in such a way that:

- the sum of the weights of the nodes assigned to the clusters is maximized;
- the sum of the weights of inter-cluster and outbound edges is minimized.

3.1   Integer Linear Programming Formulations

In this section, Integer Linear Programming models for the tagless cache management problem are presented. In the proposed models, the objectives listed at the end of the previous section are combined into a single function by introducing suitable weights.

Let
x_{ik} be a binary variable equal to 1 if node i is assigned to cluster $k \in C$, and 0 otherwise;
y_{ij} be a binary variable associated to edge $(i,j) \in A$, equal to 1 if nodes i and j are assigned to different clusters (i.e., if (i,j) is an inter-cluster edge), and 0 otherwise;
z_{ij} be a binary variable associated to edge $(i,j) \in A$, equal to 1 if node i belongs to a cluster and j does not belong to any cluster (i.e., if (i,j) is an outbound edge), and 0 otherwise;
h_{ijk} be a binary variable associated to edge $(i,j) \in E$, i.e., belonging to the static graph H, equal to 1 if nodes i and j are assigned to the same cluster $k \in C$, and 0 otherwise.

An Integer Linear Programming formulation for the problem, denoted as ILP_1, is as follows.

$$\max \sum_{i \in V} \sum_{k=1}^{K} v_i x_{ik} - \alpha \sum_{(i,j) \in A} w_{ij} y_{ij} - \beta \sum_{(i,j) \in A} w_{ij} z_{ij} \qquad (2)$$

$$\sum_{i \in V} s_i x_{ik} - \sum_{(i,j) \in E} a_{ij} h_{ijk} \leq D_k \ \ \forall k \in C \qquad (3)$$

$$\sum_{k=1}^{K} x_{ik} \leq 1 \ \forall i \in V \qquad (4)$$

$$h_{ijk} \leq x_{ik} \ \ \forall (i,j) \in E \ \ \forall k \in C \qquad (5)$$

$$h_{ijk} \leq x_{jk} \ \ \forall (i,j) \in E \ \ \forall k \in C \qquad (6)$$

$$y_{ij} + z_{ij} \leq 1 \ \ \forall (i,j) \in A \qquad (7)$$

$$x_{ik} + x_{jh} - y_{ij} \leq 1 \ \ \forall (i,j) \in A \ \ \forall k \in C \ \ \forall h \in C, \ k \neq h \qquad (8)$$

$$z_{ij} - \sum_{k=1}^{K} x_{ik} + \sum_{h=1}^{K} x_{jh} \geq 0 \ \ \forall (i,j) \in A \qquad (9)$$

$$x_{ik}, h_{ijk}, y_{ij}, z_{ij} \in \{0,1\} \qquad (10)$$

According to the goals listed in Sect. 1, the objective function (2) is composed of three terms: the total value of the nodes assigned to the clusters, the sum of the weights w_{ij} of the inter-cluster edges (i,j), and the sum of the weights w_{ij} of the outbound edges (i,j). The last two terms of the objective function are weighted by coefficients α and β, respectively, with $\alpha, \beta \geq 0$. Constraints (3) assure that the total size of nodes assigned to cluster k can not exceed the cluster size D_k. Observe that, when assigned to cluster k, the size of node i is equal to the net size as defined in relation (1). Constraints (4) state that each node can be assigned to at most one cluster. For each edge $(i,j) \in E$, Constraints (5) and (6) impose that h_{ijk} can be 1 only if node i and node j are assigned to cluster k.

Constraints (7) state that, for each edge $(i,j) \in A$, variables y_{ij} and z_{ij} can not be 1 at the same time. Constraints (8) force the variable y_{ij} to be 1 if nodes i and j belong to different clusters. Constraints (9) force the variable z_{ij} to be

1 if node i belongs to a cluster (i.e., $\sum_{k=1}^{K} x_{ik} = 1$) and node j does not belong to any cluster (i.e., $\sum_{k=1}^{K} x_{jk} = 0$).

4 Constraint Reduction and Strengthening and Variable Redefinition

In this section, new and stronger formulations for the problem are proposed.

4.1 Constraint Reduction and Strengthening

In the following we show that Constraints (7) are redundant. At this aim, observe that, Constraints (8) force the variable y_{ij} to be bigger than or equal to 1 only if two variables x_{ik} and x_{jh} equal to 1 exist, with $k \neq h$ (i.e., if the nodes i and j are assigned to different clusters). Similarly, in Constraints (9), variable z_{ij} is forced to be bigger than or equal to 1 only if $\sum_{k=1}^{K} x_{ik} = 1$ and $\sum_{k=1}^{K} x_{jk} = 0$ (recall that, by (4), $\sum_{k=1}^{K} x_{ik} \leq 1$ for all $k \in C$ and $i \in N$). As a consequence, Constraints (8) and (9) imply that z_{ij} is forced to be 1 if and only if y_{ij} is 0. On the other hand, since the objective function coefficients of variables y_{ij} and z_{ij} are all positive, y_{ij} and z_{ij} will be 1 in an optimal solution only if they are forced. Hence, Constraints (7) are redundant.

Constraints (8) can be strengthened as explained in the following. Since $\sum_{h \in C\, h \neq k} x_{jh} \leq 1$, Constraints (8) can be replaced by the following set of stronger constraints

$$x_{ik} + \sum_{h \in C\, h \neq k} x_{jh} - y_{ij} \leq 1 \quad \forall (i,j) \in A \ \forall k \in C. \tag{11}$$

Observe that, from a linear relaxation point of view, Constraints (11) are stronger than Constraints (8) and are smaller in number of a factor $K - 1$. By the above discussion, it follows that a new ILP formulation for the problem, in the following denoted as ILP_2, reads as:

$$\max \sum_{i \in V} \sum_{k=1}^{K} v_i x_{ik} - \alpha \sum_{(i,j) \in A} w_{ij} y_{ij} - \beta \sum_{(i,j) \in A} w_{ij} z_{ij} \tag{12}$$

$$\sum_{i \in V} s_i x_{ik} - \sum_{j \in V : (i,j) \in E} a_{ij} h_{ijk} \leq D_k \ \forall k \in C \tag{13}$$

$$\sum_{k=1}^{K} x_{ik} \leq 1 \ \forall i \in V \tag{14}$$

$$h_{ijk} \leq x_{ik} \ \forall (i,j) \in E \ \forall k \in C \tag{15}$$

$$h_{ijk} \leq x_{jk} \ \forall (i,j) \in E \ \forall k \in C \tag{16}$$

$$x_{ik} + \sum_{h \in C\, h \neq k} x_{jh} - y_{ij} \leq 1 \ \forall (i,j) \in A \ \forall k \in C \tag{17}$$

$$z_{ij} - \sum_{k=1}^{K} x_{ik} + \sum_{h=1}^{K} x_{jh} \geq 0 \ \forall (i,j) \in A \tag{18}$$

$$x_{ik}, h_{ijk}, y_{ij}, z_{ij} \in \{0,1\} \tag{19}$$

In the following lemma, a relation between the solutions of the linear relaxations of ILP_1 and ILP_2 is established.

Lemma 1. *Let z_{LP_1} and z_{LP_2} be the optimal solution values of the linear relaxations of ILP_1 and ILP_2, respectively. Then*

$$z_{LP_2} \leq z_{LP_1}.$$

Proof. Recall that ILP_2 has been obtained from ILP_1 by replacing Constraints (8) with (17) and by relaxing Constraints (7). Since Constraints (17) are stronger than (8), the thesis can be simply proved by showing that Constraints (7) are satisfied by the optimal solution, say (x^*, h^*, y^*, z^*), of the linear relaxation of ILP_2. By Constraints (17) and (18), we have

$$y_{ij} \geq x_{ik} + \sum_{h \in C\, h \neq k} x_{jh} - 1$$

and

$$z_{ij} \geq \sum_{k=1}^{K} x_{ik} - \sum_{h=1}^{K} x_{jh}.$$

Since variables y_{ij} and z_{ij} are not negative and do not appear in other constraints of ILP_2 and $w_{ij} > 0$ for all $(i,j) \in A$, in (x^*, h^*, y^*, z^*), it must be

$$y_{ij}^* = \max\{0; \max_{k \in C}\{x_{ik}^* + \sum_{h \in C\, h \neq k} x_{jh}^* - 1\}\}$$

and

$$z_{ij}^* = \max\{0; \sum_{k=1}^{K} x_{ik}^* - \sum_{h=1}^{K} x_{jh}^*\}.$$

Otherwise, (x^*, h^*, y^*, z^*) could not be optimal.

Summing the two relations above we obtain

$$y_{ij}^* + z_{ij}^* = \max\{0; \max_{k \in C}\{x_{ik}^* + \sum_{h \in C\, h \neq k} x_{jh}^* - 1\} + \max\{0; \sum_{k=1}^{K} x_{ik}^* - \sum_{h=1}^{K} x_{jh}^*\}.$$

Four cases can be considered: (1) $y_{ij}^* = z_{ij}^* = 0$; (2) $y_{ij}^* = \max_{k \in C}\{x_{ik}^* + \sum_{h \in C\, h \neq k} x_{jh}^* - 1\}$ and $z_{ij}^* = 0$; (3) $y_{ij}^* = 0$ and $z_{ij}^* = \sum_{k=1}^{K} x_{ik}^* - \sum_{h=1}^{K} x_{jh}^*$; (3) $y_{ij}^* = \max_{k \in C}\{x_{ik}^* + \sum_{h \in C\, h \neq k} x_{jh}^* - 1\}$ and $z_{ij}^* = \sum_{k=1}^{K} x_{ik}^* - \sum_{h=1}^{K} x_{jh}^*$.

Case (1) is trivial. In Case (2) we obtain

$$y^*_{ij} + z^*_{ij} = \max_{k \in C}\{x^*_{ik} + \sum_{h \in C\, h \neq k} x^*_{jh} - 1\} \leq 1$$

since $x^*_{ik} \leq 1$ and $\sum_{h \in C\, h \neq k} x^*_{jh} \leq 1$. Case (3) follows since $\sum_{k=1}^{K} x^*_{ik} \leq 1$ for all $i \in V$. In Case (4), we have

$$y^*_{ij} + z^*_{ij} = \max_{k \in C}\{x^*_{ik} + \sum_{h \in C\, h \neq k} x^*_{jh} - 1\} + \sum_{k=1}^{K} x^*_{ik} - \sum_{h=1}^{K} x^*_{jh}$$

Let q be the cluster in C such that

$$\max_{k \in C}\{x^*_{ik} + \sum_{h \in C\, h \neq k} x^*_{jh} - 1\} = x^*_{iq} + \sum_{h \in C\, h \neq q} x^*_{jh} - 1$$

Then

$$y^*_{ij} + z^*_{ij} = \sum_{k=1}^{K} x^*_{ik} - 1 + x^*_{iq} + \sum_{h \in C\, h \neq q} x^*_{jh} - \sum_{h=1}^{K} x^*_{jh} = \sum_{k=1}^{K} x^*_{ik} - 1 + x^*_{iq} - x^*_{jq} \leq 1$$

where the last inequality follows since $(\sum_{k=1}^{K} x^*_{ik} - 1) \leq 0$ (by Constraints (14)), $x^*_{iq} \leq 1$ and $x^*_{jq} \geq 0$. ∎

4.2 Variable Redefinition

Denote by ILP_3 the formulation obtained from ILP_2 by relaxing the integer constraints on variables y_{ij} and z_{ij}. The following lemma holds.

Lemma 2. *Let (x^*, h^*, y^*, z^*) be an optimal solution of ILP_3, then y^*_{ij} and z^*_{ij} belong to $\{0,1\}$, for all $(i,j) \in A$.*

Proof. Let us consider first the variables y_{ij}. Note that they only appear in Constraints (17). Let us consider the Constraint (17) related to a given a variable y_{ij} and a cluster $k \in C$. Since $x^*_{ik} \in \{0,1\}$ and $\sum_{h \in C\, h \neq k} x^*_{jh} \in \{0,1\}$. The following four cases can be considered: (1) $x^*_{ik} = 0$ and $\sum_{h \in C\, h \neq k} x^*_{jh} = 0$; (2) $x^*_{ik} = 1$ and $\sum_{h \in C\, h \neq k} x^*_{jh} = 0$; (3) $x^*_{ik} = 1$ and $\sum_{h \in C\, h \neq k} x^*_{jh} = 0$; (4) $x^*_{ik} = 1$ and $\sum_{h \in C\, h \neq k} x^*_{jh} = 1$. In Case (1), Constraints (17) imply that $y^*_{ij} \geq -1$, i.e., $y^*_{ij} = 0$, since $w_{ij} > 0$ for all $(i,j) \in A$ (otherwise (x^*, h^*, y^*, z^*) could not be an optimal solution). In Cases (2) and (3), Constraints (17) imply that $y^*_{ij} \geq 0$, i.e., $y^*_{ij} = 0$, too. Finally, in Case (4), by Constraints (17) we have $y^*_{ij} \geq 1$, i.e., $y^*_{ij} = 1$ (since $y_{ij} \leq 1$). Let us consider now the variables z_{ij}. Since they only appear in Constraints (18), and since $\sum_{k=1}^{K} x^*_{jk} \in \{0,1\}$ and $\sum_{h=1}^{k} x^*_{jh} \in \{0,1\}$, the thesis follows by repeating and adapting the arguments used for variables y_{ij}. ∎

According to Lemma 2, ILP_3 is a valid formulation of the problem. Now, denote by ILP_4 the formulation obtained from ILP_3 by relaxing the integer constraints on variables h_{ijk}. The following lemma holds.

Lemma 3. *Let (x^*, h^*, y^*, z^*) be an optimal solution of ILP_4, in which $h^*_{ijk} \in (0, 1)$ for some $(i, j) \in E$ and $k \in C$. Then an optimal solution of ILP_4, say $(\bar{x}, \bar{h}, \bar{y}, \bar{z})$, exists (and can be built), in which $\bar{x} = x^*$, $\bar{y} = y^*$, $\bar{z} = z^*$, $\bar{h} \in \{0, 1\}^{E \times K}$.*

Proof. Let (x^*, h', y^*, z^*) be a solution obtained from (x^*, h^*, y^*, z^*) in which $h'_{ijk} = 1$ and $h'_{uvh} = h^*_{uvh}$ for all $(u, v) \in E$ and $h \in C$, with $(u, v) \neq (i, j)$. Since the h variables do not appear in the objective function, solutions (x^*, h^*, y^*, z^*) and (x^*, h', y^*, z^*) yield the same objective function value.

We show now that (x^*, h', y^*, z^*) is a feasible solution. In fact, in ILP_4, the h variables appear in the Constraints (13), (15) and (16). Since $h'_{ijk} > h^*_{ijk}$, the solution (x^*, h', y^*, z^*) satisfies Constraints (13), and, since $x^* \in \{0, 1\}^{N \times K}$, it satisfies Constraints (15) and (16), too.   □

By Lemma 3 it follows that an optimal solution of ILP_4, (x^*, h^*, y^*, z^*), can be converted into an optimal solution of the tagless cache management problem by simply rounding up the values attained by variables h^*.

According to Lemmas 2 and 3, ILP_3 and ILP_4 are valid formulations for the tagless cache management problem. Observe that, ILP_3 and ILP_4 respectively have $2|A|$ and $2|A| + |E| \times |V|$ integer variables less than ILP_1 and ILP_2.

5   Numerical Results

In this section, experimental results are presented. The proposed ILP formulations have been tested on 12 instances arising from real world applications. The results are reported in Table 1. For each instance, columns 2 and 3 of Table 1 report the number of nodes $|V|$ and the number of edges $|A|$ of the control flow graph G. Observe that, the smallest instance has $19,436$ nodes and that the CGFs are sparse, with $|V| \simeq |A|$. In all the instances, $|C| = 8$ clusters have been considered with $D_k = 4096$ for all $k \in C$.

According to a preliminary campaign of experiments, the weights α and β of the second and third terms of the objective functions of ILP_1–ILP_4 have been set to 1 and 5, respectively, in all the instances.

The instances have been solved by CPLEX 12.5 on a 6 cores, 12 threads, PC. A time limit of 48 h has been set and the procedures to break symmetries have been enabled in CPLEX. The CPLEX's branch-and-cut has been initialized by the feasible integer solution obtained by the algorithm proposed in [8].

Table 1 reports the computational results for ILP_1–ILP_4 on the 12 instances. More precisely, the results for ILP_1 are reported in column 4, while columns 5–7, 8–10 and 11–13 report the results for ILP_2, ILP_3 and ILP_4, respectively. The last row of Table 1 reports the average values.

For each instance and ILP formulation, let f^*, f_{lp}, f_{UB} and f_h respectively be the value of the best solution found by CPLEX (within the time limit), the optimal solution value of the linear relaxation, the best upper bound obtained by CPLEX (within the time limit) and the value of the solution provided by the heuristic presented in [8]. In Table 1, for each instance: "Heu Gap" is the percentage gap between the solution provided by the algorithm presented in [8] and the best solution found by CPLEX, compuetd as $\frac{f^*-f_h}{f_h} \times 100$; "Root Gap" is the gap at the root node between the linear relaxation and the best solution found by CPLEX, computed as $\frac{f_{lp}-f^*}{f_{lp}} \times 100$; and "Opt Gap" is computed as $\frac{f_{UB}-f^*}{f_{UB}} \times 100$. As the results show, the Opt Gap of ILP_1 attained by CPLEX within the time limit is poor in terms of quality (about 22 % on average), when compared to those obtained with ILP_2–ILP_4. (And we do not report the Heu Gap and Root Gap values for ILP_1.) As it can be observed, ILP_2 and ILP_3 provide very similar results. More precisely, the Opt Gap is 1.07 % and 1.15 % on average for ILP_2 and ILP_3, respectively. Three instances out of 12 have been solved to the optimality by CPLEX on all the ILP formulations. Observe that, on these three instances, the heuristic presented in [8] provides a solution that is either optimal or very close to it. While, in the other 9 instances, the heuristic provides solution quite far from the optimum (on average at least 22 %, 22 % and 21 % far from the best solution found by ILP_2, ILP_3 and ILP_4, respectively). On the other hand, the Root Gaps of ILP_2–ILP_4 are in general quite small and close to the Opt Gap, stating the good quality of the LP relaxations of ILP_2–ILP_4. Formulation ILP_4 has a slightly worse performance with respect to ILP_2 and ILP_3. Such a behavior highlights that branching operations on the h variables performed in ILP_2 and ILP_3 are useful during the B&C search (obviously, no branching operation is performed on variables h in ILP_4, since h are continuos). Finally we point out that, since the instances have quite large dimensions, the computational time required to compute the optimal solution of the linear relaxation is big: it is about 1200 s on average, for formulations ILP_2–ILP_4, and ranges from 160 to 8,300 s.

6   Conclusions and Future Research

In this paper, an optimization problem arising in the management of a new hybrid hardware and linker-assisted approach is considered. A graph partitioning formulation is given and different ILP formulations are proposed. Some theoretical results have been presented that allow to strength the formulations or to reduce the number of integer variables. The formulations are tested on large size instances arising from real world applications. The computational results show that formulations ILP_2–ILP_4 present quite strong linear relaxations. Furthermore, a comparison of the solution provided by the algorithm presented in [8] with the best solution found by CPLEX suggests that new heuristic approaches can be investigated to find better feasible solutions for the problem.

Table 1. Results for ILP_3 and ILP_4.

Instance	|V|	|A|	ILP_1 Opt Gap	ILP_2 Heu Gap	ILP_2 Root Gap	ILP_2 Opt Gap	ILP_3 Heu Gap	ILP_3 Root Gap	ILP_3 Opt Gap	ILP_4 Heu Gap	ILP_4 Root Gap	ILP_4 Opt Gap
bzip2	20,553	17,499	0.00	0.00	0.00	0.00	0.00	0.00	0.00	0.00	0.00	0.00
cc1	94,169	109,426	-	27.78	0.18	0.17	27.64	0.29	0.27	27.66	0.27	0.26
crafty	27,713	27,805	24.48	27.24	2.69	2.22	27.41	2.56	2.13	27.25	2.68	2.20
eon	64,493	67,434	31.65	26.07	4.26	1.42	25.99	4.31	1.11	25.95	4.35	3.34
gap	45,125	45,692	63.11	55.02	6.24	4.01	53.11	7.39	5.13	41.17	14.61	14.51
gzip	20,444	17,414	0.01	0.00	0.01	0.00	0.00	0.01	0.00	0.00	0.01	0.00
mcf	19,436	16,599	0.06	0.05	0.01	0.00	0.05	0.01	0.00	0.05	0.01	0.00
parser	24,377	24,514	37.76	48.50	2.22	0.74	48.65	2.12	0.63	48.43	2.26	0.87
perlbmk	58,410	59,719	33.90	29.02	3.79	2.60	29.31	3.57	2.45	29.67	3.30	2.06
twolf	30,312	30,807	7.52	6.42	1.06	0.61	6.54	0.95	0.61	6.54	0.95	0.61
vortex	44,370	47,550	38.25	35.70	1.94	0.73	35.81	1.86	0.73	35.89	1.80	0.82
vpr	28,800	30,071	15.51	17.97	1.15	0.29	17.79	1.30	0.29	17.73	1.35	0.48
Av.	39,850.17	41,210.83	22.93	22.81	1.96	1.07	22.69	2.03	1.15	21.69	2.63	2.10

Acknowledgements. The authors would like to thank Tim M. Jones and Jonas Maebe for the discussion about the tagless cache operation and for the preparation of the input data used in this work. This work was partially supported by IT FIRB PHOTONICA project (RBFR08LE6V).

References

1. Alfieri, A., Nicosia, G., Pacifici, A.: Exact algorithms for a discrete metric labeling problem. Discrete Optim. **3**(3), 181–194 (2006)
2. Alpert, C.J., Kahng, A.B.: Recent directions in netlist partitioning: a survey. Integr. VLSI J. **19**, 1–81 (1995)
3. Grotschel, M., Wakabayashi, Y.: A cutting plane algorithm for a clustering problem. Math. Program. **45**(1), 59–96 (1989)
4. Ferreira, C.E., Martin, A., de Souza, C.C., Weismantel, R., Wolsey, L.A.: Formulations and valid inequalities for the node capacitated graph partitioning problem. Math. Program. **74**(3), 247–266 (1996)
5. Ferreira, C.E., Martin, A., de Souza, C.C., Weismantel, R., Wolsey, L.A.: The node capacitated graph partitioning problem: a computational study. Math. Prog. Series B **81**, 229–256 (1998)
6. Garey, M.R., Johnson, D.S.: Computers and Intractability - A Guide to the Theory of NP-Completeness. Freeman and co., New York (1979)
7. Holm, S., Sorensen, M.M.: The optimal graph partitioning problem. OR Spectrum **15**(1), 1–8 (1993)
8. Jones, T.M., Bartolini, S., Maebe, J., Chanet, D.: Link-time optimization for power efficiency in a tagless instruction cache. In: 2011 9th Annual IEEE/ACM International Symposium on Code Generation and Optimization (CGO), pp. 32–41 (2011)
9. Mehrotra, A., Trick, M.A.: Cliques and clustering: a combinatorial approach. Oper. Res. Lett. **22**, 1–12 (1998)
10. Nossack, J., Pesch, E.: A branch-and-bound algorithm for the acyclic problem. Comput. OR **41**, 174–184 (2014)
11. Wulf, W.A., McKee, S.A.: Hitting the memory wall: Implications of the obvious. SIGARCH Comput. Archit. News **23**(1), 20–24 (1995)

Lagrangean Decomposition for Mean-Variance Combinatorial Optimization

Frank Baumann, Christoph Buchheim, and Anna Ilyina[✉]

Fakultät für Mathematik, Technische Universität Dortmund,
Dortmund, Germany
{frank.baumann,christoph.buchheim,anna.ilyina}@tu-dortmund.de

Abstract. We address robust versions of combinatorial optimization problems, focusing on the uncorrelated ellipsoidal uncertainty case, which corresponds to so-called mean-variance optimization. We present a branch and bound-algorithm for such problems that uses lower bounds obtained from Lagrangean decomposition. This approach allows to separate the uncertainty aspect in the objective function from the combinatorial structure of the feasible set. We devise a combinatorial algorithm for solving the unrestricted binary subproblem efficiently, while the underlying combinatorial optimization problem can be addressed by any black box-solver. An experimental evaluation shows that our approach clearly outperforms other methods for mean-variance optimization when applied to robust shortest path problems and to risk-averse capital budgeting problems arising in portfolio optimization.

Keywords: Robust combinatorial optimization · Mean-risk optimization · Lagrangean decomposition

1 Introduction

Decision making under uncertainty is a challenge both from an economical and a mathematical perspective. In combinatorial optimization, where the constraints describe some problem-specific structure, the uncertainty usually appears in the objective function, i.e. the costs of the variables. We assume that a set $\mathcal{U}$ of potential cost vectors is given and aim at minimizing the value of a solution in its worst case scenario from this uncertainty set, i.e. we take a risk-averse attitude and consider the min-max criterion [1] to define solutions that are robust against variation of costs. That is we consider problems of the form

$$\min_{c \in \mathcal{U}} \max c^\top x \qquad\qquad (R)$$
$$\text{s.t. } x \in X,$$

The first author has been supported by the German Research Foundation (DFG) under grant BU 2313/2. The third author has been supported by the German Federal Ministry of Economics and Technology within the 6th Energy Research Programme.

P. Fouilhoux et al. (Eds.): ISCO 2014, LNCS 8596, pp. 62–74, 2014.
DOI: 10.1007/978-3-319-09174-7_6

with $X \subseteq \{0,1\}^n$ defining the combinatorial structure of the feasible set. We focus on problems whose deterministic versions are easy to solve, i.e. where a linear objective function can be optimized quickly over the set X.

Reasonable choices of scenario sets $\mathcal{U}$ depend on the application at hand, but also on the (theoretical and practical) tractability of the resulting problems. Among the popular types of uncertainty we find interval uncertainties, discrete scenario sets and the so-called ellipsoidal uncertainty. In the latter case the set of all possible scenarios forms an ellipsoid in $\mathbb{R}^n$ and each point in this ellipsoid represents a possible cost vector.

In this paper we address uncorrelated ellipsoidal uncertainties. Compared with interval uncertainty, which is more commonly used and easier to deal with computationally, using ellipsoidal uncertainties can avoid overly pessimistic solutions. In fact, the worst-case scenario in Problem (R) always corresponds to an extreme point of $\mathcal{U}$, so that in the interval case all coefficients are at their extremes, which is very unlikely in practice. In the ellipsoidal case this is explicitly excluded. More precisely, when assuming that the objective function coefficients are jointly normally distributed, the confidence regions form ellipsoids. Under the additional assumption that the distributions are independent, we obtain axis-parallel ellipsoids.

Interest in robust optimization under ellipsoidal uncertainty has been steadily growing in recent years, with a focus on the special case of axis-parallel ellipsoids, where the problem is equivalently reformulated to a mean-variance optimization problem. Depending on the underlying combinatorial structure efficient algorithms for Problem (R) may exist. As an example, the mean-risk *spanning tree problem* can be solved in polynomial time, as noted by Nikolova [6], who also proposes general-purpose approximation schemes. For other underlying problems, such as the *shortest path problem*, the complexity of (R) is unknown.

Another general solution approach has been to solve the problem as a general mixed-integer quadratic program. Atamtürk Narayanan [2] propose a SOCP-based branch and bound-algorithm for the robust *knapsack problem* with axis-parallel ellipsoidal uncertainty that additionally exploits the submodularity of the objective function.

In this paper we develop a new exact approach for min-max problems of type (R). We propose a branch and bound-algorithm using lower bounds obtained from Lagrangean decomposition, allowing to separate the uncertainty aspect in the objective function from the combinatorial structure of the feasible set. In particular, we present an efficient algorithm to solve (R) for $X = \{0,1\}^n$ in the case of uncorrelated ellipsoidal uncertainty. The combinatorial subproblem in the decomposition can be addressed by any black box-solver.

This paper is organized as follows. In Sect. 2 we present our Lagrangean decomposition approach for general binary nonlinear minimization problems. The special case of mean-variance combinatorial optimization is discussed in Sect. 3; we study the unconstrained binary optimization problem arising in the decomposition and devise an efficient algorithm that can deal with fixed variables. This allows us to embed the decomposition approach into a branch and bound-algorithm to compute provably optimal solutions. In Sect. 4 we evaluate our algorithm for the robust *shortest path problem* and the *risk-averse capital budgeting problem*. Extensive experimental studies show that our new algorithm clearly outperforms other approaches described in the literature.

2   A Lagrangean Decomposition Approach

Lagrangean decomposition can be considered a special case of Lagrangean relaxation, applied to a set of artificial constraints [4]. Its aim is to decompose a problem into auxiliary problems that can be easily computed. We use Lagrangean decomposition to separate the nonlinear objective function from the combinatorial constraints. Starting from the problem

$$\min f(x) \tag{P}$$
$$\text{s.t. } x \in X \subseteq \{0,1\}^n,$$

we introduce new variables $y \in \mathbb{R}^n$ along with artificial linking constraints and express the original set of combinatorial constraints in the new variables:

$$\min f(x)$$
$$\text{s.t. } x = y$$
$$x \in \{0,1\}^n$$
$$y \in X.$$

Lagrangean relaxation of the linking equations yields

$$\min f(x) + \lambda^\top (y - x)$$
$$\text{s.t. } x \in \{0,1\}^n$$
$$y \in X,$$

where $\lambda \in \mathbb{R}^n$ is the vector of Lagrangean multipliers. Since the original objective function and the set of constraints are now independent of each other, the problem decomposes into

$$\min f(x) - \lambda^\top x \quad + \quad \min \lambda^\top y \tag{$L(\lambda)$}$$
$$\text{s.t. } x \in \{0,1\}^n \qquad \text{s.t. } y \in X.$$

The two minimization problems in $(L(\lambda))$ can be solved independently. The left problem is an unconstrained nonlinear minimization problem over binary variables, whereas the problem on the right is a linear instance of the underlying combinatorial problem. For any $\lambda \in \mathbb{R}^n$, $(L(\lambda))$ is a relaxation of the original problem (P) and yields a lower bound on its optimal value. The best possible bound is obtained by computing the Lagrangean dual

$$\max_{\lambda \in \mathbb{R}^n} L(\lambda), \tag{1}$$

for example with a subgradient algorithm. In each iteration the two subproblems of $(L(\lambda))$ have to be solved for a given λ. Note that

$$L(\lambda) = \begin{cases} \min \ z - \lambda^\top x \\ \text{s.t. } (z, x) \in \text{conv}(F) \end{cases} + \begin{cases} \min \ \lambda^\top y \\ \text{s.t. } y \in \text{conv}(X) \end{cases}$$

where $F := \{(z, x) \mid x \in \{0, 1\}^n, \ z \geq f(x)\}$. By general results on Lagrangean relaxation we obtain

Lemma 1.

$$\max_{\lambda \in \mathbb{R}^n} L(\lambda) = \begin{cases} \min \ z \\ \text{s.t. } (z, x) \in \text{conv}(F) \\ \quad x \in \text{conv}(X). \end{cases}$$

Note that

$$\begin{array}{ccc} \min z & \min f(x) & \\ \text{s.t. } (z, x) \in \text{conv}(F) \quad \geq & \text{s.t. } x \in \text{conv}(\{0, 1\}^n) \quad = & \min f(x) \\ x \in \text{conv}(X) & x \in \text{conv}(X) & \text{s.t. } x \in \text{conv}(X), \end{array}$$

and that the inequality is strict in general if f is nonlinear. This is due to the fact that the objective function f is minimized over $\{0, 1\}^n$ in the left problem of $(L(\lambda))$, instead of over $[0, 1]^n$. In other words, the bounds we obtain are potentially stronger than those obtained from convexifying the feasible set in Problem (P).

Instead of solving the decomposition $(L(\lambda))$ exactly, it is possible to solve relaxations of the subproblems. This might be advantageous when the subproblems are computationally hard or no exact algorithm is known. Even if the relaxation may decrease the quality of the resulting lower bounds and hence increase the number of nodes in the branch and bound-algorithm, this effect might be compensated by the shorter time required to compute the bounds.

When embedding the computation of the Lagrangean dual into a branch and bound-algorithm, in order to obtain exact solutions, the problem solved in the root node of the branch and bound-tree is (1), but in deeper levels of the tree variable fixings have to be respected. This means that the algorithms for both the (formerly) unconstrained nonlinear subproblem and the linear combinatorial subproblem have to be adapted to handle fixed variables.

Within a branch and bound-scheme our approach can be improved significantly by reoptimization: in order to compute the Lagrangean dual (1) quickly with a subgradient method, a good starting guess for the multipliers λ is crucial. The choice of the initial multipliers in the root node should depend on the objective function, i.e. on the type of uncertainty set considered. In the remaining nodes of the branch and bound-tree, we use the optimal multipliers of the parent node for warmstart.

An important advantage of the Lagrangean decomposition approach is that we get a primal heuristic for free: each time we solve $(L(\lambda))$ we obtain a feasible solution $y \in X$ for Problem (R). In particular, we can use $f(y)$ as an upper bound in our algorithm.

In robust optimization, the function f is defined as the worst case solution quality of a vector $x \in X$ over all scenarios c in a given set $\mathcal{U}$. More formally, we consider objective functions of the form

$$f(x) := \max_{c \in \mathcal{U}} c^\top x$$

where $\mathcal{U} \subset \mathbb{R}^n$ is a compact set. In many applications, linear optimization over the feasible set X is only possible (or at least easier) if the objective function satisfies certain additional constraints such as, e.g., non-negativity or triangle inequalities. In the following, we argue that our approach also works in this case. More precisely, we claim that any homogeneous inequality that is valid for all scenarios $c \in \mathcal{U}$ can be assumed to be valid also for each vector λ appearing in the subproblem on the right in $(L(\lambda))$. To this end, one can show

Theorem 2. *Let $cone(\mathcal{U})$ denote the closed convex cone generated by $\mathcal{U}$ and let $cone(\mathcal{U})^*$ be its dual cone. Then*

$$
\begin{array}{ccc}
\min \max\limits_{c \in \mathcal{U}} c^\top x & & \min \max\limits_{c \in \mathcal{U}} c^\top x \\
s.t. \ x \in X & = & s.t. \quad x \in \{0,1\}^n \\
& & \quad y \in X \\
& & \quad x - y \in cone(\mathcal{U})^*.
\end{array}
$$

Due to space restrictions the proof is omitted here. By Theorem 2 we can apply the Lagrangean relaxation approach directly to the problem

$$
\begin{aligned}
\min \ \max_{c \in \mathcal{U}} &\ c^\top x \\
s.t. \quad & x \in \{0,1\}^n \\
& y \in X \\
& x - y \in \text{cone}(\mathcal{U})^*,
\end{aligned}
$$

meaning that the dual multipliers have to be chosen from $\text{cone}(\mathcal{U})$. It remains to investigate whether this restriction on λ yields the same bound as (1).

Theorem 3. *Assume that $\mathcal{C}$ is a polyhedral cone with $\mathcal{U} \subseteq \mathcal{C}$. Then*

$$\max_{\lambda \in \mathcal{C}} L(\lambda) = \max_{\lambda \in \mathbb{R}^n} L(\lambda).$$

Again, we have to omit the proof. By Theorem 3, any finite number of conditions on the objective function of one of the following types can be carried over from $\mathcal{U}$ to λ without weakening the lower bound:

- non-negativity or non-positivity of a given objective function coefficient;
- the triangle inequality on a given triple of coefficients;
- if variables correspond to edges of a graph, the non-negativity of the total cost of a given cycle.

Depending on the underlying combinatorial structure, such conditions may be crucial for linear optimization over X. This is true, e.g., when the underlying optimization problem asks for a *shortest path* or a *minimum cut* in a graph.

3   Uncorrelated Ellipsoidal Uncertainty

We now focus on the case of ellipsoidal uncertainty, i.e. the set $\mathcal{U}$ of all possible scenarios has the form of an ellipsoid in $\mathbb{R}^n$,

$$\mathcal{U} = \left\{ c \in \mathbb{R}^n \mid (c - c_0)^\top A^{-1} (c - c_0) \leq 1 \right\},$$

with $c_0 \in \mathbb{R}^n$ denoting the center of the ellipsoid and $A \in \mathbb{R}^{n \times n}$ being a positive definite symmetric matrix. In this case the objective function

$$f(x) = \max_{c \in \mathcal{U}} c^\top x$$

of (P) can be replaced by a closed formula: for a given $x \in \mathbb{R}^n$, the value $f(x)$ is obtained by a linear maximization over an ellipsoid. The KKT optimality conditions yield

$$f(x) = c_0^\top x + \sqrt{x^\top A x}.$$

Thereby the unconstrained min-max problem arising in the left part of problem $(L(\lambda))$ in the ellipsoidal uncertainty case reads

$$\min_{x \in \{0,1\}^n} (c_0 - \lambda)^\top x + \sqrt{x^\top A x}. \tag{2}$$

Here, c_0 and A can be interpreted as the mean values and the covariance matrix of a set of random variables.

In the following, we restrict ourselves to the case of uncorrelated random variables. In this case, the ellipsoid $\mathcal{U}$ is axis-parallel or equivalently the matrix A is diagonal. Exploiting the binarity of x we can simplify Problem (2) to

$$\min_{x \in \{0,1\}^n} (c_0 - \lambda)^\top x + \sqrt{a^\top x}, \tag{3}$$

where $A = \mathrm{Diag}(a)$ for some non-negative vector $a \in \mathbb{R}^n$.

3.1   An Efficient Algorithm for the Unconstrained Problem

Problem (3) is an unconstrained variant of the so-called *mean-risk optimization problem*. It can be solved to optimality in polynomial time since the objective function is submodular [2]. As minimization algorithms for general submodular functions are too slow to be applied in practice, we aim at a faster algorithm exploiting the special structure of Problem (3).

To this end, consider two solutions of (3) which differ in exactly one variable i. The difference between the corresponding objective values is

$$\Delta_i f(J) = (c_0 - \lambda)_i + \sqrt{\sum_{j \in J} a_j + a_i} - \sqrt{\sum_{j \in J} a_j}, \tag{4}$$

with J denoting the set of variables which are 1 in both solutions. The value (4) is also known as the discrete derivative of variable i [8]. It describes the contribution of setting variable i to 1, which clearly depends on the set J or, more precisely, on the quantity $\sum_{j \in J} a_j$. We hence define for each variable i its *contribution function* by

$$C_i(z) = (c_0 - \lambda)_i + \sqrt{z + a_i} - \sqrt{z}.$$

The functions C_i are strictly decreasing and therefore have at most one root each. The root r_i of C_i is the value which $\sum_{j \in J} a_j$ must reach such that setting variable i to 1 becomes profitable. Note that setting a variable to 1 never has a negative effect on the contributions of other variables, since the objective function of (3) is submodular.

Our basic idea for the construction of an optimal solution of (3) is that, due to the definition of r_i, a variable i cannot be 1 in an optimal solution while another variable having a smaller root is 0. This leads to an obvious sorting algorithm. However, in a first step we have to eliminate variables i for which C_i has no root, using the following observation.

Lemma 4. *There exists an optimal solution x^* of Problem (3) with the following properties:*

(i) if $(c_0 - \lambda)_i \geq 0$, then $x_i^* = 0$.
(ii) if $(c_0 - \lambda)_i \leq -\sqrt{a_i}$, then $x_i^* = 1$.

Proof. The condition in (i) implies that the function C_i is positive everywhere, as $a_i > 0$. This implies that any solution with $x_i = 1$ can be improved by setting $x_i = 0$. The condition in (ii) implies that C_i is non-positive everywhere, as

$$(c_0 - \lambda)_i + \sqrt{z + a_i} - \sqrt{z} \leq (c_0 - \lambda)_i + \sqrt{a_i} \leq 0$$

by concavity of the square-root function. The contribution of variable i to the value of an arbitrary solution is therefore non-positive, so that it may be fixed to 1 without loss of generality. □

For each i such that $-\sqrt{a_i} < (c_0 - \lambda)_i < 0$, the function C_i has exactly one positive root

$$r_i = \left(\frac{a_i - (c_0 - \lambda)_i^2}{2(c_0 - \lambda)_i} \right)^2.$$

The algorithm for solving the unconstrained problem proceeds as follows: first variables are fixed according to Lemma 4, then the remaining non-fixed variables x_i are sorted by non-decreasing roots r_i. Finally, all binary vectors where the non-fixed entries have values that are non-increasing in the resulting order are enumerated and the best such solution is reported.

Theorem 5. *Problem (3) can be solved in time $O(n \log n)$.*

Proof. The algorithm can be implemented to run in linear time except for the sorting of variables which takes $O(n \log n)$ time. It thus remains to prove correctness. By Lemma 4 we may assume that no variable is fixed, then it suffices to show that (after sorting) every optimal solution x^* satisfies $x_1^* \geq x_2^* \geq \cdots \geq x_n^*$.

Assume on contrary that x^* is an optimal solution with $x_j^* = 0$ and $x_{j+1}^* = 1$ for some $j < n$. Consider the two solutions x^0 and x^1 defined by

$$x_i^0 = \begin{cases} 0 \text{ for } i = j + 1 \\ x_i^* \text{ otherwise,} \end{cases} \qquad x_i^1 = \begin{cases} 1 \text{ for } i = j \\ x_i^* \text{ otherwise.} \end{cases}$$

By optimality of x^* we have

$$0 \geq f(x^*) - f(x^0) = C_{j+1}\left(\sum_{i \in I} a_i\right)$$

for $I = \{i \in \{1, ..., n\} \setminus \{j + 1\} \mid x_i^* = 1\}$ and hence by definition of r_{j+1} and r_j

$$\sum_{i \in I} a_i \geq r_{j+1} \geq r_j. \tag{5}$$

Then using the concavity of the square-root function we have

$$f(x^1) - f(x^*) = (c_0 - \lambda)_j + \sqrt{\sum_{i \in I} a_i + a_{j+1} + a_j} - \sqrt{\sum_{i \in I} a_i + a_{j+1}}$$

$$< (c_0 - \lambda)_j + \sqrt{\sum_{i \in I} a_i + a_j} - \sqrt{\sum_{i \in I} a_i}$$

$$\overset{(5)}{\leq} (c_0 - \lambda)_j + \sqrt{r_j + u_j} - \sqrt{r_j} = 0,$$

which contradicts the optimality of x^*. □

Note that a similar algorithm for Problem (3) using a different sorting rule has been devised by Shen et al. [7].

It is easily verified that the fixings of variables arising in the branch and bound-scheme for the min-max problem do not affect the validity of our algorithm. The roots can be computed using the same formula, because an additional constant under the root does not change the order of the roots.

3.2   A Mixed-Integer SOCP Formulation

Due to the nonlinearity of the set $\mathcal{U}$, there is no straight-forward mixed-integer linear formulation of Problem (R) in the ellipsoidal uncertainty case. However, the problem can be modeled as a mixed-integer second-order cone program (SOCP), even in the correlated case: the objective function in (2) can be modeled by an SOCP constraint, while the feasible set X has to be modeled by a polyhedral description of conv(X). In practice, mixed-integer SOCPs are much harder to solve than mixed-integer linear programs, making this approach much less competitive than similar linearization approaches used for min-max problems in the discrete scenario case. Moreover, if the polytope conv(X) does not have a compact outer description, a separation algorithm might be needed.

4   Applications

The Lagrangean decomposition approach presented in Sect. 3 is applicable to a wide range of robust combinatorial optimization problems. In the following we present numerical results for the *robust shortest path problem* and the *robust knapsack problem*. We compare the performance of the decomposition algorithm with the standard mixed-integer SOCP approach as explained in Sect. 3.2, using CPLEX 12.5 to solve the resulting programs. The left subproblem of the decomposition $(L(\lambda))$, i.e. the unconstrained nonlinear binary minimization problem, was solved with the algorithm discussed in Sect. 3.1. The initial Lagrangean multipliers were chosen as the center of the ellipsoid. To implement the optimization of the Lagrangean dual problem we used the Conic Bundle Library [5]. A very natural branching rule is choosing some variable, for which the solutions of both sides differ most. The Best-Node-First search strategy was adopted.

All experiments were carried out on a machine running SUSE Linux on an Intel Xeon CPU at 2.60 GHz. All running times are stated in CPU-seconds; the time limit for each instance was one CPU-hour.

4.1   The Shortest Path Problem with Ellipsoidal Uncertainty

For the uncorrelated ellipsoidal uncertainty variant of the *shortest path problem*, no polynomial time algorithm is known [6]. Here each arc in the graph is associated with a mean and a variance value. The uncertain part of the objective function is weighted with a parameter $\Omega \in \{1, \frac{1}{2}, \frac{1}{3}, \frac{1}{5}, \frac{1}{10}\}$, the resulting problem of minimizing

$$f(x) = c_0^\top x + \Omega \sqrt{a^\top x}$$

over the set of s, t-paths in a given directed graph falls into the class of problems considered in Sect. 3. The factor Ω leads to a scaling of the ellipsoid $\mathcal{U}$ by Ω^{-2}.

We solved the combinatorial subproblem of the decomposition $(L(\lambda))$ with the network simplex optimizer of CPLEX 12.5, allowing to deal with fixed variables easily. For the left subproblem of $(L(\lambda))$ the algorithm proposed in Sect. 3.1 is directly applicable.

All tests were done on directed grid graphs having the following form: $n \times n$ nodes are arranged on a grid, where n ranges from 100 to 500. Each node is linked by an arc to the node to the right and to the node below. The start node s is the node in the upper left corner of the grid, the end node t is in the lower right corner. In these graphs the total number of arcs is $2n(n-1)$ and each path consists of $2(n-1)$ arcs. The ellipsoid center was generated by randomly choosing coefficients in the interval $[0, 100]$, then the variances were determined as squares of randomly chosen numbers in the interval between 0 and the ellipsoid center. We generated 10 instances of each size and type.

Table 1 shows our results compared to the SOCP solver of CPLEX. Our approach could solve all but 2 instances within the time limit of 1 h while the CPLEX solver reached its limit at 500×500 grids. Instances with smaller ellipsoid volume turned out to be easier to solve in both cases, which can be seen in all

Table 1. Results for the *shortest path problem with ellipsoidal uncertainty* on $n \times n$ grid graphs with $n \in \{100, 200, 300, 400, 500\}$. The number of edges is $m = 2n(n-1)$

Vars	$\frac{1}{\Omega}$	Decomposition approach				CPLEX SOCP			
		#s	Subs	Iter	Time/s	#s	Subs	Iter	Time/s
19800	10	10	4.0	17.4	0.20	10	1.5	6774.1	8.98
	5	10	4.6	24.5	0.25	10	5.0	6874.4	11.43
	3	10	5.6	33.2	0.33	10	16.4	7033.9	12.83
	2	10	6.8	45.6	0.48	10	57.6	7337.5	15.03
	1	10	7.2	85.0	0.83	10	902.7	12405.1	35.45
79600	10	10	4.0	24.4	1.52	10	3.4	27482.8	91.51
	5	10	6.6	34.9	1.77	10	6.7	27725.9	109.95
	3	10	8.2	56.4	2.77	10	19.4	28076.3	130.95
	2	10	24.8	176.3	7.29	10	166.1	29214.9	159.11
	1	10	164.2	1165.8	42.93	8	4895.0	89498.8	637.97
179400	10	10	6.4	22.1	5.58	10	5.0	62455.9	369.34
	5	10	8.6	34.2	6.23	10	12.4	62903.1	405.86
	3	10	8.4	44.4	7.54	10	51.1	63674.3	453.39
	2	10	14.6	101.7	14.58	10	205.0	65409.2	529.34
	1	10	198.0	1390.6	144.99	5	9142.8	141693.6	2217.49
319200	10	10	5.6	25.4	14.11	10	4.3	112330.9	1168.52
	5	10	8.4	45.8	19.62	10	14.5	113040.9	1205.93
	3	10	14.6	84.5	24.75	10	67.4	114185.6	1364.80
	2	10	58.4	411.6	107.47	10	513.9	120765.1	1565.60
	1	9	323.2	2296.0	579.62	1	5324.0	160764.0	3350.32
499000	10	10	6.4	26.0	27.02	10	3.9	177395.6	2710.71
	5	10	7.8	38.6	31.54	10	21.6	178383.3	3076.77
	3	10	26.4	157.4	79.92	8	112.9	180086.5	3260.07
	2	10	92.2	638.7	285.77	2	52.5	180432.0	2970.01
	1	9	237.2	1704.0	849.53	0	–	–	–

performance indicators. While the number of subproblems is not substantially different in both approaches, CPLEX obviously spent a lot more iterations than our approach. Overall, our approach was faster than CPLEX by a factor of 10 to 100.

4.2   The Knapsack Problem with Ellipsoidal Uncertainty

In portfolio theory an important concept is to not only consider the expected return when choosing a set of investments but also take into account the risk

Table 2. Results for the *knapsack problem with ellipsoidal uncertainty*

Vars					Vars				
ε	#s	Subs	Iter	Time/s	ε	#s	Subs	Iter	Time/s
1000					**4000**				
0.10	10	11970.4	26085.8	9.15	0.10	10	87337.6	200149.1	297.48
0.05	10	11999.0	26441.9	9.21	0.05	10	76084.4	173606.0	257.95
0.03	10	18363.6	40781.0	14.29	0.03	10	129048.6	296989.3	440.18
0.02	10	17761.0	40074.8	13.98	0.02	10	156201.2	355627.5	528.33
0.01	10	17861.4	38516.4	13.65	0.01	10	170836.2	397621.9	589.41
2000					**5000**				
0.10	10	39777.8	88595.8	63.12	0.10	10	128904.8	295368.9	558.42
0.05	10	43534.4	96112.8	68.76	0.05	9	243370.6	567948.6	1071.93
0.03	10	80259.2	182641.9	129.76	0.03	9	273028.8	629101.6	1200.04
0.02	10	57073.6	126914.7	90.43	0.02	10	217465.6	512935.8	972.53
0.01	10	46486.4	106990.0	76.42	0.01	10	380360.4	894375.6	1712.76
3000					**6000**				
0.10	10	72851.0	164835.7	185.49	0.10	10	214092.4	494438.2	1143.04
0.05	10	65032.8	147170.3	165.58	0.05	8	303761.0	701917.5	1632.30
0.03	10	73101.8	167410.3	187.52	0.03	9	294969.7	690688.8	1606.99
0.02	10	198490.0	461660.2	514.55	0.02	7	327128.7	754971.9	1754.51
0.01	10	127533.0	297333.5	332.10	0.01	9	258226.6	611664.8	1421.17

associated with investments. Such *mean-risk optimization problems* can be modeled using stochastic objective functions. Potential investment decisions are represented by independent random variables that have an associated mean value as well as a variance. The mean value stands for the expected return of the investments, and the variance models the uncertainty inherent in the investment, i.e. the risk that the real return deviates from the expected. The case of continuous variables is well studied whereas the case of discrete variables has received relatively little attention yet [3].

We concentrate on the *risk-averse capital budgeting problem* with binary variables [2]. In this variant of the mean-risk optimization problem we are given a set of possible investments characterized by their costs w, expected return values c_0 and variances a, as well as a number ε. The number $\varepsilon > 0$ characterizes the level of risk the investor is willing to take. Investment decisions are binary. The only constraint is a limit on the available budget. The choice of investments guarantees that with probability $1 - \varepsilon$ the portfolio will return at least a profit of the objective value.

The corresponding nonlinear IP-model is

$$\max c_0^\top x - \sqrt{\frac{1-\varepsilon}{\varepsilon} a^\top x}$$
$$\text{s.t. } w^\top x \leq b \tag{6}$$
$$x \in \{0,1\}^n,$$

which can easily be converted into a minimization problem of the form considered in Sect. 3. In this case the underlying combinatorial optimization problem is a *knapsack problem*. Note that here the scaling factor for $0 < \varepsilon < \frac{1}{2}$ is $\Omega = \frac{1-\varepsilon}{\varepsilon} > 1$, whereas for the *shortest path problem* with ellipsoidal uncertainty it was $\Omega \leq 1$.

We generated the objective function for our instances as described in Sect. 4.1. The constraints were created as follows: the (certain) rational weights w were chosen randomly and uniformly distributed from $[0, 100]$, while the threshold b was determined as $\frac{1}{2} \sum_{i=1}^n w_i$. This choice of the right-hand side was proposed in [2] to avoid trivial instances. We generated 10 instances of each size between 1000 and 6000 and solved each instance for the values of ε given in Table 2. The legend of the table is as in Table 1.

Also here we compared the performance of the decomposition approach with the performance of the SOCP solver of CPLEX. However, we do not state the results of the SOCP solver because it was not competitive: already for $n = 75$ not all instances could be solved within the time limit of 1 h.

Atamtürk Narayanan [2] present an approach to improve the second-order cone program using additional cutting planes to strengthen the relaxation in each node of the enumeration tree. The cutting planes are derived from the submodularity of the objective function of Problem (6). Their results show that the additional cutting planes significantly improve the dual bounds and lead to a much lower number of subproblems and faster solution times. Still, their approach is not competitive with the Lagrangean decomposition approach presented here: it takes more than 800 s on average for solving the instances with $n = 100$ and $\varepsilon = 0.01$.

In the decomposition approach the dependence of the number of subproblems or the running times on the balance between the linear and the nonlinear parts of the objective function, i.e. the scaling factor ε, is nearly absent, which was not the case for the SOCP solver. If we take a closer look at the ratio between the number of calls to the combinatorial algorithms for the two parts of the decomposition and the number of subproblems in the branch and bound-tree, we see that only few calls per subproblem are necessary, showing the importance of reoptimization. For all n this ratio is less than three. Additionally, the algorithms applied to solve the subproblems are very fast in theory and in practice. In combination with strong primal bounds this leads to very moderate overall running times.

In summary we could show that the decomposition algorithm is well suited for the risk-averse capital budgeting problem. It dramatically outperforms both standard SOCP solvers and more problem-specific approaches found in the literature.

References

1. Aissi, H., Bazgan, C., Vanderpooten, D.: Min-max and min-max regret versions of combinatorial optimization problems: a survey. Eur. J. Oper. Res. **197**(2), 427–438 (2009)
2. Atamtürk, A., Narayanan, V.: Polymatroids and mean-risk minimization in discrete optimization. Oper. Res. Lett. **36**(5), 618–622 (2008)
3. Cornuéjols, G., Tütüncü, R.: Optimization Methods in Finance. Mathematics, Finance, and Risk. Cambridge University Press, Cambridge (2006)
4. Guignard, M., Kim, S.: Lagrangean decomposition: a model yielding stronger lagrangean bounds. Math. Program. **39**(2), 215–228 (1987)
5. Helmberg. C.: ConicBundle 0.3.11. Fakultät für Mathematik, Technische Universität Chemnitz (2012). http://www.tu-chemnitz.de/-helmberg/ConicBundle
6. Nikolova, E.: Approximation algorithms for reliable stochastic combinatorial optimization. In: Serna, M., Shaltiel, R., Jansen, K., Rolim, J. (eds.) APPROX 2010. LNCS, vol. 6302, pp. 338–351. Springer, Heidelberg (2010)
7. Max Shen, Z.-J., Coullard, C., Daskin, M.S.: A joint location-inventory model. Transp. Sci. **37**(1), 40–55 (2003)
8. Stobbe, P., Krause, A.: Efficient minimization of decomposable submodular functions. CoRR, abs/1010.5511 (2010)

Maximum Generalized Assignment
with Convex Costs

Marco Bender[✉] and Stephan Westphal

Institute for Numerical and Applied Mathematics, University of Göttingen,
Lotzestr. 16-18, 37083 Göttingen, Germany
{m.bender,s.westphal}@math.uni-goettingen.de

Abstract. We consider a generalization of the maximum generalized
assignment problem. We relax the hard constraints for the bin capacities, and introduce for every bin a cost function that is convex in the total
load on this bin. These costs are subtracted from the profits of assigned
items, and the task is to find an assignment maximizing the resulting
net profit.

We show that even restricted cases of this problem remain strongly
NP-complete, and identify two cases that can be solved in strongly polynomial time. Furthermore, we present a $(1 - 1/e)$-approximation algorithm for the general case. This algorithm uses a configuration based
integer programming formulation for a randomized rounding procedure.
In order to turn the rounded solution into a feasible solution, we define
appropriate estimators that linearize the convex costs.

Keywords: Generalized assignment problem · Combinatorial optimization · Complexity · Approximation algorithms

1   Introduction

The *maximum generalized assignment problem* (GAP) is a classical combinatorial
optimization problem. We are given a set of items $\mathcal{I}$ and a set of bins $\mathcal{B}$. Every
item i yields a profit of p_{ij} and has a weight w_{ij} when assigned to bin j. The
task is to assign a subset of the items to the bins such that the total profit is
maximized and every bin's capacity B_j is satisfied. This can be summarized in
the following integer program:

$$(\text{GAP}) \quad \max \quad \sum_{j \in \mathcal{B}} \sum_{i \in \mathcal{I}} p_{ij} x_{ij} \tag{1a}$$

$$\text{s.t.} \quad \sum_{j \in \mathcal{B}} x_{ij} \leq 1 \qquad \forall\, i \in \mathcal{I} \tag{1b}$$

$$\sum_{i \in \mathcal{I}} w_{ij} x_{ij} \leq B_j \qquad \forall\, j \in \mathcal{B} \tag{1c}$$

$$x_{ij} \in \{0,1\} \qquad \forall\, i \in \mathcal{I}, j \in \mathcal{B}. \tag{1d}$$

© Springer International Publishing Switzerland 2014
P. Fouilhoux et al. (Eds.): ISCO 2014, LNCS 8596, pp. 75–86, 2014.
DOI: 10.1007/978-3-319-09174-7_7

Here, the decision variable x_{ij} attains the value one if and only if item i is assigned to bin j. Constraints (1b) ensure that every item is assigned at most once, and constraints (1c) ensure that every bin's capacity is respected.

We consider a natural generalization of GAP by dropping the "hard" constraints (1c) for the bin capacities and introduce for every bin a cost function c_j that depends on the total weight of items assigned to this bin (see Sect. 1.1 for the formal definition). Then, we define the net profit of an assignment as the sum over all profits for assigned items minus the costs incurred by the bins.

If we consider arbitrary cost functions, the problem does not admit a constant-factor approximation, unless $\mathcal{P} = \mathcal{NP}$, as shown in [1] already for the case of a single bin.

We focus our attention on the class of convex cost functions. The motivation for this stems from scheduling problems concerned with energy-efficient computing environments, which is known as *speed scaling* (cf. [2]). If the workload on a processor increases, it has to run at a higher speed, and thus, its energy consumption increases. This dependence follows approximately the *cube-root rule*, which states that the energy consumption increases cubic in the speed.

Most optimization problems that have been researched in this field so far have "hard" constraints, e.g., minimizing the total energy such that all jobs are completed, or minimizing the flow time given a fixed energy budget (cf. [2]). Pruhs and Stein [3] note that it would be of more practical relevance to consider an objective function that takes both the rewards that are earned for a job (depending on when it is finished), and a convex cost function depending on the current workload, into account. This would, e.g., allow to model the situation of the operator of a large data center that has to decide whether to accept or reject requests for computation times on his servers. In our setting, the requests correspond to the items and the servers to the bins.

1.1   Problem Definition

In the course of this paper, we consider the following problem:

Definition 1. (Max-GAP with Convex Costs (GAP-CC))
We are given a set of items $\mathcal{I} = \{1, \ldots, n\}$ *and a set of* bins $\mathcal{B} = \{1, \ldots, m\}$. *If we assign item i to bin j, it yields a* profit $p_{ij} \in \mathbb{Z}_{\geq 0}$ *and increases the load of bin j by the* weight $w_{ij} \in \mathbb{Z}_{>0}$. *For each bin j there is a convex non-decreasing* cost function $c_j : \mathbb{R}_{\geq 0} \to \mathbb{R}_{\geq 0}$ *with $c_j(0) = 0$, and we assume that it can be evaluated in constant time. For pairwise disjoint subsets $I_1, \ldots, I_m \subseteq \mathcal{I}$, the assignment of items $I_1, \ldots, I_m$ to bins $1, \ldots, m$, respectively, yields a* total net profit *of*

$$\pi(I_1, \ldots, I_m) := \sum_{j \in \mathcal{B}} \left(\sum_{i \in I_j} p_{ij} - c_j \left(\sum_{i \in I_j} w_{ij} \right) \right).$$

The task is to find an assignment of disjoint subsets of the items to bins such that the total net profit is maximized.

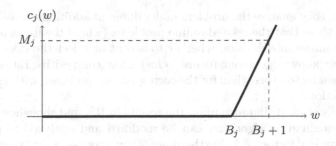

Fig. 1. Using cost functions (3) shows that GAP is a special case of GAP-CC.

This can be summarized in the following nonlinear integer program:

$$\text{(GAP-CC)} \quad \max \quad \sum_{j \in \mathcal{B}} \left(\sum_{i \in \mathcal{I}} p_{ij} x_{ij} - c_j \left(\sum_{i \in \mathcal{I}} w_{ij} x_{ij} \right) \right) \tag{2a}$$

$$\text{s.t.} \quad \sum_{j \in \mathcal{B}} x_{ij} \leq 1 \qquad\qquad\qquad \forall\, i \in \mathcal{I} \quad \text{(2b)}$$

$$x_{ij} \in \{0, 1\} \qquad\qquad\qquad \forall\, i \in \mathcal{I}, j \in \mathcal{B}. \quad \text{(2c)}$$

Here, the decision variable x_{ij} attains the value one if and only if item i is assigned to bin j, and constraints (2b) ensure that every item is assigned to at most one bin.

Note that GAP-CC contains GAP as a special case by choosing as a cost function for bin j, e.g.,

$$c_j(w) = \begin{cases} 0, & \text{if } w \leq B_j \\ M_j(w - B_j), & \text{else,} \end{cases} \tag{3}$$

as illustrated in Fig. 1. Here, the constant $M_j := \sum_{i \in \mathcal{I}} p_{ij} + 1$ ensures that in an optimal solution the load on bin j is no more than the bin capacity B_j in GAP.

1.2 Previous Work

GAP is a well-studied problem in literature [4,5]. The problem is known to be APX-hard [6], but there exists a deterministic $\frac{1}{2}$-approximation algorithm [6,7]. The currently best approximation algorithm due to Feige and Vondrak [8] achieves an approximation factor of $(1 - \frac{1}{e} + \delta)$ for some small $\delta > 0$. For the special case of fixed profits, i.e., the profit of assigning an item does not depend on the bin ($p_{ij} = p_i$), Nutov et al. [9] show how the randomized approximation algorithm of Fleischer et al. [10] can be derandomized using the method of conditional expectations (cf. [11]) yielding a deterministic $(1 - \frac{1}{e})$-approximation.

Special cases of GAP-CC have been subject to recent research. Barman et al. [12] consider the case $|\mathcal{B}| = 1$ and provide a 3-approximation algorithm that first sorts the items in non-increasing order of densities and then applies a greedy procedure.

Furthermore, they analyze the problem under different additional feasibility constraints and show that the related online problems (where the items arrive over time and an immediate decision whether to accept or reject the item has to be made without knowledge about future items) allow competitive ratios that are only a constant factor worse than for the corresponding problems without the convex cost function.

Antoniadis et al. [1] improve upon the results in [12] and show how the ideas of the 3-approximation algorithm can be modified and analyzed to guarantee an approximation factor of 2. Furthermore, they propose a dynamic program and show how this can be scaled in order to obtain an FPTAS. For the case of concave cost functions, they show that the problem can be solved optimally in polynomial time by a greedy procedure.

2 Complexity

We have already seen in Sect. 1.1 that GAP-CC contains GAP as a special case. In this section, we show strong $\mathcal{NP}$-completeness for restricted cases of GAP-CC that use, in particular, more realistic cost functions.

Theorem 1. GAP-CC *is strongly $\mathcal{NP}$-complete even if the cost functions of all bins are identical ($c_j = c$) and quadratic, the weight of an item is the same for each bin ($w_{ij} = w_i$), and either*

(i) the weight of each item i equals its profit ($w_i = p_i$), or
(ii) all items have the same profit ($p_{ij} \equiv p$).

Proof. Membership in $\mathcal{NP}$ holds since the objective function can be evaluated in polynomial time.

To show $\mathcal{NP}$-hardness, we perform a reduction from 3-PARTITION which is known to be strongly $\mathcal{NP}$-complete (cf. [13]). An instance of 3-PARTITION is given by non-negative integers $a_1, \ldots, a_{3m}, B \in \mathbb{Z}_{\geq 0}$ such that $\frac{B}{4} < a_i < \frac{B}{2}$ for all $i \in \{1, \ldots, 3m\}$, and $\sum_{i=1}^{3m} a_i = mB$. The task is to decide whether there exists a partition into sets of three elements each, i.e., $U_1, \ldots, U_m \subseteq \{1, \ldots, 3m\}$ such that $\cup_{i=1}^m U_i = \{1, \ldots, 3m\}$ and $U_i \cap U_j = \emptyset$ for all $i \neq j$, such that $\sum_{i \in U_j} a_i = B$ for all $j \in \{1, \ldots, m\}$.

Given an instance of 3-PARTITION, we construct an instance of GAP-CC as follows: every bin j has the same convex, quadratic cost function $c_j(w) := \frac{w^2}{2B}$.

In case (i), where $w_i = p_i$, we define an item with $w_i := p_i := a_i$ for all $i \in \{1, \ldots, 3m\}$. The net profit when assigning $I_1, \ldots, I_m \subset \mathcal{I}$ to bins $1, \ldots, m$, respectively, is then given as

$$\pi(I_1, \ldots, I_m) = \sum_{j=1}^m \left(\sum_{i \in I_j} a_i - \frac{\left(\sum_{i \in I_j} a_i\right)^2}{2B} \right),$$

and attains the value $\frac{mB}{2}$ if and only if there exists a 3-PARTITION.

It is easy to see that if there is a 3-PARTITION, then the corresponding assignment yields a net profit of $\frac{mB}{2}$. Vice versa, if there is an assignment with net profit $\frac{mB}{2}$, the contribution of every bin to the net profit is $\frac{B}{2}$ and this implies that there is a 3-PARTITION.[1]

In case (ii), where $p_{ij} \equiv p$, we define an item with $w_i := a_i$ and $p_i := p := c(n \max_i w_i) + 1$ for all $i \in \{1, \ldots, 3m\}$. Note that we assume that the cost function can be evaluated in constant time, and hence, p can be determined in constant time. This "large" profit ensures that all items have to be assigned to some bin in an optimal solution, and there exists an assignment with net profit $\pi^* := np - \frac{mB}{2}$ if and only if there exists a 3-PARTITION.

If there exists an assignment with net profit π^*, convexity of the cost functions imply that the load on every bin j is exactly B.[2] Thus, there exists a 3-PARTITION. The other implication follows immediately from the definition. □

Note that, using a similar reduction from PARTITION instead of 3-PARTITION, we can see that the problem is also (weakly) $\mathcal{NP}$-complete under the assumptions of the previous theorem even if we have only two bins. In particular, the case of constant profits which can be solved in polynomial time for the single bin case, becomes weakly $\mathcal{NP}$-complete for two bins.

3 Polynomially Solvable Cases

In this section, we show that special cases of GAP-CC can be solved in polynomial time.

3.1 A Round-Robin Algorithm for $w_{ij} \equiv w, p_{ij} = p_i, c_j = c$

We have seen that GAP-CC is strongly $\mathcal{NP}$-complete under the assumptions of Theorem 1. If we restrict the problem slightly more and assume that the profits do not depend on the bin ($p_{ij} = p_i$), all cost functions are identical ($c_j = c$), and the weights are constant ($w_{ij} \equiv w$), the problem can be solved in $\mathcal{O}(n \log n)$.

The idea of Algorithm 1 is to sort the items in non-increasing order of profits, and then keep on adding them to the bins in a round-robin manner, i.e., item i is assigned to bin i mod m, as long as the net profit increases.

Theorem 2. GAP-CC *can be solved in* $\mathcal{O}(n \log n)$ *if the profits do not depend on the bin* ($p_{ij} = p_i$), *the weights are constant* ($w_{ij} \equiv w$), *and all bins have the same cost function* ($c_j = c$).

---

[1]Observe that the function $h(w) := w - \frac{w^2}{2B}$ attains its global maximum $\frac{B}{2}$ for $w = B$.

[2]Let $B + \delta_j$ be the load on bin j, where $\sum_{j \in \mathcal{B}} \delta_j = 0$. Then, the total costs are $\sum_{j \in \mathcal{B}} c_j(B + \delta_j) = \frac{1}{2B}\left(mB^2 + \sum_{j \in \mathcal{B}} \delta_j^2\right) = \frac{mB}{2}$, which implies that $\delta_j = 0$ for all j.

Algorithm 1. Round-Robin Algorithm for $p_{ij} = p_i, w_{ij} \equiv w, c_j = c$

1: Sort items $\mathcal{I}$ in non-increasing order of profits such that $p_1 \geq p_2 \geq \ldots \geq p_n$.
2: **for** $i \in \mathcal{I}$ **do**
3:    **if** net profit increases when item i is assigned to bin i mod m **then**
4:       assign item i to bin i mod m.
5:    **end if**
6: **end for**

Proof. We assume w.l.o.g. $w_{ij} \equiv 1$. Observe that an optimal solution assigns items to the bins such that the loads of the bins differ by at most one. Otherwise, items could be reassigned increasing the net profit since the profits do not change and the cost functions are convex and identical for all bins. Hence, we can assume that an optimal solution is given in a round robin-manner.

Since the net profit function can be evaluated in constant time, the running time is dominated by the time needed for the sorting of the items, which can be done in $\mathcal{O}(n \log n)$.   □

3.2   A Minimum-Cost Flow Algorithm for $w_{ij} = w_j$

The problem can also be solved in polynomial time if we restrict GAP-CC to have weights that do not depend on the items, i.e., $w_{ij} = w_j$. In this case, we can use a minimum-cost flow algorithm.

Therefore, consider the following network: there is a node for every item i and for every bin j, an additional source s, and a sink t. There are arcs from s to all items with costs 0. Furthermore, there is an arc between item i and bin j with costs $-p_{ij}$, and there are n parallel arcs connecting bin j with the sink t that have costs of $c_j(a) - c_j(a-1)$ for $a \in \{1, \ldots, n\}$, respectively, representing the additional costs incurred by the a-th unit of weight on bin j. All of these arcs have unit capacity. Finally, there is an arc connecting t with s at cost 0, and capacity ∞. This network is illustrated in Fig. 2.

Observe that we can assume all weights to be 1 (if they were different, we could just scale them appropriately by modifying the cost function), and since the cost functions are convex and non-decreasing we have that $c_j(a) - c_j(a-1) \leq c_j(b) - c_j(b-1)$ for all $a \leq b$. Thus, a minimum-cost circulation then corresponds to a maximum net profit assignment of items to bins.

Hence, in order to solve GAP-CC for $w_{ij} = w_j$, we simply need to compute a minimum-cost circulation in the given network. Since the minimum cost flow problem can be solved in strongly polynomial time (cf. [14]), and since the graph has $n + m + 2 = \mathcal{O}(n + m)$ nodes and $n + 2nm + 1 = \mathcal{O}(mn)$ arcs, we obtain the following result:

Theorem 3. GAP-CC *can be solved in strongly polynomial time if the weights only depend on the bin* ($w_{ij} = w_j$).

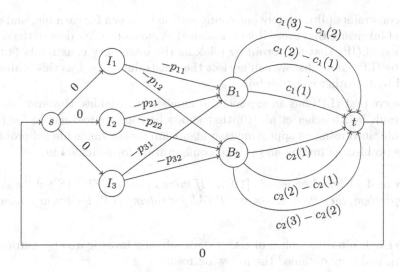

Fig. 2. Illustration of the network for three items and two bins. The labels on the arcs correspond to the costs. Arc (t, s) has capacity ∞, all other arcs have capacity 1.

4 Approximation Algorithm for the General Case

In this section, we design an approximation algorithm for the general case based on the ideas of Fleischer et al. [10] for the classical GAP.

4.1 A Configuration Based Formulation

Besides the straight-forward formulation (2), GAP-CC can also be formulated using a configuration based integer linear program. A *configuration* of a bin j is a subset of the items that may be assigned to the bin.

Although it might lead to negative net profits, it is allowed to assign any subset of the items to each of the bins. Hence, the set of feasible configurations for bin j is given by $T(j) := 2^{\mathcal{I}}$. We denote by $T := \bigcup_{j \in \mathcal{B}} T(j)$ the disjoint union of all bins' configurations. The net profit of a configuration $t \in T(j)$ is given by $\pi_t := \sum_{i \in t} p_{ij} - c_j \left(\sum_{i \in t} w_{ij} \right)$.

We can then write the problem as the following integer linear program:

$$\text{(IP)} \quad \max \quad \sum_{t \in T} \pi_t x_t \tag{4a}$$

$$\text{s.t.} \quad \sum_{t \in T(j)} x_t = 1 \qquad \forall j \in \mathcal{B} \tag{4b}$$

$$\sum_{t \in T : i \in t} x_t \leq 1 \qquad \forall i \in \mathcal{I} \tag{4c}$$

$$x_t \in \{0, 1\} \qquad \forall t \in T. \tag{4d}$$

By constraints (4b), exactly one configuration is chosen for each bin, and constraints (4c) ensure that every item is assigned at most once. We denote the linear relaxation of (IP) that we obtain by relaxing the integrality constraints (4d) to $x_t \geq 0$ by (LP) (note that we can neglect the constraints $x_t \leq 1$ as this is already implied by the other constraints).

Observe that (LP) has an exponential number of variables. However, we can use a result by Fleischer et al. [10] that shows how an approximation for (LP) can be obtained using an approximation algorithm for the single bin subproblem, i.e., the problem of finding an optimal configuration for a single bin.

Theorem 4 (Fleischer et al. [10]). *If there exists an FPTAS for the single bin subproblem, then there exists an FPTAS for solving (LP) the linear relaxation of (4).*

The single bin subproblem of GAP-CC was already investigated by Antoniadis et al. [1], and they obtained the following result:

Theorem 5 (Antoniadis et al. [1]). *For the single bin subproblem of GAP-CC there exists an FPTAS.*

Combining the two previous results, we obtain:

Corollary 1. *There exists an FPTAS for solving (LP).*

4.2 The Randomized Rounding Procedure

Our randomized rounding procedure then works as follows: For a fixed $\epsilon > 0$, we first solve (LP) approximately using the FPTAS as described above and obtain a fractional solution $x^{\mathrm{LP}} \in [0, 1]^{|T|}$ in time polynomial in the encoding length of the problem and $1/\epsilon$. Then, for each bin j, we choose a configuration independently at random, where configuration $t \in T(j)$ is chosen with probability x_t^{LP}. Note that, by constraints (4b), the values $\{x_t^{\mathrm{LP}} : t \in T(j)\}$ define a probability distribution for every bin j.

We denote the solution obtained by this procedure by $x^{\mathrm{IP}} \in \{0, 1\}^{|T|}$, where $x_t^{\mathrm{IP}} = 1$ if and only if configuration t is chosen. The expected net profit of the rounded solution x^{IP} is then given as

$$\mathbb{E}\left[\sum_{t \in T} \pi_t x_t^{\mathrm{IP}}\right] = \sum_{t \in T} \pi_t \mathbb{E}\left[x_t^{\mathrm{IP}}\right] = \sum_{t \in T} \pi_t x_t^{\mathrm{LP}} \geq (1 - \epsilon) \cdot \mathrm{OPT}_{\mathrm{LP}}. \tag{5}$$

By construction, x^{IP} satisfies (4b). However, it does not, in general, fulfill (4c) since an item might be assigned to more than one bin.

4.3   Obtaining a Feasible Solution

In order to turn x^{IP} into a feasible solution of (IP), we need to remove multiple copies of items from the bins they are assigned to. In classical GAP, the straightforward way of doing this is to delete each item from all bins it is assigned to except for the one where it yields highest profit (cf. [10]). Due to the nonlinear cost functions c_j, one also has to take the cost increase caused by the items' weights into account. One natural way to do this is to "linearize" the costs by charging every item an amount of the costs proportional to its weight. This motivates the following definition:

Definition 2. *For a configuration* $t \in T(j)$, *we define the* net profit contribution *of item* i *with respect to* t *as*

$$\mu_i^t := \begin{cases} p_{ij} - \dfrac{w_{ij}}{\sum\limits_{k \in t} w_{kj}} c_j \left(\sum\limits_{k \in t} w_{kj} \right), & \textit{if } i \in t \\ 0, & \textit{else.} \end{cases}$$

Note that, for each $t \in T$, the net profit contributions of all items with respect to t sum up to the net profit of configuration t, i.e., $\sum_{i \in t} \mu_i^t = \pi_t$. Moreover, the following result shows that we can use the net profit contributions with respect to a configuration in order to obtain a lower bound on the net profit of any subset of the configuration:

Lemma 1. *For every configuration* $t' \subseteq t$, *we have*

$$\sum_{i \in t'} \mu_i^t \leq \pi_{t'}.$$

Proof. For $t' = t$, the result holds as $\sum_{i \in t} \mu_i^t = \pi_t$ by definition of the μ_i^t. For $t' \subsetneq t$, let j denote the bin to which configuration t corresponds, i.e., $t \in T(j)$. Observe that, for $0 \leq x < y$, the convexity of c_j and the assumption that $c_j(0) = 0$ imply that

$$c_j(x) = c_j \left(\left(1 - \frac{x}{y} \right) \cdot 0 + \frac{x}{y} \cdot y \right) \leq \left(1 - \frac{x}{y} \right) \cdot \underbrace{c_j(0)}_{=0} + \frac{x}{y} \cdot c(y) = \frac{x}{y} \cdot c(y).$$

If we choose $x := \sum_{i \in t'} w_{ij}$ and $y := \sum_{i \in t} w_{ij}$ (which satisfy $0 \leq x < y$ since $w_{ij} > 0$ for all i, j and $t' \subsetneq t$), this shows that

$$c_j \left(\sum_{i \in t'} w_{ij} \right) \leq \frac{\sum_{i \in t'} w_{ij}}{\sum_{i \in t} w_{ij}} \cdot c_j \left(\sum_{i \in t} w_{ij} \right).$$

Using this inequality, we obtain

$$\sum_{i \in t'} \mu_i^t = \sum_{i \in t'} p_{ij} - \frac{\sum_{i \in t'} w_{ij}}{\sum_{k \in t} w_{kj}} c_j \left(\sum_{k \in t} w_{kj} \right) \leq \sum_{i \in t'} p_{ij} - c_j \left(\sum_{i \in t'} w_{ij} \right) = \pi_{t'}.$$

$\square$

In order to decide to which bin a multiply assigned item is assigned, we now use the net profit contributions. If an item i is assigned to multiple bins, we delete it from all bins except for the one where the expression

$$E_{ij} := \sum_{t \in T(j): i \in t} \frac{x_t^{LP}}{y_j} \cdot \mu_i^t \tag{6}$$

is maximal. Here, E_{ij} denotes the conditional expectation of the profit contribution of item i in bin j given that the configuration chosen for bin j contains i. We denote by $y_j := \sum_{t \in T(j): i \in t} x_t^{LP}$ the probability that item i is assigned to bin j. Note that for the classical GAP, $\mu_i^t = p_{ij}$ for all $t \in T(j)$ and thus, $E_{ij} = p_{ij}$.

In the following, let $i \in \mathcal{I}$ be an arbitrary, but fixed item, and number the bins in non-increasing order of (6) such that $E_{i1} \geq E_{i2} \geq \ldots \geq E_{im}$.

With probability y_1 the configuration that is chosen for bin 1 contains item i. In this case, item i is assigned to bin 1 and its expected profit contribution in bin 1 is at least E_{i1}. With probability $(1 - y_1)y_2$ the configuration chosen for bin 1 does not contain item i, but the one for bin 2 does (this holds since the configurations for the bins are chosen independently). In this case, item i is assigned to bin 2 and its expected profit contribution in bin 2 is at least E_{i2}.

This argument can be applied in the same way for the following bins, and overall, we obtain that the expected profit contribution of item i is at least

$$y_1 E_{i1} + (1 - y_1)y_2 E_{i2} + \ldots + \left(\prod_{j=1}^{m-1} (1 - y_j) \right) y_m E_{im}$$

$$= \sum_{j=1}^{m} \prod_{k=1}^{j-1} (1 - y_k)\, y_j\, E_{ij}$$

$$\geq \left(1 - \left(1 - \frac{1}{m} \right)^m \right) \cdot \sum_{j=1}^{m} y_j E_{ij}$$

$$\geq \left(1 - \frac{1}{e} \right) \cdot \sum_{j=1}^{m} y_j E_{ij}$$

$$= \left(1 - \frac{1}{e} \right) \cdot \sum_{j=1}^{m} \sum_{t \in T(j): i \in t} x_t^{LP} \mu_i^t.$$

Here, we used the arithmetic-geometric mean inequality for the first inequality, and the fact that $(1 - 1/k)^k \leq e^{-1}$ for all $k \geq 1$ for the second inequality (cf. [15]).

If we do this for every multiply assigned item, we choose subsets of the configurations that have been selected in the rounding procedure, and by Lemma 1 we know that using the net profit contributions for the chosen items we underestimate the actual net profit. Hence, the total expected net profit we obtain is at least the sum of the expected profit contributions of all items that remain in the solution after removing multiple copies.

Summing up over all items $i \in \mathcal{I}$, this shows that we obtain a solution with expected net profit at least

$$\sum_{i \in \mathcal{I}} \left(1 - \frac{1}{e}\right) \cdot \sum_{t \in T : i \in t} \mu_i^t x_t^{\mathrm{LP}} = \left(1 - \frac{1}{e}\right) \cdot \sum_{t \in T} \underbrace{\sum_{i \in \mathcal{I}} \mu_i^t x_t^{\mathrm{LP}}}_{= \pi_t}$$

$$\overset{(5)}{\geq} \left(1 - \frac{1}{e}\right) \cdot (1 - \epsilon) \cdot \mathrm{OPT}_{\mathrm{LP}}$$

$$\geq \left(1 - \frac{1}{e} - \epsilon\right) \cdot \mathrm{OPT}_{\mathrm{LP}}. \qquad (7)$$

Hence, we obtain the following result:

Lemma 2. *For every $\epsilon > 0$, there exists a randomized $\left(1 - \frac{1}{e} - \epsilon\right)$-approximation algorithm for* GAP-CC *whose running time is polynomial in the encoding length of the problem and $1/\epsilon$.*

In fact, one can even slightly improve upon the previous result. If we use instead of the arithmetic-geometric mean inequality, the sharper estimate

$$1 - (1 - 1/k)^k \geq 1 - \frac{1}{e} + \frac{1}{32k^2} \text{ for all } k \geq 1$$

due to Nutov et al. [9], we can bound the expected profit we obtain similar to (7) from below by

$$\left(1 - \frac{1}{e} + \frac{1}{32m^2} - \epsilon\right) \cdot \mathrm{OPT}_{\mathrm{LP}}.$$

If we then choose $\epsilon = 1/32m^2$ (which is polynomial in the encoding length of the problem), we obtain the following result:

Theorem 6. *There exists a randomized $\left(1 - \frac{1}{e}\right)$-approximation algorithm for* GAP-CC.

5   Open Problems

For classical GAP with fixed profits ($p_{ij} = p_i$), the approximation algorithm from [10] can be derandomized [9]. It remains an open problem whether such a derandomization can also be performed in our setting for some cases, or whether there exist other deterministic approximation algorithms.

Acknowledgments. This research was partially supported by the German Research Foundation (DFG), grant GRK 1703/1 for the Research Training Group "Resource Efficiency in Interorganizational Networks – Planning Methods to Utilize Renewable Resources".

References

1. Antoniadis, A., Huang, C.-C., Ott, S., Verschae, J.: How to pack your items when you have to buy your knapsack. In: Chatterjee, K., Sgall, J. (eds.) MFCS 2013. LNCS, vol. 8087, pp. 62–73. Springer, Heidelberg (2013)
2. Albers, S.: Energy-efficient algorithms. Commun. ACM **53**(5), 86–96 (2010)
3. Pruhs, K., Stein, C.: How to schedule when you have to buy your energy. In: Serna, M., Shaltiel, R., Jansen, K., Rolim, J. (eds.) APPROX and RANDOM 2010. LNCS, vol. 6302, pp. 352–365. Springer, Heidelberg (2010)
4. Cattrysse, D.G., Van Wassenhove, L.N.: A survey of algorithms for the generalized assignment problem. Eur. J. Oper. Res. **60**(3), 260–272 (1992)
5. Pentico, D.W.: Assignment problems: a golden anniversary survey. Eur. J. Oper. Res. **176**(2), 774–793 (2007)
6. Chekuri, C., Khanna, S.: A PTAS for the multiple knapsack problem. SIAM J. Comput. **35**(3), 713–728 (2006)
7. Shmoys, D.B., Tardos, É.: An approximation algorithm for the generalized assignment problem. Math. Program. **62**, 461–474 (1993)
8. Feige, U., Vondrák, J.: Approximation algorithms for allocation problems: improving the factor of $1-1/e$. In: Proceedings of the 47th Annual IEEE Symposium on the Foundations of Computer Science (FOCS), pp. 667–676 (2006)
9. Nutov, Z., Beniaminy, I., Yuster, R.: A $(1 - 1/e)$-approximation algorithm for the maximum generalized assignment problem with fixed profits. Oper. Res. Lett. **34**, 283–288 (2006)
10. Fleischer, L., Goemans, M.X., Mirrokni, V.S., Sviridenko, M.: Tight approximation algorithms for maximum general assignment problems. Math. Oper. Res. **36**(3), 416–431 (2011)
11. Alon, N., Spencer, J.H.: The Probabilistic Method. John Wiley and Sons, New York (1992)
12. Barman, S., Umboh, S., Chawla, S., Malec, D.: Secretary problems with convex costs. In: Czumaj, A., Mehlhorn, K., Pitts, A., Wattenhofer, R. (eds.) ICALP 2012, Part I. LNCS, vol. 7391, pp. 75–87. Springer, Heidelberg (2012)
13. Garey, M.R., Johnson, D.S.: Computers and Intractability (A Guide to the Theory of NP-Completeness). W.H. Freeman and Company, New York (1979)
14. Ahuja, R.K., Magnanti, T.L., Orlin, J.B.: Network Flows. Prentice Hall, Englewood Cliffs (1993)
15. Goemans, M.X., Williamson, D.P.: New 3/4-approximation algorithms for the maximum satisfiability problem. SIAM J. Discrete Math. **7**, 656–666 (1994)

An Integer Programming Formulation for the Maximum k-Subset Intersection Problem

Eduardo T. Bogue, Cid C. de Souza, Eduardo C. Xavier,
and Alexandre S. Freire[✉]

Institute of Computing, University of Campinas, Campinas, Brazil
afreire@ime.usp.br

Abstract. In this paper, we study the MAXIMUM k-SUBSET INTERSECTION (MkSI) problem. Given an integer k, a ground set U and a collection $\mathcal{S}$ of subsets of U, the MkSI problem is to select k subsets $S_1, S_2, \ldots, S_k$ in $\mathcal{S}$ whose intersection size $|S_1 \cap S_2 \cap \cdots \cap S_k|$ is maximum. The MkSI problem is NP-hard and hard to approximate. Some applications of the MkSI problem can be found in the literature and, to the best of our knowledge, no exact method was proposed to solve this problem. In this work, we introduce a very effective preprocessing procedure to reduce the size of the input, introduce a GRASP heuristic which was able to find solutions very close to be optimal ones, propose an integer programming formulation for the problem and present computational experiments made with instances that come from an application.

Keywords: Maximum k-subset intersection · Integer programming · GRASP

1 Introduction

In this paper[1], we study the MAXIMUM k-SUBSET INTERSECTION (MkSI) problem. Given an integer k, a ground set U and a collection $\mathcal{S}$ of subsets of U, the MkSI problem is to select k subsets $S_1, S_2, \ldots, S_k$ in $\mathcal{S}$ whose intersection size $|S_1 \cap S_2 \cap \cdots \cap S_k|$ is maximum. Using graph theory terminology, the MkSI problem can be stated as follows: given a bipartite graph $G = (V = L \cup R, E)$, find a *biclique* (a complete bipartite graph) B, subgraph of G, such that $|V(B) \cap L| = k$ and $|V(B) \cap R|$ is maximized. It is worth noticing that the constraint $|V(B) \cap L| = k$ can be relaxed to $|V(B) \cap L| \geq k$. In fact, given an optimal biclique B such that $|V(B) \cap L| \geq k$, we can remove $|V(B) \cap L| - k$ vertices from $(V(B) \cap L)$ and obtain a feasible solution of the MkSI problem with the same objective value $|V(B) \cap R|$.

The MkSI problem was introduced by Vinterbo in [9]. In [9] and [10], it is shown that the problem is NP-hard and Xavier shows in [10] that the MkSI

[1] This work was partially founded by CNPq (grants 477692/2012-5, 302804/2010-2 and 302067/2011-6) and FAPESP (grants 2012/17585-9 and 2013/03447-6).

© Springer International Publishing Switzerland 2014
P. Fouilhoux et al. (Eds.): ISCO 2014, LNCS 8596, pp. 87–99, 2014.
DOI: 10.1007/978-3-319-09174-7_8

problem is hard to approximate. A related NP-hard problem is the MAXIMUM EDGE BICLIQUE (MEB) problem, which is to find a biclique in a graph G with maximum number of edges. A natural idea for solving the MEB problem is to solve the MkSI, for $k = 1, 2, \ldots, |L|$, and return a solution with the maximum number of edges. Intuitively, this process leads to an optimal solution of the MEB problem. In fact, in [10] it is shown that the MEB problem can be reduced to the MkSI problem. In [1], Acuña et al. introduced a Mixed Integer Linear Programming (MILP) formulation for the MEB problem and mention some applications of this problem in bioinformatics and other fields.

A possible way to tackle the MkSI problem is simply to enumerate all $\binom{|L|}{k}$ subsets of L with size k and return one solution which maximizes the objective function. This observation suggests that, if k is bounded by a constant, then the problem can be solved in polynomial time. In [3], Ganter et al. proposed a polynomial delay algorithm for enumerating all closed sets, which can be used to enumerate all maximal bicliques of a given bipartite graph. A (L, R)-bipartite graph is said to be *convex* if there is a total order σ of R, such that, for each $u \in L$, we have that the vertices adjacent to u are consecutive in σ. In [6], D. Nussbaum et al. showed that a *convex* bipartite graph has a polynomial number of maximal bicliques, which implies that the MEB and MkSI problems can be solved in polynomial time for this special class of graphs. Apart from these two special cases, to the best of our knowledge, there is no other special case in which the MkSI problem can be solved in polynomial time.

In [9], an application of the MkSI problem in patients' data privacy control is proposed. In this application, each vertex in L represents a patient, each vertex in R represents an attribute, while an edge indicates that a certain patient has a certain attribute. The objective is to publish the maximum amount of patients' data (to be used in statistical analysis, for instance) without violating patients' privacy. For that purpose, a set of attributes can be published only if at least k patients have all those attributes in common. In [6], an application of the MkSI problem in DNA Microarray analysis is mentioned, in which the objective is to select k genes with the greatest number of characteristics in common. Although some applications have been previously proposed, we could not find in the literature any exact method for solving the MkSI problem.

In this work, we consider the following application of the MkSI problem: each vertex in L represents a musical artist (singer, band, etc.), each vertex in R represents a person, while an edge indicates that a certain person is a fan of a certain musical artist. The objective is to find a set of k musical artists (to be part of a jazz festival, for instance) with maximum number of fans in common. In this way, it is expected that most of the people attending the concert will enjoy all the bands. We obtained real data for this application from the *Last FM* database. Our main contributions can be summarized as follows: we introduce a very effective preprocessing procedure to reduce the size of the input; we introduce a GRASP heuristic that yielded an optimal solution for most of the tested instances (for all of them it yielded primal bounds very close to the optimum); we propose an Integer Linear Programming (ILP) formulation with which we solved instances of relatively large size; and we present computational experiments made with instances obtained from the *Last FM* database.

The next sections are organized as follows. In Sect. 2, we introduce the notation and definitions used throughout this paper. In Sect. 3, we present an ILP formulation for the MkSI, study the associated polytope, establish its relation with the so-called *stable set* polytope and propose a class of facet-defining inequalities. In Sect. 4, we describe a preprocessing procedure, a GRASP heuristic and discuss some other issues that we considered in our implementation. In Sect. 5, we present computational experiments with instances obtained from the *Last FM* database. Finally, in Sect. 6, we draw concluding remarks and point out some possible future work.

2   Notation and Definitions

This section introduces the notation and definitions used throughout this paper.

Let $G = (V, E)$ be a bipartite graph and let L and R be the two parts of the bipartition of V (i.e., $V = L \cup R$, $L \cap R = \emptyset$ and $E \subseteq L \times R$). For simplicity, we say that G is a (L, R)-bipartite graph and, given an edge $\{u, v\} \in E$, we assume, as a convention, that $u \in L$, $v \in R$ and write uv without braces (i.e., the label in the left side corresponds to the vertex in L, and the label in the right side corresponds to the vertex in R). Given a subgraph B of G, we denote its vertex set by V_B, its edge set by E_B and we say that B is a (L_B, R_B)-bipartite graph, where $L_B = V_B \cap L$ and $R_B = V_B \cap R$. Given a subset of vertices $S \subset V$, the subgraph of G *induced* by S is the graph $G' = (V', E')$, such that $V' := S$ and $E' = \{uv \in E \mid u \in V' \text{ and } v \in V'\}$. Given a vertex u in V, let $N(u) = \{v \in V \mid uv \in E\}$ be the set of *neighbours* of u and let $\overline{N}(u)$ be the set of non-neighbours of u in the opposite part (i.e., if $u \in L$ then $\overline{N}(u) = R \setminus N(u)$, otherwise $\overline{N}(u) = L \setminus N(u)$). Given a subset S of V, let $\Gamma(S) = \bigcap_{u \in S} N(u)$ be the set of neighbours that all vertices in S have in common. We denote by $\overline{E} = \{uv \mid u \in L \text{ and } v \in \overline{N}(u)\}$ the set of edges that are missing for G to be a complete bipartite graph. We say that B is a *biclique* if $uv \in E_B$, for each $u \in L_B$ and $v \in R_B$. We denote by $\chi(B)$ the *incidence* vector of B, where $\chi(B) \in \{0, 1\}^{|V|}$ and $\chi(B)_u = 1$ if and only if $u \in V_B$. In this paper, we study the problem stated below.

Problem 1. Given a (L, R)-bipartite graph $G = (V, E)$ and a positive integer $k \leq |L|$, the MAXIMUM k-SUBSET INTERSECTION (MkSI) problem is to find a biclique B, subgraph of G, such that $|L_B| = k$ and $|R_B|$ is maximum.

We denote by $opt(G, k)$ the optimum of an instance (G, k) of the MkSI problem.

3   Formulation and Preliminary Polyhedral Study

In this section, we present an Integer Linear Programming (ILP) formulation for the MkSI, study the associated polytope, establish its relation with the so-called *stable set* polytope and propose a class of facet-defining inequalities. To ease the discussions that follow, we state the problem in a slightly different way.

Suppose that we have a lower bound λ on $opt(G, k)$. Let $\mathcal{L}(\lambda)$ be the set of pairs of vertices from L that have at most $\lambda - 1$ neighbours in common, i.e. $\mathcal{L}(\lambda) = \{\{u, v\} \subset L \text{ s.t. } |\Gamma(\{u, v\})| < \lambda\}$. Analogously, for an integer t, we define $\mathcal{R}(t) = \{\{u, v\} \subset R \text{ s.t. } |\Gamma(\{u, v\})| < t\}$, the set of pair of vertices in R whose neighborhoods intersect in less than t vertices. Let $\mathcal{F}(t, \lambda) = \mathcal{L}(\lambda) \cup \mathcal{R}(t) \cup \overline{E}$. By construction, none of the pairs in $\mathcal{F}(k, \lambda)$ can be part of an optimal solution of MkSI.

That said, we introduce an ILP formulation for a slightly different variant of the MkSI problem, denoted by MkSI(λ), in which we consider only the feasible solutions which do not contain pairs in $\mathcal{F}(k, \lambda)$. Since an optimal solution for the MkSI does not contain such pairs, this integer program can be used to solve the latter problem. In this model, to each vertex u in V, we associate a binary variable x_u which is set to one if and only if u is in the computed biclique.

$$\max z(x) = \sum_{v \in R} x_v$$

$$\text{(IP1)} \qquad \text{s.t.} \sum_{u \in L} x_u = k \qquad\qquad\qquad (1)$$

$$x_u + x_v \leq 1, \qquad \text{for each } \{u, v\} \in \mathcal{F}(k, \lambda) \qquad (2)$$

$$x \in \{0, 1\}^{|V|} \qquad\qquad\qquad\qquad (3)$$

We first argue informally that (IP1) is a formulation for the MkSI(λ) problem, and latter prove it formally. Given an optimal solution x^* of (IP1), we obtain a subgraph B^* of G, such that $\chi(B^*) = x^*$. Constraints (1) and (3) together assure that $|L_{B^*}| = k$. Constraints (2) and (3) assure that B^* does not contain any pair in $\overline{E}$, which implies that B^* is a biclique. From the objective function, we have that $z(x^*) = |R_{B^*}|$. Therefore, (IP1) is a formulation for the MkSI(λ) problem. It is worth noticing that a more natural formulation for MkSI is derived from (IP1) by replacing $\mathcal{F}(k, \lambda)$ in constraint (2) by $\mathcal{F}(0, 0) = \overline{E}$. Clearly, in this way no knowledge is assumed of a primal bound λ and no pairs of vertices of R are eliminated *a priori* from the solution, leaving this latter task to the ILP solver. We name this natural formulation as the M$_{\text{EDGE}}$ model.

Let $\mathcal{P}_{\text{M}k\text{SI}}(G, k, \lambda) = \text{conv}\{\chi(B) \in \{0, 1\}^{|V|} \text{ s.t. } B \text{ is a biclique in } G, |L_B| = k \text{ and } |\Gamma(\{u, v\})| \geq \lambda, \text{ for each } u \text{ and } v \text{ in } L_B\}$ be the polytope associated with the MkSI(λ) problem, which corresponds to the convex hull of all feasible solutions of the problem. Since λ is a valid lower bound, we have that there is at least one biclique B in G with $|R_B| \geq \lambda$, thus $\mathcal{P}_{\text{M}k\text{SI}}(G, k, \lambda) \neq \emptyset$. Let $\mathcal{P}_{\text{IP1}}(G, k, \lambda) = \text{conv}\{x \in \mathbb{R}^{|V|} \mid x \text{ satisfies } (1), (2) \text{ and } (3)\}$ be the polytope associated with the formulation (IP1). We now show the equivalence between $\mathcal{P}_{\text{IP1}}(G, k, \lambda)$ and $\mathcal{P}_{\text{M}k\text{SI}}(G, k, \lambda)$, then we use this fact to prove that (IP1) is a formulation for the MkSI(λ) problem.

Lemma 1. $\mathcal{P}_{\mathrm{MkSI}}(G, k, \lambda) = \mathcal{P}_{\mathrm{IP1}}(G, k, \lambda)$.

Proof. Clearly it suffices to check that the set of integral vectors in both polytopes are the same. So, given a vector $\chi(B)$ in $\mathcal{P}_{\mathrm{MkSI}}(G, k, \lambda)$, let $x := \chi(B)$. We now show that x is a feasible solution of (IP1). Since $|L_B| = k$, we have that constraint (1) is satisfied by x. Since B is a biclique, such that $|L_B| = k$ and $|\Gamma(\{u, v\})| \geq \lambda$, for each u and v in V_B, then we have that $\{u, v\} \notin \mathcal{L}(\lambda) \cup \mathcal{R}(k) \cup \overline{E}$. Thus, constraints (2) are satisfied by x. Hence, $\mathcal{P}_{\mathrm{MkSI}}(G, k) \subseteq \mathcal{P}_{\mathrm{IP1}}(G, k)$.

Conversely, let x' be a feasible solution of (IP1) and let B' be the subgraph of G induced by $V_{B'}$, where $V_{B'} = \{u \in V \mid \chi(B')_u = 1\}$. By constraint (1), we have that $|L_{B'}| = k$. By constraints (2), for each $u \in L_{B'}$ and $v \in R_{B'}$, we have that $\{u, v\} \notin \mathcal{L}(\lambda) \cup \overline{E}$ and, thus, B' is a biclique such that $|\Gamma(\{u, v\})| \geq \lambda$, for each u and v in $L_{B'}$. Hence $\mathcal{P}_{\mathrm{IP1}}(G, k, \lambda) \subseteq \mathcal{P}_{\mathrm{MkSI}}(G, k, \lambda)$. $\qquad\square$

Lemma 2. (IP1) *is a formulation for the* MkSI(λ) *problem.*

Proof. Given a feasible solution x of (IP1), let B be the subgraph of G with $\chi(B) = x$. Clearly, we have that $|R_B| = z(x)$. Together with Lemma 1, this implies that (IP1) is a formulation for the MkSI(λ) problem. $\qquad\square$

Due to the equality constraint (1), $\mathcal{P}_{\mathrm{IP1}}(G, k, \lambda)$ is not full dimensional. As a result, for technical convenience, we study the monotone polytope defined as

$$\mathcal{P}_{\mathrm{IP2}}(G, k, \lambda) = \mathrm{conv}\{x \in \mathbb{R}^{|V|} \mid x \text{ satisfies } (2), (3) \text{ and } \sum_{u \in L} x_u \leq k\}.$$

In order to derive a formulation for the MkSI(λ) problem from $\mathcal{P}_{\mathrm{IP2}}(G, k, \lambda)$, the objective function is modified to:

$$z'(x) = (|R| + 1) \sum_{u \in L} x_u + \sum_{v \in R} x_v.$$

Since $\mathcal{P}_{\mathrm{IP1}}(G, k, \lambda) \subseteq \mathcal{P}_{\mathrm{IP2}}(G, k, \lambda)$, the high costs assigned to the vertices in L forces that any optimal solution for (IP2) must be in (IP1). In other words, a solution x^* which maximizes $z'(x^*)$ also maximizes $z(x^*) = z'(x^*) - k(|R| + 1)$. For computational purposes, there is no reason to deal with $\mathcal{P}_{\mathrm{IP2}}(G, k, \lambda)$ instead of $\mathcal{P}_{\mathrm{IP1}}(G, k, \lambda)$. However, from a theoretical point of view, it is more convenient to work with a full dimensional polytope, which is the case of $\mathcal{P}_{\mathrm{IP2}}(G, k, \lambda)$, as shown in Lemma 3. In the sequel, for each $u \in V$, we denote by B_u be the biclique which contains only the vertex u and let $\mathcal{B} = \{\chi(B_u) \in \{0, 1\}^{|V|} \mid u \in V\}$.

Lemma 3. $\mathcal{P}_{\mathrm{IP2}}(G, k, \lambda)$ *is full dimensional.*

Proof. Note that, for each $u \in V$, we have that $\chi(B_u) \in \mathcal{P}_{\mathrm{IP2}}(G, k, \lambda)$. Moreover, the null vector also belongs to $\mathcal{P}_{\mathrm{IP2}}(G, k, \lambda)$. These $|V| + 1$ vectors are affinely independent and, since $\mathcal{P}_{\mathrm{IP2}}(G, k, \lambda) \subset \mathbb{R}^{|V|}$, we have that $\mathcal{P}_{\mathrm{IP2}}(G, k, \lambda)$ is full dimensional. $\qquad\square$

Let $G' = (V', E')$ be the graph obtained as follows: $V' := V$ and $E' := \mathcal{F}(k, \lambda)$. Note that each feasible solution of (IP2) corresponds to a *stable set* (a set with pairwise non-adjacent vertices) in G', but the converse is not true, since the stable sets in G' with more than k vertices in L do not correspond to any feasible solution of (IP2). For a survey on branch-and-cut algorithms for the MAXIMUM STABLE SET problem, as well as a polyhedral study of the so-called *stable set* polytope, we refer to the work of S. Rebennack et al. [7].

A natural question that arises in this context is: which properties of the stable set polytope corresponding to G' hold for $\mathcal{P}_{\mathrm{IP2}}(G, k, \lambda)$? We give a partial answer to this question by showing that some inequalities that are known to define facets of the stable set polytope also have this same property with respect to $\mathcal{P}_{\mathrm{IP2}}(G, k, \lambda)$. To this end, we first consider the non-negativity inequalities.

Lemma 4. *For each $u \in V$, the inequality $x_u \geq 0$ defines a facet of $\mathcal{P}_{\mathrm{IP2}}(G, k, \lambda)$.*

Proof. Let u be a vertex in V and consider the face $F = \{x \in \mathcal{P}_{\mathrm{IP2}}(G, k, \lambda) \mid x_u = 0\}$. Note that, for each $v \in V \setminus \{u\}$, we have that $\chi(B_v) \in F$. Since the null vector is also in F, we have found $dim(\mathcal{P}_{\mathrm{IP2}}(G, k, \lambda))$ affinely independent vectors in this face. Hence, F is a facet of $\mathcal{P}_{\mathrm{IP2}}(G, k, \lambda)$.   □

Now, suppose that there are three vertices u, v and w in V, such that $\{u, v\}$, $\{u, w\}$ and $\{v, w\}$ are in $\mathcal{F}(k, \lambda)$. It can be easily checked that, in this case, $x_u + x_v + x_w \leq 1$ is a valid inequality for $\mathcal{P}_{\mathrm{IP2}}(G, k, \lambda)$. This argument can be generalized to show that, in general, constraints (2) are dominated by *clique inequalities* and, as a consequence, can not define facets of $\mathcal{P}_{\mathrm{IP2}}(G, k, \lambda)$. For the graph G', the clique inequalities are defined as follows.

Let $\mathcal{C}(G')$ be the set of all cliques in G' with at least two vertices. The clique inequalities for G' are given by

$$\sum_{u \in C} x_u \leq 1, \text{ for each } C \in \mathcal{C}(G') \tag{4}$$

Clique inequalities are known to be valid for the stable set polytope and, only for maximal cliques, they are facet-defining. Next, we show that the same property holds for $\mathcal{P}_{\mathrm{IP2}}(G, k, \lambda)$.

Lemma 5. *Inequalities (4) are valid for $\mathcal{P}_{\mathrm{IP2}}(G, k, \lambda)$.*

Proof. Let C be a clique in $\mathcal{C}(G')$. We now show, by induction in $|C|$, that $\sum_{u \in C} x_u \leq 1$ is valid for $\mathcal{P}_{\mathrm{IP2}}(G, k, \lambda)$. If $|C| = 2$ then the corresponding constraint (4) is explicitly added in (IP2), and therefore, valid. Suppose now that $|C| > 2$. Then, by the induction hypothesis, we have that the inequality $\sum_{u \in C'} x_u \leq 1$ is valid for $\mathcal{P}_{\mathrm{IP2}}(G, k, \lambda)$, for each $C' \subset C$, such that $|C'| = |C| - 1$. Then, adding all these inequalities we have that $\sum_{u \in C}(n - 1)x_u \leq n$ is valid for $\mathcal{P}_{\mathrm{IP2}}(G, k, \lambda)$, where $n = |C|$. Since all feasible solutions of (IP2) are integral, then we have that $\sum_{u \in C} x_u \leq \left\lfloor \frac{n}{n-1} \right\rfloor = 1$ is valid for $\mathcal{P}_{\mathrm{IP2}}(G, k, \lambda)$.   □

Lemma 6. *For each maximal clique C in $\mathcal{C}(G')$, we have that the inequality $\sum_{u \in C} x_u \leq 1$ defines a facet of $\mathcal{P}_{IP2}(G, k, \lambda)$.*

Proof. Let C be a maximal clique in $\mathcal{C}(G')$ and let $F = \{x \in \mathcal{P}_{IP2}(G, k, \lambda) \mid \sum_{u \in C} x_u = 1\}$. Note that, for each $u \in C$, we have that $\chi(B_u) \in F$. If $C = V$ then the proof is complete. Suppose that $C \neq V$. Since C is maximal, given a vertex $u \in V \setminus C$, we have that there is at least one vertex $\phi(u)$ in C, such that $\{u, \phi(u)\} \notin E'$. For each $u \in V \setminus C$, let $\check{B}_u$ be a subgraph of G induced by $\{u, \phi(u)\}$ and let $\check{\mathcal{B}} = \{\chi(\check{B}_u) \in \{0, 1\}^{|V|} \mid u \in V \setminus C\}$. Note that $\chi(\check{B}_u) \in F$, for each $\check{B}_u \in \check{\mathcal{B}}$, and the square matrix M composed by the vectors in $\check{\mathcal{B}} \cup \mathcal{B}$ and its inverse can be written as below:

$$M = \begin{pmatrix} I_{n \times n} & A \\ \mathbf{0} & I_{m \times m} \end{pmatrix} \text{ and } M^{-1} = \begin{pmatrix} I_{n \times n} & -A \\ \mathbf{0} & I_{m \times m} \end{pmatrix},$$

where $n = |C|$ and $m = |V| - |C|$. Since $rank(M) = |V|$ then we have that the vectors in $\check{\mathcal{B}} \cup \mathcal{B}$ are affinely independent. Therefore, $dim(F) = |V| - 1$ and F is a facet of $\mathcal{P}_{IP2}(G, k, \lambda)$. $\square$

We leave for further investigation the question of identifying which other inequalities, valid for the stable set polytope, are also valid for $\mathcal{P}_{IP2}(G, k, \lambda)$, as well as under which conditions they are facet-defining for $\mathcal{P}_{IP2}(G, k, \lambda)$.

4 Implementation Issues

This section describes a preprocessing procedure, a GRASP heuristic and the implementation of our integer programming formulation for solving the MkSI problem. All the algorithms were implemented in the C++ programming language.

A Preprocessing Procedure. The goal of the preprocessing is to reduce the instance size by eliminating vertices from L and R that can not be part of a feasible solution with cost higher than the primal bound λ. This is done as follows. For each vertex u in L, let $\mathcal{L}_u(\lambda) = \{v \in L \mid \{u, v\} \in \mathcal{L}(\lambda)\}$ be the set of vertices from L that can not be together with u in any feasible point of $\mathcal{P}_{IP2}(G, k, \lambda)$. Besides, for each vertex v in R, let $\mathcal{R}_v(k) = \{w \in R \mid \{v, w\} \in \mathcal{R}(k)\}$ be the set of vertices from R that can not be together with v in any feasible point of $\mathcal{P}_{IP2}(G, k, \lambda)$. Note that each vertex u in L with $|N(u)| < \lambda$ or with $|L| - |\mathcal{L}_u(\lambda)| < k$ can be removed from G. Similarly, each vertex u in R with $|N(u)| < k$ or with $|R| - |\mathcal{R}_u(k)| < \lambda$ can be removed from G. After removing such vertices, as well as their incident edges, it may happen that new vertices with "low degree" appear in the resulting graph. Thus, we reiterate the process until no further vertices are removed from the graph. Of course, this procedure can be applied every time a new lower bound λ' is found, such that $\lambda' > \lambda$. In Sect. 5, we show the effectiveness of this preprocessing procedure in practice.

A GRASP Heuristic. We now address the question of how to find a good lower bound λ. For this purpose, we developed a *Greedy Randomized Adaptive Search Procedure* (GRASP) heuristic.

The constructive phase of our GRASP begins with an empty set L', with $\lambda = 0$ and constructs a set $S \subseteq (L \setminus L')$ of *candidate* vertices. In order to explain how we set the maximum size of S, let us introduce some additional notation. For each $u \in L \setminus L'$, let $c(u) = |\Gamma(L' \cup \{u\})|$ be the *incremental value* of u. Let $c_{\min} = \min_{u \in L \setminus L'} c(u)$ and $c_{\max} = \max_{u \in L \setminus L'} c(u)$, and let α be a parameter, such that $\alpha \in \{0.0, 0.1, 0.2, \ldots, 1.0\}$. The candidates included in S are the ones with incremental value inside the interval $[c_{\min} + \alpha \times (c_{\max} - c_{\min}), c_{\max}]$. If $\alpha = 0$ then we have that all vertices in $L \setminus L'$ are candidates. As the value of α increases, only the vertices with the highest incremental value are candidates and, as a consequence, the size of the candidate list tends to decrease. In our implementation we change the value of α at each iteration using a probabilistic criterium. This strategy is known as *Reactive* GRASP [8]. At each iteration, the algorithm randomly chooses a candidate to be removed from S and adds this vertex to L'. This process is repeated until $|L'| = k$. After that, the algorithm performs a *local search* until a local optimal solution is found, where the *neighbourhood* of a solution B contains all bicliques obtained by switching one vertex of L_B with a vertex in $L \setminus L_B$. The algorithm performs 500 iterations.

Finally, after executing the procedure described above, we apply a *path-relinking* [4] procedure among pairs of the best solutions found so far, which are kept in a set $\mathcal{S}^*$. In order to get better results, the solutions selected to be in $\mathcal{B}^*$ must be well diversified. At each iteration of the path-relinking phase, the algorithm randomly chooses two solutions B_1 and B_2 from $\mathcal{B}^*$ and explores a *path* of intermediate solutions connecting B_1 to B_2. The *path* is formed by a sequence of solutions, starting in B_1 and ending in B_2, and such that each of them can be obtained from the previous one through a local search operation. The local search and the path-relinking phases are executed only if the value of the current solution is at least 80 % of the value of the best solution found so far.

Strengthening the Integer Programming Model. After computing a lower bound λ with our GRASP heuristic and applying the preprocessing procedure, we solve a strengthened version of (IP1). In this model we include the non-negativity constraints and, for each $\{u, v\} \in \mathcal{F}(k, \lambda)$, a clique inequality for an arbitrarily chosen maximal clique in G' containing both u and v. In other words, we take each constraint (2) of model (IP1) and lift it to a facet-defining inequality of $\mathcal{P}_{\text{IP2}}(k, \lambda)$. We refer to this model as the M_{CLIQUE} model.

Initially, we tested with a *branch-and-cut* algorithm which started with the M_{CLIQUE} model and separated clique inequalities using the heuristics developed in [5] to that purpose. But, in preliminary experiments, we observed that smaller computing times were attained by turning off our separation routine. Notice, however, that the latter can not be considered a pure branch-and-bound algorithm since modern ILP solvers have their own separation routines for generating cuts throughout the computation. Anyway, we decided to use the M_{CLIQUE} model as input for the ILP solver without adding cuts from our own separation routine.

5   Computational Experiments

In this section, we present computational experiments made with instances coming from the application discussed in Sect. 1. We compare the M_{CLIQUE} model (see previous section), against the M_{EDGE} model (see Sect. 3), aiming at assess the benefits of using a stronger formulation.

The instances used in our tests were generated from a dataset made available (at http://www.dtic.upf.edu/~ocelma/MusicRecommendationDataset/lastfm-1K.html – last accessed in December, 2013) by Celma in his work [2] on music recommendation and obtained from *Last FM* (http://www.last.fm). The dataset has 176.948 bands, 992 listeners and 19.150.868 listener-band pairs (edges) indicating listening habits of users until May, 2009. Observing the degree distribution of vertices representing bands, we could see that it follows a power law. As a consequence, there are a few bands with a high number of listeners and most of the vertices representing bands have degree close to zero.

From the described dataset, we have generated 6 classes of instances containing 50, 70, 90, 110, 130 and 150 bands. The bands play the role of the vertices in L whereas the listeners are associated to vertices in R. Obviously, an edge exists between the vertex representing band u and listener v if v is a fan of u. To generate an instance with n bands, we selected the $2n$ most listened bands and randomly chose n bands among them. For each class, we generated 10 instances and each instance was tested with $k = 2, 3, 4, 5, 6, 7$. In all tables presented in this section we show the arithmetic mean values followed by the standard deviation values in parenthesis. We used IBM CPLEX 12 as the ILP solver, with all primal heuristics and all general purpose cutting-plane's generation routines (except the CLIQUE one) provided by CPLEX turned off. We have executed our code in single thread mode in a machine with the following configurations: Intel(R) Xeon(R) 2.40 GHz processor and 8 GB of RAM memory.

In the first two tables we show the effect of the preprocessing procedure. Table 1 displays the percentage of listeners, bands and edges removed, for different values of k. Table 2 exhibits the number of listeners, bands and edges, before and after the preprocessing, for each class of instances. According to Table 1, for $2 \leq k \leq 4$, the preprocessing removed most of the listeners and edges. For all values of k, the percentage of listeners removed was much larger than that of bands. As can be noticed, the effectiveness of the preprocessing decreases as the value of k increases. This can be explained by the fact that an increase in the value of k leads to a decrease in the value of λ (the size of the solution found by the GRASP heuristic). Intuitively, if we consider very large values for k we will observe the same phenomenon but in the opposite "direction" (i.e., almost all bands will be removed by the preprocessing).

One can see from Table 2 that, as the number of bands increases, the number of listeners, bands and edges remain almost the same in the preprocessed graphs. This can be explained by the fact that, as mentioned before, there are a few bands with a high number of listeners and most of the vertices representing bands have degree close to zero. Thus, even if we select a very large number of bands, most of them will be removed by the preprocessing because they have a small number

Table 1. The effect of the preprocessing for different values of k.

k	Percentage removed		
	Bands	Listeners	Edges
2	48.3 (4.2)	97.7 (0.9)	97.2 (0.8)
3	18.4 (9.3)	84.8 (10.0)	80.5 (11.2)
4	9.8 (2.0)	64.3 (15.7)	57.9 (14.2)
5	8.4 (1.9)	44.4 (22.9)	40.3 (18.9)
6	7.5 (1.2)	24.8 (20.8)	24.6 (16.2)
7	7.7 (1.7)	10.7 (13.2)	14.7 (10.1)

Table 2. The effect of the preprocessing for different number of bands.

Original graph			Preprocessed graph		
# bands	# listeners	# edges	# bands	# listeners	# edges
50	967 (2)	23191 (505)	30 (19)	793 (136)	13889 (8420)
70	974 (2)	30112 (558)	41 (26)	820 (160)	17787 (10351)
90	975 (2)	36616 (510)	44 (31)	820 (151)	18346 (11784)
110	977 (1)	42788 (308)	43 (34)	821 (144)	18178 (12356)
130	982 (1)	48001 (697)	47 (37)	818 (154)	18993 (13133)
150	982 (1)	53495 (768)	43 (36)	806 (151)	17812 (12931)

of listeners. In fact, for graphs with up to 4000 bands generated from the *Last FM* database, the figures of the preprocessed graph remain about the same as those displayed in Table 2.

Once the importance of the preprocessing has been established, we turn our attention to the choice of the ILP model. In the experiments reported so far, a time limit of 1800 s was set to solve each instance. In Table 3 we give the percentage of instances solved to optimality by each model and the *duality gaps* calculated in the following way. Given a model (M_{CLIQUE} or M_{EDGE}), let LP be the value of its linear relaxation and LB and UB be the values of the best lower and upper bounds found during the execution, respectively. Thus, the values in column "LP/Best-LB Gap" are given by $\frac{LP-LB}{LP} \times 100$ and those in column "Final Gap" are given by $\frac{UB-LB}{UB} \times 100$. Column "% Opt. Found" shows the percentage of solutions solved to optimality.

As column "LP/Best-LB Gap" shows, the model M_{CLIQUE} is much stronger than the model M_{EDGE} for these instances. As the value of k increases, the gaps obtained by both models increase as well and the percentage of solutions solved to optimality decreases. The model M_{CLIQUE} could solve more instances to optimality than model M_{EDGE} and, for the instances not solved to optimality, the final gaps obtained by the M_{CLIQUE} model were much smaller than the gaps obtained by the M_{EDGE} model. We observed that when we do not set any time limit, the model M_{CLIQUE} could solve all instances with $k = 7$ in at most 1h20m of running time.

Table 3. Percentage of instances solved to optimality and duality gaps.

k	LP/best-LB gap		Final gap		% opt. found	
	M_{CLIQUE}	M_{EDGE}	M_{CLIQUE}	M_{EDGE}	M_{CLIQUE}	M_{EDGE}
2	0.0 (0.0)	0.0 (0.0)	0.0 (0.0)	0.0 (0.0)	100.0	100.0
3	0.0 (0.0)	33.2 (17.4)	0.0 (0.0)	0.0 (0.0)	100.0	100.0
4	11.4 (9.8)	56.2 (4.7)	0.0 (0.0)	0.0 (0.0)	100.0	100.0
5	29.3 (8.4)	63.8 (3.8)	0.0 (0.0)	0.4 (3.2)	100.0	98.3
6	40.3 (6.2)	68.9 (2.6)	8.3 (1.2)	23.3 (9.4)	96.6	53.3
7	46.9 (5.1)	72.1 (2.3)	14.1 (5.3)	39.2 (13.9)	70.0	23.3

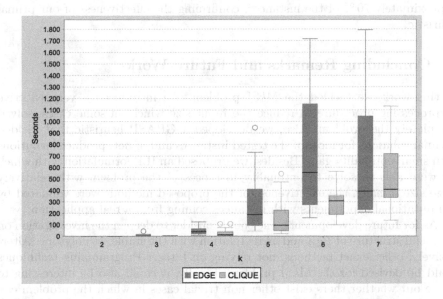

Fig. 1. Comparing the running times of M_{CLIQUE} and M_{EDGE} models.

In Fig. 1 we show in a *box plot* the running times obtained by the M_{CLIQUE} and M_{EDGE} models, considering only the instances solved to optimality by both models. The bottom and top of the box are the first and third quartiles, respectively, and the band inside the box is the second quartile (the median). The *whiskers*, i.e. the ends of the lines extending vertically from the boxes, are calculated as follows: maximum value lesser than $q_3 + 1,5 \times (q_3 - q_1)$ for the upper end and minimum value greater than $q_1 - 1,5 \times (q_3 - q_1)$ for the lower end, where q_1, q_2 and q_3 are the first, second and third quartiles, respectively. Each *outlier*, i.e. each point not included between the whiskers, is drawn as a small circle.

As seen in Fig. 1 and Table 3, for $k \leq 4$, both models solve each instance in less than 2 minutes. Independently of the model, as the value of k increases, the running time increases as well. We highlight that the data in Fig. 1 refer only to the instances solved to optimality and, as shown in Table 3, the model M_{EDGE}

found the optimum only for 23.3 % of the instances with $k = 7$. Thus, for $k = 7$, the instances considered in Fig. 1 are the 23.3 % easiest ones. This explains the somewhat unexpected decay in the running time needed to solve the problem with model M_{EDGE}, from $k = 6$ to $k = 7$. Extending the analysis to the entire set of instances, would confirm that the running time actually increases as k increases. However, some preliminary experiments that we carried out showed that after a threshold, the running time diminishes as the value of k augments.

As for the primal bound, our GRASP encounters an optimal solution in about 94 % of the instances, in less than 7 s in average. For the 6 % remaining ones, the average optimality gap was 13.5 %, with a maximum of 21.7 %. For comparison, we also implemented a simple greedy heuristic which found the optimum in approximately 70 % of the instances, confirming the effectiveness of our primal heuristic.

6   Concluding Remarks and Future Work

In this work, we tackled the MkSI problem. We introduced a very effective preprocessing procedure to reduce the input size which, in some cases, solved completely the problem. Also, we developed a GRASP heuristic that yielded optimal solutions for most of the tested instances and, if not, produced solutions with small optimality gaps. Besides, we proposed an ILP formulation with which we were able to solve to optimality instances of the problem with relatively large size. Finally, the effectiveness of the proposed methods were validated by computational experiments with instances coming from a real application.

As for future developments, one could consider to deepen the investigation on the facial structure of $\mathcal{P}_{IP2}$ and on its relation with the stable set polytope. Alternatively, other exact methods, not relying on Integer Programming techniques could be devised for the MkSI problem. Finally, it could also be interesting to figure out whether there exist other non trivial cases in which the problem can be solved exactly, or well approximated, in polynomial time. One such situation occurs, for example, when the input graph is convex, as observed in Sect. 1.

References

1. Acuña, V., Ferreira, C.E., Freire, A.S., Moreno, E.: Solving the maximum edge biclique packing problem on unbalanced bipartite graphs. Discrete Appl. Math. (2011, in press). doi:10.1016/j.dam.2011.09.019
2. Celma, O.: Music Recommendation and Discovery in the Long Tail. Springer, Berlin (2010)
3. Ganter, B., Reuter, K.: Finding all closed sets: a general approach. Order 8, 283–290 (1991). doi:10.1007/BF00383449
4. Glover, F.: Tabu search and adaptive memory programing advances, applications and challenges. In: Barr, R.S., et al. (eds.) Interfaces in Computer Science and Operations Research, pp. 1–75. Kluwer, Dordrecht (1996)
5. Nemhauser, G.L., Sigismondi, G.: A strong cutting plane/branch-and-bound algorithm for node packing. J. Oper. Res. Soc. 43(5), 443–457 (1992)

6. Nussbaum, D., Pu, S., Sack, J.-R., Uno, T., Zarrabi-Zadeh, H.: Finding maximum edge bicliques in convex bipartite graphs. In: Thai, M.T., Sahni, S. (eds.) COCOON 2010. LNCS, vol. 6196, pp. 140–149. Springer, Heidelberg (2010)
7. Rebennack, S., Reinelt, G., Pardalos, P.M.: A tutorial on branch and cut algorithms for the maximum stable set problem. Int. Trans. Oper. Res. **19**, 161–199 (2012)
8. Resende, M.G.C., Ribeiro, C.C.: Greedy randomized adaptive search procedures. In: Glover, F., Kochenberger, G.A. (eds.) Handbook of Metaheuristics, pp. 219–249. Springer, New York (2003)
9. Vinterbo, S.A.: A note on the hardness of the k-ambiguity problem. Technical report, Harvard Medical School, Boston, MA, USA (2002)
10. Xavier, E.C.: A note on a maximum k-subset intersection problem. Inf. Process. Lett. **112**, 471–472 (2012)

b-Coloring is NP-Hard on Co-Bipartite Graphs and Polytime Solvable on Tree-Cographs

Flavia Bonomo[1]([⊠]), Oliver Schaudt[2], Maya Stein[3],
and Mario Valencia-Pabon[4]

[1] CONICET and Dep. de Computación, Facultad de Ciencias Exactas y Naturales,
Universidad de Buenos Aires, Buenos Aires, Argentina
fbonomo@dc.uba.ar
[2] Institut de Mathématiques de Jussieu, CNRS UMR7586,
Université Pierre et Marie Curie (Paris 6), Paris, France
schaudt@math.jussieu.fr
[3] Centro de Mod. Mat., Universidad de Chile, Santiago, Chile
mstein@dim.uchile.cl
[4] Université Paris 13, Sorbonne Paris Cité, LIPN, CNRS UMR7030,
Villetaneuse, France
valencia@lipn.univ-paris13.fr

Abstract. A *b-coloring* of a graph is a proper coloring such that every color class contains a vertex that is adjacent to all other color classes. The *b-chromatic number* of a graph G, denoted by $\chi_b(G)$, is the maximum number t such that G admits a b-coloring with t colors. A graph G is called *b-continuous* if it admits a b-coloring with t colors, for every $t = \chi(G), \ldots, \chi_b(G)$, and *b-monotonic* if $\chi_b(H_1) \geq \chi_b(H_2)$ for every induced subgraph H_1 of G, and every induced subgraph H_2 of H_1.
We investigate the b-chromatic number of graphs with stability number two. These are exactly the complements of triangle-free graphs, thus including all complements of bipartite graphs. The main results of this work are the following:

1. We characterize the b-colorings of a graph with stability number two in terms of matchings with no augmenting paths of length one or three. We derive that graphs with stability number two are b-continuous and b-monotonic.
2. We prove that it is NP-complete to decide whether the b-chromatic number of a co-bipartite graph is at most a given threshold.
3. We describe a polynomial time dynamic programming algorithm to compute the b-chromatic number of co-trees.
4. Extending several previous results, we show that there is a polynomial time dynamic programming algorithm for computing the b-chromatic number of tree-cographs. Moreover, we show that tree-cographs are b-continuous and b-monotonic.

Partially supported by UBACyT Grant 20020100100980, PICT ANPCyT 2012-1324 and CONICET PIP 112-200901-00178 and 112-201201-00450CO (Argentina) and MathAmSud Project 13MATH-07 (Argentina–Brazil–Chile–France).
M. Valencia-Pabon – Currently Invited at INRIA Nancy - Grand Est.

P. Fouilhoux et al. (Eds.): ISCO 2014, LNCS 8596, pp. 100–111, 2014.
DOI: 10.1007/978-3-319-09174-7_9

1   Introduction

A *b-coloring* of a graph G by k colors is a proper coloring of the vertices of G such that every color class contains a vertex that is adjacent to all the other $k - 1$ color classes. Such a vertex will be called a *dominant vertex*.

The *b-chromatic number* of a graph G, denoted by $\chi_b(G)$, is the maximum number k such that G admits a b-coloring with k colors. Clearly, $\chi_b(G) \leq \Delta(G) + 1$ where $\Delta(G)$ denotes the maximum degree of G. The b-chromatic number was introduced in [11]. The motivation, similarly as the well known achromatic number (cf. e.g., [2,7] and ref. therein), comes from algorithmic graph theory. Suppose one colors a given graph properly, but in an arbitrary way. After all vertices are colored, one would wish to perform some simple operations to reduce the number of colors. A simple operation consists in recoloring all the vertices in one color class with a possible different color. Such recoloring is impossible if each color class contains a dominant vertex. Hence, the b-chromatic number of the graph serves as the tight upper bound for the number of colors used by this coloring heuristic. From this point of view, both complexity results and polynomial time algorithms for particular graph families are interesting. And, in particular, this arguments show that any coloring of a graph G with $\chi(G)$ many colors is a b-coloring (as usual, we denote by $\chi(G)$ the minimum number of colors needed for a proper coloring of the vertices of a graph).

Assume that the vertices $v_1, v_2, \ldots, v_n$ of a graph G are ordered such that $d(v_1) \geq d(v_2) \geq \ldots \geq d(v_n)$, where $d(x)$ denotes the degree of vertex x in G. Let

$$m(G) := \max\{i : d(v_i) \geq i - 1\}$$

be the maximum number i such that G contains at least i vertices of degree $\geq i - 1$. It is clear that $m(G) \leq \Delta(G) + 1$. Irving and Manlove [11] show that this parameter bounds the b-chromatic number:

Proposition 1. *For every graph G, $\chi(G) \leq \chi_b(G) \leq m(G)$.*

Irving and Manlove [11] also show that determining $\chi_b(G)$ is NP-complete for general graphs, but polynomial-time solvable for trees. Kratochvíl, Tuza and Voigt [13] prove that the problem of determining if $\chi_b(G) = m(G)$ is NP-complete even for connected bipartite graphs G with $m(G) = \Delta(G)+1$. A graph G is *tight* if it has exactly $m(G)$ *dense* vertices (a vertex v of a graph G is dense if $d(v) \geq m(G)-1$), each of which has degree exactly $m(G)-1$. Havet, Linhares-Sales and Sampaio [8] recently investigated the problem on tight graphs. They proved that the problem of determining if a tight graph G has $\chi_b(G) = m(G)$ is NP-complete for bipartite graphs and ptolemaic graphs, but polynomial-time solvable for complements of bipartite graphs, split graphs and block graphs.

In last years, several related concepts concerning b-colorings of graphs have been studied in [6,9,10,12]. A graph G is defined to be *b-continuous* [6] if it admits a b-coloring with t colors, for every $t = \chi(G), \ldots, \chi_b(G)$. In [12] (see also [6]) it is proved that chordal graphs and some planar graphs are b-continuous. A graph G is defined to be *b-monotonic* [3] if $\chi_b(H_1) \geq \chi_b(H_2)$ for every induced subgraph H_1 of G, and every induced subgraph H_2 of H_1. They prove that P_4-sparse graphs (and, in particular, cographs) are b-continuous and b-monotonic. Besides, they give a dynamic programming algorithm to compute the b-chromatic number in polynomial time within these graph classes.

Our paper is organized as follows. In the next section, we characterize b-colorings of graphs with stability number two in terms of matchings with no augmenting paths of length one or three. In Sect. 3, we prove that graphs with stability at most two are both b-continuous and b-monotonic. In Sect. 4, we prove that computing the b-chromatic number of co-bipartite graphs is an NP-complete problem. Finally, in Sect. 5, first we describe a polynomial-time dynamic programming algorithm to compute the b-chromatic number of co-trees. Next, we extend our results to the family of tree-cographs by showing that there is a polynomial time dynamic programming algorithm for computing the b-chromatic number of graphs in this family and that these are also b-continuous and b-monotonic.

Most of our results are given without proof due to lack of space.

2   b-Colorings and Matchings

The *stability* of a graph G is defined as the maximum cardinality of a subset of pairwise non-adjacent vertices in G (i.e. a *stable set* or *independent set*). Given a graph G, we denote by $\overline{G}$ the complement graph of G, which is the graph on the same set of vertices as G that has an edge between two different vertices u and v if and only if u and v are non-adjacent in G. It is not difficult to see that G is a graph with stability one if and only if it is complete, and G is a graph with stability at most two if and only if $\overline{G}$ is a triangle-free graph. In this section, we will see that matchings in triangle-free graphs are very important when we deal with b-colorings of graphs with stability at most two.

Let M be a matching of a graph G. Denote by $V(M)$ the set of all vertices covered by M. An *augmenting path* for M is a path starting and ending outside $V(M)$ whose edges alternate between $E(G) - M$ and M. Usually, M is called *maximal* if no further edge can be included in M. In other words, G does not contain an augmenting path of length one with respect to M. Following this terminology we call M *strongly maximal* if G does not contain augmenting paths of length one or three with respect to M. By definition, maximum matchings are strongly maximal, and strongly maximal matchings are maximal. Our next lemma shows why strongly maximal matchings are important in our setting.

Lemma 1. *Let G be a graph of stability at most two and let c be a proper coloring of G. Then c is a b-coloring if and only if the set*

$$M = \{uv : u, v \in V, u \neq v \text{ and } c(u) = c(v)\}$$

is a strongly maximal matching in $\overline{G}$. Moreover, the number of colors c uses is $|V(G)| - |M|$.

Proof. First, observe that M is a (possibly empty) matching of $\overline{G}$ because G has stability at most two. Now, suppose that $\overline{G}$ contains an augmenting path P of length 1 or 3 for M. If P consists of only one edge uv, then in G, the vertices u and v are non-adjacent, and each makes up a singleton colour class. Thus c is not a b-coloring. If P has three edges, then for each of the endvertices of its middle edge uv there is a singleton color class which it does not see in G. So the color class $\{u,v\}$ witnesses the fact that c is not a b-coloring.

Next, suppose that c is not a b-coloring. Note that, as G has stability at most two, every vertex of G is adjacent (in G) to at least one vertex of any given color class of size 2. So, the witness for c not being a b-coloring is one of the following two: either it is a singleton color class whose vertex is non-adjacent to another singleton color class, or it is a color class $\{u,v\}$ of size two, such that u is non-adjacent to some singleton color class, and v is non-adjacent to a different singleton color class. The first situation corresponds to an augmenting path of M on one edge, and the second situation corresponds to an augmenting path of M on three edges. □

Observe that coloring c from Lemma 1 is a maximum (minimum) b-coloring of G if and only if M is a minimum (maximum) strongly maximal matching of $\overline{G}$.

3   b-Continuity and b-Monotonicity of Graphs with Stability at most Two

In order to prove the b-continuity of graphs with stability at most two, we need the following result.

Lemma 2. *Let M be a strongly maximal matching of a graph G and let P be a minimum length augmenting path in G with respect to M. Then, the matching $M' = (M \backslash E(P)) \cup (E(P) \backslash M)$ is a strongly maximal matching of G, and $|M'| = |M| + 1$.*

Proof. Let $P = (x_1, x_2, \ldots, x_k)$. By basic results from matching theory, the only thing we need to prove is that M' is again strongly maximal. Since the maximality of M' is clear, suppose for contradiction that there is an augmenting path of length 3, say $Q = (u, v, w, x)$. Necessarily vw is an edge of $M' \backslash M$, and thus w.l.o.g. there is some $i \in \{1, 2, \ldots, k-1\}$ with $v = x_i$ and $w = x_{i+1}$. Moreover, $u, x \notin V(M)$. Thus both paths $(x_1, x_2, \ldots, x_i, u)$ and $(x, x_{i+1}, x_{i+2}, \ldots, x_k)$ are augmenting paths for M and at least one of these paths is shorter than P. This is a contradiction to the choice of P. □

By Lemma 1, any b-coloring using $k > \chi(G)$ colors of a graph G of stability at most two corresponds to a strongly maximal matching M that is not maximum. By Berge's lemma [1], there is an augmenting path for M. Using Lemma 2 we obtain a strongly maximal matching M' of cardinality $|M| + 1$, which, again by Lemma 1, corresponds to a b-coloring with $k - 1$ colors. Repeatedly applying this argument gives the following result.

Theorem 1. *Graphs of stability at most two are b-continuous.*

Given a maximum b-coloring of a graph G of stability at most two, we can thus find b-colorings for all values between $\chi(G)$ and $\chi_b(G)$. Moreover, we can do this in polynomial time, provided we can find a minimum length augmenting path for a given matching in polynomial time. This is the aim of the following lemma that can be derived by a slight modification of Edmonds' blossom algorithm [5].

Lemma 3. *Let M be a matching in a graph G. Then, a minimum length augmenting path P in G with respect to M can be computed in polynomial time.*

Lemma 3 together with the proof of Theorem 1 implies that given a graph G of stability at most two, and a b-coloring of G using $k > \chi(G)$ colors, we can compute in polynomial time a b-coloring for G with $k - 1$ colors. Notice that the converse is not necessarily true, i.e., if we have a b-coloring of G using $k < \chi_b(G)$ colors, we do not know how to compute in polynomial time a b-coloring for G with $k + 1$ colors. Indeed, we will prove in the next section that the problem of computing the b-chromatic number of a graph with stability at most two is NP-complete, even restricted to the smaller class of co-bipartite graphs.

We now turn to the b-monotonicity of graphs of stability at most two.

Theorem 2. *Graphs of stability at most two are b-monotonic.*

Proof. (Sketch). The class of graphs of stability at most two is closed under taking induced subgraphs. Thus we only have to prove that χ_b is monotonously decreasing under the deletion of a vertex. In view of Lemma 1, it is sufficient to show that given a graph G of stability 2 and some vertex $v \in V(G)$ the following holds: If there is a strongly maximal matching of $\overline{G - v}$ of size k, then there is a strongly maximal matching of $\overline{G}$ of size at most $k + 1$. This implies $\chi_b(G) \geq \chi_b(G - v)$. $\qquad\square$

4   NP-Hardness Result for Co-Bipartite Graphs

As mentioned in Sect. 1, Havet, Linhares-Sales and Sampaio [8] proved that the problem of determining if a tight co-bipartite graph G has $\chi_b(G) = m(G)$ is polynomial-time solvable. However, the computational complexity of χ_b in the class of co-bipartite graphs was left open. In the next theorem, we prove that b-coloring general co-bipartite graphs is a hard problem.

Theorem 3. *Given a co-bipartite graph G and a natural number t, it is NP-complete to decide whether G admits a b-coloring with at least t colors.*

Proof. By Lemma 1, it suffices to prove that it is NP-complete to decide whether a bipartite graph G admits a strongly maximal matching containing at most k edges, when G and k are given input.

Our reduction is from the minimum maximal matching problem which is to decide whether a given graph admits a maximal matching of at most k edges, for given k. This problem is NP-complete even if the instances are restricted to bipartite graphs, as shown by Yannakakis and Gavril [15].

Given a bipartite graph G with m edges, we define a new graph H_G as follows. For each edge $uv \in E(G)$ we introduce a set of new vertices

$$X_{uv} = \{x_{uv}^1, x_{uv}^2, x_{uv}^3, x_{uv}^4, x_{vu}^1, x_{vu}^2, x_{vu}^3, x_{vu}^4\}$$

and edges

$$F_{uv} = \{ux_{uv}^1, x_{uv}^1 x_{uv}^2, x_{uv}^2 x_{uv}^3, x_{uv}^3 x_{uv}^4, x_{uv}^1 x_{vu}^1, x_{vu}^1 x_{vu}^2, x_{vu}^2 x_{vu}^3, x_{vu}^3 x_{vu}^4, vx_{vu}^1\}.$$

Note that $X_{uv} = X_{vu}$ and $F_{uv} = F_{vu}$. Then H_G is defined by

$$V(H_G) = V(G) \cup \bigcup_{uv \in E(G)} X_{uv},$$

$$E(H_G) = \bigcup_{uv \in E(G)} F_{uv}.$$

Clearly, H_G can be computed in polynomial time. Moreover, H_G is bipartite since G is. For each edge $uv \in E(G)$, we define the following auxiliary sets of edges in H_G:

$$F_{uv}^{\in} = \{ux_{uv}^1, x_{uv}^2 x_{uv}^3, x_{vu}^2 x_{vu}^3, vx_{vu}^1\} \text{ and } F_{uv}^{\notin} = \{x_{uv}^1 x_{vu}^1, x_{uv}^2 x_{uv}^3, x_{vu}^2 x_{vu}^3\}$$

We claim the following:

Claim. There exists a minimum strongly maximal matching M of H_G such that

$$x_{uv}^3 x_{uv}^4 \notin M \text{ for each edge } uv \in E(G).$$

Moreover, M can be obtained from any minimum strongly maximal matching of H_G in polynomial time.

In order to prove this claim, we proceed by contradiction. Assume that every minimum strongly maximal matching of H_G contains at least an edge $x_{uv}^3 x_{uv}^4$ for some edge $uv \in E(G)$, and let M be a minimum strongly maximal matching of H_G having a minimum number of edges of the form $x_{uv}^3 x_{uv}^4$. Note that the choice of M implies that for every edge $uv \in E(G)$ we have that

(i) $x_{uv}^3 x_{uv}^4 \in M$ if and only if $x_{uv}^1 x_{uv}^2 \in M$. If $x_{uv}^1 x_{uv}^2 \in M$ then $x_{uv}^3 x_{uv}^4 \in M$, otherwise, M is not maximal. If $x_{uv}^3 x_{uv}^4 \in M$ then $x_{uv}^1 x_{uv}^2 \in M$, otherwise, we could replace $x_{uv}^3 x_{uv}^4$ by $x_{uv}^2 x_{uv}^3$ in M (the resulting matching is strongly maximal as M is), contradicting the choice of M.

(ii) If the edges $x_{uv}^3 x_{uv}^4$ and $x_{uv}^1 x_{uv}^2$ are in M, then we have that vertices u and x_{vu}^1 are each matched by M. Otherwise, if u is unmatched, we can replace $x_{uv}^1 x_{uv}^2, x_{uv}^3 x_{uv}^4 \in M$ with the edges $x_{uv}^2 x_{uv}^3, ux_{uv}^1$. This again yields a strongly maximal matching (since u has no neighbors unmatched by M), contradicting the choice of M. We can use the same argument in the case x_{vu}^1 is unmatched.

These are also some of the steps in order to transform any minimum strongly maximal matching into the desired one.

Now, let uv be an edge in the graph G such that $x_{uv}^3 x_{uv}^4 \in M$. By (i) and (ii), we can deduce that $|M \cap F_{uv}| = 4$. Consider the matching

$$\tilde{M} := (M \backslash F_{uv}) \cup F_{uv}^{\notin}$$

We claim that $\tilde{M}$ is strongly maximal. As $\tilde{M}$ is smaller than M, we thus obtain the desired contradiction. So assume $\tilde{M}$ is not strongly maximal. Then, as u is matched, there is an augmenting path P of length 1 or 3 starting at v.

Now, observe that all neighbors of v are of the form x_{vw}^1 (for some $w \in V(G)$), and thus, as neither $x_{vw}^1 x_{vw}^2$ nor $x_{vw}^1 x_{vw}^2 x_{vw}^3 x_{vw}^4$ is an augmenting path for the strongly maximal matching M, all neighbors of v are matched by M.

So, P has length 3, and it is easy to see that P has to end in some (unmatched) vertex $w \in V(G) \backslash \{u, v\}$ (by the maximality of M, every vertex x_{wz}^3 is matched by M, and by the choice of M, every vertex x_{wz}^2 is matched by M). By (i) and (ii), we know that $F_{vw} \cap M = F_{vw}^{\notin}$. Consider the matching

$$(\tilde{M} \backslash F_{vw}^{\notin}) \cup F_{vw}^{\in}.$$

It can be seen that this matching is strongly maximal, and has fewer edges of the form $x_{uv}^3 x_{uv}^4$, contradicting the choice of M. (And this is the remaining step in order to transform any minimum strongly maximal matching into the desired one.) This ends the proof of this claim.

Therefore, by the previous claim, we have that there is a minimum strongly maximal M' in H_G that verifies either $F_{uv} \cap M' = F_{uv}^{\in}$ or $F_{uv} \cap M' = F_{uv}^{\notin}$ for each edge $uv \in E(G)$. Next we show that if M is a minimum maximal matching of G and M' is a minimum strongly maximal matching of H_G, $|M| = |M'| - 3\,\mathrm{m}$. As explained above, this completes the proof.

Let M be a minimum maximal matching of G. Using the auxiliary sets $F_{uv}^{\in}$ and $F_{uv}^{\notin}$, we define a strongly maximal matching M' of H_G by

$$M' = \bigcup_{uv \in M} F_{uv}^{\in} \cup \bigcup_{uv \notin M} F_{uv}^{\notin}.$$

Note that $|M'| = |M| + 3|E(G)|$.

Now, let M' be a minimum strongly maximal matching of H_G that verifies either $F_{uv} \cap M' = F_{uv}^{\in}$ or $F_{uv} \cap M' = F_{uv}^{\notin}$ for each edge $uv \in E(G)$. We define a maximal matching M of G by setting

$$M = \{uv : uv \in E(G), \ F_{uv} \cap M' = F_{uv}^{\in}\}.$$

Clearly, $|M| = |M'| - 3|E(G)|$, which completes the proof.   □

5   b-Coloring Co-Trees and Tree-Cographs

5.1   Co-Trees

Theorem 4. *In the class of co-trees, χ_b can be computed in polynomial time.*

Proof. According to Lemma 1, the problem is equivalent to finding a minimum strongly maximal matching (MSMM) in a tree. We will find such a matching by dynamic programming. In order to do so, we will define five functions $F_i(r, s)$, $i = 1, \ldots, 5$, for a nontrivial tree T_{rs} rooted at a leaf r with neighbor s. As we will apply them to the subtrees of a tree, we will assume that r can have neighbors outside T_{rs}.

- $F_1(r, s)$: cardinality of an MSMM of T_{rs} such that r is unmatched, and ∞ if it does not exist.
- $F_2(r, s)$: cardinality of an MSMM of T_{rs} that uses the edge rs and such that s may or may not have an unmatched neighbor (this case will apply when r has no unmatched neighbor outside T_{rs}), and ∞ if it does not exist.
- $F_3(r, s)$: cardinality of an MSMM of T_{rs} that uses the edge rs and such that s does not have an unmatched neighbor (this case will apply when r has already an unmatched neighbor outside T_{rs}, so an unmatched neighbor of s will complete an augmenting path of length 3 in the whole tree), and ∞ if it does not exist.
- $F_4(r, s)$: cardinality of an MSMM of T_{rs} such that the vertex s is matched with some vertex different from r and the vertex r is considered as "already matched" (this case will apply when r is already matched with a vertex outside T_{rs}), and ∞ if it does not exist.
- $F_5(r, s)$: cardinality of an MSMM of T_{rs} such that the vertex s remains unmatched and the vertex r is considered as "already matched", and ∞ if it does not exist.

With these definitions, for the base case in which $V(T_{rs}) = \{r, s\}$, we have

- $F_1(r, s) = \infty$ (if r is unmatched and s has no further neighbors, the matching will never be maximal)
- $F_2(r, s) = 1$ (precisely, the edge rs)
- $F_3(r, s) = 1$ (precisely, the edge rs)
- $F_4(r, s) = \infty$ (it is not feasible because s has no further neighbors)
- $F_5(r, s) = 0$

For the case in which s has children $v_1, \ldots, v_k$, we have

- $F_1(r, s) = \min_{i=1,\ldots,k}\{F_3(s, v_i) + \sum_{j=1,\ldots,k;j \neq i} \min\{F_4(s, v_j), F_5(s, v_j)\}\}$. In order to obtain a maximal matching, we need to match s with one of its children, say v_i. Since r will be unmatched, v_i should not have an unmatched neighbor, in order to prevent an augmenting path of length 3. When considering the trees T_{sv_j} for $j \neq i$, the vertex s will have the status of "already matched". Furthermore, since we are already assuming that s has an unmatched neighbor, we do not need to care about the vertices v_j being matched or not.
- $F_2(r, s) = 1 + \sum_{i=1,\ldots,k} \min\{F_4(s, v_i), F_5(s, v_i)\}$. We will use the edge rs, and then when considering the trees T_{sv_i} for $i = 1, \ldots, k$, the vertex s will have the status of "already matched". Furthermore, since s may or may not have an unmatched neighbor, we can take the minimum over F_4 and F_5 for each of the trees T_{sv_i}.

- $F_3(r,s) = 1 + \sum_{i=1,...,k} F_4(s,v_i)$. This case is similar to the previous one, but now the vertex s cannot have unmatched neighbors, so we will just consider F_4 for each of the trees T_{sv_i}.

- $F_4(r,s) = \min\{\min_{i=1,...,k}\{F_2(s,v_i) + \sum_{j=1,...,k;j\neq i} F_4(s,v_j)\}, \min_{i=1,...,k}\{F_3(s,v_i) + \sum_{j=1,...,k;j\neq i} \min\{F_4(s,v_j), F_5(s,v_j)\}\}\}$. As in the first case, we need to match s with one of its children, say v_i. But now, since r is assumed to be matched, s may or may not have an unmatched neighbor, depending on the matching status of the vertices v_j with $j \neq i$. So we will take the minimum among allowing v_i to have an unmatched neighbor and forcing v_j, $j \neq i$, to be matched, or forbidding v_i to have an unmatched neighbor and allowing v_j, $j \neq i$, to be either matched or not.

- $F_5(r,s) = \sum_{i=1,...,k} F_1(s,v_i)$. This last case is quite clear.

In this way, in order to obtain the cardinality of a minimum strongly maximal matching of a nontrivial tree T, we can root it at a leaf r whose neighbor is s and compute $\min\{F_1(r,s), F_2(r,s)\}$. By keeping some extra information, we can also obtain in polynomial time the matching itself. □

5.2 Tree-Cographs

A graph is a *tree-cograph* if it can be constructed from trees by disjoint union and complement operations. Tree-cographs have been introduced by Tinhofer [14] as a generalization of trees and cographs. Let $G_1 = (V_1, E_1)$ and $G_2 = (V_2, E_2)$ be two graphs with $V_1 \cap V_2 = \emptyset$. The *union* of G_1 and G_2 is the graph $G_1 \cup G_2 = (V_1 \cup V_2, E_1 \cup E_2)$, and the *join* of G_1 and G_2 is the graph $G_1 \vee G_2 = (V_1 \cup V_2, E_1 \cup E_2 \cup V_1 \times V_2)$. Note that $\overline{G_1 \vee G_2} = \overline{G_1} \cup \overline{G_2}$. Tree-cographs can be recursively defined as follows: a graph G is a tree-cograph if and only if

(i) G is a tree or a co-tree, or
(ii) G is the union of two tree-cographs G_1 and G_2, or
(iii) G is the join of two tree-cographs G_1 and G_2.

Notice that if (i) in the above definition is replaced by "G is a single vertex" then, the obtained graph is a cograph. The notion of *dominance sequence* has been introduced in [3] in order to compute the b-chromatic number of P_4-sparse graphs and, in particular, cographs. Formally, given a graph G, the dominance sequence $\mathrm{dom}_G \in \mathbb{Z}^{\mathbb{N}_{\geq \chi(G)}}$, is defined such that $\mathrm{dom}_G[t]$ is the maximum number of distinct color classes admitting dominant vertices in any coloring of G with t colors, for every $t \geq \chi(G)$. Note that it suffices to consider this sequence until $t = |V(G)|$, since $\mathrm{dom}_G[t] = 0$ for $t > |V(G)|$. Therefore, in the sequel we shall consider only the *dominance vector* $(\mathrm{dom}_G[\chi(G)], \ldots, \mathrm{dom}_G[|V(G)|])$. Notice that a graph G admits a b-coloring with t colors if and only if $\mathrm{dom}_G[t] = t$. Moreover, it is clear that $\mathrm{dom}_G[\chi(G)] = \chi(G)$. The following results given in [3] are very important in order to compute the b-chromatic number of graphs that can be decomposed recursively in modules via disjoint union or join operations.

Theorem 5 [3]. *Let $G_1 = (V_1, E_1)$ and $G_2 = (V_2, E_2)$ be two graphs such that $V_1 \cap V_2 = \emptyset$. If $G = G_1 \cup G_2$ and $t \geq \chi(G)$, then*

$$\text{dom}_G[t] = \min\{t, \text{dom}_{G_1}[t] + \text{dom}_{G_2}[t]\}.$$

Theorem 6 [3]. *Let $G_1 = (V_1, E_1)$ and $G_2 = (V_2, E_2)$ be two graphs such that $V_1 \cap V_2 = \emptyset$. Let $G = G_1 \vee G_2$ and $\chi(G) \leq t \leq |V(G)|$. Let $a = \max\{\chi(G_1), t - |V(G_2)|\}$ and $b = \min\{|V(G_1)|, t - \chi(G_2)\}$. Then $a \leq b$ and*

$$\text{dom}_G[t] = \max_{a \leq j \leq b}\{\text{dom}_{G_1}[j] + \text{dom}_{G_2}[t - j]\}.$$

In order to compute the dominance vector of a tree-cograph and its corresponding b-chromatic number, by Theorems 5 and 6, it is sufficient to compute the dominance vector for both trees and co-trees.

Dominance Vector for Trees. Irving and Manlove [11] shown that the b-chromatic number of any tree T is equal to $m(T) - 1$ or $m(T)$, depending on the existence of a unique vertex in T called a *pivot*. A vertex v of T is called *dense* if $d(v) \geq m(T) - 1$. Based on Irving and Manlove's results, we are able to prove the following result.

Theorem 7. *If G is tree, then dom_G can be computed in polynomial time.*

Dominance Vector for Co-Trees. Let G be a graph and M be a matching of it. Let $S_1(G, M)$ be the number of unmatched vertices that have at least one unmatched neighbor and let $S_2(G, M)$ be the number of edges of M that are the center of an augmenting path of length 3 for M. Let $F(\overline{G}, k)$ be the minimum value of $S_1(G, M) + S_2(G, M)$ over all matchings M of G with $|M| = k$.

Now, let G be a graph with stability at most two and consider a coloring of it. Let M be the matching of $\overline{G}$ corresponding to that coloring. The number of color classes without a dominant vertex are exactly $S_1(\overline{G}, M) + S_2(\overline{G}, M)$. So, for $\chi(G) \leq i \leq |V(G)|$, $\text{dom}_G[i] = i - F(\overline{G}, |V(G)| - i)$. We will show how to compute $F(T, k)$ for a tree T and a nonnegative integer k in polynomial time.

Theorem 8. *If G is a co-tree, then dom_G can be computed in polynomial time.*

Proof. (Sketch) As we noticed above, the problem is equivalent to computing $F(\overline{G}, k)$. We use dynamic programming. In order to do so, and in a similar fashion as in Theorem 4, we will define seven functions $F_i(r, s, k)$, $i = 1, \ldots, 7$, for a nontrivial tree T_{rs} rooted at a leaf r with neighbor s and a nonnegative integer k. As we will apply them to the subtrees of a tree, we will assume that r can have neighbors outside T_{rs}. Nevertheless, we will count for S_2 just the edges of $M \cap E(T_{rs})$ and for S_1 the vertices of $V(T_{rs})$, with the exception of r when it is unmatched but has already an unmatched neighbor outside T_{rs}, in order to avoid double counting.

For $i = 1, \ldots, 7$, $F_i(r, s, k)$ will be the minimum of $S_1(T_{rs}, M) + S_2(T_{rs}, M)$ over all the matchings M with $|M| = k$ such that:

- $F_1(r, s, k)$: r is unmatched and s is matched by M with some vertex different from r.
- $F_2(r, s, k)$: M uses the edge rs and r has no unmatched neighbor outside T_{rs}.
- $F_3(r, s, k)$: M uses the edge rs and r has an unmatched neighbor outside T_{rs}.
- $F_4(r, s, k)$: the vertex s is matched by M with some vertex different from r and the vertex r is already matched with a vertex outside T_{rs}.
- $F_5(r, s, k)$: the vertex s remains unmatched and the vertex r is already matched with a vertex outside T_{rs}.
- $F_6(r, s, k)$: r is unmatched, s remains unmatched, and r has no unmatched neighbor outside T_{rs}.
- $F_7(r, s, k)$: r is unmatched, s remains unmatched, and r has an unmatched neighbor outside T_{rs} (we will not count r for S_1 as we assume it is already counted).

In all cases, the value will be ∞ if no such M does exist.

Notice that as the values of the functions F_i are bounded by the number of vertices of the corresponding tree, and k is also bounded by that number, taking the minimum over $k_1 + \cdots + k_\ell = k$ of some combination of these F_i is equivalent to solving a polynomially bounded number of knapsack problems where both the weights and the utilities are polynomially bounded as well, so this can be done by dynamic programming in polynomial time [4].

In this way, in order to obtain $F(T, k)$ for a nontrivial tree T, we can root it at a leaf r whose neighbor is s and compute $\min\{F_1(r, s, k), F_2(r, s, k), F_6(r, s, k)\}$. By keeping some extra information, we can also obtain in polynomial time the matching itself. $\square$

b-Continuity and b-Monotonicity of Tree-Cographs. The following result was proved for union and join of graphs.

Lemma 4 [3]. *Let $G_1 = (V_1, E_1)$ and $G_2 = (V_2, E_2)$ be two graphs such that $V_1 \cap V_2 = \emptyset$. If G_1 and G_2 are b-continuous, then $G_1 \cup G_2$ and $G_1 \vee G_2$ are b-continuous.*

As a corollary of the lemma, Theorem 1, and the b-continuity of chordal graphs [6,12], we have the following result.

Theorem 9. *Tree-cographs are b-continuous.*

Concerning the b-monotonicity, the following results are known for general graphs and for union and join of graphs.

Lemma 5 [3]. *Let G be a graph. The maximum value of $\mathrm{dom}_G[t]$ is attained in $t = \chi_b(G)$.*

Lemma 6 [3]. *Let $G_1 = (V_1, E_1)$ and $G_2 = (V_2, E_2)$ be two b-continuous graphs such that $V_1 \cap V_2 = \emptyset$, and let $G = G_1 \cup G_2$ (resp., let $G = G_1 \vee G_2$). Assume that for every $t \geq \chi(G_i)$ and every induced subgraph H of G_i we have $\mathrm{dom}_H[t] \leq \mathrm{dom}_{G_i}[t]$, for $i = 1, 2$. Then, for every $t \geq \chi(G)$ and every induced subgraph H of G, $\mathrm{dom}_H[t] \leq \mathrm{dom}_G[t]$ holds.*

In order to prove the b-monotonicity of tree-cographs, we need the following two lemmas.

Lemma 7. *Let T be a tree and H an induced subgraph of T. Then for every $t \geq 2$, $\mathrm{dom}_H[t] \leq \mathrm{dom}_T[t]$.*

Lemma 8. *Let G be a graph with stability at most two and H an induced subgraph of G. Then for every $t \geq \chi(G)$, $\mathrm{dom}_H[t] \leq \mathrm{dom}_G[t]$.*

So, we can conclude the following.

Theorem 10. *Tree-cographs are b-monotonic.*

References

1. Berge, C.: Two theorems in graph theory. Proc. Natl. Acad. Sci. U.S.A. **43**, 842–844 (1957)
2. Bodlaender, H.L.: Achromatic number is NP-complete for cographs and interval graphs. Inf. Process. Lett. **31**, 135–138 (1989)
3. Bonomo, F., Durán, G., Maffray, F., Marenco, J., Valencia-Pabon, M.: On the b-coloring of cographs and P_4-sparse graphs. Graphs Comb. **25**(2), 153–167 (2009)
4. Dantzig, G.B.: Discrete-variable extremum problems. Oper. Res. **5**, 266–277 (1957)
5. Edmonds, J.: Paths, trees and flowers. Can. J. Math. **17**, 449–467 (1965)
6. Faik, T.: La b-continuité des b-colorations: complexité, propriétés structurelles et algorithmes. Ph.D. thesis, L.R.I., Université Paris-Sud, Orsay, France (2005)
7. Harary, F., Hedetniemi, S.: The achromatic number of a graph. J. Comb. Theor. **8**, 154–161 (1970)
8. Havet, F., Linhares-Sales, C., Sampaio, L.: b-coloring of tight graphs. Discrete Appl. Math. **160**(18), 2709–2715 (2012)
9. Hoàng, C.T., Kouider, M.: On the b-dominating coloring of graphs. Discrete Appl. Math. **152**, 176–186 (2005)
10. Hoàng, C.T., Linhares Sales, C., Maffray, F.: On minimally b-imperfect graphs. Discrete Appl. Math. **157**(17), 3519–3530 (2009)
11. Irving, R.W., Manlove, D.F.: The b-chromatic number of a graph. Discrete Appl. Math. **91**, 127–141 (1999)
12. Kára, J., Kratochvíl, J., Voigt, M.: b-continuity. Technical report M 14/04, Technical University Ilmenau, Faculty of Mathematics and Natural Sciences (2004)
13. Kratochvíl, J., Tuza, Z., Voigt, M.: On the b-Chromatic number of graphs. In: Kučera, L. (ed.) WG 2002. LNCS, vol. 2573, pp. 310–320. Springer, Heidelberg (2002)
14. Tinhofer, G.: Strong tree-cographs are Birkoff graphs. Discrete Appl. Math. **22**(3), 275–288 (1989)
15. Yannakakis, M., Gavril, F.: Edge dominating sets in graphs. SIAM J. Appl. Math. **38**(3), 364–372 (1980)

Proactive Reactive Scheduling in Resource Constrained Projects with Flexibility and Quality Robustness Requirements

Mario Brčić[✉], Damir Kalpić, and Marija Katić

Faculty of Electrical Engineering and Computing,
University of Zagreb, Zagreb, Croatia
{mario.brcic,damir.kalpic,marija.katic}@fer.hr

Abstract. This paper presents a new approach to proactive reactive scheduling of stochastic resource-constrained project scheduling problems with known probability distributions of activity durations. To facilitate the search for cost-flexible proactive schedules that are adjustable and incur lower expected cost of future rescheduling, a new family of cost-based flexibility measures is introduced. Under these measures, cost is incurred on each rescheduling while taking into account the temporal distance of changes in the baseline schedule. We propose a new model that describes the integrated approach using the proposed cost-based flexibility measures where, in each stage, reactive scheduling can adjust the baseline schedule to accommodate flexibility and quality requirements. The model is based on bounded stochastic shortest path with finite state and action spaces. The commonly used schedule stability measure is put in the context of proposed family of flexibility measures and contrasted to them in the terms of project execution system properties.

Keywords: Project scheduling · Risk · Proactive reactive scheduling · Stochastic dynamic programming · Cost-based flexibility

1 Introduction

Stochastic Resource Constrained Project Scheduling Problem (SRCPSP) is a generalization of the classical family of deterministic scheduling problems with complete information [1]. It introduces uncertainty by using random variables to model some of its data. In this paper, we shall focus on uncertain activity durations with known probability distributions. There are three main approaches to solving SRCPSP: predictive, reactive and proactive procedures [1] and combinations of the main approaches.

Predictive approach ignores stochasticity of the problem and uses point estimations, most usually expectation or median, instead of random variables. This has been shown to underestimate project cost/duration [2]. Reactive procedures make scheduling decisions during the project run-time. They can work with baseline schedule, as schedule repair procedures, and without baseline schedule

© Springer International Publishing Switzerland 2014
P. Fouilhoux et al. (Eds.): ISCO 2014, LNCS 8596, pp. 112–124, 2014.
DOI: 10.1007/978-3-319-09174-7_10

as completely online procedures. In the latter case they see the project as a multi-stage decision making process and dynamically create schedule in stages, using policies. Proactive project scheduling for SRCPSP is interested in creating a baseline schedule of increased robustness to unexpected outcomes (according to the used robustness measure explained below) such as longer than anticipated activity duration. In such a way, it can remain feasible under various conditions. The two most commonly used notions of robustness in proactive scheduling are: quality and stability robustness. Quality robustness pertains to maximizing the probability of completing the project on time. Stability robustness aims to make schedule stable with respect to possible disruptions, so it does not change much during the execution. Solution robustness or schedule stability is justified by examples where several separate entities cooperate on the project and need to synchronize their actions. Also, in cases with in-house project running, schedule stability increases the setup efficiency. Quality and stability are most commonly two competing criteria and problems containing both are bi-objective. However, they are implicitly converted to single-objective problems by parameterization into monetary costs, scalarizing the two objectives into one. If the monetary cost is the only interest, then such approach is valid and we can continue our work with that assumption.

Although protected against some future disruptions, proactive baseline schedule can become infeasible during the execution due to unanticipated disturbances. In that case rescheduling needs to be done. At this point, reactive schedule repair procedures are used [3]. Such a combination of proactive and reactive procedures to SRCPSP is called a proactive-reactive approach. Current rescheduling procedures mostly focus on restoring the schedule feasibility by starting activities with the least rescheduling cost w.r.t. the first baseline and/or they do not produce proactive schedules that are hedged against future unexpected outcomes in the same way as it is done for the baseline schedule [4]. This paper explores the problem of proactive-reactive scheduling of SRCPSP where changes to baseline schedule can be made in advance at a lesser cost than if being done at the activity start, hence yielding a new proactive baseline using all the information available up to that moment.

The main contributions of the paper are:

- A new family of flexibility measures based on realistic assumptions on cost functions. Total idleness is shown to be potentially optimal behaviour in certain situations. A bound is put on the worst-case performance of optimal policy, which ensures the termination.
- A model capturing the aspects of the general problem is presented.
- Commonly used stability measure is put into the relation with the model and compared to our family of flexibility measures.

The organization of this paper is as follows: in Sect. 2, we lay out the overview of the related work done in this area. Section 3 shortly presents the problem and Sect. 4 presents the family of flexibility measures. In Sect. 5, we present the stochastic dynamic programming model and in Sect. 6 we put commonly used stability measure into the relation to our model. Finally, Sect. 7 gives conclusions and future work.

2   Related Work

Authors in [1] have offered the survey of resource-constrained project scheduling under uncertainty. In this section we describe related work in pure reactive approaches based on stochastic dynamic programming and proactive-reactive approaches.

The most influential work on pure reactive based methods is given by Möhring et al. [5,6]. They modelled a general stochastic scheduling problem with regular performance measures (measures that are non-decreasing in activity completion times) as a stochastic dynamic program. Their theoretical results are built on the fact that the total idleness is non-optimal behaviour for their problem. Stork, based on [5,6], dealt with different scheduling policy families in [7]. There are several works on pure reactive scheduling [8–10] that use Markov Decision Process (MDP). Tai in [8] used dynamic programming and authors in [9,10] used reinforcement learning to find solutions. The work listed above focused on regular performance measures and do not use baseline schedules.

Of proactive-reactive approaches, Leus and Herroelen [11] proposed stability measure expressed as the weighted sum of absolute differences between baseline and realized schedule activity start times. This measure is used in the majority of the project scheduling literature [3], including the works listed below. Van de Vonder et al. in [12] describe Starting Time Criticality + Descent (STC+D) heuristic with surrogate measure and simplifying assumptions in approximations for generation of proactive schedules with time buffering on locked resource flows and predefined policy family. Van de Vonder et al. in [4] used robust schedule generation schemes with priority lists in basic sampling approach and with time windows, where point estimates of duration times were used. Deblaere et al. in [13], based on ideas in STC+D, proposed a family of proactive policies that use activity priority list and release times in parameterization. The final schedule is not necessarily resource-feasible, but it minimizes the combination of expected deviation and due date exceeding costs. In all the works above, the policies used are starting activities in every stage of schedule creation, using variants of parallel/serial scheduling schemes, and perform only just-in-time rescheduling of the activities that are about to be run. Lambrechts in his PhD thesis [14] developed a tabu-search based method that does bi-objective optimization for proactive rescheduling with respect to uncertainty of resource availability. The method is, at the same time, keeping the new schedule close to the schedule obtained in the previous phase using the deviation measure. The author used scalarization to transform bi-objective into a single objective problem. This method achieved moderate results [3].

Although there are various interesting approaches to the problem or resource constrained project scheduling under uncertainty, there is no approach that considers scheduling where rescheduling can be done in advance with smaller cost than if done at activity start times. In this paper we give such a proposal where we approach to the problem as a stochastic dynamic program in a similar way as Möhring et al. [5,6]. However, in order to model proactiveness we allow for a special family of performance measures (not guaranteed to be regular) on SRCPSP.

3   The Problem Definition

The problem under consideration in this paper is the single mode non-preemptive stochastic resource-constrained project scheduling problem with quality and flexibility robustness requirements, uncertain activity durations and with known probability distribution of activity durations. Let H be the space of all such problems. Each $h \in H$ is a combinatorial optimization problem defined as a tuple $(V, E, p, R, B, D, \delta, c)$. $V = \{0, ...n + 1\}$ is a set of $n + 2$ activities where 0 and $n + 1$ are dummy activities that represent project beginning and end respectively. Let $V' = V \setminus \{0, n + 1\}$. Precedence relation between activities is defined as transitive closure of relation $E \subset V \times V$, where 0 precedes and $n + 1$ succeeds all other activities in V. Precedence relation must be asymmetric. Let $\Delta(\mathbb{N}_0^{n+2})$ be the space of all discrete probability distributions defined over $\mathbb{N}_0^{n+2}$ with bounded support. $p \in \Delta(\mathbb{N}_0^{n+2})$ is a joint probability distribution of activity durations represented by a random vector d, where p_0 and p_{n+1}, marginal distributions for dummy activities, have all the mass on duration of 0. Also, $\forall a \in V$ either $p_a(0) = 1$ or $p_a(0) = 0$. $R = \{R_1, ..., R_r\}$ defines a set of r renewable resources and $B \in \mathbb{N}^r$ is a vector of resource availabilities. Activity demands on resources are given in the matrix $D \in \mathbb{N}_0^{(n+2) \times r}$, where $(\forall i \in \{a \in V | p_a(0) = 1\}), (\forall r \in R) D_{i,r} = 0$ and $(\forall r \in R)(\forall i \in V) D_{i,r} \leq B_r$. $\delta \in \mathbb{N}_0$ is the project due date. Before defining the objective function c, we need to define some necessary intermediate objects.

Let S be the countable space of all project states, where each state $x \in S$ stores all relevant information about the project during the execution. This information includes the global project time (the time elapsed from the start of the project execution), statuses of activities, durations of finished activities and the current schedule. Schedule is a vector in $\mathbb{N}_0^{n+2}$, where i-th component is the scheduled start time of activity i. The start time of activity 0 is in each schedule equal to 0. In order to extract the schedule from the state, let us define the *schedule extraction function* $L : S \rightarrow \mathbb{N}_0^{n+2}$. Let C be the countable space of all controls (decisions) that control project's execution. Controls start the execution of activities, change the current baseline schedule or do nothing, at any control point of project. As this is a dynamic optimization problem where the objective function c will be defined as expected total cost, where the next project state and stage cost depend only on current state and control, the solution is a randomized Markov *policy* $\mu^* : S \rightarrow \Delta(C)$ [15] in policy space Ξ_{rand}. Let $\Xi_{\text{det}} \subset \Xi_{\text{rand}}$ be the space of all deterministic Markov policies $\mu : S \rightarrow C$. In the rest of the paper we shall focus on deterministic Markov policies.

Let $N : \mathbb{N}_0^{n+2} \times \Xi_{\text{det}} \rightarrow \overline{\mathbb{N}}_0$, i.e. $N(\gamma, \mu)$ for *realized activity duration vector (scenario)* γ and policy μ, be the number of stages where decisions take place before the end of project execution. When it is clear from the context, we omit the dependency and only write N. For each stage $k = 1..N(\gamma, \mu)$ let $x_k^{\gamma,\mu} \in S$ be the project state reached under duration vector γ and policy μ and let $x_k^{\gamma,\mu|t} \in \mathbb{N}_0$ be the global project time of that state. Let $s_k^{\gamma,\mu,i} \in \mathbb{N}_0$ be the scheduled start time of activity i in the schedule $L(x_k^{\gamma,\mu})$. Let $\Pi \subset \Xi_{\text{det}}$ be the space of admissible policies, i.e. all policies that respect 0-lag precedence, resource, non-anticipativity, non-retroactiveness, non-prematureness constraints and condition of project terminability defined below. We search for the solution in the space Π. Let $Z : \Pi \to \mathbb{N}_0$ be the worst-case schedule duration given the problem $h \in H$ and policy $\mu \in \Pi$, i.e. $Z(\mu) = \max_{\gamma \in \text{supp}(p)} s_N^{\gamma,\mu,n+1}$. 0-lag precedence constraints are defined as:

$$s_N^{\gamma,\mu,j} \geq s_N^{\gamma,\mu,i} + \gamma_i, \forall (i,j) \in E, \forall \mu \in \Pi, \forall \gamma \in \text{supp}(p) \ .$$

The set of concurrent activities at timepoint $t \in \mathbb{N}_0$ under policy μ with the vector of activity durations γ is:

$$\Lambda_{t,\gamma,\mu} = \{i \in V | t - s_N^{\gamma,\mu,i} < \gamma_i\} \ .$$

Using the set of concurrent activities, resource constraints are defined as:

$$\sum_{i \in \Lambda_{t,\gamma,\mu}} D_{ij} \leq B_j, \forall t \geq 0, \forall R_j \in R, \forall \mu \in \Pi, \forall \gamma \in \text{supp}(p) \ .$$

Non-anticipativity ensures that for all scenarios $\gamma \in \mathbb{N}_0^{n+2}$, policies $\mu \in \Pi$, and all timepoints $t \in \mathbb{N}_0$, the behaviour of policy μ at any stage k depends only on the history of γ w.r.t. μ up to t. These constraints are formally described in [5]. Non-retroactiveness constraints for all timepoints t disallow rescheduling of activities started at any timepoints $t' < t$. Also, starting activities at any timepoint $t' < t$ or rescheduling starts of not yet started activities to any timepoint $t' < t$ is forbidden. Non-prematureness constraint $\forall t < 0, \forall i \in V$ prohibits start of activity i at timepoint t. Terminability condition means that $(\forall \mu \in \Pi)(\exists M \in \mathbb{N})(\forall \gamma \in \text{supp}(p))$ project execution finishes at least until M under μ and γ. Let $\Pi_{rand} \subset \Xi_{\text{rand}}$ be the set of admissible randomized Markov policies with similar constraints as described above, generalized to the case of randomized policies.

Definition 1. *Function* $c_d : \mathbb{N}_0 \to \mathbb{R}_+$ *is the* quality robustness penalty. *It is defined as* $c_d(x) = \beta_d \cdot \max(0, x - \delta), \beta_d > 0$, *incurred in stages as stage-cost* $c_{d,s}(x) = \beta_d \cdot \mathbf{1}_{x > \delta}$.

Function $c_s : \mathbb{N}_0^{n+2} \times \mathbb{N}_0^{n+2} \times \mathbb{N} \to \mathbb{R}_+$ is a **rescheduling cost function** defined in the next section as part of a cost-based flexibility measure.

The objective function is $c : S \times \Pi \to \mathbb{R}_+$:

$$c(x_{-1}, \mu) = \mathbb{E}_{d \sim p}^{\mu} \left[\sum_{k=1}^{N(\mathbf{d},\mu)} \left(c_s(L(x_{k-1}^{d,\mu}), L(x_k^{d,\mu}), x_{k-1}^{d,\mu|t}) + c_{d,s}(x_{k-1}^{d,\mu|t}) \right) \right] \quad (1)$$

where x_{-1} is the initial empty state at the timepoint -1 and x_0^μ is the initial running schedule state that contains baseline schedule created by the policy μ offline at stage -1 from the empty schedule in state x_{-1}, at no cost. The effects of applied policy control at stage k are first visible in the state at the next stage.

4   Family of Cost-Based Flexibility Measures

Introduction of Cost-based Flexibility (CBF) measure enables modelling situations where rescheduling in advance might be opportunistic due to lower costs. This enables search for flexible proactive schedules that are adjustable and incur minimal total costs for rescheduling and due date exceeding.

Let, $\forall i \in V'$, $c_{s,i} : \mathbb{N}_0 \times \mathbb{N}_0 \times \mathbb{N}_0 \to \mathbb{R}_+$, $c_{s,i}(x,y,t)$, be the **activity rescheduling cost function** defined on each point of the domain. It is monotonically non-decreasing in $|x-y|$ for each constant $\min(x,y)-t$ and for each constant $|x-y|$ it is monotonically non-increasing in positive $\min(x,y)-t$, i.e. the distance of the schedule change from the current timepoint t. Also, $\forall i \forall t\, (x = y \Rightarrow c_{s,i}\,(x,y,t) = 0)$.

Rescheduling cost function $c_s : \mathbb{N}_0^{n+2} \times \mathbb{N}_0^{n+2} \times \mathbb{N} \to \mathbb{R}_+$ measures the difference between two successive schedules $L(x_{k-1}^{\gamma,\mu})$ and $L(x_k^{\gamma,\mu})$:

$$c_s\left(L\left(x_{k-1}^{\gamma,\mu}\right), L\left(x_k^{\gamma,\mu}\right), x_{k-1}^{\gamma,\mu|t}\right) = \sum_{i\in V'} c_{s,i}(s_{k-1}^{\gamma,\mu,i}, s_k^{\gamma,\mu,i}, x_{k-1}^{\gamma,\mu|t}) \ .$$

Definition 2. *Cost-based flexibility measure is the function* $c_f : \Pi \times \mathbb{N}_0^{n+2} \to \mathbb{R}_+$ *of the form:*

$$c_f(\mu,\gamma) = \sum_{k=1}^{N(\gamma,\mu)} c_s\left(L\left(x_{k-1}^{\gamma,\mu}\right), L\left(x_k^{\gamma,\mu}\right), x_{k-1}^{\gamma,\mu|t}\right) \ . \tag{2}$$

Lemma 1. $(\forall h \in H)(\forall \mu' \in \Pi_{rand})(\exists \mu \in \Pi)c(x_{-1},\mu) \leq c(x_{-1},\mu')$

Definition 3. $\forall h \in H, \forall t \in \mathbb{N}_0$ *period* $[t, t+1)$ *is total idleness period under policy* μ *and vector of activity durations* γ *if and only if* $\Lambda_{t,\gamma,\mu} = \emptyset$.

Lemma 2. $\exists h \in H$ *such that for* $\mu^* \in \Pi, \exists \gamma \in \mathrm{supp}(p)$ *where there is at least one total idleness period.*

Lemma 2 causes the departure from previous theoretical results laid in [5,6]. We build results that put bound on the amount of total idleness in optimal policy. That enables the creation of model that can be solved using standard methods.

Theorem 1. *For arbitrary problem* $h \in H$, $\exists \mu_r \in \Pi$ *such that* $c(x_{-1},\mu_r) = c_d\left(\sum_{i\in V} \max(\mathrm{supp}(p_i))\right)$. *Also,*

$$(\forall \mu \in \Pi)\left(c(x_{-1},\mu) \leq c(x_{-1},\mu_r) \Rightarrow c_d\left(Z\left(\mu\right)\right) \leq \frac{c_d\left(\sum_{i\in V} \max(\mathrm{supp}(p_i))\right)}{\min_{d\in\mathrm{supp}(p)} p(d)}\right) \ .$$

Corollary 1. *In the case $\delta < \sum_{i \in V} \max(\text{supp}(p_i))$, the bound ζ on the worst case project duration $Z(\mu^*)$ under optimal policy μ^* can be uniquely inferred from the cost bound given in Theorem 1 due to the properties of c_d. Otherwise, the solution to the problem is trivial and the bound ζ is set to $\sum_{i \in V} \max(\text{supp}(p_i))$.*

Corollary 2. *For non-trivial values of δ, we can search for optimal policy only in the set $\{\mu \in \Pi | (\forall \gamma \in \text{supp}(p)) s_N^{\gamma,\mu,n+1} \leq \zeta\}$. Optimal policy $\mu^* \in \Pi$ exists.*

The proofs of Lemmas 1, 2, Theorem 1 and Corollary 2 are given in the appendix. The bound in Theorem 1 is loose, depending on the continuous parameter of probability distribution. Stricter bounds can be found using more information about SRCPSP instance at hand. Relying on the Corollary 2, we can model the problem using finite horizon dynamic programming (DP). Also, using finiteness of the worst case performance of the optimal policy, we can model the problem as a stochastic shortest path with finite state and control spaces, using the undiscounted objective function (1) and infinite horizon DP theory.

5   The Model

In this section, we introduce a new stochastic dynamic programming model for the defined problem based on theoretical results from the previous section. The given problem is modelled as a stochastic shortest path (SSP) problem [16] and it is a Markov Decision Process (MDP) with variable number of stages as is the case in the project scheduling.

Definition 4. *Finite Markov Decision Process is a 7-tuple (S, C, W, U, P, f, g) where S is the finite discrete state space, C is the finite discrete control space, W is the finite sample space modelling elementary random information we can receive, $U : S \rightarrow 2^C$ is the control availability function. $P(\omega \in 2^W | x \in S, u \in C)$ is the distribution of random information conditioned on the last state and applied control. $f : S \times C \times W \rightarrow S$ is the state transition function, and $g : S \times C \rightarrow \mathbb{R}$ is the immediate cost function. Transition probabilities $T_{i,u,j}$ between the states $i, j \in S$ under applying control $u \in U(i)$ can be obtained by using the transition function f: $T_{i,u,j} = P\{\{\omega \in W | f(i, u, \omega) = j\} | i, u\}$.*

To make the state and the control space finite, we are using the upper bound on duration of project execution, Θ. Θ can be determined heuristically, or a conservative upper bound based on Corollary 1 can be used. In the rest of the paper we assume conservative upper bound. Let $O = \{i \in \mathbb{N}_0 | i \leq \Theta\}$.

Let $F : S \rightarrow 2^V$ define the set of resource and precedence feasible activities that have not been started yet. Let $F^s(x_k) \subset 2^{F(x_k)}$ be the set of sets of resource and precedence feasible combinations of activities for each state $x_k \in S$.

State Space. Each state $x_k \in S$ is a tuple (v, K, b, t).

- v - r-tuple $(v_1, ..., v_r)$ containing resource availabilities at state x_k, $\forall j \in R$
- K - state information about project activities; $(\forall a \in V \setminus \{0\}) K_a = (\phi_a, s_a, d_a)$. Information for activity 0 are not stored as it is under all policies always started at timepoint 0.
  - $\phi_a \in \{$'inactive','started', 'finished'$\}$ - the status of activity a
  - s_a - start time of activity a in the current schedule. If the activity has not been started yet, s_a is predicted/scheduled start time.
  - d_a realized duration of activity a, known only if the activity is completed. Otherwise, it is 0 and has no meaning. This information is useful only for activities whose activity durations are not independent.
- $b \in$ 'bounded', 'non-bounded'$\}$ - the status of project state used to terminate all executions of projects surpassing duration of Θ
- t - the current time of the project global clock, measured in discrete units.

Each element of state x_k is marked using the superscript notation, for example $x_k^{\phi_a}$ is the status of activity a in state x_k. The terminal state $x_k \in S$ is the one in which $x_k^{\phi_{n+1}} =$'finished' or $x_k^b =$'bounded'. The initial state x_{-1} has all resources free, all activities inactive, their scheduled start times set to 0, state marked as non-bounded and the global project time set to -1.

Control Space and Available Controls. $C \subset (O \cup \{$'start','empty'$\})^{n+1}$ is the set of controls that present decisions for rescheduling of activity scheduled start times and for starting any set of activities. $(\forall \mu \in \Pi)(\forall x_k \in S)\mu(x_k) \in U(x_k)$, where set of available controls in each state has to satisfy next conditions:

1. $(\forall x_k \in S)U(x_k) = \{[u_1, ..., u_{n+1}] | \{i | u_i = $ 'start'$\} \in F^s(x_k)\}$
2. $(\forall i \in V \setminus \{0\})(\forall x_k \in S) \left(x_k^{\phi_i} \neq$ 'inactive' $\Rightarrow u_i(x_k) = $ 'empty'$\right)$
3. $(\forall i \in V \setminus \{0\})(\forall x_k \in S) (u_i(x_k) \notin \{$'start', 'empty'$\} \Rightarrow u_i(x_k) \geq x_k^t)$
4. $(\forall x_k \in S) (x_k^t < 0 \Rightarrow (\forall i \in V \setminus \{0\})u_i(x_k) \neq $ 'start'$)$.

Random Information. The timing of the next stage is $t_{k+1} := t_k + 1$. Sample space is defined as $W = \text{supp}(p)$. Random information distribution function P is based on activity duration probability distribution p, with the probabilities of random information conditioned on durations of finished activities and on current execution times of running activities. All of the information that the conditioning is done upon is contained in the state and last applied control. Non-anticipativity of policies is ensured by the fact that after (state, action) pair (x, u) we receive information only on running activities in $[x^t, x^t + 1)$ where some subset of activities is finished at timepoint $x^t + 1$ and the rest continues with execution, in a similar way as in [5].

Transition Function. The model uses a notion of *post-decision state*. The transition function f is a composition $f = \sigma \circ \psi$ of the post-decision transition function $\psi : S \times C \to S$ and the stochastic transition function $\sigma : S \times W \to S$. The post-decision transition function ψ updates the state with the applied control, but without receiving new random information. The state we transition to after applying ψ is the post-decision state, and no control can be applied to it. The stochastic transition function σ updates the post-decision state with the random information. The result is a new pre-decision state at the next timepoint at which we can apply the new control. Transition function $f = \sigma \circ \psi$ is defined algorithmically as:

function $\phi(x_k, u_k)$
    if $x_k^{\phi_{n+1}} = $ 'finished' $\lor x_k^b = $ 'bounded' **then return** x_k
    $x_k^{u_k} \leftarrow \text{COPY}(x_k)$
    if $x_k^t = \Theta \land x_k^{\phi_{n+1}} \neq $ 'finished' $\land u_{k,n+1} \neq $ 'start' **then**
        $x_k^{u_k,b} \leftarrow $ 'bounded'
        return x_k^u
    for all $i \in \{i | u_{k,i} = $ 'start'$\}$ **do**
        Take up resources in $x_k^{u_k}$ for activity i
        $x_k^{u_k,s_i} \leftarrow x_k^{u_k,t}$, $x_k^{u_k,\phi_i} \leftarrow $ 'started'
    for all $i \in \{i | u_{k,i} = $ 'start' $\land p_i(0) = 1\}$ **do** $x_k^{u_k,\phi_i} \leftarrow $ 'finished'
        for all $i \in \{i | u_{k,i} \in O\}$ **do** $x_k^{u_k,s_i} \leftarrow u_{k,i}$
    return $x_k^{u_k}$
function $\sigma(x_k^{u_k}, \omega_k)$
    if $x_k^{u_k,\phi_{n+1}} = $ 'finished' $\lor x_k^{u_k,b} = $ 'bounded' **then return** $x_k^{u_k}$
    $x_{k+1} \leftarrow \text{COPY}(x_k^{u_k})$, $x_{k+1}^t \leftarrow x_k^t + 1$
    for all $i \in \{i | i$ finishes in x_{k+1}^t according to $\omega_k\}$ **do**
        release resources in x_{k+1} for i
        $x_{k+1}^{\phi_i} \leftarrow $ 'finished', $x_{k+1}^{d_i} \leftarrow x_{k+1}^t - x_{k+1}^{s_i}$
    return x_{k+1}

Cost Function

$$g(x_k, u_k) := \begin{cases} 0, & \text{if } x_k^{\phi_{n+1}} = \text{'finished'} \\ & \lor x_k^b = \text{'bounded'} \\ c_\text{s}(L(x_k), L(\psi(x_k, u_k)), x_k^t) + c_{\text{d},s}(x_k^t), & \text{if } x_k^t \geq 0 \\ M, & \text{if } x_k^t = \Theta \land u_{k,n+1} \neq \text{'start'} \\ 0, & \text{otherwise} \end{cases}$$

where M is a sufficiently large penalty for not terminating the project before exceeding the bound Θ. Cost bound from Theorem 1 can be used as a conservative basis for setting M.

The proposed model proceeds through all unit timesteps and at each stage searches for optimal controls in high dimensional discrete space, where the dimension depends on the number of activities. Similarly to [13], resource feasibility is not explicitly enforced in the non-started part of schedule. In our model the same holds for precedence feasibility as well.

Since the proposed model is SSP with finite state and control spaces with all admissible policies terminating (some are artificially made proper by bounding and penalizing), there is a unique optimal cost-to-go function $J^* : S \to \mathbb{R}_+$ that satisfies Bellman's optimality equations [16]:

$$J^*(x) = \min_{u \in U(x)} \mathbb{E}_{\omega \sim W(x,u)} \left[g(x,u) + J^* \left(f(x,u,\omega) \right) \right], \forall x \in S \ . \tag{3}$$

Standard methods for solving MDP such as value iteration, policy iteration or linear programming converge to the solution for this model [16] and can generally be used for solving. The proposed model is nearly acyclic and the solution can be found using simple adaptation to the shortest path algorithm for Directed Acyclic Graphs (DAG). Such an adaptation sets optimal costs-to-go value of terminal states to 0, resolving the only cycles in the graph, and uses expectations in calculations of distances in stochastic transitions.

6   Stability vs. Cost-Based Flexibility

The research based on the schedule stability measure as defined by [11] a priori assumes that the baseline schedule is static and that changes on scheduled activity start times between the creation of the baseline and the realized activity start times are forbidden or of no benefit and are ignored. We consider that the costs due to (in)stability have their root in inflexibilities in the project executing system and that these costs should be lower with increased temporal distance of changes from the current timepoint. That gives us the incentive to switch the baseline schedule in order to reduce anticipated costs. Here we present the conditions under which CBF reduces to the stability measure and when the search for the optimal policy can be done in simpler policy subspace.

Definition 5. $\forall h \in \Pi, \Pi_{st} \subset \Pi$ is the space of policies that offline, before the start of project execution, create a baseline schedule while online, $\forall \gamma \in \text{supp}(p)$, in each stage only perform rescheduling of activities at their start and leave the non-started part of the schedule unchanged.

Lemma 3. $\forall h \in H$ it holds:

$$(\forall \mu \in \Pi_{st})(\forall \gamma \in \text{supp}(p)) \left(c_f(\mu, \gamma) = \sum_{i \in V'} c_{s,i}(s_0^{\gamma,\mu,i}, s_N^{\gamma,\mu,i}, s_N^{\gamma,\mu,i}) \right) \ .$$

Theorem 2. $\forall h \in H, \min_{\mu \in \Pi} c(x_{-1}, \mu) = \min_{\mu' \in \Pi_{st}} c(x_{-1}, \mu')$ if $(\forall i \in V') c_{s,i}(x, y, t)$ have the following properties:

1. $(\forall x, y) \min_t c_{s,i}(x, y, t) = c_{s,i}(x, y, \min(x, y))$,
2. $(\forall x, y, z) [c_{s,i}(x, y, \min(x, y)) + c_{s,i}(y, z, \min(y, z)) \geq c_{s,i}(x, z, \min(x, z))]$,
3. $(\forall x, y) [x \leq y \Rightarrow (c_{s,i}(x, y, x) = c_{s,i}(x, y, y))]$.

The proofs of Lemma 3 and Theorem 2 are given in the appendix.

Corollary 3. *Let $h \in H$ be such that the expression for c_f under Lemma 3, $c_{f,st}$, has the form of Leus and Herroelen's stability measure. Let h' be the problem identical to h except that c_f in the objective function of h' is replaced by $c_{f,st}$. If h satisfies conditions of the Theorem 2 then $\min_{\mu \in \Pi} c(x_{-1}, \mu) = \min_{\mu' \in \Pi_{st}} c'(x_{-1}, \mu')$.*

Corollary 3 shows that a priori decision on search in Π_{st} and using stability measure results in no loss of optimality if there is obviously no advantage in rescheduling in advance and if the project executing system is fairly inflexible.

Let $\tau(x, y, t) = \max\{0, \min(x, y) - t\}$. An example of CBF measure is defined by the following rescheduling cost function:

$$c_s \left(L(x_{k-1}^{d,\mu}), L(x_k^{d,\mu}), x_{k-1}^{d,\mu|t} \right) = \sum_{i \in V'} b_i \alpha_i^{\tau(s_{k-1}^{d,\mu,i}, s_k^{d,\mu,i}, x_{k-1}^{d,\mu|t})} \cdot |s_{k-1}^{d,\mu,i} - s_k^{d,\mu,i}| \quad (4)$$

where $b_i \in \mathbb{R}$ is the activity-specific basic cost of rescheduling while $\alpha_i \in (0, 1]$ is activity-specific discount factor. α_i does not model the economic discounting, but the inflexibility of the project execution system included in execution of that activity. For example, the system is inflexible if the discount factor is very close to 1 as there is small or no benefit to reschedulings with advance notice. Obviously, under minimization of the problem that has rescheduling cost function (4) with $\forall i \in V'(\alpha_i > 1)$ we can restrict the search for the solution to Π_{st}. Using Corollary 3, the solution to the problem $h \in H$ with the rescheduling cost function (4), where $\forall i \in V'(\alpha_i = 1)$, can be found by solving the problem h' with the flexibility cost of the form of stability measure, with restriction to search in Π_{st}.

7   Conclusions and Future Work

In this paper, a new approach to proactive-reactive scheduling in SRCPSP has been introduced. To the best of our knowledge, this is the first work that approaches modelling of proactive scheduling with proactive reschedules. There are three main contributions discussed. Firstly, a new family of cost based flexibility measures is proposed in order to measure flexibility robustness. Flexibility measures are integrated into the sequential decision making procedure in order to obtain reactive approach with proactive reschedules. Furthermore, we presented a bounded stochastic shortest path based model with factored state representations that captures important aspects of the given problem. The model is finite MDP with variable number of stages. Standard solving methods, including simple adaptation to the shortest path algorithm for DAGs, could be applied to it using Bellman's optimality equation. Optimal solutions can be obtained only for small projects. The proactive-reactive optimization is done within a single framework of dynamic programming. Our third contribution refers to the commonly used stability measure. We compared it with the proposed family of flexibility measures in the context of proactive scheduling. Schedule stability is shown to be attained under special conditions on the flexibility measure when the project execution system does not show more flexibility with the advance notice or even forbids the advance notice.

Possible future work refers to finding better upper bounds on the worst case schedule duration for optimal policies with bounds depending on discrete parameters of the project. Characterizing the trade-off between the expected performance and different imputed worst case bounds is important for the development of solving procedures. The creation of approximation methods, possibly exploiting near-acyclicity, in order to scale the application scope onto bigger projects is viable venue. For example, approximate dynamic programming can be used on the model with heuristically determined bound on the worst case schedule duration. Finding rescheduling cost function, i.e. sub-elements of flexibility measure, between the consecutive schedules that would balance realistic modelling and computational costs of solving is also an interesting research topic. Finally, research into policy families that would be appropriate for solving problems with special classes of cost-based flexibility measures, could bring the results closer to the application domain.

References

1. Brcic, M., Kalpic, D., Fertalj, K.: Resource constrained project scheduling under uncertainty: a survey. In: Proceedings of the 23rd Central European Conference on Information and Intelligent Systems, pp. 401–409 (2012)
2. Klingel, A.R.: Bias in PERT project completion time calculations for a real network. Manag. Sci. **13**(4), 194–201 (1966)
3. Demeulemeester, E., Herroelen, W.: Robust Project Scheduling. Now Publishers, Boston (2011)
4. Van de Vonder, S., Ballestin, F., Demeulemeester, E., Herroelen, W.: Heuristic procedures for reactive project scheduling. Comput. Ind. Eng. **52**(1), 11–28 (2007)
5. Möhring, R.H., Radermacher, F.J., Weiss, G.: Stochastic scheduling problems I general strategies. Z. Oper. Res. **28**(7), 193–260 (1984)
6. Möhring, R.H., Radermacher, F.J.: Introduction to stochastic scheduling problems. In: Neumann, P.D.K., Pallaschke, P.D.D. (eds.) Contributions to Operations Research. Lecture Notes in Economics and Mathematical Systems, vol. 240, pp. 72–130. Springer, Heidelberg (1985)
7. Stork, F.: Stochastic resource-constrained project scheduling. Ph.D. thesis, Technical University at Berlin, Berlin, Germany (2001)
8. Tai, C.H.C.: A stochastic project scheduling problem with resource constraints. Ph.D. thesis, University of Missouri-Columbia (1997)
9. Choi, J., Realff, M.J., Lee, J.H.: A Q-learning-based method applied to stochastic resource constrained project scheduling with new project arrivals. Int. J. Robust Nonlinear Control **17**(13), 1214–1231 (2007)

10. Csáji, B., Monostori, L.: Adaptive stochastic resource control: a machine learning approach. J. Artif. Intel. Res. **32**(1), 453–486 (2008)
11. Leus, R., Herroelen, W.: Stability and resource allocation in project planning. IIE Trans. **36**(7), 667–682 (2004)
12. Van de Vonder, S., Demeulemeester, E., Leus, R., Herroelen, W.: Proactive-reactive project scheduling trade-offs and procedures. In: Jzefowska, J., Weglarz, J. (eds.) Perspectives in Modern Project Scheduling, vol. 92, pp. 25–51. Springer, New York (2006)
13. Deblaere, F., Demeulemeester, E., Herroelen, W.: Proactive policies for the stochastic resource-constrained project scheduling problem. Eur. J. Oper. Res. **214**(2), 308–316 (2011)
14. Lambrechts, O.: Robust project scheduling subject to resource breakdowns. Ph.D. thesis, Faculty of Business and Economics, Katholieke Universiteit Leuven, Leuven, Belgium (2007)
15. Puterman, M.L.: Markov Decision Processes: Discrete Stochastic Dynamic Programming, 1st edn. Wiley-Interscience, New York (1994)
16. Bertsekas, D.P.: Dynamic Programming and Optimal Control, vol. 1, 3rd edn. Athena Scientific, Belmont (2005)

Active Set Methods with Reoptimization for Convex Quadratic Integer Programming

Christoph Buchheim and Long Trieu[✉]

Fakultät für Mathematik, Technische Universität Dortmund,
Dortmund, Germany
{christoph.buchheim,long.trieu}@math.tu-dortmund.de

Abstract. We present a fast branch-and-bound algorithm for solving convex quadratic integer programs with few linear constraints. In each node, we solve the dual problem of the continuous relaxation using an infeasible active set method proposed by Kunisch and Rendl [10] to get a lower bound; this active set algorithm is well suited for reoptimization. Our algorithm generalizes a branch-and-bound approach for unconstrained convex quadratic integer programming proposed by Buchheim, Caprara and Lodi [4] to the presence of linear constraints. The main feature of the latter approach consists in a sophisticated preprocessing phase, leading to a fast enumeration of the branch-and-bound nodes. Experimental results for randomly generated instances are presented. The new approach significantly outperforms the MIQP solver of CPLEX 12.4 for instances with a small number of constraints.

1 Introduction

We consider integer optimization problems with strictly convex quadratic objective functions and linear constraints,

$$\min \quad f(x) = x^\top Q x + L^\top x + c$$
$$\text{s.t.} \quad Ax \le b \qquad\qquad\qquad\text{(QIP)}$$
$$x \in \mathbb{Z}^n,$$

where $Q \in \mathbb{R}^{n \times n}$ is positive definite, $L \in \mathbb{R}^n$, $c \in \mathbb{R}$, $A \in \mathbb{R}^{m \times n}$ and $b \in \mathbb{R}^m$. All approaches discussed in this paper can be generalized to the presence of additional linear equations in a straightforward way. Quadratic Integer Programming (QIP) is a special case of Integer Nonlinear Programming. While convex Quadratic Programming (QP) can be handled efficiently by the ellipsoid method [9], adding integrality makes the problem NP-hard. In fact, Problem (QIP) remains NP-hard even in the unconstrained case [13], or in the case where the objective function is linear. Up to now, very few publications consider exact solution methods for (QIP); a survey is given by Volkovich, Roshchin, and Sergienko [14]. However, numerical experiments by Fletcher and Leyffer [6] showed that branch-and-bound is the most effective approach out of all the common methods for solving (QIP), because the QP relaxation is very easy to

© Springer International Publishing Switzerland 2014
P. Fouilhoux et al. (Eds.): ISCO 2014, LNCS 8596, pp. 125–136, 2014.
DOI: 10.1007/978-3-319-09174-7_11

solve. Standard solvers that can handle (QIP) include CPLEX [1], SCIP [2] and BONMIN [3].

Many optimization problems in real world applications can be formulated as quadratic integer programs with few linear constraints, e.g., the classical Mean-Variance Optimization (MVO) for the selection of portfolios of assets [5].

1.1   Basic Ideas

The main idea of our approach is to use a fast branch-and-bound scheme, enumerating nodes very quickly. By fixing the branching order in advance, we lose the possibility of choosing sophisticated branching strategies but gain the advantage of shifting expensive computations into the preprocessing phase. Each node is explored using a specialized active set method to solve the continuous relaxation in order to determine the local lower bound. Another key ingredient of our algorithm is that we solve the dual problem of the continuous relaxation, which again is a convex QP, while we keep branching on the primal variables. Since all constraints of the continuous relaxation of (QIP) are affine, strong duality holds if the primal problem is feasible. Furthermore, the dual has only non-negativity constraints; this allows us to use the tailored active set method proposed by Kunisch and Rendl [10]. By considering the dual problem, it suffices to find an approximate solution, as each feasible solution yields a valid lower bound. We can thus prune the branch-and-bound node as soon as the current upper bound is exceeded by the lower bound and the current iterate is feasible in the active set process. Finally, we make use of warmstarts: after each fixing we use the optimal active set from the parent node as an initial guess for the optimal active set in the child nodes. In our algorithm, the overall running time per node is linear in the dimension n if the number of constraints is fixed, it can therefore be seen as an extension of the algorithm devised by Buchheim et al. [4].

1.2   Organization of the Paper

Section 2 presents an outline of the new algorithm, starting with some recapitulation of the main ideas of general active set methods for solving standard convex QPs. Section 2.2 discusses the advantages of considering the corresponding dual problem instead of the primal one. In Sect. 2.3, we explain the idea of reoptimization, using warmstarts within the branch-and-bound scheme. The next Sect. 2.4 deals with some tricks to speed up the algorithm by using incremental computations and preprocessing. In the last part of Sect. 2 the active set method of Kunisch and Rendl [10] is described. In Sect. 3 we present computational results and compare the performance of the proposed algorithm, applied to randomly generated instances, to the MIQP solver of CPLEX 12.4. To point out the benefits of the different ingredients of the algorithm, we also compared it to several versions of our implementation, which differ only in the consideration of the primal or dual relaxation and in the use of warmstarts.

2 Active Set Methods within Branch-and-Bound

For a better understanding of the branch-and-bound algorithm, we shortly summarize its key ingredients; some ideas are taken from [4] where the unconstrained case is addressed. Consider the convex quadratic integer program (QIP). As our branch-and-bound scheme aims at a fast enumeration of the nodes, we focus on bounds that can be computed quickly. A straightforward choice for determining lower bounds is to solve the continuous relaxation of (QIP). For this, we use an active set method presented by Kunisch and Rendl [10], shortly described in Sect. 2.5. Instead of applying it to the relaxation directly, we solve its dual formulation, which again is a convex QP. The solution can be used as a lower bound for f over all feasible integer points and is as strong as the lower bound obtained by solving the primal problem, since strong duality holds. As an initial upper bound, we simply compute an integer point which is feasible, i.e., some $x \in \mathbb{Z}^n$ satisfying $Ax \leq b$. This can be done by using a phase-1 approach, similar to the linear programming case. Note that this poor upper bound gets improved very quickly, because we use a depth-first branching strategy.

In our branch-and-bound scheme, we branch by fixing the *primal* variables in increasing distance to their values in the solution of the continuous relaxation. For a scalar $a \in \mathbb{R}$, let $\lfloor a \rceil$ denote the integer value closest to a. Then by exploiting the convexity of f and its symmetry with respect to the continuous minimizer $\bar{x}$, all consecutive lower bounds obtained by fixing x_i to integer values in increasing distance to $\lfloor x_i \rceil$ are non-decreasing. Thus, we can cut off the current node of the tree and all its siblings as soon as we fix a variable to some value for which the resulting lower bound exceeds the current best known upper bound. Using these ingredients we get a straightforward branch-and-bound algorithm, sketched in Algorithm 1. Using a predetermined branching order, some of the expensive calculations can be moved into a preprocessing phase, as described in Sect. 2.4.

2.1 Basic Active Set Strategy for Quadratic Programming

The continuous relaxation of (QIP) is the QP

$$
\begin{aligned}
\min \quad & f(x) = x^\top Q x + L^\top x + c \\
\text{s.t.} \quad & Ax \leq b \\
& x \in \mathbb{R}^n.
\end{aligned}
\tag{QP}
$$

Essentially two classes of algorithmic approaches have been developed for solving convex quadratic programs of the form (QP), namely active set methods (ASM) [7] and interior point methods (IPM) [8,11]. For a detailed comparison, we refer to [12]. In practice, active set methods are preferable to interior point methods if the number of constraints is rather small, so we will focus on ASM in the following.

Algorithm 1. Basic Branch-and-Bound Scheme

---

input : a strictly convex function $f : \mathbb{R}^n \to \mathbb{R}, x^\top Q x + L^\top x + c, A \in \mathbb{R}^{m \times n}, b \in \mathbb{R}^m$
output: integer minimizer x^* of f such that $Ax \leq b$

determine a variable order $x_1, \ldots, x_n$; set $d := 0, ub := \infty$;
while $d \geq 0$ **do**

> define $\bar{f} : \mathbb{R}^{n-d} \to \mathbb{R}$ by $\bar{f}(x) := f(r_1, \ldots, r_d, x_1, \ldots, x_{n-d})$;
> compute $\bar{L}$ and $\bar{c}$ such that $\bar{f}(x) = x^\top Q_d x + \bar{L}^\top x + \bar{c}$;
> // compute lower bound
> construct and solve the dual problem (2)
> reconstruct $\bar{x}$ from (1) and set lb $:= \bar{f}(\bar{x})$;
> // compute upper bound and update solution
> set $r_j := \lfloor \bar{x}_{j-d} \rfloor$ for $j = d+1, \ldots, n$ to form $r \in \mathbb{Z}^n$;
> **if** $Ar \leq b$ **and** $\bar{f}(r_{d+1}, \ldots, r_n) < ub$ **then**
>
> > set $r^* = r$;
> > set $ub = \bar{f}(r_{d+1}, \ldots, r_n)$;
>
> // prepare next node
> **if** $lb < ub$ **then**
>
> > // branch on variable x_{d+1}
> > set $d := d + 1$;
> > set $r_d := \lfloor \bar{x}_1 \rfloor$;
>
> **else**
>
> > // prune current node
> > set $d := d - 1$;
> > **if** $d > 0$ **then**
> >
> > > // go to next node
> > > increment r_d by increasing distance to the continuous relaxation;

The main idea of active set methods is to solve a sequence of equality constrained quadratic programs by just considering the active constraints at the current iterate. The active and inactive set for some iterate x^k are denoted by

$$A(x^k) := \{i \in \{1, \ldots, m\} \mid a_i^\top x^k = b_i\} \quad \text{and} \quad I(x^k) := \{1, \ldots, m\} \setminus A(x^k).$$

For any $A \subseteq \{1, \ldots, n\}$ and $x \in \mathbb{R}^n$ we write x_A for the components of x indexed by A, i.e. $x_A := (x_i)_{i \in A}$. For a matrix Q and two index sets $A, I \subseteq \{1, \ldots, n\}$, $Q_{A,I}$ denotes the submatrix of Q consisting of the rows indexed by A and columns indexed by I. The basic idea of general ASM for convex QP relies on the following observation: if x^* is the minimizer of (QP), we have

$$x^* = \operatorname{argmin} q(x) \text{ s.t. } Ax \leq b$$
$$= \operatorname{argmin} q(x) \text{ s.t. } a_i^\top x = b_i \ \forall i \in A(x^*).$$

The approach starts by choosing a subset $W_k \subseteq \{1, \ldots, m\}$ and computes

$$x_{k+1} = \operatorname{argmin} q(x) \text{ s.t. } a_i^\top x = b_i \ \forall i \in W_k.$$

If the new iterate x_{k+1} is not the minimizer of (QP), the set W_k is replaced by a new guess W_{k+1} essentially using KKT-conditions and the updated equality constrained QP is solved again. This iterative process is repeated until an optimal solution to (QP) is obtained.

It is common to distinguish between primal and dual ASM. While a primal ASM ensures primal feasibility of all iterates, a dual ASM ensures dual feasibility.

2.2 Dual Approach

In the following, we derive the dual problem of (QP) and point out some advantages when using the dual approach in the branch-and-bound framework. The dual can be computed by first forming the Lagrangian

$$\mathscr{L}(x, \lambda) = x^\top Q x + L^\top x + c + \lambda^\top (Ax - b)$$

and then, for fixed λ, minimizing $\mathscr{L}$ with respect to the primal variables x. As Q is assumed to be positive definite, the unique minimizer can be computed from the first order optimality condition

$$\nabla_x \mathscr{L}(x, \lambda) = 2Qx + L + A^\top \lambda = 0 \iff x = -\frac{1}{2} Q^{-1}(L + A^\top \lambda). \qquad (1)$$

Inserting this into the Lagrangian $\mathscr{L}$ yields the following dual function

$$\mathscr{L}(\lambda) = \lambda^\top \left(-\frac{1}{4} A Q^{-1} A^\top \right) \lambda - \left(b^\top + \frac{1}{2} L^\top Q^{-1} A^\top \right) \lambda - \frac{1}{4} L^\top Q^{-1} L + c.$$

Defining $\widetilde{Q} := \frac{1}{4} A Q^{-1} A^\top$, $\widetilde{L} := \frac{1}{2} A Q^{-1} L + b$ and $\tilde{c} := \frac{1}{4} L^\top Q^{-1} L - c$, we can thus write the dual of (QP) as

$$
\begin{aligned}
-\min \quad & \lambda^\top \widetilde{Q} \lambda + \widetilde{L}^\top \lambda + \tilde{c} \\
\text{s.t.} \quad & \lambda \geq 0 \\
& \lambda \in \mathbb{R}^m.
\end{aligned}
\qquad (2)
$$

Note that (2) is again a convex QP, since $\widetilde{Q}$ is positive semidefinite. The first crucial difference in considering the dual problem is that its dimension changed from n to m, which is beneficial if $m \ll n$. The second one is that $\lambda = 0$ is always feasible for (2). Finally, note that having the optimal solution $\lambda^* \in \mathbb{R}^m$ of (2), it is easy to reconstruct the corresponding optimal primal solution $x^* \in \mathbb{R}^n$ using the first order optimality condition (1).

Within a branch-and-bound framework, a special feature of the dual approach is that we can stop the iteration process and prune the node as soon as the current iterate λ_k is feasible and its objective function value exceeds the current upper bound, since each dual feasible solution yields a valid bound. As feasibility in the dual problem is equivalent to non-negativity of the variables, this can be checked quickly. The additional time spent to evaluate the objective function for bound comparison is little when m is small. Note however that, in case we cannot prune, an optimal solution of the dual problem is required, since it is needed for the computation of the corresponding primal solution x^* which in turn is needed to decide the enumeration order in the branch-and-bound scheme.

2.3 Reoptimization

At every node of the branch-and-bound tree, we use an active set algorithm for solving (2). A crucial advantage of ASM is the possibility of using warmstarts, i.e., of passing information on the optimal active set from a parent node to its children. In the dual approach the dimension of the subproblems is always m, independently of the depth d in the branch-and-bound tree. When fixing a variable, only the objective function changes, given by the adapted data $\widetilde{Q}$, $\widetilde{L}$ and $\widetilde{c}$. So instead of choosing $A = \{1, \ldots, m\}$ as the starting guess in a child node, we start with $A = A(\lambda^*)$, i.e., we use the active set of the optimal solution λ^* of the parent node.

This approach also works in the primal setting. However, as our experimental results show, it seems that warmstarting is only effective for the dual approach, since maintaining primal feasibility requires solving an LP at each branch-and-bound node for finding some starting point x satisfying

$$a_i^\top x = b_i \ \forall i \in A(x^*)$$
$$a_i^\top x \leq b_i \ \forall i \in I(x^*),$$

where x^* denotes the optimal solution of the primal QP in the parent node. On the other hand, in the dual approach, any feasible solution λ^* of the parent node remains feasible for the child nodes and it corresponds to the same active set.

2.4   Incremental Computations and Preprocessing

A remarkable speed-up can be achieved by exploiting the fact that the subproblems enumerated in the branch-and-bound tree are closely related to each other. Let $d \in \{0, \ldots, n-1\}$ be the current depth in the branch-and-bound tree and recall that after fixing the first d variables, the problem reduces to the minimization of

$$\bar{f} \colon \mathbb{Z}^{n-d} \to \mathbb{R}, \ x \mapsto x^\top Q_d x + \bar{L}^\top x + \bar{c}$$

over the feasible region $\bar{\mathcal{F}} = \{x \in \mathbb{Z}^{n-d} \mid A_d x \leq \bar{b}\}$, where $Q_d \succ 0$ is obtained by deleting the corresponding d rows and columns of Q and $\bar{L}$ and $\bar{c}$ are adapted properly; see [4]. Similarly, A_d is obtained by deleting the corresponding d columns of A and the reduced right hand side $\bar{b}$ is updated according to the current fixing. In particular, since we predetermine the order in which variables are fixed, the reduced matrices Q_d and A_d only depend on the depth d, but not on the specific fixings. Similarly to the reduced matrix Q_d, the quadratic part of the reduced dual objective function $\widetilde{Q}_d$ and its inverse $\widetilde{Q}_d^{-1}$ can also be computed in the preprocessing phase, because they only depend on Q_d and A_d.

Concerning the linear part $\widetilde{L}$ and the constant part $\widetilde{c}$ of the dual reduced problem, it is possible to compute both of them incrementally in linear time per node: let $r = (r_1, \ldots, r_d) \in \mathbb{Z}^d$ be the current fixing at depth d. By definition of $\widetilde{L}$, we have

$$\widetilde{L}(r) = \frac{1}{2} A_d Q_d^{-1} L(r) + b(r),$$

where the suffix (r) always denotes the corresponding data after fixing the first d variables to r.

Theorem 1. *After a polynomial time preprocessing, the vector $\widetilde{L}(r)$ can be constructed incrementally in $O(n - d + m)$ time per node.*

Proof. Defining $y(r) := -\frac{1}{2}Q_d^{-1}L(r)$, we have

$$\frac{1}{2}A_d Q_d^{-1}L(r) = -A_d \cdot y(r).$$

Note that $y(r)$ is the unconstrained continuous minimizer of $f(r)$. In [4], it was shown that $y(r)$ can be computed incrementally by

$$y(r) := [y(r') + \alpha z^{d-1}]_{1,\ldots,n-d} \in \mathbb{R}^{n-d}$$

for some vector $z^{d-1} \in \mathbb{R}^{n-d+1}$ and $\alpha := r_d - y(r')_d \in \mathbb{R}$, where $r' = (r_1, \ldots, r_{d-1})$ is the fixing at the parent node. This is due to the observation that the continuous minima according to all possible fixings of the next variable lie on a line, for which z^{d-1} is the direction. The vectors z^{d-1} only depend on the depth d and can be computed in the preprocessing [4]. Updating y thus takes $O(n - d)$ time. We now have

$$\widetilde{L}(r) = -A_d[y(r') + \alpha z^{d-1}]_{1,\ldots,n-d} + b(r)$$
$$= -A_d[y(r')]_{1,\ldots,n-d} - \alpha A_d[z^{d-1}]_{1,\ldots,n-d} + b(r)$$
$$= -(A_{d-1}y(r') - y(r')_{n-d+1} \cdot A_{\bullet,n-d+1}) - \alpha A_d[z^{d-1}]_{1,\ldots,n-d} + b(r).$$

In the last equation, we used the fact that the first part of the computation can be taken over from the parent node by subtracting column $n - d + 1$ of A, scaled by the last component of $y(r')$, from $A_{d-1}y(r')$, which takes $O(m)$ time. The second part $A_d[z^{d-1}]_{1,\ldots,n-d}$ can again be computed in the preprocessing. The result then follows from the fact that also $b(r)$ can easily be computed incrementally from $b(r')$ in $O(m)$ time.   □

Lemma 1. *After a polynomial time preprocessing, the scalar $\tilde{c}(r)$ can be constructed incrementally in $O(n - d)$ time per node.*

Proof. Recalling that

$$\tilde{c}(r) = \frac{1}{4}L(r)^\top Q_d^{-1}L(r) - c(r),$$

this follows from the fact that $y(r) = -\frac{1}{2}Q_d^{-1}L(r)$ and $c(r)$ can be computed in $O(n - d)$ time per node [4].   □

Corollary 1. *After a polynomial time preprocessing, the dual problem (2) can be constructed in $O(n - d + m)$ time per node.*

Besides the effort for solving the QP with the active set method, computing the optimal solution of the primal problem from the dual solution is the most time consuming task in each node. The following observation is used to speed up its computation.

Theorem 2. *After a polynomial time preprocessing, the optimal primal solution $x^*(r)$ can be computed from the optimal dual solution $\lambda^*(r)$ in $O(m \cdot (n-d))$ time per node.*

Proof. From (1) we derive

$$
x^*(r) = -\frac{1}{2}Q_d^{-1}\left(\sum_{i=1}^{m}\lambda^*(r)_i a_i + L(r)\right)
$$

$$
= y(r) + \sum_{i=1}^{m}\lambda^*(r)_i\left(-\frac{1}{2}Q_d^{-1}a_i\right).
$$

The first part can again be computed incrementally in $O(n-d)$ time per node. For the second part, we observe that $-\frac{1}{2}Q_d^{-1}a_i$ can be computed in the preprocessing phase for all $i = 1, \dots, m$. ☐

The above results show that the total running time per node is linear in $n-d$ when the number m of constraints is considered a constant: the construction of the dual problem takes $O(n-d+m)$ time, the running time for solving the dual problem is constant as its dimension is m, and the optimal primal solution can be obtained from the dual solution in $O(m \cdot (n-d))$ time.

2.5   Infeasible Primal-Dual Active Set Method

The ASM we discuss in this section is an infeasible primal-dual active set method, meaning that given a primal-dual pair $(x, s) \in \mathbb{R}^n \times \mathbb{R}^m$, neither primal nor dual feasibility is required during the iterations but only the first order optimality condition and the complementary slackness conditions have to be satisfied in the KKT system. The active set is adjusted until (x, s) is feasible. Kunisch and Rendl [10] have derived sufficient conditions for the algorithm to stop in a finite number of steps with an optimal pair (x^*, s^*). From their computational experiments, they concluded that their approach requires only a few (less than 10) iterations on average to find the optimal solution.

The main advantage of this algorithm is the opportunity to change more than one active/inactive constraint per iteration to quickly reach the optimal active set, while standard active set methods usually just add or delete one single constraint from the current active set.

It is well known that the KKT system corresponding to (2) is

$$
2\widetilde{Q}\lambda + \widetilde{L} - s = 0 \tag{3}
$$

$$
-s_i\lambda_i = 0 \ \forall i = 1, \dots, m \tag{4}
$$

$$
-\lambda \leq 0 \tag{5}
$$

$$
s \geq 0 \tag{6}
$$

and a pair (λ^*, s^*) is a solution of that system if and only if λ^* is a global minimizer of (2). The active set method of [10] starts by relaxing the constraints

(5) and (6) and choosing $\lambda_A = 0$ and $s_I = 0$. This a valid choice for the non-linear Eq. (4), so that a system of linear equations remains. By partitioning (3) according to the active and inactive variables, we get

$$\begin{pmatrix} 2\widetilde{Q}_{A,A} & 2\widetilde{Q}_{A,I} \\ 2\widetilde{Q}_{I,A} & 2\widetilde{Q}_{I,I} \end{pmatrix} \begin{pmatrix} \lambda_A \\ \lambda_I \end{pmatrix} + \begin{pmatrix} \widetilde{L}_A \\ \widetilde{L}_I \end{pmatrix} + \begin{pmatrix} -s_A \\ -s_I \end{pmatrix} = 0. \tag{7}$$

Then we solve this system for λ_I and s_A, yielding

$$\lambda_I = -\frac{1}{2}\widetilde{Q}_{I,I}^{-1}\widetilde{L}_I \tag{8}$$

and

$$s_A = \widetilde{L}_A + 2\widetilde{Q}_{A,I}\lambda_I.$$

Before resolving the updated reduced system (7), we compute a new guess for the active set A^+ as follows:

$$A^+ = \{i \in \{1,\ldots,m\} \mid (\lambda_i < 0) \vee (s_i > 0)\}.$$

The choice of A^+ can be justified by the following reasoning. On the one hand, if $s_i > 0$, the solution is dual feasible, so our previous choice $i \in A$ is confirmed and we keep the index in the subsequent active set. On the other hand, if $\lambda_i < 0$, the solution is primal infeasible and we add the index to our subsequent active set. For the starting guess of A, we can choose $A = \{1,\ldots,m\}$, which gives $\lambda = 0$ and $s = \widetilde{L}$. Alternatively, we could use $A = \emptyset$, such that $s = 0$ and $\lambda = -\frac{1}{2}\widetilde{Q}^{-1}\widetilde{L}$, or simply choose A randomly.

As the starting guess A can be chosen arbitrarily, this method is very well-suited for reoptimization within a branch-and-bound scheme. Even in a primal approach, the active set of the parent node can be used to initialize the active set method in the children nodes. However, this approach only allows non-negativity constraints, so that we can only apply it to our dual problem (2). It cannot be adapted directly to QPs with linear equations $Bx = d$, since the resulting system (7) may not be solvable, even if B has full rank. In particular, modeling general linear inequalities by slack variables is also not possible. Nevertheless, if (QIP) contains also linear equations, the dual of its continuous relaxation is again a QP of the form (2) with an additional vector of unrestricted multipliers, preserving the simple structure of the dual problem.

Note that in a single iteration of this algorithm the main computational effort consists in the solution of a system of linear equations of dimension $|I|$ for computing λ_I according to (8). For a more detailed description of the algorithm and convergence results, see [10].

3   Experimental Results

To show the potential of our algorithm, called CQIP, we implemented it in C++ and compared it to CPLEX 12.4 [1] and four other algorithms based on ASM:

(P) The branch-and-bound Algorithm 1 with a standard ASM for solving the QP relaxations

(P+w) as (P) but with reoptimization, as described in Sect. 2.3

(D) as (P) but considering the *dual* problems of the QP relaxations, as described in Sect. 2.2

(D+w) as (D) but with reoptimization

For both (P) and (D), we implemented a standard ASM; see e.g. [12]. Whenever applicable, the incremental computations described in Sect. 2.4 have been used in all approaches. The difference between (D+w) and (CQIP) consists in the fact that (CQIP) uses the infeasible ASM discussed in Sect. 2.5. All experiments were carried out on Intel Xeon processors running at 2.60 GHz.

We randomly generated 200 instances for Problem (QIP), namely 10 for each combination of $n \in \{50, 55, 60, 65, 70\}$ and $m \in \{1, 5, 10, 15\}$. For generating the positive definite matrix Q, we chose n eigenvalues λ_i uniformly at random from $[0, 1]$ and orthonormalized n random vectors v_i, each chosen uniformly at random from $[-1, 1]^n$, then we set $Q = \sum_{i=1}^{n} \lambda_i v_i v_i^\top$. All entries of L, b, and A were chosen uniformly at random from $[-1, 1]$, moreover we set $c = 0$. The time-limit for each instance was set to three cpu-hours for each algorithm and we used an absolute optimality tolerance of 10^{-6}. The results for these instances can be found in Table 1. Running times are measured in cpu seconds.

From the results we can see that instances with up to five linear inequalities and up to 65 variables can be effectively solved to optimality by the dual approach using the tailored active set method, while the simple primal approach as well as CPLEX both suffer from the increasing number of variables. As expected, for larger number of inequalities the running times of our algorithm increase rapidly, it is generally outperformed by CPLEX for $m \geq 10$. In Table 1, we left out the results for the algorithm (P+w), i.e. the primal approach with a standard active set method including reoptimization, because there was no improvement in running times compared to (P), see Sect. 2.3.

By comparing the running times of (P), (D), (D+w), and (CQIP), one can observe the stepwise improvement obtained by considering the dual problem, by reoptimization, and by using the special active set method of Kunisch and Rendl. It is also worth to mention that the average number of systems of linear equations needed to be solved at each node stays very small, independently of n and m. Again, a decrease in the number of such systems to be solved is observed in each of the improvement steps.

Note that the average number of branch-and-bound nodes in our dual approach is approximately 30 times greater than that needed by CPLEX. Nevertheless the overall running times of our approach are much faster for moderate sizes of m, emphasizing both the quick enumeration process within the branch-and-bound tree and the benefit of using reoptimization.

Table 1. Numbers of instances solved within the time limit, average running times, average numbers of linear equation systems per node solved and average numbers of branch-and-bound nodes on randomly generated instances of type (QIP). All averages are taken over the set of instances solved within the time limit.

| inst | | CQIP | | | | Active Set Based Methods | | | | | | | | | | | | CPLEX 12.4 | | |
| | | | | | | P | | | | D | | | | D+w | | | | | | |
n	m	#	time	sle/node	nodes	#	time	sle/node	nodes	#	time	sle/node	nodes	#	time	sle/node	nodes	#	time	nodes
50	1	10	9.98	0.50	9.27e+6	10	926.58	2.98	9.27e+6	10	43.55	1.45	9.27e+6	10	29.25	1.05	9.27e+6	10	55.17	5.54e+5
50	5	10	50.05	1.37	1.57e+7	10	3288.73	5.44	1.57e+7	10	427.25	4.66	1.57e+7	10	158.03	1.67	1.57e+7	10	92.15	8.66e+5
50	10	10	238.54	1.72	3.16e+7	7	5050.32	9.73	1.17e+7	9	2165.73	8.85	1.87e+7	10	871.08	2.63	3.16e+7	10	170.10	1.47e+6
50	15	10	988.98	2.06	6.70e+7	1	4822.78	13.64	6.61e+6	2	5131.80	11.34	1.64e+7	9	4455.05	3.98	6.70e+7	10	504.43	3.78e+6
55	1	10	38.03	0.60	3.63e+7	9	4353.64	2.58	3.41e+7	10	166.54	1.62	3.63e+7	10	113.78	1.02	3.63e+7	10	160.70	1.40e+6
55	5	10	147.91	1.17	5.67e+7	1	9869.31	6.37	3.24e+7	10	1363.31	3.86	5.67e+7	10	497.70	1.32	5.67e+7	10	289.76	2.34e+6
55	10	10	1243.21	1.53	1.46e+8	1	8289.16	8.61	1.50e+7	6	4813.29	6.91	5.14e+7	9	2150.02	2.02	8.98e+7	10	1123.09	7.89e+6
55	15	9	3702.92	1.72	3.00e+8	0	-	-	-	0	-	-	-	3	8357.10	2.15	1.54e+8	9	2691.91	1.70e+7
60	1	10	79.86	0.21	9.65e+7	5	7642.13	3.22	4.04e+7	10	5514.38	6.93	9.65e+7	10	319.31	1.04	9.65e+7	10	431.75	3.22e+6
60	5	10	1078.43	1.21	3.26e+8	0	-	-	-	6	3058.93	4.02	1.21e+8	10	2689.96	1.35	3.26e+8	10	1581.32	1.10e+7
60	10	7	3786.00	1.62	4.83e+8	0	-	-	-	1	5514.38	6.93	6.07e+7	4	7433.36	2.43	2.70e+8	8	4036.91	2.47e+7
60	15	4	6806.86	1.77	5.69e+8	0	-	-	-	0	-	-	-	0	-	-	-	8	4738.51	2.62e+7
65	1	10	455.82	0.41	4.16e+8	2	7016.44	2.38	3.90e+7	10	1857.82	1.35	4.16e+8	10	1384.08	1.05	4.16e+8	10	1830.83	1.17e+7
65	5	10	965.18	1.27	3.62e+8	0	-	-	-	7	4805.10	3.84	2.29e+8	10	3470.16	1.51	3.62e+8	9	1480.94	9.27e+6
65	10	6	5834.21	1.46	1.08e+9	0	-	-	-	0	-	-	-	0	-	-	-	4	5502.69	3.02e+7
65	15	4	4600.28	1.54	6.28e+8	0	-	-	-	0	-	-	-	0	-	-	-	5	6261.84	3.19e+7
70	1	10	1330.68	0.50	1.29e+9	0	-	-	-	8	8981.48	1.48	8.97e+8	9	3455.93	1.03	1.05e+9	8	5114.02	2.84e+7
70	5	6	5602.57	1.32	1.75e+9	0	-	-	-	0	-	-	-	1	10101.19	1.43	1.04e+9	4	7531.69	4.09e+7
70	10	1	10391.83	1.24	2.92e+9	0	-	-	-	0	-	-	-	0	-	-	-	1	9709.13	6.11e+7
70	15	0	-	-	-	0	-	-	-	0	-	-	-	0	-	-	-	0	-	-

References

1. IBM ILOG CPLEX Optimizer 12.4 (2013). http://www-01.ibm.com/software/integration/optimization/cplex-optimizer/
2. Achterberg, T.: SCIP: solving constraint integer programs. Math. Program. Comput. **1**(1), 1–41 (2009)
3. Bonami, P., Biegler, L.T., Conn, A.R., Cornuéjols, G., Grossmann, I.E., Laird, C.D., Lee, J., Lodi, A., Margot, F., Sawaya, N., Wächter, A.: An algorithmic framework for convex mixed integer nonlinear programs. Discrete Optim. **5**(2), 186–204 (2008)
4. Buchheim, C., Caprara, A., Lodi, A.: An effective branch-and-bound algorithm for convex quadratic integer programming. Math. Program. **135**, 369–395 (2012)
5. Cornuéjols, G., Tütüncü, R.: Optimization Methods in Finance. Mathematics, Finance, and Risk. Cambridge University Press, Cambridge (2006)
6. Fletcher, R., Leyffer, S.: Numerical experience with lower bounds for MIQP branch-and-bound. SIAM J. Optim. **8**(2), 604–616 (1998)
7. Goldfarb, D., Idnani, A.: A numerically stable dual method for solving strictly convex quadratic programs. Math. Program. **27**(1), 1–33 (1983)
8. Goldfarb, D., Liu, S.: An $O(n^3 L)$ primal interior point algorithm for convex quadratic programming. Math. Program. **49**(1–3), 325–340 (1990)
9. Kozlov, M.K., Tarasov, S.P., Khachiyan, L.G.: The polynomial solvability of convex quadratic programming. USSR Comput. Math. Math. Phys. **20**(5), 223–228 (1980)
10. Kunisch, K., Rendl, F.: An infeasible active set method for quadratic problems with simple bounds. SIAM J. Optim. **14**(1), 35–52 (2003)
11. Mehrotra, S., Sun, J.: An algorithm for convex quadratic programming that requires $O(n^{3.5} L)$ arithmetic operations. Math. Oper. Res. **15**(2), 342–363 (1990)
12. Nocedal, J., Wright, S.J.: Numerical Optimization. Springer, New York (2006)
13. Van Emde Boas, P.: Another NP-complete problem and the complexity of computing short vectors in a lattice. Technical report 81–04, Department of Mathematics, University of Amsterdam (1981)
14. Volkovich, O.V., Roshchin, V.A., Sergienko, I.V.: Models and methods of solution of quadratic integer programming problems. Cybern. **23**, 289–305 (1987)

Fixed-Parameter Algorithms for Scaffold Filling

Laurent Bulteau[1], Anna Paola Carrieri[2], and Riccardo Dondi[3]([✉])

[1] Department of Software Engineering and Theoretical Computer Science,
Technische Universität Berlin, Berlin, Germany
l.bulteau@gmail.com

[2] Dipartimento di Informatica, Sistemistica e Comunicazione,
Università degli Studi di Milano-Bicocca, Milano, Italy
annapaola.carrieri@disco.unimib.it

[3] Dipartimento di Scienze Umane e Sociali, Università degli Studi di Bergamo,
Bergamo, Italy
riccardo.dondi@unibg.it

Abstract. In this paper we consider two combinatorial problems related to genome comparison. The two problems, starting from possibly incomplete genomes produced from sequencing data, aim to reconstruct the complete genomes by inserting a collection of missing genes. More precisely, in the first problem, called *One-sided scaffold filling*, we are given an incomplete genome B and a complete genome A, and we look for the insertion of missing genes into B with the goal of maximizing the common adjacencies between the resulting genome B' and A. In the second problem, called *Two-sided scaffold filling*, we are given two incomplete genomes A, B, and we look for the insertion of missing genes into both genomes so that the resulting genomes A' and B' have the same multi-set of genes, with the goal of maximizing the common adjacencies between A' and B'. While both problems are known to be NP-hard, their parameterized complexity when parameterized by the number of common adjacencies of the resulting genomes is still open. In this paper, we settle this open problem and we present fixed-parameter algorithms for the *One-sided scaffold filling* problem and the *Two-sided scaffold filling* problem.

1 Introduction

Genome comparison is a fundamental problem in bioinformatics, and it aims to identify differences and similarities among genomes, with the goal of understanding their function and evolutionary history. In this context several interesting combinatorial problems have been introduced (see for example [10]).

The introduction of new sequencing techniques (Next Generation Sequencing technologies, NGS) has led to a huge increase of the amount of DNA/RNA and protein sequences available for genomic and trascriptomic analyses [4]. These high-throughput sequencing technologies produce millions of short DNA/RNA reads that are joined together into longer sequences by means of assembly algorithms.

© Springer International Publishing Switzerland 2014
P. Fouilhoux et al. (Eds.): ISCO 2014, LNCS 8596, pp. 137–148, 2014.
DOI: 10.1007/978-3-319-09174-7_12

However, due to limitations of the NGS technologies, the cost of finishing a genome is still high compared to the cost of sequencing, hence most of the released genomes are unfinished and incomplete [4].

The use of incomplete draft genomes (called *scaffolds*) in genomic analyses may introduce errors. Hence, a relevant combinatorial problem is to fill the scaffolds with missing genes in order to obtain complete genomes that are as similar as possible to a given reference genome. Recently in [14] it has been introduced the *One-sided scaffold filling problem* that consists of filling a scaffold B in order to obtain a complete genome B' such that the Double-Cut and Join (DCJ) distance [16] (the minimum number of allowed rearrangement operations transforming one genome into the other) between B' and the reference genome A is minimized. Moreover, Jiang et al. in [11] considered the *Two-sided scaffold filling problem*, where the second genome A (on which the comparison is based) is incomplete as well.

In this paper we consider a different similarity measure, that is *the maximum number of common adjacencies* between two genomes, which has been introduced for the One-sided/Two-sided scaffold filling problems in [5]. Both problems are NP-hard under this similarity measure [12]. However, it has been shown that both problems admit constant factor approximation algorithms. In [12] it has been given a factor $\frac{4}{3}$ approximation algorithm for the One-sided scaffold filling problem and a factor 2 approximation algorithm for the Two-sided scaffold filling problem. The former approximation factor has been recently improved in [13], where it has been presented an approximation algorithm of factor $\frac{5}{4}$ for the One-sided scaffold filling problem.

In this paper, we focus on the parameterized complexity of the two scaffold filling problems. Parameterized complexity aims to characterize the complexity of a problem with respect to interesting parameters, with the goal of understanding if the exponential explosion of an exact algorithm can be confined only to the considered parameters. For an introduction to parameterized complexity we refer the reader to [8,15].

A first step in the analysis of the parameterized complexity of the One-sided scaffold filling problem has started in [12]. The authors presented two Fixed Parameter Tractable (FPT) algorithms for two special cases of the One-sided scaffold filling problem. In the first case, the number k of common adjacencies between a filled genome B' and a reference genome A, and the maximal number d of occurrences of a gene are considered as parameters, and it is presented an FPT algorithm of time complexity $O((2d)^{2k}poly(|A||B|))$. In the second case, the authors consider, as parameters, the number k of common adjacencies between a filled genome B' and a reference genome A and the size c of the set of symbols (genes) and they give an FPT algorithm that runs in time $O(c^{2k}poly(|A||B|))$. However, the parameterized complexity of the One/Two-sided scaffold filling problems, when parameterized only by the maximum number of common adjacencies k, has been left open in [12].

Our Contribution. In this paper we present two FTP-algorithms for both problems, thus answering an open question in [12]. More precisely, we present an algorithm of time complexity $2^{O(k)}poly(|A||B|)$ for One-sided scaffold filling and an algorithm of time complexity $2^{O(k \cdot \log k)}poly(|A||B|)$ for Two-sided scaffold filling.

The rest of the paper is organized as follows. First, in Sect. 2 we introduce some preliminary definitions and we formally define the two combinatorial problems we are interested in. Then, in Sect. 3, we describe the FPT algorithm for the One-sided case, while in Sect. 4 we present the FPT algorithm for the Two-sided case. We conclude the paper with some open problems. Some of the proofs are omitted due to page limit.

2 Preliminaries

Let Σ be a non-empty finite set of symbols. An (unsigned) unichromosomal genome A is represented as a string over an alphabet Σ, where the symbols in A (that are genes) form a multiset $[A]$ on Σ. For example $A = abcdabcdaa$ with $\Sigma = \{a, b, c, d\}$ and $[A] = \{a, a, a, a, b, b, c, c, d, d\}$. We write $A[i]$ for the symbol of A in i-th position, and we write $A[i \ldots j]$ for the substring of A between positions i and j.

Given a string A, an *adjacency* is an unordered pair of consecutive elements of A ($A[i]A[i+1]$ or $A[i+1]A[i]$). A position i induces an adjacency ab, if either $A[i] = a$, $A[i+1] = b$ or $A[i] = b$ and $A[i+1] = a$. The endpoints of A, are the first and the last position of A.

We write $[\![A]\!]$ for the multi-set of adjacencies of A (i.e., if $A = abcdabcdaa$, then $[\![A]\!] = \{aa, ab, ab, ad, ad, bc, bc, cd, cd\}$).

In order to deal with endpoints of the two strings, we assume that given a string A, with $|A| = n$, $A[1] = A[n] = \natural$, where $\natural$ is a dummy symbol. The dummy symbols are not considered when defining the set of adjacencies $[\![A]\!]$, i.e. $\natural A[1]$ and $A[n]\natural$ are not in $[\![A]\!]$.

Comparing two strings A and B, we denote by $X = [A] \setminus [B]$ the multi-set of symbols of A missing in B, and by $Y = [B] \setminus [A]$ the multi-set of symbols of B missing in A. Given a multi-set of symbols $[G]$ over alphabet Σ, a *scaffold* is a string on $[G]$ with some missing elements.

The two scaffold filling problems we will deal with are based on the definition of *common adjacency* between two genomes (strings).

Definition 1. *Consider two strings A, B on alphabet Σ. The multi-set of common adjacencies between A, B is defined as $[\![A]\!] \cap [\![B]\!]$. A matching M of the adjacencies of A and the adjacencies of B is a relation between the positions of A and the positions of B such that:*

– *for each position i of A or B, there exists at most one pair in M containing i;*
– *for each position i of A and j of B, $(i, j) \in M$ if and only if position i and position j induces the same adjacency;*
– *each position that induces a common adjacency belongs to some pair of M.*

A position of A or B is *matched*, if it belongs to a pair of M. Informally, M relates the positions inducing common adjacencies.

Given a scaffold B and a multi-set of symbols X, a string B' is a *filling* of B with X if (1) $[B'] = [B] \cup X$, and (2) B is a subsequence of B' such that the first and last symbols of B' are respectively the first and last symbols of B.

In the following we give the definitions of the two Scaffold Filling problems (parameterized versions) investigated in this paper.

One-sided Scaffold Filling to Maximize the Number of common String Adjacencies (One-sided SF-MNSA)

Input: Two strings A and B, such that $[B] \subseteq [A]$.
Output: A filling B' of B with $X = [A] \setminus [B]$ such that A and B' have at least k common adjacencies.
Parameter: k.

Two-sided Scaffold Filling to Maximize the Number of common String Adjacencies (Two-sided SF-MNSA)

Input: Two strings A and B.
Output: A filling B' of B with $X = [A] \setminus [B]$ and a filling A' of A with $Y = [B] \setminus [A]$ such that A' and B' have at least k common adjacencies.
Parameter: k.

Notice that the restriction of Two-sided SF-MNSA with $Y = \emptyset$ is exactly the One-sided SF-MNSA problem.

Now, we discuss some properties that will be useful to design our FPT-algorithms. First, we present the following property for the parameter k, proved in [12].

Lemma 1 [12]. *Let A and B two strings on an alphabet Σ, $X = [A] \setminus [B]$, and $Y = [B] \setminus [A]$. Let k be the optimal number of common adjacencies for Two-sided SF-MNSA between two fillings A' and B'. Then $|X|, |Y| \leq k$.*

Notice that Lemma 1 holds also for One-sided SF-MNSA, that is when $Y = \emptyset$, it holds $|X| \leq k$.

Let A and B be two strings of symbols over an alphabet Σ, which are input of One-sided SF-MNSA or Two-sided SF-MNSA. Consider now the set AD of common adjacencies between A and B. Notice that we can assume that $|AD| < k$, otherwise we already know that One-sided SF-MNSA/Two-sided SF-MNSA admits a solution consisting of at least k common adjacencies. Now, we can compute a partition of AD into two subsets as follows:

- the set $AD_{pr} \subseteq AD$ of common adjacencies that are preserved after the filling of B and/or A;
- the set $AD_{br} \subseteq AD$ of common adjacencies that are broken by inserting symbols of $X = [A] \setminus [B]$ (of $Y = [B] \setminus [A]$ respectively) into B (into A respectively).

Then the following easy property holds.

Property 1. *Let A and B be two strings of symbols over alphabet Σ and let AD be the set of common adjacencies between A and B. Then, if there exists a solution for the One-sided SF-MNSA/Two-sided SFMNSA that partitions the set AD into the sets AD_{pr}, AD_{br}, we can compute the partition of AD into the two subsets AD_{pr} and AD_{br} in time $O(2^k)$.*

This property is implicitly used in the two fixed-parameter algorithms to guess which adjacencies of the set AD will be preserved, that is those adjacencies induced by positions where no insertion is possible when computing a filling of an input string. Hence, in what follows, we assume that when a string is inserted into A or B, then it is not inserted in a position associated with an adjacency in AD_{pr}.

Color Coding. The FPT-algorithms we present are mainly based on the *color-coding* technique and on the perfect family of hash functions [1]. Color-coding is a well-known technique for designing fixed-parameter algorithms, and it has been applied to several combinatorial problems, for example for the longest path problem [1], for the graph motif problem [2,6,7,9] and for problems on strings [3].

Informally, given a set U of size n, color-coding aims to identify a subset $S \subseteq U$ of size k by coloring the elements of U with k colors, so that each element in S is associated with a distinct color. While enumerating the subsets having size k of U takes time $O(n^k)$, by means of the coloring and using combinatorial properties of the problem, in some cases it is possible to compute whether a solution of size k exists in time $f(k)poly(n)$, thus leading to an FPT algorithm.

We now introduce the definition of a perfect family of hash functions, which are used to compute the coloring.

Definition 2. *Let I be a set, a family F of hash functions from I to $\{c_1, \ldots, c_k\}$ is called* perfect *if for any subset $I' \subseteq I$, with $|I'| = k$, there exists a function $f \in F$ which is injective on I'.*

A perfect family F of hash functions from I to $\{c_1, \ldots, c_k\}$, having size $O(\log |I| 2^{O(k)})$, can be constructed in time $O(2^{O(k)} |I| \log |I|)$ [1].

3   An FPT Algorithm for One-Sided SF-MNSA

In this section we present an FPT algorithm for One-sided SF-MNSA parameterized by k, the number of common adjacencies between the input string A and the filling B' of B with the multi-set X of symbols of A missing in B. Recall that, by Lemma 1, it holds $|X| \leq k$. Furthermore, we assume that we have already computed the subset AD_{pr} of AD (the common adjacencies of A, B) where no insertion is possible during the filling (see Property 1).

Let $C_A = \{c_1, \ldots, c_k\}$ be a set of *colors*. Consider a family F of perfect hash functions from the positions inducing the adjacencies of A in AD_{br} to colors in C_A. Informally, the coloring is used to identify a matching of the positions of A and the positions of B that induce new adjacencies due to the insertion of symbols in X.

In the following, we assume that the coloring of the positions of A is induced by some injective function $f \in F$. Given a string S, S is *colorful* for C_A if there exist $\{s_c \mid c \in C_A\} \subseteq [\![S]\!]$ such that for each $c \in C_A$ there is a position of A colored by c which induces the adjacency s_c. Our objective is thus to compute a filling of B colorful for C_A.

We first focus on inserting a set of elements at one given position of B. Given j a position in B, $X_j \subseteq X$ and $C_j \subseteq C_A$, define $\mathtt{Ins}_j (X_j, C_j)$ as follows:

$$\mathtt{Ins}_j (X_j, C_j) = \begin{cases} 1 \text{ if there exists a filling of } B[j-1, j] \text{ with } X_j \text{ which is} \\ \quad \text{colorful for } C_j, \\ 0 \text{ else.} \end{cases}$$

Using dynamic programming, $\mathtt{Ins}$ can be computed in time $O(2^{2k}k^2)$ yielding the following lemma.

Lemma 2. *Let $X_j \subseteq X$, $C_j \subseteq C_A$, such that $|X_j|, |C_j| \le k$ and j be an integer s.t. $j \le |B|$. Then we can compute $\mathtt{Ins}_j (X_j, C_j)$ in time $O(2^{2k}k^2)$.*

We now define a table $\mathtt{Fill}_j (X', C_A')$ computed by the following recurrence. The objective, as stated in Lemma 3, is to determine whether a prefix of B can be filled with any given subset of X so as to be colorful for any given subset of C_A.

Recurrence 1. *Let $X' \subseteq X$, $C_A' \subseteq C_A$.*

- *For $j = 1$, let $\mathtt{Fill}_1 (X', C_A') = \mathtt{Ins}_1 (X', C_A')$.*
- *For all $j \ge 2$, let:*

$$\mathtt{Fill}_j (X', C_A') = \max_{X_j \subseteq X', C_j \subseteq C_A'} \begin{cases} \mathtt{Fill}_{j-1} (X' \setminus X_j, C_A' \setminus C_j) \\ \wedge \mathtt{Ins}_j (X_j, C_j) \end{cases}$$

In the following, we prove that $\mathtt{Fill}_{|B|} (X, C_A)$ allows us to determine whether B admits a filling with k common adjacencies.

Lemma 3. *Let (A, B) be an instance of One-sided Scaffold Filling, $X = [A] \setminus [B]$, k be an integer, C_A be a set of k colors, and F be a perfect family of hash functions from the positions of A to C_A. Then the following propositions are equivalent:*
(i) There exists a filling B' of B with X such that A and B' have k common adjacencies;
(ii) There exists a coloring $f \in F$ for which $\mathtt{Fill}_{|B|} (X, C_A) = 1$.

Next, we show how the recurrence described in Recurrence 1 yields a dynamic programming algorithm to solve One-sided SF-MNSA.

Theorem 1. *Let A, B be two strings of symbols on an alphabet Σ and let $X = [A] \setminus [B]$ be the multiset of symbols missing in B. It is possible to compute a solution of One-sided SF-MNSA in time $O(2^{O(k)}poly(|A| + |B|))$.*

Proof. Recurrence 1 yields a dynamic programming algorithm: for each j from 1 to $n + 1$, compute the entry $\mathtt{Fill}_j (X', C_A')$ for each set $X' \subseteq X$ and $C_A' \subseteq C_A$. Then, by Lemma 3, there exists a filling B' of B creating $|C_A| = k$ common adjacencies if and only if $\mathtt{Fill}_{|B|} (X, C_A) = 1$.

Now, we consider the time complexity of the algorithm. Write $n = |A| + |B|$. First, a perfect family of hash functions that color-codes the positions of A

can be computed in time $2^{O(k)}poly(n)$. Once the family is computed, there are $2^{O(k)}\log(n)$ color codings to iterate through. For each color coding, the table $\text{Fill}_j(X', C'_A)$ is computed in time $O(2^{2k}k^2n)$ (see Lemma 2). Then the $O(2^{2k}n)$ entries of table $\text{Fill}_j(X', C'_A)$ are computed, where each entry requires $O(2^{2k})$ look-ups, depending on the choice of X_j and C_j. Thus Recurrence 2 requires $O(2^{4k}n)$ to compute table $\text{Fill}_j(X', C'_A)$. Finally, the overall complexity is indeed $2^{O(k)}poly(n)$. $\square$

4   An FPT Algorithm for Two-Sided SF-MNSA

In this section, we consider the Two-sided SF-MNSA problem and we give a fixed-parameter tractable algorithm for it. As for the One-sided case, the algorithm is based on color-coding and dynamic programming. However, new challenges appear which make the problem more complicated. First, there exist a new kind of common adjacencies: with adjacencies that are created in the fillings although they never appear as such in the input strings. Also, unlike the One-sided case, it is not known a priori whether a given adjacency may be used in a common adjacency or should be split to insert a substring. We deal with the first issue by bounding (and enumerating) the possible arrangements of such rare adjacencies, and with the second by introducing "insertion" colors, where corresponding adjacencies can only be used to insert a substring, not to create a common adjacency.

Given two strings A and B over alphabet Σ, denote by k the number of common adjacencies between two fillings A' and B' of A, B respectively. Let $X = [A] \setminus [B]$ be the multi-set of symbols of A missing in B and let $Y = [B] \setminus [A]$ be the multi-set of symbols of B missing in A, where $X, Y \neq \emptyset$ (otherwise the problem is equivalent to One-sided SF-MNSA) and $X \cap Y = \emptyset$ (by the definition of sets X and Y for the Two-sided SF-MNSA).

Recall that, by Lemma 1, the following property holds: $|X|, |Y| \leq k$. Furthermore, as in the previous section, we assume that we have already computed the subset AD_{pr} of $[A] \cap [B]$, that is those common adjacencies of A, B, that must be preserved during the filling (see Property 1).

Before giving the details of the FPT-algorithm, we present an (informal) overview. A filling B' (A' respectively) of B (of A respectively) consists of inserting substrings over alphabet X (over alphabet Y respectively) into B (into A respectively). In the first step, the algorithm "guesses" how these inserted strings are formed from X and Y (since $|X|, |Y| \leq k$, the number of cases to try depends only on a function of k, see Property 2).

We now identify two kinds of common adjacencies for two fillings A', B'. In the first kind, one adjacency appears already in $[A]$ or $[B]$: this case can be dealt with as in the one-sided algorithm. In the second kind, both adjacencies have been created during the filling, using one element from X in B' and one from Y in A'. They are called (X, Y)-adjacencies. Since $X \cap Y = \emptyset$, such adjacencies use *exactly* one element of X and Y, hence they consist of an endpoint of an inserted string as well as a letter already present in the original strings A and B.

The second step of the algorithm consists in identifying and matching the end-points of strings which correspond to such (X, Y)-adjacencies (see Definition 4 and Property 3).

In Step 3 the algorithm opportunely color-codes the positions of A and B in order to (i) match non (X, Y)-adjacencies (like in the previous algorithm), and (ii) identify the positions of A and B where an insertion is possible (we will show that the number of these positions is bounded by k in Property 2). This allows, in Step 4, to finally insert the strings into A and B by dynamic programming while creating the remaining adjacencies (see Recurrence 2).

We can now present the details of the algorithm.

Step 1: *Compute inserted strings.*

Let S_X and S_Y be the two multi-sets of strings over the multi-sets X and Y that have to be inserted in B and A respectively in an optimal solution. The algorithm simply iterates through all such pairs (S_X, S_Y) of multi-sets of strings over (X, Y): in some iteration, the correct pair (S_X, S_Y) is clearly considered. The following property bounds both the number of possible pairs (S_X, S_Y) and the number of positions where strings can be inserted in A and B.

Property 2. *Let X, Y be two multi-sets of symbols to be inserted into the strings B and A respectively. Then (1) the number of positions in each of A, B where a string of S_Y, S_X is inserted is bounded by k and (2) the number of possible multi-sets S_X and S_Y of strings over X, Y to be inserted into B and A respectively is bounded by $O(k^{2k})$.*

Step 2: *Identify (X, Y)-adjacencies.*

We first define formally the concept of (X, Y)-*adjacency* (see Fig. 1 for an example).

Definition 3. *Consider a filling B' of B with X and a filling A' of A with Y. A common adjacency $z \in [\![A']\!] \cup [\![B']\!]$ is an (X, Y)-adjacency if it is induced by positions i, j of A', B' respectively, and either $A'[i]$ or $A'[i+1]$ is the endpoint of an inserted string $s_x \in S_X$, and either $B'[j]$ or $B'[j+1]$ is the endpoint of an inserted string $s_y \in S_Y$.*

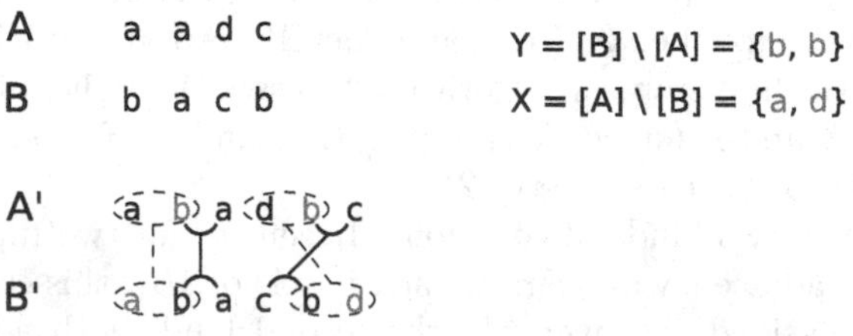

Fig. 1. An instance of the Two-sided SFMNA problem. Given two scaffolds A and B, we obtain the filled genomes A' and B' by inserting symbols X in B and Y in A (inserted symbols are in red). Lines connect common adjacencies, dotted lines connect (X,Y)-adjacencies (Color figure online).

Notice that, since $X \cap Y = \emptyset$, it follows that any new common adjacency of A' (of B' respectively) is either not involved in an insertion (hence, in one string, it is induced by a position where no string is inserted), or it is an (X, Y)-adjacency.

Now, the algorithm defines which endpoints of the strings in S_X, S_Y induce a common (X, Y)-adjacency. Denote by E_X (E_Y respectively), the set of endpoints of the strings in set S_X (in set S_Y respectively). We consider a procedure, called *number assignment*, that associates with each endpoint in E_X and E_Y a number which identifies the (X, Y)-adjacency, if any, which uses this endpoint.

Definition 4. *A number assignment for the strings in $S_X \cup S_Y$ is the data of an integer k' and of a function from $E_X \cup E_Y$ to $\{0, 1, \ldots, k'\}$, where each number $\{1, \ldots, k'\}$ is assigned to exactly one endpoint in E_X and one endpoint in E_Y.*

Consider a solution, a corresponding number assignment is obtained as follows. Let k' be the number of (X, Y)-adjacencies. Consider an endpoint $e_z \in E_X \cup E_Y$, then:

- Endpoint e_z is associated with 0 iff it is not involved in an (X, Y)-adjacency;
- Endpoint e_z is associated with a number $i \in \{1, \ldots, k'\}$ iff it is involved in the i-th (X, Y)-adjacency.

The set $E'_X \subseteq E_X$ ($E'_Y \subseteq E_Y$) denotes the set of endpoints of E_X (of E_Y respectively) associated with a positive number.

The following property gives an easy upper bound on the number of such assignments.

Property 3. *There are at most $(2k)^{k+1}$ number assignments.*

Hence, in what follows assume that the algorithm guesses the correct number assignment to $E_X \cup E_Y$. Now, we show how we can bound the possible symbols that are adjacent to an endpoint in $E'_X \cup E'_Y$. First, we introduce the following definition.

Definition 5. *Consider a string $s_x \in S_X$ ($s_y \in S_Y$ respectively). Let $e_x \in E'_X$ ($e_y \in E'_Y$ respectively) be an endpoint of s_x (of s_y respectively). Then, $v(e_x)$ ($v(e_y)$ respectively) is the symbol of Y (of X respectively) adjacent to e_x in B' (to e_y in A' respectively).*

Notice that the number assignment immediately defines the values $v(e_x)$, $v(e_y)$, for each $e_x \in E'_X$, $e_y \in E'_Y$. Indeed, if $e_x \in E'_X$ and $e_y \in E'_Y$ are associated with the same number i, then $v(e_x)$ must be the symbol contained in $s_y[e_y]$, while $v(e_y)$ must be the symbol contained in $s_x[e_x]$.

Remark 1. *A number assignment uniquely determines the value $v(e_z)$ for $e_z \in E'_X \cup E'_Y$.*

Using this value, the algorithm creates the following table which tells whether or not, according to (X, Y)-adjacencies, a string can be inserted at a certain position. Let Z be an input string among A, B, $s \in S_Z$, and $j \in \{1, \ldots, |Z|\}$. Write s_l and s_r for the left and right endpoints of s respectively:

$$\texttt{XY} - \texttt{Fits}_j (Z, s) = \begin{cases} 0 & \text{if } (s_l \in E'_Z \text{ and } Z[j-1] \neq v(s_l)) \\ & \text{or } (s_r \in E'_Z \text{ and } Z[j] \neq v(s_r)) \\ 1 & \text{otherwise.} \end{cases}$$

Step 3: *Color-code the positions in A and B.*

We are now able to define the color-coding of the positions of A and B. Consider a coloring f of the positions of A and B with a set C of z, $z \leq 2k$, colors. Moreover, we partition C into disjoint subsets $C_{M,A}$, $C_{M,B}$, $C_{I,A}$, $C_{I,B}$ defined as follows:

- Let $C_{M,B}$ ($C_{M,A}$ respectively) be a set of colors associated with positions of B (of A respectively) that matches positions of A' (of B' respectively). Notice that in a position colored by $C_{M,A}$ ($C_{M,B}$ respectively) a string of S_X (of S_Y respectively) cannot be inserted.
- Let $C_{I,B}$ ($C_{I,A}$ respectively) be a set of colors assigned to positions in B (in A respectively) where insertions of strings of S_X (of S_Y respectively) are allowed.

Note that, since S_X, S_Y and the number assignment with k' (X,Y)-adjacencies are fixed, we only consider partitions where $|C_{I,A}| = |S_Y|$, $|C_{I,B}| = |S_X|$, and $|C_{M,A}| + |C_{M,B}| + k' = k$.

There are k values of z to test. For each z, there are $O(2^{O(z)} \log n)$ colorings [1], and for each coloring, 4^z ways of partitioning C into $C_{M,A}$, $C_{M,B}$, $C_{I,A}$, $C_{I,B}$. Overall, there are thus $O(2^{O(k)} \log n)$ cases to consider.

Step 4: *Insert strings by dynamic programming.*

Now, we can define the dynamic programming recurrence. Similarly to the One-sided case, we define $\texttt{Ins}_{Z,j} (s, C_{M,j})$, where $C'_{M,j} \subseteq C_{M,W}$ and W, Z are different strings of $\{A, B\}$, as follows:

$$\texttt{Ins}_{Z,j} (s, C_{M,j}) = \begin{cases} 1 & \text{if } \texttt{XY} - \texttt{Fits}_j (Z, s) = 1 \text{ and the string } Z[j-1]sZ[j] \\ & \text{is colorful for } C_{M,j}, \\ 0 & \text{otherwise.} \end{cases}$$

Similary to Property 2, any entry $\texttt{Ins}_{Z,j} (s, C'_{M,W})$ can be computed in time $O(2^{2k}n)$.

Lemma 4. *Let $C'_{M,W} \subseteq C_{M,W}$, and j be an integer s.t. $j \leq |W|$. Then we can compute $\texttt{Ins}_{Z,j} (s, C'_{M,W})$ in time $O(2^{2k}n)$.*

We can now compute a filling of B satisfying all the above constraints. We define the following table $\texttt{Fill} - \texttt{B}_j (S'_X, C'_{M,A}, C'_{I,B})$ for each $S'_X \subseteq S_X$, $C'_{M,A} \subseteq C_{M,A}$, $C'_{I,B} \subseteq C_{I,B}$ and $0 \leq j \leq |B|$.

Recurrence 2. *Let* $S'_X \subseteq S_X$, $C'_{M,A} \subseteq C_{M,A}$, $C'_{I,B} \subseteq C_{I,B}$

- *For* $j = 0$, $\mathtt{Fill-B}_j\left(S'_X, C'_{M,A}, C'_{I,B}\right) = 1$ *iff* $S'_X = C'_{M,A} = C'_{I,B} = \emptyset$.
- *For all* $j \geq 1$, $\mathtt{Fill-B}_j\left(S'_X, C'_{M,A}, C'_{I,B}\right) = 1$ *iff one of the following is true:*
  - $\mathtt{Fill-B}_{j-1}\left(S'_X, C'_{M,A}, C'_{I,B}\right) = 1$.
  - $f(j) \in C'_{I,B}$ *and* $\exists s_x \in S'_X, C_{M,j} \subseteq C'_{M,A}$ *such that*
  $$\begin{cases} \mathtt{Fill-B}_{j-1}\left(S'_X \setminus \{s_x\}, C'_{M,A} \setminus C_{M,j}, C'_{I,B} \setminus \{f(j)\}\right) \\ \wedge \, \mathtt{Ins}_{B,j}\left(s_x, C_{M,j}\right) \end{cases}$$

A filling of A is computed using a table $\mathtt{Fill-A}_j\left(S'_Y, C_{M,B}, C_{I,A}\right)$ defined similarly.

Lemma 5. *Let* (A, B) *be an instance of One-sided Scaffold Filling,* $X = [A] \setminus [B]$, $Y = [B] \setminus [A]$, k *be an integer,* C *be a set of colors, and* F *be a perfect family of hash functions from the positions of* A *and* B *to* C. *Then the following propositions are equivalent:*
(i) There exists a filling A' *of* A *with* Y *and a filling* B' *of* B *with* X *such that* A' *and* B' *have* k *common adjacencies;*
(ii) There exist two multi-sets of strings S_X *and* S_Y *over* X, Y, *a number assignment, a color-coding* $f \in F$ *and a partition* $C = C_{M,A} \cup C_{M,B} \cup C_{I,A} \cup C_{I,B}$ *such that* $\mathtt{Fill-A}_{|A|}\left(S_Y, C_{M,B}, C_{I,A}\right) = \mathtt{Fill-B}_{|B|}\left(S_X, C_{M,A}, C_{I,B}\right) = 1$.

We present now the main result of this section.

Theorem 2. *Let* A, B *be two strings over alphabet* Σ *and let* $X = [A] \setminus [B]$ *be the multiset of symbols of* A *missing in* B *and* $Y = [B] \setminus [A]$ *the multiset of symbols of* B *missing in* A. *It is possible to compute a solution of Two-sided SFMNSA over instance* (A, B) *in time* $2^{O(k \log k)} poly(n)$.

Proof. The correctness of the algorithm is directly given by Lemma 5: once a perfect family of hash functions F is fixed and two multi-sets of strings S_X and S_Y over X, Y, a number assignment, a color-coding $f \in F$ and a partition $C = C_{M,A} \cup C_{M,B} \cup C_{I,A} \cup C_{I,B}$ are selected by exhaustive branching, it suffices to compute the entries $\mathtt{Fill-A}_{|A|}\left(S_Y, C_{M,B}, C_{I,A}\right)$ and $\mathtt{Fill-B}_{|B|}\left(S_X, C_{M,A}, C_{I,B}\right)$, and return the corresponding fillings of A and B if both entries are equal to 1.

The time complexity of the algorithm is dominated by the iteration over all possible pairs (S_X, S_Y) and of the number assignment. The number of possible sets S_X, S_Y is bounded by k^{2k} from Property 2. By Property 3 there are $O(2k^{k+1})$ number assignments to iterate through. The dynamic programming recurrence requires time $O(2^{4k}n)$. Since a family of perfect hash function of size $O(2^{O(k)} poly(n))$ can be computed in time $O(2^{O(k)} poly(n))$ [1], and the possible partitions of C into sets $C_{M,A}$, $C_{M,B}$, $C_{I,A}$, $C_{M,B}$ are less than 2^{4k} (including the constraint $|C_{M,A}| + |C_{M,B}| + k' = k$), it follows that the overall time complexity of the algorithm is bounded by $O((2k)^{2k+1} 2^{O(k)} poly(n)) = 2^{O(k \log k)} poly(n)$. $\square$

5   Conclusion

In this paper we presented two FPT algorithms for the One-sided SF-MNSA problem and the Two-sided SF-MNSA problem. There are some interesting open problems from an algorithmic perspective. First, it would be interesting to improve upon the time complexity of the algorithms we presented. Moreover, the approximation complexity of the Scaffold Filling problems, in particular of the Two-sided case, should be further investigated. An interesting open problem in this direction is whether it is possible to design an approximation algorithm for Two-sided SF-MNSA with approximation factor better than 2.

References

1. Alon, N., Yuster, R., Zwick, U.: Color-coding. J. ACM **42**(4), 844–856 (1995)
2. Betzler, N., van Bevern, R., Fellows, M.R., Komusiewicz, C., Niedermeier, R.: Parameterized algorithmics for finding connected motifs in biological networks. IEEE/ACM Trans. Comput. Biology Bioinform. **8**(5), 1296–1308 (2011)
3. Bonizzoni, P., Della Vedova, G., Dondi, R., Pirola, Y.: Variants of constrained longest common subsequence. Inf. Process. Lett. **110**(20), 877–881 (2010)
4. Chain, P., Grafham, D., Fulton, R., Fitzgerald, M., Hostetler, J., Muzny, D., Ali, J., et al.: Genome project standards in a new era of sequencing. Science **326**, 236–237 (2009)
5. Chen, Z., Fu, B., Xu, J., Yang, B., Zhao, Z., Zhu, B.: Non-breaking similarity of genomes with gene repetitions. In: Ma, B., Zhang, K. (eds.) CPM 2007. LNCS, vol. 4580, pp. 119–130. Springer, Heidelberg (2007)
6. Dondi, R., Fertin, G., Vialette, S.: Complexity issues in vertex-colored graph pattern matching. J. Discrete Algorithms **9**(1), 82–99 (2011)
7. Dondi, R., Fertin, G., Vialette, S.: Finding approximate and constrained motifs in graphs. Theor. Comput. Sci. **483**, 10–21 (2013)
8. Downey, R., Fellows, M.: Parameterized Complexity. Springer, New York (1999)
9. Fellows, M.R., Fertin, G., Hermelin, D., Vialette, S.: Upper and lower bounds for finding connected motifs in vertex-colored graphs. J. Comput. Syst. Sci. **77**(4), 799–811 (2011)
10. Fertin, G., Labarre, A., Rusu, I., Tannier, E., Vialette, S.: Combinatorics of Genome Rearrangements. The MIT Press, Cambridge (2009)
11. Jiang, H., Zheng, C., Sankoff, D., Zhu, B.: Scaffold filling under the breakpoint distance. In: Tannier, E. (ed.) RECOMB-CG 2010. LNCS, vol. 6398, pp. 83–92. Springer, Heidelberg (2010)
12. Jiang, H., Zheng, C., Sankoff, D., Zhu, B.: Scaffold filling under the breakpoint and related distances. IEEE/ACM Trans. Comput. Biology Bioinform **9**(4), 1220–1229 (2012)
13. Liu, N., Jiang, H., Zhu, D., Zhu, B.: An improved approximation algorithm for scaffold filling to maximize the common adjacencies. In: Du, D.-Z., Zhang, G. (eds.) COCOON 2013. LNCS, vol. 7936, pp. 397–408. Springer, Heidelberg (2013)
14. Muñoz, A., Zheng, C., Zhu, Q., Albert, V., Rounsley, S., Sankoff, D.: Scaffold filling, contig fusion and gene order comparison. BMC Bioinform. **11**, 304 (2010)
15. Niedermeier, R.: Invitation to Fixed-Parameter Algorithms. Oxford University Press, Oxford (2006)
16. Yancopoulos, S., Attie, O., Friedberg, R.: Efficient sorting of genomic permutations by translocation, inversion and block interchange. Bioinformatics **21**, 3340–3346 (2005)

Finding Totally Independent Spanning Trees with Linear Integer Programming

Alexandre Salles da Cunha[1] and Fernanda Sumika Hojo de Souza[2(✉)]

[1] Departamento de Ciência da Computação,
Universidade Federal de Minas Gerais,
Belo Horizonte, Brazil
acunha@dcc.ufmg.br
[2] Departamento de Ciência da Computação,
Universidade Federal de São João del-Rei,
São João del-Rei, Brazil
fsumika@ufsj.edu.br

Abstract. Two spanning trees of an undirected graph are totally independent if they are edge disjoint and if the unique paths that connect any pair of vertices in these trees are also node disjoint. Accordingly, $K \geq 2$ spanning trees are totally independent if they are pairwise totally independent. The problem of finding K totally independent spanning trees (KTIST) or proving that no such trees do exist is NP-Complete. We investigate KTIST and an optimization problem which consists of finding K totally independent spanning trees with the minimum possible number of central nodes. Both problems have applications in the design of interconnection networks. We propose an integer programming formulation, valid inequalities and a Branch-and-cut algorithm to solve them. We also present an experimental evaluation of such an algorithm.

Keywords: Combinatorial optimization · Totally independent spanning trees · Packing connected dominating sets · Branch-and-cut algorithms

1 Introduction

Let $G = (V, E)$ be a connected and undirected graph with $n = |V|$ vertices and $m = |E|$ edges. Two spanning trees $T^1 = (V, E^1)$ and $T^2 = (V, E^2)$ of G are *r-independent* if the paths that connect $r \in V$ to any $v \in V \setminus \{r\}$ in (V, E^1) and in (V, E^2) are internally edge and vertex disjoint, i.e., the paths do not share edges or vertices other than r and v. Trees T^1 and T^2 are *totally independent* if they are r-independent for any choice of $r \in V$ and if they are edge disjoint. Accordingly, K spanning trees are totally independent if they are pairwise totally independent. The problem of deciding whether or not a general graph has $K \geq 2$ totally independent spanning trees, denoted here KTIST, is NP-complete [14].

This research is partially funded by CNPq grants 477863/2010-8, 305423/2012-6, 471464/2013-9 and FAPEMIG grant PPM-VII-00164-13.

© Springer International Publishing Switzerland 2014
P. Fouilhoux et al. (Eds.): ISCO 2014, LNCS 8596, pp. 149–160, 2014.
DOI: 10.1007/978-3-319-09174-7_13

Spanning trees are fundamental structures used to disseminate information in several types of communication networks [5]. Quite often, requirements other than connectivity need to be imposed on network topologies such that they guarantee a certain level of quality of service parameters like system reliability, latency, throughput and many others. Multi-trees [17] are one of the possible approaches to overcome drawbacks found in networks based solely on single (constrained or not) spanning trees. Totally independent spanning trees find thus applications in the design of communication networks based on multi-trees, for example, in fault tolerant broadcasting networks in parallel computing [13–15]. The idea is that K copies of a message are sent from a processor r to the others, through r-independent spanning trees. Even if $K - 1$ processors do not work properly, the message is still received by the others. Totally independent spanning trees also inherit applications from closely related problems like finding edge disjoint spanning trees [21,25,29] and packing dominating trees [1,2].

In this paper, we discuss an integer programming formulation and a Branch-and-cut algorithm for KTIST and for an optimization version of it, denoted K-Packing Totally Independent Spanning Trees (KPTIST). The latter seeks for a packing of G into K totally independent spanning trees, whose set of central nodes has minimum cardinality. In this sense, KPTIST also relates to the Minimum Connected Dominating Set Problem (MCDS) [4,9,12,26] and the Max-Leaf Spanning Tree Problem (MLST) [7,8,20].

Our aim is to conduct a polyhedral investigation of the convex hull of K totally independent spanning trees and to use valid inequalities for that polytope in a Branch-and-cut algorithm to solve KTIST and KPTIST. To the best of our knowledge, there are no exact algorithms addressing KTIST or KPTIST; the later being actually introduced here. The remaining of the paper is organized as follows. In Sect. 2, we point out some of the most important results and applications of totally independent spanning trees and related structures. An integer programming formulation along with some valid inequalities are discussed next, in Sect. 3. A Branch-and-cut algorithm is introduced in Sect. 4. Preliminary computational results for that algorithm are discussed in Sect. 5. We close the paper in Sect. 6, indicating future steps of our research.

2   Literature Review

The applications reported in [13–15] bring KTIST close to combinatorial optimization problems where network survivability is a primary goal. One example is the generalized Steiner problem [31]. Given an edge weighted undirected graph $G = (V, E)$ and a symmetric $|V| \times |V|$ matrix $R = [r_{st}]$, the goal is to find a minimum cost subgraph of G such that r_{st} node (edge) disjoint paths are required to connect every pair of vertices s and t of V. Another closely related problem was proposed in [10] and imposes that, connecting two vertices s, t, there must be at least r_{st} node (edge) disjoint paths, after the deletion of any node subset of cardinality k_{st}, where the later is also a given design parameter. In a third type of survivability model, nodes are divided in a hierarchical way, according to their importance to the network. For every node $i \in V$, a connectivity parameter $\rho_i \in \mathbb{Z}_+$ denotes the hierarchical level of that node. The connectivity requirements impose that $\min\{\rho_s, \rho_t\}$ node (edge) disjoint paths must be selected to connect every pair of nodes s, t in the network. For an in-depth review of these and related problems, we quote references [3,11,19,24,27] and those therein.

Because of their applications in interconnection networks, independent and totally independent spanning trees have deserved the attention of several studies. Examples of studies whose aim is to characterize the existence and algorithms for finding such trees in particular types of graphs could be found in [15], for torus networks, in [32], for hypercubes, in [18], for chordal rings, and finally, in [16,32], for de Bruijn and Kautz graphs.

The problem of finding edge disjoint spanning trees has also been investigated for quite a long time, in both weighted [25] and non-weighted [21,29] versions. The former, which seeks for a minimum cost set of K edge disjoint spanning trees of an edge weighted graph, was proven to be polynomially solvable in $O(m \log m + k^2 n^2)$ time, by a greedy matroid algorithm introduced in [25]. Close to it is the Minimum Congestion Spanning Tree Problem [30], which looks for a minimum cost set of K spanning trees, not necessarily edge disjoint, penalizing the cost whenever an edge appears in more than one tree.

A well known result due to Nash-Williams [21] states that every $2K$ edge connected graph has K edge disjoint spanning trees. Based on that, Hasunuma [14] conjectured that a $2K$ connected graph has K totally independent spanning trees. The conjecture was proven correct for maximal planar graphs [14]. Recently, Péterfalvi [23] has shown K-connected graphs, for any $K \geq 2$, that do not have two totally independent spanning trees, proving thus that Hasunuma's conjecture does not hold true in general. More recently, Pai et al. [22] has shown other types of graphs, for which the conjecture is true. More precisely, it was proven that there are $\lfloor \frac{n}{2} \rfloor$ completely independent spanning trees in complete graphs with $n \geq 4$ and in complete bipartite graphs with at least four vertices in each partition. In addition, they have shown the existence of $\lfloor \frac{n_1+n_2}{2} \rfloor$ completely independent spanning trees in complete tripartite graphs, where n_1 and n_2 are the number of vertices in the two smallest partitions.

The following nice and useful characterization of totally independent spanning trees was given by Hasunuma [13]:

Property 1. [Theorem 2.1 in [13]] $\{T^k = (V, E^k) : k = 1, \dots, K\}$ are K totally independent spanning trees of G if and only if: (a) they are edge-disjoint, and, (b) given any vertex $i \in V$, there is at most one spanning tree T^k such that $|\delta(i) \cap E^k| > 1$, i.e., no vertex is a central node in more than one tree.

Another application for KTIST and KPTIST, different from those reported in [13–15], follows from the fact that, due to Property 1, totally independent spanning trees are very closely related to connected dominating set of G [1,2]. A dominating set of G is a set $S \subseteq V$ such that every vertex of V either belongs to S or has a neighbor in it. A connected dominating set is a dominating set that implies a connected subgraph of G. Accordingly, a connected dominating set packing of size K (KCDSP) is a collection of K disjoint connected dominating sets. According to [2], packings of dominating sets generalize edge disjoint spanning trees, in the context of vertex connectivity.

It should be clear that, given K totally independent spanning trees, one can easily obtain a KCDSP, by simply removing their leaves. The resulting sets of vertices are disjoint, connected and dominating. However, the opposite does not always hold true, since K totally independent spanning trees must not share edges. As an example, assume that G is defined by $V = \{1, \ldots, 4\}$ and $E = \{\{1,2\}, \{2,3\}, \{3,4\}, \{1,4\}\}$. While $C^1 = \{2,3\}$ and $C^2 = \{1,4\}$ form a 2CDSP, they do not lead two totally independent spanning trees.

Multi-trees whose internal nodes form a KCDSP are important structures in synchronous communication networks, where message coding is not employed [1,2]. In such networks, nodes send their own messages and forward those they receive, without packing them into larger ones. During each time interval, each node can send a message (of a bounded size) to all its neighbors. It was shown in [2] that the network throughput is maximized when communications are organized in terms of the largest size of a connected dominating set packing of G. In other words, when the maximum size of a connected dominating set packing is used, the throughput is maximized.

KTIST can thus be seen as a constrained version of KCDSP, where aiming at reducing network congestion, one imposes that tree edges must be disjoint. Finally, when one seeks for a KTIST solution with the minimum number of central nodes, the network latency decreases, since the number of hops between any two pairs, in any of the trees, tend to be small.

3   Integer Programming Formulation

Property 1 allows KPTIST and KTIST to be formulated using any spanning tree representation along with leaf defining decision variables. The formulation discussed next makes use of directed cutsets to impose connectivity on each spanning tree. As such, the formulation requires a specific vertex r as the root of the K arborescences.

In order to present the model, consider the following notation used in the remaining of the paper. Given a set $S \subseteq V$, its complement in V is denoted $\overline{S}$. The set of edges of E with both endpoints in S is denoted $E(S)$. For any disjoint sets of vertices S and W, let $\delta(S, W) \subseteq E$ denote the edges in E with one endpoint in S and another in W. For simplicity, we denote $\delta(S, V \setminus S)$ by $\delta(S)$. Whenever S has a single vertex, say i, we replace $\delta(\{i\})$ by $\delta(i)$.

The formulation discussed next is based on orientations of edges in E. Thus, we denote by $D = (V, A)$ the directed graph obtained by orienting the edges of E, i.e., $A = \{(i,j) \cup (j,i) : \{i,j\} \in E\}$. Sets $\delta^+(S) = \{(i,j) \in A : i \in S, j \in \overline{S}\}$ and $\delta^-(S) = \{(i,j) \in A : i \in \overline{S}, j \in S\}$ respectively denote the arcs pointing outwards and inwards of $S \subset V$. Whenever S has a single vertex i, we also replace $\delta^-(\{i\})$ and $\delta^+(\{i\})$ respectively by $\delta^-(i)$ and $\delta^+(i)$. For any real valued function $f : Q \to \mathbb{R}$ defined over a finite domain Q, $f(Q')$ denote $\sum_{q \in Q'} f_q$.

Consider the following decision variables:

- $\{y_i^k \in \mathbb{B} : i \in V, k = 1, \ldots, K\}$ to indicate whether or not vertex i is a leaf in tree T^k. In case i is a leaf, $y_i^k = 1$. Otherwise, $y_i^k = 0$.
- $\{x_{ij}^k \in \mathbb{B} : (i,j) \in A\}$ to indicate whether or not arc (i,j) is included in the the k-th arborescence. In case it is included, $x_{ij}^k = 1$. Otherwise, $x_{ij}^k = 0$.

Now, consider the polyhedral region $\mathcal{P}_d$ defined by (1), (2) and (3):

$$x^k(A) = n - 1, \ k = 1, \ldots, K \tag{1a}$$

$$x^k(\delta^-(i)) = 1, \ k = 1, \ldots, K, i \in V \setminus \{r\} \tag{1b}$$

$$x^k(\delta^+(S)) \geq 1, \ k = 1, \ldots, K, S \subset V, r \in S \tag{1c}$$

$$x_{ij}^k \geq 0, \ k = 1, \ldots, K, (i,j) \in A \tag{1d}$$

$$x^k(\delta^+(i)) \leq (|\delta(i)| - K)(1 - y_i^k), \ k = 1, \ldots, K, i \in V \setminus \{r\} \tag{2a}$$

$$x^k(\delta^+(r)) \leq (|\delta(r)| - K)(1 - y_r^k) + 1, \ k = 1, \ldots, K \tag{2b}$$

$$y_i^k + x^k(\delta^+(i)) \geq 1, \ k = 1, \ldots, K, i \in V \setminus \{r\} \tag{2c}$$

$$y_i^k \geq 0, \ k = 1, \ldots, K, i \in V \tag{2d}$$

$$y_i^k \leq 1, \ k = 1, \ldots, K, i \in V \tag{2e}$$

$$\sum_{k-1}^{K}(x_{ij}^k + x_{ji}^k) \leq 1, \{i,j\} \in E \tag{3a}$$

$$\sum_{k=1}^{K} y_i^k \geq K - 1, i \subset V \tag{3b}$$

Constraints (1a) provide the convex hull of spanning arborescences of D [28]. Constraints (1b) guarantee that precisely one arc of D is incident to i. Cutset constraints (1c) guarantee that there is a path connecting the root r to every other node in each arborescence.

Constraints (2c) enforce that either $i \in V \setminus \{r\}$ is a leaf or at least one arc must point outwards of it in arborescence k. Constraints (2a) impose that, whenever a non-root vertex i is a leaf, no arc can point outwards of i in arborescence k. For spanning arborescences in general, at most $|\delta(i)| - 1$ arcs are allowed to point outwards of a central node $i \neq r$. Note, however, that (2a) enforce that at most $|\delta(i)| - K$ arcs may leave i. That applies, since if more than $|\delta(i)| - K$ edges were incident to i in T^k, at least one of them would certainly be included in another tree. Similar conditions are enforced by (2b), for the root vertex r. However, since no arcs point inwards of r, the right-hand-side of (2b) is increased by one. Finally, inequalities (3a) enforce that the spanning trees are edge disjoint while (3b) guarantee that no vertex can be a central node twice or more times.

Define $\mathcal{X} := \{(x,y) \in (\mathbb{B}^{2Km} \times \mathbb{B}^{Kn}) : (x,y) \in \mathcal{P}_d\}$ and let $\text{conv}(\mathcal{X})$ denote the convex hull of solutions in $\mathcal{X}$. KTIST consists of deciding whether or not $\mathcal{X} \neq \emptyset$. Accordingly, a formulation for KPTIST is

$$\min\left\{\sum_{k=1}^{K}\sum_{i\in V}(1-y_i^k) : (x,y)\in\mathcal{P}_d\cap(\mathbb{B}^{2Km}\times\mathbb{B}^{Kn})\right\}. \tag{4}$$

3.1 Valid Inequalities for $\mathcal{X}$ Coming from the Max-Leaf Spanning Tree and Minimum Connected Dominating Set Polytopes

Formulation $\mathcal{P}_d$ can be strengthened by exploiting different structures found in solutions in $\mathcal{X}$. The first set of valid inequalities used to that aim are stated in (5) and come from the MLST polytope [6,8].

$$x_{ij}^k + y_i^k \le 1, \ k=1,\dots,K, (i,j)\in A, i\neq r \tag{5a}$$

$$x_{rj}^k + y_r^k + y_j^k \le 2, \ k=1,\dots,K, (r,j)\in\delta^+(r) \tag{5b}$$

$$\sum_{\{i,j\}\in F}(x_{ij}^k + x_{ji}^k) + (|F|-1)y_i^k \le |F|, k=1,\dots,K, i\in V, F\subseteq\delta(i), |F|\ge 2. \tag{5c}$$

Constraints (5a) enforce that $i\in V\setminus\{r\}$ cannot be a leaf node if any arc (i,j) pointing outwards of it is included in arborescence k. For the root, (5b) guarantee that r and j cannot simultaneously be leaves in the same arborescence, if arc (r,j) is included in that arborescence. These inequalities proved quite important to reinforce the MLST linear programming upper bounds in [20].

Constraints (5c) state that, whenever i is a leaf in arborescence k, at most one arc out of the orientations of edges in $F\subseteq\delta(i):|F|\ge 2$ can be included in that arborescence. Undirected versions of (5c) were shown to be facet defining for the MLST polytope [8].

Another set of valid inequalities comes from the fact that no solution in $\mathcal{X}$ may involve a star of G. Thus, at least one neighbor on each vertex must be a central node. Such observation translates to

$$\sum_{j:\{i,j\}\in\delta(i)} y_j^k \le |\delta(i)|-1, \ \forall i\in V, \ \forall k=1,\dots,K. \tag{6}$$

Although such an idea was already explored in (2a)–(2b), constraints (6) indeed help strengthening $\mathcal{P}_d$. An inequality similar to (6) was used in the MCDS solution approaches in [9], to rule out minimum connected dominating sets of cardinality one, whose existence in G could be evaluated in polynomial time.

3.2 Lifted Node Cutset Constraints

Since $(V,\bigcup_{k=1}^{K}E^k)$ is a uniform K node and edge connected subgraph of G [27], lower bounds for KPTIST can be reinforced with valid inequalities for those polytopes. One of such valid inequalities are node cutsets [27]:

$$\sum_{k=1}^{K}\sum_{\{i,j\}\in\delta(S,V\setminus(S\cup U))}(x_{ij}^k + x_{ji}^k) \ge K-|U|, S\subset(V\setminus U), S\neq\emptyset, U\subset V, |U|\le K. \tag{7}$$

They enforce that $(V, \bigcup_{k=1}^{K} E^k)$ must remain $K - |U|$ connected after removing vertices in $U : |U| \leq K$ and edges in $\delta(U) \cup E(U)$ from it. Since our formulation is indexed by k and uses leaf-defining variables, inequalities (7) can be made stronger. To that aim, let $\{S^k : k = 1, \ldots, K\}$ denote K subsets of vertices such that $S^{k_1} \cap S^{k_2} \neq \emptyset$ and $(V \setminus (S^{k_1} \cup U)) \cap (V \setminus (S^{k_2} \cup U)) \neq \emptyset$, for any pair $k_1, k_2 \in \{1, \ldots, K\}$, such that $k_1 \neq k_2$. Then we have:

$$\sum_{k=1}^{K} \sum_{\{i,j\} \in \delta(S^k, V \setminus (S^k \cup U))} (x_{ij}^k + x_{ji}^k) \geq K - |U|. \tag{8}$$

Validity of (8) comes from the fact that any $s \in \bigcap_{k=1}^{K} S^k$ and $t \in \bigcap_{k=1}^{K} (V \setminus (S^k \cup U))$ must remain $K - |U|$ connected after removing U and $\delta(U) \cup E(U)$ from $(V, \bigcup_{k=1}^{K} E^k)$. Since no vertex can be a central node in more than one tree, removing U cannot disconnect s and t in more than $|U|$ trees. Thus, the connectivity of $(V, \bigcup_{k=1}^{K} E^k)$ can only drop from K to K minus the number of vertices in U that are central nodes in one of the K trees. Such an observation actually allows one to lift node cutset constraints to the stronger form:

$$\sum_{k=1}^{K} \sum_{\{i,j\} \in \delta(S^k, V \setminus (S^k \cup U))} (x_{ij}^k + x_{ji}^k) \geq K - \sum_{k=1}^{K} \sum_{i \in U} (1 - y_i^k) \tag{9}$$

3.3 Symmetry Breaking Constraints

One drawback of formulation $\mathcal{P}_d$ is its symmetry. As an attempt to go around the negative computational side effects of that, we impose the following symmetry breaking constraints:

$$y_i^k = 1, \; i = 1, \ldots, K - 1, \; k = i + 1, \ldots, K. \tag{10}$$

Since no vertex can be central in two or more spanning trees, constraints (10) simply define the indices of the trees for which K vertices are allowed to be central nodes. According to (10), node $i = 1$ is forbidden of being a central node in those trees of index $k \geq 2$. Likewise, vertex $i = 2$, is necessarily a leaf node in trees from index $k = 3$ up to $k = K$. Such a vertex is allowed to be a central node either in the same tree of vertex 1 or else in the subsequent tree, of index $k = 2$. Similar reasonings define the indices of trees where the remaining $K - 2$ nodes are allowed to be central. For the use of constraints (10), we assume that $V = \{1, \ldots, n\}$ and that $|\delta(1)| \geq |\delta(2)| \geq \cdots \geq |\delta(n)|$.

The main motivation behind the use of symmetry breaking constraints like (10) in linear programming (LP) based Branch-and-bound algorithms is to avoid the investigation of enumeration tree branches that essentially lead to the same optimal solutions. However, as we shall see in the computational results, constraints (10) also improved the LP bounds provided by our model.

4   Branch-and-Cut Algorithm for KPTIST

Our Branch-and-cut algorithm (BC) first solves the LP relaxation $\min\{\sum_{k=1}^{K}\sum_{i\in V}(1-y_i^k) : (x,y) \in \overline{P}\}$, where $\overline{P}$ is implied by (1a),(1b),(1d),(2), (3),(5a),(5b), (6) and (10). Let $(\overline{x},\overline{y}) = (\overline{x}^1,\ldots,\overline{x}^K,\overline{y}^1,\ldots,\overline{y}^K)$ be the solution to such a LP. If $(\overline{x}^k,\overline{y}^k) \in \mathbb{B}^{2m+n}$ for every $k = 1,\ldots,K$ and if each of these vectors imply arborescences of D, $(\overline{x},\overline{y})$ also solves KPTIST. Otherwise, we look for violated directed cutset constraints (1c) and Fujie's inequalities (5c) to reinforce $\overline{P}$.

Cutset constraints (1c) are separated in $O(Kn^4)$ time complexity as follows. For a given $k \in \{1,\ldots,K\}$, let $\overline{D}^k = (V,\overline{A}^k)$ ($\overline{A}^k := \{(i,j) \in A : \overline{x}_{ij}^k > 0\}$) be the support directed graph associated to $(\overline{x}^k,\overline{y}^k)$. For each $i \in V\setminus\{r\}$, compute a minimum cut $\delta^+(S) : r \in S$, $i \in \overline{S}$, that separates r and i in the network given by $\overline{D}^k$ and arc capacities $\{\overline{x}_{ij}^k : (i,j) \in \overline{A}^k\}$. If $\overline{x}^k(\delta^+(S)) < 1$, a cutset (1c) is violated and is appended in the relaxation. The exact separation of directed cutset constraints is conducted for every $k = 1,\ldots,K$.

Attempting to speed-up the cutset separation procedure above, we settled for computing the min-cut separating r and $i \in V\setminus\{r\}$ only when i is not included in any set S associated with a previously identified violated cutset inequality, for the particular network $\overline{D}^k$ at hand. This strategy guarantees that at least one violated cutset inequality separating r and every $i \neq r$, for each k, is appended to the relaxation, provided they exist. One possible advantage of using such a procedure is that it also attempts to reduce the number of violated inequalities being used and the number of calls of the min-cut algorithm.

For identifying violated Fujie's cuts (5c), we use the exact separation procedure introduced in [20], that works as follows. For each k and $i \in V$, we compute $\overline{z}_{ij}^k := \overline{x}_{ij}^k + \overline{x}_{ji}^k : \{i,j\} \in \delta(i)$. Then, we sort the edges in $\delta(i)$ in a non-increasing order of the values $\overline{z}_{ij}^k$. Assume that $\delta(i) = \{e_1,e_2,\ldots,e_{|\delta(i)|}\}$ and that $\overline{z}_{e_1}^k \geq \overline{z}_{e_2}^k \geq \cdots \geq \overline{z}_{e_{|\delta(i)|}}^k$. For every $l = 2,\ldots,|\delta(i)|$, whenever $\sum_{p=1}^{l}\overline{z}_{e_p}^k > l+(1-l)\overline{y}_i^k$, a violated Fujie's cut defined for i,k and $F = \{e_1,\ldots,e_l\}$ is violated and appended into the LP relaxation.

The separation of (1c) and (5c) is carried out until no violated cuts are found. At that moment, if the solution to the relaxation is not integer feasible, BC branches on variables. BC was implemented with calls to XPRESS Mixed Integer Programming solver (23.01.06) callback routines. The node selection policy was set to *depth-first*. Apart from the heuristics, cut generation and pre-processing procedures that were turned off, all other default XPRESS settings were used.

Lifted node cutset constraints (9) can also be exactly separated in polynomial time, by solving $O(n^{K-1})$ max-flow problems, in conveniently defined networks (see [27] for details). Despite the fact they are not redundant with respect to $\mathcal{P}_d$, even after valid inequalities (5), (6) and (10) are appended to that polytope, BC does not separate (9). We decided to do so since our computational experience indicated that the LP relaxations for the instances in our test bed did not violate them. Our experience also indicated that these inequalities are likely to be violated for LP relaxations of edge weighted versions of KTIST.

5   Preliminary Computational Experiments

Since no instances are available in the KTIST literature, we used MLST instances introduced in [20], in order to evaluate BC. Instances in [20] were generated by a two-step procedure, as follows. For a given instance size n, the first set of edges included in G are those in a Hamiltonian path, built by randomly choosing a permutation of the vertices of the input graph, and connecting consecutive vertices in the permutation. Additional edges are included in G, according to a uniform probability, until a desired a priori defined graph density is achieved.

Depending of the values of $n \in \{30, 50, 70, 100, 120, 150, 200\}$, instances in [20] have graph densities in the interval between 5 % and 70 %. In our computational study, only a subset of the instances in [20] were actually tested. Some instances in [20] were left out of our experiments because they are not K connected and because they are out of reach for the current implementation of BC. In total, our test bed consists of 23 instances, with values of $n \in \{30, 50, 70, 100\}$ for $K = 2$ and $n \in \{30, 50, 70\}$ for $K = 3$. Instances in [20] whose graph densities are greater than 50 % were not considered here.

All algorithms were implemented in C and all computational results reported here were obtained with an Intel XEON E5645Core TMi7-980 hexa-core machine, running at 2.4GHz, with 24 GB of shared RAM memory. No multi-threading was allowed. A time limit of 1 CPU hour was imposed on the execution of BC.

In Table 1, we report on computational results for all instances considered here. Each one is indicated in the first column of the table, by a word that starts with n, followed by _d and then by the graph density (in % figures). For example, 30_d50 is the instance in [20] with $n = 30$ and $d = 50$ % graph density. In the second column, we report the value of $K \in \{2, 3\}$ considered for each instance.

The next five columns in the table provide KPTIST LP lower bounds; the first of them being the bound implied by formulation $\mathcal{P}_d$. Subsequent lower bounds in the table are those obtained when polytope $\mathcal{P}_d$ is reinforced with each set of valid inequalities: logical constraints (5a)–(5b), Fujie's cuts (5c), stars preventing inequalities (6) and, finally, symmetry breaking constraints (10). For the computation of each of the four last lower bounds, each of these groups of inequalities are sequentially added to the previous formulation, one group at a time, according to this order. For example, the second lower bound is obtained when constraints (5a)–(5b) are appended to $\mathcal{P}_d$. For the the third, $\mathcal{P}_d$ was also reinforced with (5c). Therefore, for the fifth lower bound indicated in the table, all inequalities were taken into account. The bounds presented here are not symmetrical with respect to the vertex that acts as the root of the arborescences. All of them were evaluated with $r = 1$ as the root.

In the last columns of the table, BC results are given: the best lower (BLB) and upper (BUB) bounds found after the search was finished or when the CPU time limit was hit, the number of enumeration tree nodes investigated, followed by $t(s)$, the total CPU time (in seconds) needed to conclude the search. The last column in the table gives $t_f(s)$, the time (also in seconds) needed to find the first integer feasible solution. Values presented for BLB in the table are the best lower bounds rounded up. Whenever an instance was not solved to proven optimality, an indication "-" is given for the corresponding CPU time entries.

Table 1. Linear programming and Branch-and-cut results.

| Instance | K | $\mathcal{P}_d$ | $\mathcal{P}_d$ strengthened with | | | | BLB | BUB | nodes | $t(s)$ | $t_f(s)$ |
			(5a)–(5b)	(5)	(5)–(6)	(5)–(6) +(10)					
30_d30	2	6.127	6.632	6.659	6.844	7.070	9	9	91	2.6	0.8
30_d50	2	3.454	4.000	4.000	4.024	4.033	6	6	118	2.1	0.8
50_d10	2	22.633	24.614	24.743	24.743	24.754	28	28	14185	1059.0	2.7
50_d20	2	7.803	10.541	10.552	10.554	10.758	14	14	457	21.8	1.5
50_d30	2	5.443	6.813	6.813	6.913	6.998	10	10	1225	145.6	2.0
50_d50	2	3.265	4.020	4.020	4.095	4.095	6	6	139	19.8	2.5
70_d10	2	17.893	22.640	22.667	22.879	23.001	26	29	2594	-	6.7
70_d20	2	8.275	10.293	10.294	10.588	10.664	13	15	2954	-	11.1
70_d30	2	5.050	6.706	6.706	6.749	6.749	9	11	1749	-	2.9
70_d50	2	3.129	4.055	4.055	4.155	4.198	6	6	263	41.3	3.8
100_d5	2	48.083	49.894	50.080	50.104	50.129	52	64	5355	-	34.0
100_d10	2	16.395	21.744	21.753	21.835	21.971	23	33	219	-	1144.9
100_d20	2	7.629	10.491	10.491	10.694	10.721	12	16	880	-	8.5
100_d30	2	5.509	6.756	6.756	6.862	6.875	8	12	1019	-	212.7
100_d50	2	3.412	4.213	4.213	4.246	4.248	6	8	1202	-	51.3
30_d30	3	11.238	11.480	11.491	11.571	11.759	15	15	2924	365.1	3.6
30_d50	3	5.780	6.143	6.143	6.184	6.252	9	9	1763	92.7	2.1
50_d20	3	13.956	17.126	17.136	17.136	17.436	20	23	3372	-	16.4
50_d30	3	9.101	10.281	10.281	10.374	10.683	13	16	3692	-	5.9
50_d50	3	5.190	6.042	6.042	6.146	6.176	8	10	6107	-	4.2
70_d20	3	13.659	15.859	15.859	16.041	16.185	17	26	568	-	511.2
70_d30	3	8.167	10.059	10.059	10.124	10.253	11	18	109	-	14.1
70_d50	3	4.935	6.083	6.083	6.232	6.336	8	11	1968	-	7.6

Computational results indicated in the table suggest that constraints (5a)–(5b) were the most important in strengthening the lower bounds. Another interesting result is that symmetry breaking constraints (10) also helped in improving the formulation. Despite the lower bound improvements we report in the table, KTIST and KPTIST are still very difficult to solve. Certificates for the feasibility problem KTIST were obtained in all 23 cases, sometimes after several CPU minutes. For only 9 out of 23 cases, optimality certificates were provided for KPTIST. Quite often, duality gaps obtained at the end of the imposed CPU time limit go beyond 50 %.

6   Conclusions and Future Research

In this paper, we proposed an integer programming formulation, valid inequalities and a Branch-and-cut algorithm to find K totally independent spanning trees. Our aim is to proceed with the polyhedral investigation of the problem, attempting to better approximate the convex hull of K totally independent spanning trees of a graph. From an algorithmic point of view, we also plan to investigate other exact solution approaches, like Combinatorial Benders Decomposition, that provided good results for related problems (see [9], for instance).

We are also currently studying the problem of finding the maximum value of K such that G has K totally independent trees. For such a problem, an integer programming formulation that involves one binary decision variable for each spanning tree of G is considered. The formulation naturally leads to a Branch-and-price algorithm, whose pricing problem consists in finding a minimum weight spanning tree, where weights, assigned to edges and to the central nodes in the tree, respectively come from the dual variables assigned to constraints (3a) and (3b) in the restricted linear programming master. Preliminary computational results provided by such a method confirm our claim that finding K totally independent spanning trees is indeed very difficult in practice.

References

1. Censor-Hillel, K., Ghaffari, M., Kuhn, F.: Distributed connectivity decomposition. Technical report, Cornell University Library (2013). arXiv:1311.5317
2. Censor-Hillel, K., Ghaffari, M., Kuhn, F.: A new perspective on vertex connectivity. In: Proceedings of ACM-SIAM Symposium on Discrete Algorithms (2014)
3. Chimani, M., Kandyba, M., Ljubić, I., Mutzel, P.: Orientation-based models for 0,1,2-survivable network design: theory and practice. Mathe. Program. **124**, 413–439 (2010)
4. Dai, F., Wu, J.: On constructing k-connected k-dominating set in wireless Ad hoc and sensor networks. J. Parallel Distrib. Comp. **66**, 947–958 (2006)
5. England, D., Veeravalli, B., Weissman, J.: A robust spanning tree topology for data collection and dissemination in distributed environments. IEEE Trans. Parallel Distrib. Syst. **18**(5), 608–620 (2007)
6. Fernandes, M., Gouveia, L.: Minimal spanning trees with a constraint on the number of leaves. Eur. J. Oper. Res. **104**, 250–261 (1998)
7. Fujie, T.: An exact algorithm for the maximum leaf spanning tree problem. Comput. Oper. Res. **30**(13), 1931–1944 (2003)
8. Fujie, T.: The maximum-leaf spanning tree problem: formulations and facets. Networks **43**(4), 212–223 (2004)
9. Gendron, B., Lucena, A., da Cunha, A., Simonetti, L.: Benders decomposition, Branch-and-cut and hybrid algorithms for the minimum connected dominating set problem. INFORMS J. Comp. (2013). http://dx.doi.org/10.1287/ijoc.2013.0589, http://pubsonline.informs.org/page/terms-and-conditions
10. Grötschel, M., Monma, C.: Integer polyhedra associated with certain network design problem with connectivity constraints. SIAM J. Discrete Math. **3**, 502–523 (1990)
11. Grötschel, M., Monma, C., Stoer, M.: Design of survivable networks. In: Ball, M., Magnanti, T., Monma, C. (eds.) Network Models, Handbooks in Operations Research and Management Science, vol. 7, pp. 617–672. Elsevier, Amsterdam (1995)
12. Guha, S., Khuller, S.: Approximation algorithms for connected dominating sets. Algorithmica **20**(4), 374–387 (1998)

13. Hasunuma, T.: Completely independent spanning trees in the underlying graph of a line digraph. Discrete Math. **234**, 149–157 (2001)
14. Hasunuma, T.: Completely independent spanning trees in maximal planar graphs. In: Kučera, L. (ed.) WG 2002. LNCS, vol. 2573, pp. 235–245. Springer, Heidelberg (2002)
15. Hasunuma, T., Morisaka, C.: Completely independent spanning trees in torus networks. Networks **60**, 59–69 (2012)
16. Hasunuma, T., Nagamochi, H.: Independent spanning trees with small depths in iterated line digraphs. Discrete Appl. Math. **10**, 189–211 (2001)
17. Itai, A., Rodeh, M.: The multi-tree approach to reliability in distributed networks. Inform. Comput. **79**, 43–59 (1984)
18. Iwasaki, Y., Kajiwara, Y., Obokata, K., Igarashi, Y.: Independent spanning trees of chordal rings. Inf. Process. Lett. **69**, 155–160 (1999)
19. Kerivin, H., Mahjoub, A.: Design of survivable networks: a survey. Networks **46**, 1–21 (2005)
20. Lucena, A., Maculan, N., Simonetti, L.: Reformulation and solution algorithms for the maximum leaf spanning tree problem. Comput. Manag. Sci. **7**(3), 289–311 (2010)
21. Nash-Williams, C.: Edge-disjoint spanning trees of finite graphs. J. Lond. Math. Soc. **36**, 445–450 (1961)
22. Pai, K., Tang, S., Chang, J., Yang, J.: Completely independent spanning trees on complete graphs, complete bipartite graphs and complete tripartite graphs. Advances in Intelligent Systems & Applications. SIST, vol. 20, pp. 107–113. Springer, Heidelberg (2013)
23. Péterfalvi, F.: Two counterexamples on totally independent spanning trees. Discrete Math. **312**, 808–810 (2012)
24. Raghavan, S.: Formulations and algorithms for network design problems with connectivity requirements. Ph.D. thesis, Department of Electrical Engineering and Computer Science, Massachusetts Institute of Technology (1994)
25. Roskind, J., Tarjan, R.: A note on finding minimum cost edge-disjoint spanning trees. Math. Oper. Res. **10**(4), 701–708 (1985)
26. Simonetti, L., da Cunha, A.S., Lucena, A.: The minimum cConnected dominating set problem: formulation, valid inequalities and a branch-and-cut algorithm. In: Pahl, J., Reiners, T., Voß, S. (eds.) INOC 2011. LNCS, vol. 6701, pp. 162–169. Springer, Heidelberg (2011)
27. Stoer, M.: Design of survivable networks. Lecture Notes in Mathematics, vol. 1531. Springer, Berlin, New York (1992)
28. Magnanti, T.L., Wolsey, L.: Optimal trees. In: Ball, M.O. (ed.) Handbooks in OR and MS, vol. 7, pp. 503–615. North-Holland, Amsterdam (1995)
29. Tutte, W.: On the problem of decomposing a graph into n connected factors. J. Lond. Math. Soc. **36**, 445–450 (1961)
30. Werneck, R., Setubal, J., da Conceição, A.: Finding minimum congestion spanning trees. J. Exper. Algor. **5** (2000)
31. Winter, P.: Steiner problem in networks: a survey. Networks **17**, 129–167 (1987)
32. Yang, J., Tang, S., Chang, J., Wang, Y.: Parallel construction of optimal independent spanning trees on hypercubes. Parallel Comput. **33**, 73–79 (2007)

Coupled-Tasks in Presence of Bipartite Compatibilities Graphs

Benoit Darties[1][(✉)], Gilles Simonin[2], Rodolphe Giroudeau[3],
and Jean-Claude König[3]

[1] LE2I-CNRS-UMR 6306-8, Rue Alain Savary, 21000 Dijon, France
benoit.darties@u-bourgogne.fr
[2] Insight Centre for Data Analytics, University College Cork, Cork, Ireland
[3] LIRMM-CNRS-UMR 5506-161, rue Ada, 34090 Montpellier, France

Abstract. We tackle the makespan minimization coupled-tasks problem in presence of incompatibility constraints. In particular, we focus on stretched coupled-tasks, i.e. coupled-tasks having the same sub-tasks execution time and idle time duration. We study several problems in the framework of classic complexity and approximation for which the compatibility graph is bipartite (star, chain, ...). In such context, we design efficient polynomial-time approximation algorithms according to different parameters of the scheduling problem.

1 Introduction

We consider a non-preemptive coupled-tasks scheduling problem in presence of incompatibility constraint on a single processor. From the point of view of scheduling theory, the problem is also defined as a scheduling problem with exact delays on single machine. In this article, we will show the close relationship between coupled-task in presence of incompatibility constraint and the classic bin packing problem in the framework of complexity and approximation.

The coupled-tasks model, was first introduced by Shapiro [13] in order to model some data acquisition processes i.e. radar sensors: a sensor emits a radio pulse (first sub-task), and finally listen for an echo reply (second sub-task). Between these two instants (emission and reception), clearly there is an idle time due to the propagation, in both sides, of radio pulse. Therefore, a coupled-task is constituted by the triplet: the two sub-tasks and the idle between them. Thus, in order to minimise the makespan (schedule length), it is necessary to execute one or several different sub-tasks during the idle time of a coupled-task. Therefore, the aim is to find a best packing of coupled-tasks in which the sum of idle times is minimised. Notice that in the basic model, all coupled-tasks may be executed in each other according to processing time of sub-tasks and the duration of the idle time. Hereafter, we consider a relaxation of the previous model in which for a fixed coupled-task $\mathcal{A}$ there are only a subset of coupled-tasks compatible to be processed in the idle time of $\mathcal{A}$. This model is motivated by the problem of data acquisition in presence of incompatibility constraint in a submarine torpedo.

© Springer International Publishing Switzerland 2014
P. Fouilhoux et al. (Eds.): ISCO 2014, LNCS 8596, pp. 161–172, 2014.
DOI: 10.1007/978-3-319-09174-7_14

A collection of sensors acquires data for the torpedo. The incompatibility constraint is expressed to prevent interference issues caused by tasks using sensors working at the same frequency. So, the constraints are represented by a compatibility graph in which vertices are the coupled-tasks and edges represent compatibility between two tasks. In this article the variation of the complexity according to severals structural parameters are considered and some efficient polynomial-time approximation results on $\mathcal{NP}$-hard instances are presented without omitting the relationship to bin packing problems.

Above all, we will show the close relationship between the studied problem and four packing-related problems, for which known approximation will be used as routine for scheduling coupled-tasks problem:

1. The SUBSET SUM (SS) problem: given a set $\mathcal{S}$ of n positive values and $v \in \mathbb{N}$, the aim is to find a subset $\mathcal{S}^* \subseteq \mathcal{S}$ such that $\sum_{i \in \mathcal{S}^*} i = v$. This problem is known to be $\mathcal{NP}$-complete (see [8]). The optimization version is sometimes viewed as a KNAPSACK problem, where each item profit and weight coincide to a value in $\mathcal{S}$, the knapsack capacity is v, and the aim is to find the set of packable items with maximum profit.
2. The MULTIPLE SUBSET SUM (MSS) problem: variant of BIN PACKING in which a number of identical bins are given and one aims to maximize the overall weight of the items packed in the bins without violating the constraint on the capacity of each bin. The problem is a special case of the MULTIPLE KNAPSACK problem in which all knapsacks have the same capacity and the item profits and weights coincide. MSS admits a $\mathcal{PTAS}$ [2] and a $\frac{3}{4}$−approximation algorithm [3], but does not admit a $\mathcal{FPTAS}$ even for only two knapsacks.
3. MULTIPLE SUBSET SUM WITH DIFFERENT KNAPSACK CAPACITIES (MSSDC) [1] is an extension of MSS considering different bin capacities. MSSDC also admits a $\mathcal{PTAS}$ [1].
4. As a generalization of MSSDC, MULTIPLE KNAPSACK ASSIGNMENT RESTRICTION (MKAR) problem consists in packing weighted items into non-identical capacity-constrained bins, with the additional constraint that each item can be packed into some bins only. Each item as a profit, the objective here is to maximize the sum of profits of packed items. Considering that the profit of each item is equal to its weight, [5] proposed a $\frac{1}{2}$-approximation.

2   Presentation of Coupled-Tasks and Related Work

We model a task $\mathcal{A}_i$ with a triplet (a_i, L_i, b_i), where a_i (resp. b_i) is the duration of the first (resp. second) sub-task, and L_i the idle time to respect between the execution of sub-tasks. We note $\mathcal{A}$ the set of tasks, and describe the incompatibility constraint between tasks with a graph $G_c = (\mathcal{A}, E)$. There is an edge $(\mathcal{A}_i, \mathcal{A}_j) \in E$ iff a (or both) sub-task from $\mathcal{A}_i$ may be scheduled during the idle time of $\mathcal{A}_j$ or reciprocally. In a valid schedule, we said that $\mathcal{A}_i$ is *packed* into $\mathcal{A}_j$ if the entire task $\mathcal{A}_i$ is scheduled during the idle time of $\mathcal{A}_j$. This is only possible when $a_i + L_i + b_i \leq L_j$. We call *stretched coupled-task* a task $\mathcal{A}_i$ such that $a_i = L_i = b_i = \alpha(A_i)$, where $\alpha(A_i)$ is the *stretch factor* of task A_i. And for any set W of tasks, we define $seq(W) = 3 \sum_{x \in W} \alpha(x)$.

Due to the combinatorial nature of the parameters of the problem, we use the Graham's notation scheme $\alpha|\beta|\gamma$ [9] (respectively the machine environment, job characteristic and objective function) to characterize the problems related to coupled-tasks. The job characteristics summarizes the conditions made on the values of a_i, L_i, b_i (independent between tasks, or equal to a constant), and the shape of the compatibility graph G. The coupled-tasks scheduling problems under incompatibility constraints has been studied in the framework of classic complexity and approximation in [7,12].

3   Stretched Coupled-Task: Model and Contribution

3.1   Model

This paper focuses on *stretched* coupled-tasks.

In the rest of the paper, all tasks are always stretched coupled-tasks, and, for two compatible tasks A_j and A_j to be scheduled in parallel, one of the following conditions must hold:

1. $\alpha(A_i) = \alpha(A_j)$: the idle time of one task is fully exploited to schedule a sub-task from the other (i.e. b_i is scheduled during L_j, and a_j is scheduled during L_i), and the completion of the two tasks is done without idle time.
2. $3\alpha(A_i) \leq \alpha(A_j)$: task A_i is fully executed during the idle time L_j of A_j.

From this observation, one can obtain from the compatibility graph $G = (\mathcal{A}, E)$ a directed compatibility graph $G_c = (\mathcal{A}, E_c)$ by assigning a direction to each edge E from the task with the lowest stretch factor to the task with the highest one. If two compatible tasks x and y have the same stretch factor, then E_c contains both the arc (x, y) and the arc inverted (y, x). Remark that if for any pair of compatible tasks x and y we have $\alpha(x) \neq \alpha(y)$, then G_c is a directed acyclic graph.

We note $N_G(v)$ the neighbourhood of v in G. We note $d_G(v) = |N(x)|$ the degree of v in G, and Δ_G the maximum degree of G. As we focus our work on bipartite graphs, we recall that a k-*stage bipartite graph* is a digraph $G = (V_0 \cup \cdots \cup V_k, E_1 \cup \cdots \cup E_k)$ where $V_0 \ldots V_k$ are disjoint vertex sets, and each arc in E_i is from a vertex in V_i to a vertex in V_{i+1}. The vertices of V_i are said to be at rank i, and the subgraph $G_i = (V_{i-1} \cup V_i, E_i)$ is called the i-th stage of G. For clarity, 1-stage bipartite graphs can be referred as triplet (X, Y, E) instead of (V_0, V_1, E).

4   Computational Complexity

In this section, we present several $\mathcal{NP}$-complete and polynomial results. We first show the problem is $\mathcal{NP}$-hard even when the compatibility graph is a star (Theorem 1), but solvable with an $O(n^3)$ time complexity algorithm when G is a chain (Theorem 2). Then we focus our analysis when G_c is a 1-stage bipartite graph. We prove the problem is solvable with an $O(n^3)$ polynomial algorithm if $\Delta_G = 2$ (Theorem 3), but becomes $\mathcal{NP}$-hard when $\Delta_G = 3$ (Theorem 4).

Theorem 1. *The problem* $1|a_i = L_i = b_i = \alpha(A_i), G = star|C_{max}$ *is*

- *polynomial if the central node admits at least one outcoming arc.*
- $\mathcal{NP}$*-hard if the central node admits only incoming arcs.*

Proof. If there exists at least one outgoing arc $(x, y) \in G_c$ from the central node x, then the optimal solution consists in executing the x-task into the y-task, then in processing sequentially the remaining tasks after the completion of the y-task.

In the case where the central node admits only incoming arcs, first one can easily see that $1|\alpha(A_i) = a_i = L_i = b_i, G = star|C_{max}$ is $\mathcal{NP}$. Second, we propose the following polynomial construction from an instance of SS to an instance of our problem: $\forall i \in S$ we add a coupled-task x with $\alpha(x) = i$; let $\mathcal{T}$ be the set of these tasks; we add a task y with $\alpha_y = a_y = L_y = b_y = 3 \times v$; we define an incompatibility constraint between each task $x \in \mathcal{T}$ and y modelled by the compatibility graph G. In brief, G is a star with y as the central node. From this transformation, one can easily show the reduction between both problems. □

Theorem 2. *The problem* $1|a_i = L_i = b_i = \alpha(A_i), G = chain|C_{max}$ *admits a polynomial-time algorithm.*

Sketch of proof. Due to space limitation, we give only the main idea of the proof. Nevertheless, one can find the entire proof in the technical report [4].

The proof consists first in simplifying the original instance by defining some elements of an optimal solution, in order to obtain a sub-instance where in any solution at most one task can be packed in another. This new instance of the problem can be solved in polynomial time by reducing it to the search of a minimum weighted perfect matching. Basically, this reduction consists in duplicating the compatibility graph G, then in linking each node to its clone with an edge. From this new graph, we add for each edge $\{x, y\}$ a weight $w(\{x, y\})$ corresponding to half the execution time of these two tasks, and to the processing time of x if y is the clone of x. Then we perform a minimum weighted perfect matching in $O(n^2 m)$ by [6].

In following, we study the variation of the complexity in presence of a 1-stage bipartite graph according to the different values.

Theorem 3. *The problem of deciding whether an instance of* $1|a_i = L_i = b_i = \alpha(A_i), G_c = 1-stage\ bipartite, \Delta_{G_c} = 2|C_{max}$ *is polynomial.*

Proof. Let $G_c = (X, Y, E)$ be a 1-stage bipartite compatibility graph. Y-tasks will always be scheduled sequentially. The aim is to fill their idle time with a maximum of tasks of X, while the remained tasks will be executed after the Y-tasks. We just have to minimize the length of the remained tasks. Note that $d_{G_c}(y) \leq 2$. The algorithm use three steps:

1. for each task $y \in Y$ such that $3 \times \alpha(x_1) + 3 \times \alpha(x_2) \leq \alpha(y)$ where x_1 and x_2 are the only two neighbors of Y, we add y to the schedule and execute $x1$ and x_2 sequentially during the idle time of y. Then we remove y, x_1 and x_2 from the instance.
2. Each remaining task $y \in Y$ admits at most two incoming arcs (x_1, y) and/or (x_2, y). We add a weight $\alpha(x)$ to the arc (x, y) for each $x \in N(y)$, then perform a maximum weight matching on G_c in order to minimize the length of the remained tasks of X. Thus, the matched coupled-tasks are executed, and these tasks are removed from G_c.
3. Then, remaining tasks from X are allotted sequentially after the other tasks.

The complexity of an algorithm is $O(n^3)$.

Theorem 4. *The problem of deciding whether an instance of $1|a_i = L_i = b_i = \alpha(A_i), G_c =1-stage\ bipartite, \Delta_{G_c} = 3|C_{max}$ has a schedule of length at most $54n$ is $\mathcal{NP}$-complete with n the number of tasks.*

Proof. It is easy to see that our problem is in $\mathcal{NP}$. Our proof is based on a reduction from ONE-IN-$(2,3)$SAT$(2,\bar{1})$: does there exist an assignment of a set $\mathcal{V}$ of n boolean variables with $n \mod 3 \equiv 0$, a set of n clauses of cardinality two and $n/3$ clauses of cardinality three such that:

- Each clause of cardinality 2 is equal to $(x \vee \bar{y})$ for some x, $y \in \mathcal{V}$ with $x \neq y$.
- Each of the n literals x (resp. of the literals $\bar{x}$) for $x \in \mathcal{V}$ belongs to one of the n clauses of cardinality 2, thus to only one of them.
- Each of the n (positive) literals x belongs to one of the $n/3$ clauses of cardinality 3, thus to only one of them.
- Whenever $(x \vee \bar{y})$ is a clause of cardinality 2 for some x, $y \in \mathcal{V}$, then x and y belong to different clauses of cardinality 3.

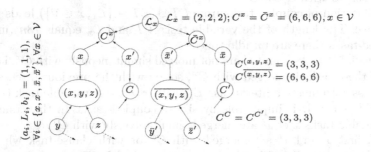

Fig. 1. A partial compatibility graph for the $\mathcal{NP}$-completeness of the scheduling problem $1|bipartite\ of\ depth\ one, d(G_c) \leq 3, \alpha_i = a_i = L_i = b_i|C_{max}$

We construct an instance π of our problem in following way (see Fig. 1):

1. For all $x \in \mathcal{V}$, we introduce four variable-tasks: x, x', $\bar{x}$ and $\bar{x}'$ with $(a_i, L_i, b_i) = (1, 1, 1), \forall i \in \{x, x', \bar{x}, \bar{x}'\}$. This variable-tasks set is noted $\mathcal{VT}$.

2. For all $x \in \mathcal{V}$, we introduce three literal-tasks $\mathcal{L}_x, C^x$ and $\bar{C}^x$ with $\mathcal{L}_x = (2, 2, 2); C^x = \bar{C}^x = (6, 6, 6)$. The set of literal-tasks is denoted $\mathcal{LT}$.
3. For all clauses with a length of three, we introduce two clause-tasks C^i and $\bar{C}^i$ with $C^i = (3, 3, 3)$ and $\bar{C}^i = (6, 6, 6)$.
4. For all clauses with a length of two, we introduce one clause-task C^i with $C^i = (3, 3, 3)$. The set of clause-tasks is denoted $\mathcal{CT}$.
5. The following arcs model the incompatibility constraints:
   (a) For all boolean variables $x \in \mathcal{V}$, we add the arcs $(\mathcal{L}_x, C^x)$ and $(\mathcal{L}_x, \bar{C}^x)$
   (b) For all clauses with a length of three denoted $C_i = (y \vee z \vee t)$, we add the arcs $(y, C^i), (z, C^i), (t, C^i)$ and $(\bar{y}', \bar{C}^i), (\bar{z}', \bar{C}^i), (\bar{t}', \bar{C}^i)$.
   (c) For all clauses with a length of two denoted $C_i = (x \vee \bar{y})$, we add the arcs (x', C^i) and $(\bar{y}, C^i)$.
   (d) Finally, we add the arcs $(x, C^x), (x', C^x)$ and $(\bar{x}, \bar{C}^x)$ and $(\bar{x}', \bar{C}^x)$.

This transformation can be computed clearly in polynomial time. The proposed compatibility graph is 1-stage bipartite and $d_{G_c}(x) \leq 3, \forall x \in \mathcal{VT} \cup \mathcal{LT} \cup \mathcal{CT}$.

In follows, we say that a task x is merged to a task y, if it exists a incompatibility constraint from x to y; i.e. the coupled-task x may be executed during the idle of coupled-task y.

$\Rightarrow$ Let us first assume that there is a schedule with length of $54n$ at most. We prove that there is a truth assignment $I : \mathcal{V} \rightarrow \{0, 1\}$ such that each clause in $\mathcal{C}$ has exactly one true literal. We make some essentials remarks:

1. The length of the schedule is given by an execution time of the coupled-tasks admitting only incoming arcs, and the value is $54n = 3\alpha_{\mathcal{CT}}|\mathcal{CT}| + \alpha_{\mathcal{LT}}(|\mathcal{LT}| - |\{\mathcal{L}_x, x \in \mathcal{V}\}|) = 9|\{C^i \in \mathcal{CT}$ of length 2 and $3\}| + 18|\{\bar{C}^i \in \mathcal{CT}\}| + 18|\{C^x$ and $\bar{C}^x \in \mathcal{LT}\}| = 9 \times \frac{4n}{3} + 18 \times \frac{n}{3} + 18 \times 2n$.
   Thus, all tasks from $\mathcal{VT} \cup \{\mathcal{L}_x, x \in \mathcal{V}\}$ must be merged with tasks from $\mathcal{CT} \cup (\mathcal{LT} - \{\mathcal{L}_x, x \in \mathcal{V}\})$.
2. By the construction, at most three tasks can be merged together.
3. $\mathcal{L}_x$ is merged with C^x or $\bar{C}^x$.
4. The allocation of coupled-tasks from $\mathcal{CT} \cup (\mathcal{LT} - \{\mathcal{L}_x, x \in \mathcal{V}\})$ leads to $18n$ idle time. The length of the variable-tasks $\mathcal{VT}$ and $\mathcal{L}_x$ equals $18n$ (in these coupled-tasks there are $6n$ idle times).
5. If the variable-tasks x and x' are not merged simultaneously with C^x, i.e. only one of these tasks is merged with C^x, so, by with the previous discussion, it is necessary to merge a literal-task $\mathcal{L}_y$, with $x \neq y$ one variable-task ($\bar{y}$ or $\bar{y}'$) with C^y or $\bar{C}^y$. It is impossible by size of coupled-tasks. In the same ways, the variable-tasks $\bar{x}$ et $\bar{x}'$ are merged simultaneously with $\bar{C}^x$.
6. Hence, first x and x' are merged with C^x or with clause-task where the variable x occurs. Second, $\bar{x}$ and $\bar{x}'$ are merged with $\bar{C}^x$ or a clause-task.

So, we affect the value "true" to the variable l iff the variable-task l is merged with clause-task(s) corresponding to the clause where the variable l occurs. It is obvious to see that in the clause of length three and two we have one and only one literal equal to "true".

$\Leftarrow$ Conversely, we suppose that there is a truth assignment $I : \mathcal{V} \rightarrow \{0, 1\}$, such that each clause in $\mathcal{C}$ has exactly one true literal.

1. If the variable $x = true$ then we merged the vertices $\mathcal{L}_x$ with C^x; x with the clause-task C^i corresponding to the clause of length three which x occurs; x' with the clause-task C^i corresponding to the clause of length two which x occurs; and $\bar{x}, \bar{x}'$ with $\bar{C}^x$.
2. If the variable $x = false$ then we merged the vertices $\mathcal{L}_x$ with $\bar{C}^x$; $\bar{x}$ with the clause-task corresponding to the clause of length two which $\bar{x}$ occurs; $\bar{x}'$ with the clause-task $\bar{C}^i$ corresponding to the clause (C) of length three which x occurs; and x, x' with C^x.

For a feasible schedule, it is sufficient to merge vertices which are in the same partition. Thus, the length of the schedule is at most $54n$.

5   Polynomial-Time Approximation Algorithms

5.1   Star Graph

Theorem 5. $1|a_i = L_i = b_i = \alpha(A_i), G = star|C_{max}$ admits a $\mathcal{FPTAS}$.

Proof. We may use the solution given by the SUBSET SUM (SS) (see [10,11]). Indeed, the schedule is follows: first the central node is executed, second during its idle time we process the coupled-tasks chosen by an $\mathcal{FPTAS}$ algorithm from SS, and finally the remaining tasks are processed after the completion of the central node.

5.2   1−Stage Bipartite Graph

Scheduling coupled-tasks during the idle time of others tasks can be related to packing problems, especially when the compatibility graph G_c is a bipartite graph. In the following, we propose several approximation results when G_c is a 1−stage bipartite graph.

Lemma 1. Let $\mathcal{P}$ be a problem with $\mathcal{P} \in \{$MKAR MSSDC, MSS$\}$ such that $\mathcal{P}$ admits a ρ-approximation, then the following problems

1. $1|a_i = L_i = b_i = \alpha(A_i), G_c = 1$-stage bipartite$|C_{max}$,
2. $1|\alpha_i = a_i = L_i = b_i, complete$ 1-stage bipartite$|C_{max}$
3. $1|\alpha_i = a_i = L_i = b_i, complete$ 1-stage bipartite$|C_{max}$ where the compatibility graph is a complete bipartite $G = (X, Y)$, and all the tasks from Y have the same $\alpha(y)$.

are approximable to a factor $1 + \frac{(1-\rho)}{3}$.

Proof. 1. Let consider an instance of $1|\alpha_i = a_i = L_i = b_i, G_c = 1-$stage bipartite$|C_{max}$ with $G_c = (X, Y, E)$ and a stretch factor function $\alpha : X \cup Y \rightarrow$ IN. In such instance, any valid schedule consists in finding for each task $y \in Y$ a subset of compatible tasks $X_y \subseteq X$ to pack into $y \in Y$, each task of x being packed at most once. Let $X_p = \cup_{y \in Y} X_y$ be the union of tasks of X

packed into a task from Y, and let $X_{\bar{p}}$ be the set of remaining tasks, with $X_{\bar{p}} = X/X_p$. Obviously, we have:

$$seq(X_p) + seq(X_{\bar{p}}) = seq(X) \qquad (1)$$

As Y is an independent set in G_c, tasks from Y have to be scheduled sequentially in any (optimal) solution. The length of any schedule S is then the processing time of Y-tasks plus the execution time of the $X_{\bar{p}}$-tasks. Formally:

$$\begin{aligned} C_{max}(S) &= seq(Y) + seq(X_{\bar{p}}) \\ &= seq(Y) + seq(X) - seq(X_p). \end{aligned} \qquad (2)$$

We use here a reduction to MKAR: each task x from X is an item having a weight $3.\alpha(x)$, each task from Y is a bin with capacity $\alpha(y)$, and each item x can be packed on y if and only if the edge $\{x, y\}$ belongs to G_c.
Using algorithms and results from the literature, one can compute the set X_p of packed items. The cost of the solution for the MKAR problem is $seq(X_p)$. If MKAR is approximable to a factor ρ, then we have:

$$seq(X_p) \geq \rho \times seq(X_p^*), \qquad (3)$$

where X_p^* is the set of packable items with the maximum profit. Combining Eqs. (2) and (3), we obtain a schedule S with a length equal to:

$$C_{max}(S) \leq seq(Y) + seq(X) - \rho \times seq(X_p^*) \qquad (4)$$

As X and Y are two fixed sets, an optimal solution S^* with minimal length $C_{max}(S^*)$ is obtained when $seq(X_p)$ is maximum, i.e. when $X_p = X_p^*$. Therefore, the ratio obtained between our solution S and the optimal one S^* is:

$$\frac{C_{max}(S)}{C_{max}(S^*)} \leq \frac{seq(Y) + seq(X) - \rho \times seq(X_p^*)}{seq(Y) + seq(X) - seq(X_p^*)} \leq 1 + \frac{(1-\rho) \times seq(X_p^*)}{seq(Y) + seq(X) - seq(X_p^*)} \qquad (5)$$

By definition, $X_p^* \subseteq X$. Moreover, as the processing time of X_p^* cannot excess the idle time of tasks from Y, we obtain: $seq(X_p^*) \leq \frac{1}{3} seq(Y)$. And thus combined to Eq. (5), we obtain the desired upper bound:

$$\frac{C_{max}(S)}{C_{max}(S^*)} \leq 1 + \frac{(1-\rho)}{3} \qquad (6)$$

2. For the problem $1|\alpha_i = a_i = L_i = b_i, complete\ 1-stage\ bipartite|C_{max}$, the proof is identical using MSSDC as a special case of MKAR where each item can be packed in any bin.

3. For the problem $1|\alpha_i = a_i = L_i = b_i, complete\ 1-stage\ bipartite|C_{max}$ where all the tasks from Y have the same stretch factor $\alpha(y)$, the proof is identical as previously since MSSDC is a generalisation of MSS. $\square$

Theorem 6. *These problems admit a polynomial-time approximation algorithm:*

1. *The problem* $1|a_i = L_i = b_i = \alpha(A_i), G_c = 1\text{-stage bipartite}|C_{max}$ *is approximable to a factor* $\frac{7}{6}$.
2. *The problem* $1|a_i = L_i = b_i = \alpha(A_i), G_c = complete\ 1\text{-stage bipartite}|C_{max}$ *admits a* $\mathcal{PTAS}$.
3. *The problem* $1|a_i = L_i = b_i = \alpha(A_i), G_c = complete\ 1\text{-stage bipartite}|C_{max}$, *where all the tasks from* Y *have the same stretch factor* $\alpha(y)$:
   (a) is approximable to a factor $\frac{13}{12}$.
   (b) admits a $\mathcal{PTAS}$.

Proof. 1. Authors from [5] proposed a $\rho = \frac{1}{2}$−approximation algorithm for MKAR. Reusing this result with Lemma 1, we obtain a $\frac{7}{6}$−approximation.
  2. We know that MSSDC admits a $\mathcal{PTAS}$ [1], i.e. $\rho = 1 - \epsilon$. Using this algorithm to compute such a $\mathcal{PTAS}$ and the Lemma 1, we obtain an approximation ratio of $1 + \frac{\epsilon}{3}$ for this problem.
  3. (a) Authors from [3] proposed a $\rho = \frac{3}{4}$−approximation algorithm for MSS. Reusing this result and the Lemma 1, we obtain a $\frac{13}{12}$−approximation.
  (b) They also proved that MSS admits a $\mathcal{PTAS}$ [2], i.e. $\rho = 1 - \epsilon$. Using the algorithm to compute such a $\mathcal{PTAS}$ and the Lemma 1, we obtain an approximation ratio of $1 + \frac{\epsilon}{3}$. $\square$

5.3 2−Stage Bipartite Graph

Theorem 7. *The problem* $1|a_i = L_i = b_i = \alpha(A_i), G_c = 2\text{-stage bipartite}|C_{max}$ *is approximable to a factor* $\frac{13}{9}$.

Proof. We consider an instance of the problem with $G_c = (V_0 \cup V_1 \cup V_2, E_1 \cup E_2)$, where each arc in E_i is oriented from a vertex in V_i to another one in V_{i+1}, for $i \in 1, 2$.

Before presenting our heuristic and the analyse of its approximation factor, we will give several notations, properties and equations in relation with the specificities of this instance, in any (optimal) solution:

- $\forall i = 0, 1$, let V_{ip} (p=packed), (resp. V_{ia} (a=alone)) be the set of tasks merged (resp. remaining) into any task from V_{i+1} in a solution S, and V_{ib} (b=box) the set of tasks scheduled with some tasks from V_{i-1} merged into it. This notation is extended to an optimal solution S^* by adding a star in the involved variables.
- Given any solution S to the problem and considering the specificities of the instance, note that $\{V_{0p}, V_{0a}\}$ is a partition of V_0, G_c, $\{V_{1p}, V_{1a}, V_{1b}\}$ is a partition of V_1, and G_c, $\{V_{2a}, V_{2b}\}$ is a partition of V_2.
- Any solution would consists in scheduling first each task with at least one task merged into it, then to schedule the remaining tasks (alone). Given an optimal solution S^*, the length of S^* is given by the following equation:

$$S^* = seq(V_{1b}^*) + seq(V_2 b) + seq(V_{0a}^*) + seq(V_{1a}^*) + seq(V_{2a}^*)$$
$$S^* = seq(V_2) + seq(V_{1b}^*) + seq(V_{0a}^*) + seq(V_{1a}^*) \tag{7}$$

One can remark that V_{0p}^* and V_{1p}^* are not part of the equation, as they are scheduled during the idle time of V_{1b}^* and V_{2b}^*.

– Let consider an restricted instance of G_c to a sub-graph $G_0 = G_c[V_0 \cup V_1]$ (resp. $G_1 = G_c[V_1 \cup V_2]$) which is the 1-th (resp. 2-th) stage of G_c.

Let $S[G_0]$ (resp. $S^*[G_0]$) be any (an optimal) solution on G_0, $V_{0p}[G_0]$ (resp. $V_{0p}^*[G_0]$) is the set of tasks from V_0 packed into tasks from V_1 in $S[G_0]$ (resp. $S^*[G_0]$), and $V_{0a}[G_0]$ (resp. $V_{0a}^*[G_0]$) the set of remaining tasks. In addition tho these notation, let $V_{1b}[G_0]$ be the set of tasks from V_1 with at least one task from V_0 merged into them, and $V_{1a}[G_0]$ the remaining tasks. A first observation gives for G_0:

$$S^*[G_0] = seq(V_1) + V_{0a}^*[G_0] \tag{8}$$

– From Theorem 6, Lemma 1, and the demonstration presented in their proof from [5], several equations can be computed for a solution $S[G_0]$:

$$seq(V_{0p}[G_0]) \geq \frac{1}{2} seq(V_{0p}^*[G_0]) \tag{9}$$

$$seq(V_{0a}^*[G_1]) \leq seq(V_{0a}^*) \tag{10}$$

$$seq(V_{0p}[G_0]) + seq(V_{0a}[G_0]) = seq(V_{0p}^*[G_0]) + seq(V_{0a}^*[G_0]) = seq(V_0) \tag{11}$$

$$seq(V_{0a}[G_0]) \leq seq(V_{0a}^*[G_0]) + \frac{1}{2} seq(V_{0p}^*[G_0]) \leq seq(V_{0a}^*) + \frac{1}{2} seq(V_{0p}^*[G_0]) \tag{12}$$

– We use an analog reasoning on the sub-graph G_1 with equivalent notations for V_1 and V_2, and we obtain:

$$seq(V_{1p}[G_1]) \geq \frac{1}{2} seq(V_{1p}^*[G_1]) \tag{13}$$

$$seq(V_{1a}[G_1]) \leq seq(V_{1a}^*[G_1]) + 1/2 seq(V_{1p}^*[G_1]) \leq seq(V_{1a}^*) + 1/2 seq(V_{1p}^*[G_1]) \tag{14}$$

From this notations and observation, we can propose a good heuristic. We design the feasible solution S for G_c as follows:

– We compute a solution $S[G_1]$ on G_1, then we add to S each task from V_2 and the tasks from V_1 merged into them (i.e. $V_{1p}[G_1]$) in $S[G_1]$.
– Then we compute a solution $S[G_0]$ on G_0, then we add to S each task v from $V_{1b}[G_0]/V_{1p}[G_1]$ and the tasks from V_0 merged into them.
– Tasks $V_{1a}[G_1]/V_{1b}[G_0]$ and $V_{0a}[G_0]$ are added to S sequentially.
– We note $V_{conflict}$ the set of remaining tasks, i.e. the set of tasks from V_0 which are merged into a task $v \in V_1$ in $S[G_0]$, thus that v is merged into a task from V_2 in $S[G_1]$.

Remark that:

$$seq(V_{1b}[G_0]/V_{1p}[G_1]) + seq(V_{1a}[G_1]/V_{1b}[G_0]) = V_{1a}[G_1] \tag{15}$$

Thus the cost of our solution S is

$$S = seq(V_2) + seq(V_{1a}[G_1]) + seq(V_{0a}[G_0]) + seq(V_{conflict}) \tag{16}$$

It is also clear that:

$$seq(V_{conflict}) \le \frac{1}{3} seq(V_{1p}[G_1]) \le \frac{1}{3} seq(V_{1p}^*[G_1]) \tag{17}$$

Using Eqs. (12), (14) and (17) in Eq. (16), we obtain

$$S \le seq(V_2) + seq(V_{1a}^*) + \frac{5}{6} seq(V_{1p}^*[G_1]) + seq(V_{0a}^*) + \frac{1}{2} seq(V_{0p}^*[G_0]) \tag{18}$$

$$\le S^* + \frac{5}{6} seq(V_{1p}^*[G_1]) + \frac{1}{2} seq(V_{0p}^*[G_0]), \text{ using Eq. (7)} \tag{19}$$

We know that $S^* \ge seq(V_2)$ and $S^* \ge seq(V_1)$, as V_1 is an independent set of G_c. We also know that tasks from $(V_{1p}^*[G_1])$ (resp. $(V_{0p}^*[G_0])$) must be merged into tasks from V_2 (resp. V_1) and cannot exceed the idle time of V_2 (resp. V_1), implying that $seq(V_{1p}^*[G_1])) \le \frac{1}{3} seq(V_2)$ (resp. $seq(V_{0p}^*[G_0])) \le \frac{1}{3} seq(V_1)$). One can write the following :

$$\frac{\frac{5}{6} seq(V_{1p}^*[G_1])}{S^*} \le \frac{\frac{5}{6} \times \frac{1}{3} seq(V_2)}{seq(V_2)} \le \frac{5}{18} \tag{20}$$

$$\frac{\frac{1}{2} seq(V_{0p}^*[G_0])}{S^*} \le \frac{\frac{1}{2} \cdot \frac{1}{3} seq(V_1)}{seq(V_1)} \le \frac{1}{6} \tag{21}$$

Finally, from Eqs. (19), (20) and (21) the proof is concluded:

$$\frac{S}{S^*} \le 1 + \frac{5}{18} + \frac{1}{6} = \frac{13}{9}$$

□

6   Conclusion

The results proposed in this paper are summarised in Table 1. New presented results suggest the main problem of coupled tasks scheduling remains difficult even for restrictive instances, here stretched coupled-tasks when the constraint graph is a bipartite graph. When we consider stretched coupled-tasks, the maximum degree Δ_G seems to play an important role on the problem complexity, as the problem is already $\mathcal{NP}$-Hard to solve when the constraint graph is a star. Approximation results presented in this paper show the problem can be approximated with interesting constant ratio on $k-$stage bipartite graphs for $k = 1$ or 2. The presented approach suggests a generalisation is possible for $k \ge 3$. This part constitutes one perspective of this work. Other perspective would consists to study coupled-tasks on other significant topologies, including degree-bounded trees, or regular topologies like the grid.

Table 1. Complexity and approximation results.

Topology	Complexity	Approximation
$uug(G_c)$=Star graph	$\mathcal{NP} - \mathcal{C}$ (Theorem 1)	$\mathcal{FPTAS}$ (Theorem 5)
$uug(G_c)$=Chain graph	$O(n^3)$ (Theorem 2)	
G_c= 1-stage bipartite, $\Delta(G_c) = 2$	$O(n^3)$ (Theorem 3)	
G_c= 1-stage bipartite, $\Delta(G_c) = 3$	$\mathcal{NP} - \mathcal{C}$ (Theorem 4)	$\frac{7}{6}$-$\mathcal{APX}$(Theorem 6)
G_c= complete 1-stage bipartite	$\mathcal{NP} - \mathcal{C}$ (see [12])	$\mathcal{PTAS}$ (Theorem 6)
G_c= complete 1-stage bipartite	$\mathcal{NP} - \mathcal{C}$ (see [12])	$\mathcal{PTAS}$ (Theorem 6)
with constraint $\alpha(x) = \alpha(y)$, $\forall x, y \in X_1$		$\frac{13}{12}$-$\mathcal{APX}$(Theorem 6)
G_c= 2-stage bipartite	$\mathcal{NP} - \mathcal{C}$ (Theorem 4)	$\frac{13}{9}$-$\mathcal{APX}$ (Theorem 7)

Acknowledgment. This work has been funded by the regional council of Burgundy.

References

1. Caprara, A., Kellerer, H., Pferschy, U.: A PTAS for the multiple subset sum problem with different knapsack capacities. Inf. Process. Lett. **73**(3–4), 111–118 (2000)
2. Caprara, A., Kellerer, H., Pferschy, U.: The multiple subset sum problem. Siam J. Optim. **11**(2), 308–319 (2000)
3. Caprara, A., Kellerer, H., Pferschy, U.: A 3/4-approximation algorithm for multiple subset sum. J. Heuristics **9**(2), 99–111 (2003)
4. Darties, B., Simonin, G., Giroudeau, R., König, J.-C.: Scheduling stretched coupled-tasks with compatibilities constraints: model, complexity and approximation results for some class of graphs. Report, February 2014
5. Dawande, M., Kalagnanam, J., Keskinocak, P., Salman, F.S., Ravi, R.: Approximation algorithms for the multiple knapsack problem with assignment restrictions. J. Comb. Optim. **4**(2), 171–186 (2000)
6. Edmonds, J.: Maximum matching and a polyhedron with 0, 1 vertices. J. Res. Natl. Bur. Stand. **69B**, 125–130 (1965)
7. Simonin, G., Darties, B., Giroudeau, R., König, J.C.: Isomorphic coupled-task scheduling problem with compatibility constraints on a single processor. J. Sched. **14**(5), 501–509 (2011)
8. Garey, M.R., Johnson, D.S.: Computers and Intractability: A Guide to the Theory of NP-Completeness. W. H. Freeman & Co., New York (1979)
9. Graham, R.L., Lawler, E.L., Lenstra, J.K., Kan, A.H.G.R.: Optimization and approximation in deterministic sequencing and scheduling: a survey. Ann. Discret. Math. **5**, 287–326 (1979)
10. Ibarra, O.H., Kim, C.E.: Fast approximation algorithms for the knapsack and sum of subset problems. J. ACM **22**(4), 463–468 (1975)
11. Kellerer, H., Mansini, R., Pferschy, U., Speranza, M.G.: An efficient fully polynomial approximation scheme for the subset-sum problem. J. Comput. Syst. Sci. **66**(2), 349–370 (2003)
12. König, J.-C., Simonin, G., Giroudeau, R.: Complexity and approximation for scheduling problem for coupled-tasks in presence of compatibility tasks. In: Project Management and Scheduling (2010)
13. Shapiro, R.D.: Scheduling coupled tasks. Naval Res. Logist. Q. **27**, 477–481 (1980)

The Computational Complexity
of Stochastic Optimization

Cassio Polpo de Campos[4], Georgios Stamoulis[1,3], and Dennis Weyland[1,2](✉)

[1] Università della Svizzera italiana, Lugano, Switzerland
{stamoulis.georgios,dennisweyland}@gmail.com
[2] Università degli Studi di Brescia, Brescia, Italy
[3] Lamsade, Université Paris Dauphine, Paris, France
[4] Dalle Molle Institute for Artificial Intelligence, Lugano, Switzerland
cassio@idsia.ch

Abstract. This paper presents an investigation on the computational
complexity of stochastic optimization problems. We discuss a scenario-
based model which captures the important classes of two-stage stochastic combinatorial optimization, two-stage stochastic linear programming,
and two-stage stochastic integer linear programming. This model can also
be used to handle chance constraints, which are used in many stochastic
optimization problems. We derive general upper bounds for the complexity of computational problems related to this model, which hold under
very mild conditions. Additionally, we show that these upper bounds are
matched for some stochastic combinatorial optimization problems arising
in the field of transportation and logistics.

Keywords: Stochastic combinatorial optimization · Computational
complexity · Chance constraints · Stochastic vehicle routing

1 Introduction

Stochastic optimization problems have received increasing attention in recent
years. While these problems are used extensively in practice, our theoretical
understanding of their complexity is far from complete. Hardness results have
been obtained in the context of two-stage stochastic linear programming and
two-stage stochastic integer linear programming [3]. It has been shown that just
the evaluation of the objective function is already #P-hard, where #P denotes
the famous class of counting problems originally introduced in [7]. The same
hardness results could have been derived for the corresponding decision and
optimization variants. Similar lower bounds have been obtained in the context
of stochastic combinatorial optimization problems for a widely used stochastic
vehicle routing problem [8]. Analogously, it can be shown that the evaluation of
stochastic/chance constraints is #P-hard as well. On the other hand, we do not
have strong upper bounds for the computational complexity of stochastic optimization problems. Some attempts have been done in [3], but the corresponding

© Springer International Publishing Switzerland 2014
P. Fouilhoux et al. (Eds.): ISCO 2014, LNCS 8596, pp. 173–185, 2014.
DOI: 10.1007/978-3-319-09174-7_15

results are controversial, since the equality of $P^{\#P}$ and $NP^{\#P}$ is assumed, which is to the best of our knowledge still an open problem.

In this paper we investigate a very general scenario-based model for stochastic optimization problems. This model includes, among others, the above mentioned classes of two-stage stochastic linear programming and two-stage stochastic combinatorial optimization. Our main results are general upper bounds, which hold under very mild assumptions, and lower bounds derived for a very plausible stochastic vehicle routing problem. We show that the evaluation of the objective function is in $FP^{\#P[1]}$ and that the evaluation of constraints is in PP. Furthermore, the decision variant of such problems resides in $NP^{\#P[1]}$ and the optimization variant can be solved with multiple calls to the corresponding decision variant and is therefore in $FP^{NP^{\#P[1]}}$. We then show that these bounds are actually matched by an existing stochastic vehicle routing problem, where the objective function is #P-hard and the decision and optimization variants are both $NP^{\#P[1]}$-hard.

The remaining part of this paper is organized as follows. We start with a discussion of the model used in this paper in Sect. 2, and explain why this model captures many important stochastic optimization problems. In Sect. 3 we derive general upper bounds for some computational tasks related to this model, which hold under very mild conditions. We then continue to show that these upper bounds are actually matched for a (non artificial) stochastic vehicle routing problem (Sect. 4). Finally, we conclude the paper with a short discussion of the results and their implications in Sect. 5.

2   The Stochastic Optimization Model

We discuss a scenario-based model for stochastic optimization problems that is very general and captures, among others, two-stage stochastic combinatorial optimization, two-stage stochastic linear programming and two-stage stochastic integer linear programming. Additionally, this model can also handle chance constraints which are used in many stochastic optimization problems.

The basic assumption of our model is that we can describe the objective function and the constraints in a scenario-based way. For a given problem instance, we have a set $\mathcal{X}$ of solutions and a set $\mathcal{S}$ of scenarios. There is a mass function $p : \mathcal{S} \rightarrow \mathbb{R}^+$ representing the probabilities of the scenarios and a function $k : \mathcal{X} \times \mathcal{S} \rightarrow \mathbb{R}^+$ representing the costs of a solution under a specific scenario. The objective function $f : \mathcal{X} \rightarrow \mathbb{R}^+$ is then the expectation of the costs over the scenarios, that is, for a given solution $x \in \mathcal{X}$, we have $f(x) = \sum_{s \in \mathcal{S}} p(s)k(x, s)$. Constraints representing simple predicates on the solution space are allowed and divided into the two sets $\mathcal{C}$ and $\mathcal{D}$. Constraints $c \in \mathcal{C}$ are computable predicates $c : \mathcal{X} \rightarrow \{\text{false}, \text{true}\}$ and correspond to non-stochastic constraints. Constraints in $\mathcal{D}$ are defined in a similar way as the objective function and correspond to stochastic constraints. Each $d \in \mathcal{D}$ is associated with two functions $p_d : \mathcal{S} \rightarrow \mathbb{R}^+$ and $k_d : \mathcal{X} \times \mathcal{S} \rightarrow \mathbb{R}$ and a bound $b_d \in \mathbb{R}$. For a given solution $x \in \mathcal{X}$, the predicate d is simply $\sum_{s \in \mathcal{S}} p_d(s)k_d(x, s) \leq b_d$.

For the model to be meaningful, we need the following additional assumptions. First of all, we assume that the input, including all the functions' specifications, is shortly encoded in the sense that the sets $\mathcal{X}$ and $\mathcal{S}$ are of at most exponential size with respect to the input. Additionally, we require that the size of these sets can be efficiently computed and that these two sets can be efficiently enumerated. Furthermore, we require p and k to provide efficiently computable numbers[1]. Analogously, the functions p_d and k_d associated with the constraints $d \in \mathcal{D}$ are required to provide efficiently computable numbers. Finally, the predicates $c \in \mathcal{C}$ are required to be computable in polynomial time.

This model is very powerful and it is easy to verify that it captures the important classes of two-stage stochastic combinatorial optimization, two-stage stochastic linear programming, two-stage stochastic integer linear programming (with continuous variables in the second stage) and chance-constrained programming. For more information about these classes we refer to [1].

3   General Upper Bounds

The relation between stochastic optimization problems [1] and counting problems [7] has proven to be very useful in order to derive results about the computational complexity of stochastic optimization problems [3, 8]. In this work we use the recently introduced framework of weighted counting [2] to derive upper bounds for the computational complexity of our model. For the sake of clarity, we first give an overview about the framework of weighted counting and the results which are needed in the context of this paper. We then discuss the very mild technical condition to our stochastic optimization model and give formulations for the corresponding computational problems. After that, we derive upper bounds for these computational problems using the results about weighted counting.

Weighted counting problems are a natural generalization of conventional counting problems [2]. The computational variant and the decision variant of weighted counting problems can be defined as follows.

Definition 1 (Weighted Counting Problem). *We are given a polynomial p and a function $w : \{0,1\}^\star \times \{0,1\}^\star \to \mathbb{R}$ that can be approximated by a polynomial time (in the size of the first two arguments and the third argument) computable function $v : \{0,1\}^\star \times \{0,1\}^\star \times \mathbb{N} \to \mathbb{Z}$, such that $|w(x,u) - v(x,u,b)/2^b| \leq 2^{-b}$ for all $x \in \{0,1\}^\star, u \in \{0,1\}^\star, b \in \mathbb{N}$. The weighted counting problem associated with p and w is to compute for $x \in \{0,1\}^\star$ the function*

$$f(x) = \sum_{u \in \{0,1\}^{p(|x|)}} w(x,u).$$

---

[1] That means p and k have one additional input which specifies the number of output bits that are required. In this way p and k are providing the numbers to any desired accuracy in polynomial time with respect to the input and the number of output bits.

Definition 2 (Weighted Counting Problem, Decision Variant). *We are given a weighted counting problem defined by a polynomial p and a function $w : \{0,1\}^\star \times \{0,1\}^\star \to \mathbb{R}$ as well as a threshold value $t \in \mathbb{Q}$. The corresponding decision problem is to decide for $x \in \{0,1\}^\star$ whether $f(x) \geq t$ or not.*

Here the variable x is the "input" and w is a function which assigns an efficiently computable weight to each of the exponentially many values of u. At a first glance, the relationship to our scenario-based model is apparent. The important observation is that we are able to describe the objective function and the stochastic constraints of stochastic optimization problems using the scenario-based model in terms of weighted counting. In fact, the computation of the objective function and the evaluation of the stochastic constraints can be seen as weighted counting problems themselves. By exploiting this fact, we derive upper bounds for the complexity of computational tasks related to stochastic optimization problems. The (slightly adapted) results regarding weighted counting [2] that are important for this work can be stated as follows.

Theorem 1. *We are given a weighted counting problem defined by a polynomial p and a function $w : \{0,1\}^\star \times \{0,1\}^\star \to \mathbb{R}$. If the size of the output (eventually encoded as a fraction) is bounded by a polynomial $q(|x|)$, then the given weighted counting problem is in $FP^{\#P[1]}$.*

Theorem 2. *We are given the decision variant of a weighted counting problem defined by a polynomial p, a function $w : \{0,1\}^\star \times \{0,1\}^\star \to \mathbb{R}$ and a threshold value $t \in \mathbb{Q}$. If the size of the output (of the computational variant, eventually encoded as a fraction) is bounded by a polynomial $q(|x|)$, then the problem is in PP.*

It is evident that in order to use these results we have to make an additional, mild condition to our scenario-based model: We assume that the size of the output of the objective function and the (computational variants of the) constraints is polynomially bounded. That means that we require the output to be limited to a polynomial number of bits and this is indeed an extremely mild condition[2]. With this assumption, we can immediately obtain the following results.

Theorem 3. *The following statements hold for stochastic optimization problems using the scenario-based model.*

(i) The evaluation of the objective function is in $FP^{\#P[1]}$.
(ii) Deciding whether a given solution has costs of at most $t \in \mathbb{Q}$ is in PP.
(iii) The evaluation of a stochastic constraint is in PP.
(iv) Checking the feasibility of a given solution is in PP.

---

[2] In fact, we can even handle problems which do not fulfill this assumption by truncating the objective function and the constraints after a certain number of at most polynomially many bits. The resulting problem is then a slight perturbation of the original problem which should not incur any difference for practical purposes.

(v) The task of computing the objective function in case the given solution is feasible, or returning some arbitrarily fixed value in case the given solution is not feasible is in $FP^{\#P[1]}$.

Proof. (i), (ii) and (iii) follow directly from the fact that we can write the objective function and the stochastic constraints as weighted counting problems. Since PP is closed under multiple non-adaptive/independent calls [4] and since the constraints can be checked independently, (iv) follows. For (v) we have to combine the computation of the objective function and the verification of the constraints. It is clear that we can perform this task in polynomial time with 2 calls to a #P-oracle. Since these oracle calls are independent, they can be combined into a single call, which shows that the task is in $FP^{\#P[1]}$.   □

Using this result for the evaluation of solutions, we can derive the following upper bounds for the optimization and decision variants of stochastic optimization problems using the scenario-based model.

Theorem 4. *We have given a stochastic optimization problem using the scenario-based model and a bound $t \in \mathbb{Q}$. The problem to decide whether a solution with costs at most t exists or not is in $NP^{\#P[1]}$.*

Proof. Here the idea is to create all possible solutions for the given problem in a nondeterministic way. According to Theorem 3 the objective function and the constraints can be evaluated within the NP machine using only a single call to a #P-oracle. Finally, a solution is accepted if it is feasible and has costs of at most t. That means our $NP^{\#P[1]}$ accepts the input if and only if there is at least one feasible solution which obeys the cost bound.   □

Theorem 5. *We have given a stochastic optimization problem using the scenario-based model. The task to compute an optimal solution (if it exists) or to return some arbitrarily fixed value in the case where no feasible solution exists can be solved in polynomial time using an oracle for the decision version. In other words, this problem is in $FP^{NP^{\#P[1]}}$.*

Proof. We have to show that we can solve the optimization variant in polynomial time with oracle access to the decision variant. We start with a binary search to determine the costs of an optimal solution. Since the size of the output of the objective function is polynomially bounded, we can do this with polynomially many oracle calls. Once we know the value of an optimal solution, we perform a second binary search on the set of solutions. This can be done efficiently by dividing the set of solutions that is enumerated in the NP part of the oracle into two parts. We then continue with any of the two sets of solutions that still contains an optimal solution. Since the number of possible solutions is at most exponentially large, this step can also be performed with at most polynomially many oracle calls.   □

4   Hardness Results for the Dependent PTSPD

In this section we will complement the general upper bounds for our model of stochastic optimization problems with lower bounds for a specific stochastic vehicle routing problem. It seems fairly easy to prove such lower bounds for artificially created stochastic optimization problems. The strength of our results is based on the fact that we are able to show strong lower bounds for a practical stochastic combinatorial optimization problem.

We focus on the Dependent Probabilistic Traveling Salesman Problem with Deadlines (Dependent PTSPD, [8]). As a generalization of the Probabilistic Traveling Salesman Problem with Deadlines, the objective function is #P-hard [8] and therefore basically matches the lower bound derived in the previous section. The Dependent PTSPD also inherits the #P-hardness for the decision and optimization variants from the PTSPD. We strengthen these hardness results and show that the optimization and decision variants of the Dependent PTSPD are both in fact $NP^{\#P[1]}$-hard. For this purpose we use reductions from the $NP^{\#P[1]}$-complete problem E-MINORITY-SAT (explained later on). We first give the formal definitions of the problems used in this section. After that we present the reduction from E-MINORITY-SAT to the decision variant of the Dependent PTSPD in detail. At the end, we show how this reduction can be modified for the optimization variant of the Dependent PTSPD.

4.1   Problem Definitions

In [6] it has been shown that the problem E-MAJ-SAT is $NP^{\#P[1]}$-complete. For this problem we have given a formula in conjunctive normal form. The variables are partitioned into two sets. The question is if there exists an assignment for the first set of variables such that at least half of the assignments for the second set of variables (together with the assignment for the first set of variables) satisfy the given formula. For our proof we use a variant of this problem called E-MINORITY-SAT, which is defined analogously but asks for *at most* half of the assignments to satisfy the formula.

Problem 1 (E-MINORITY-SAT). We have given a boolean formula over n variables $x_1, x_2, \ldots, x_n$ in conjunctive normal form with m clauses and a number $k \in \{0, 1, \ldots, n\}$. The task is to decide if there exists an assignment of the first k variables, such that at most half of the assignments of the remaining variables (together with the assignment of the first k variables) satisfy the given formula.

E-MINORITY-SAT is $NP^{\#P[1]}$-complete. The proof is analogous to the one for E-MAJ-SAT [6] and makes use of PP closure properties [4].

The formal definition of the Dependent Probabilistic Traveling Salesman Problem (Dependent PTSPD, [8]) is more intricate. Here we refer as V a set of n locations, including the special starting point $v_0 \in V$. We have given distances / travel times between the locations which are represented by a function $d : V \times V \to \mathbb{Q}^+$. Deadlines for different customers are modeled using a function

$t : V \rightarrow \mathbb{Q}^+$ and penalty values for different customers are modeled using a function $h : V \rightarrow \mathbb{Q}^+$. For simplicity we also define these values for the starting point v_0, although we meet the deadline at v_0 nevertheless, since we start the tour there. Additionally, the presence of customers is modeled in a stochastic way. We allow certain kinds of dependencies between the presence of different customers. Two customers $v_i, v_j \in V$ can be bonded in the following way: (1) the presence of v_i and v_j is independent, (2) v_i is present if and only if v_j is present, or (3) v_i is present if and only if v_j is absent. These dependencies can be efficiently modeled by defining sets of paired customers and their associated bonds. We avoid further details for the sake of clarity, since we explicitly point out the necessary dependencies used in our reduction. The probabilities for the customers' presence are represented by a function $p : V \rightarrow [0, 1]$. Obviously, p is assumed to respect the dependencies.

A solution can now be represented by a permutation $\tau : [n] \rightarrow V$ with $\tau_1 = v_0$. For a specific realization of the customers' presence we use this solution to derive a second-stage solution just by skipping customers which are absent. The costs for the second-stage solution are then the sum of the travel times plus the penalties for missed deadlines. We assume that a customer specific fixed penalty of $h(v)$ for customer $v \in V$ occurs in case the deadline is missed, independently of the actual delay. The costs for a solution τ are the expected costs of the second-stage solutions derived from τ over the different realizations of the customers' presence.

Let $\tau : [n] \rightarrow V$ with $\tau_1 = v_0$ be a solution. For all $v \in V$, let A_v be a random variable indicating the arrival time at customer v. Since the travel times of the second-stage solutions are identical to those of the Probabilistic Traveling Salesman Problem (PTSP, [5,9]), the costs of τ can be expressed as

$$f_{\text{ptspd}}(\tau) = f_{\text{ptsp}}(\tau) + \sum_{i=1}^{n} \Pr(A_{\tau_i} \geq t(\tau_i)) \, h(\tau_i).$$

The first part of the costs is the (polynomial time computable) objective function of the PTSP and represents the expected travel times over the second-stage solutions. The second part represents the penalties for missed deadlines. With this expression for the costs of a solution, we define the decision and optimization variants of the Dependent PTSPD in the following way.

Problem 2 (Dependent PTSPD - Decision Variant). Given a set V of size n with a special element $v_0 \in V$, a function $d : V \times V \rightarrow \mathbb{Q}^+$, sets of pairs of V defining the customers' bonds, a function $p : V \rightarrow [0, 1]$ respecting the dependencies imposed by the partition, a function $t : V \rightarrow \mathbb{Q}^+$, a function $h : V \rightarrow \mathbb{Q}^+$, and a bound $b \in \mathbb{Q}$, the problem is to decide if there exists a solution $\tau : [n] \rightarrow V$ with $\tau_1 = v_0$ such that $f_{\text{ptspd}}(\tau) \leq b$.

Problem 3 (Dependent PTSPD - Optimization Variant). Given a set V of size n with a special element $v_0 \in V$, a function $d : V \times V \rightarrow \mathbb{Q}^+$, sets of pairs of V defining the customers' bonds, a function $p : V \rightarrow [0, 1]$ respecting the dependencies imposed by the partition, a function $t : V \to \mathbb{Q}^+$ and a function $h : V \to \mathbb{Q}^+$, the problem is to compute a permutation $\tau^\star : [n] \to V$ with $\tau_1^\star = v_0$ such that $f_{\mathrm{ptspd}}(\tau^\star) \leq f_{\mathrm{ptspd}}(\tau)$ for any permutation $\tau : [n] \to V$ with $\tau_1 = v_0$.

4.2 Hardness of the Decision Variant

We now present a reduction from E-MINORITY-SAT to the decision variant of the Dependent PTSPD. First, we give the general idea behind the reduction, and then we show step by step how for any given instance of E-MINORITY-SAT an instance of the Dependent PTSPD can be constructed. Based on this construction we conclude that the Dependent PTSPD is $NP^{\#P[1]}$-hard. Later we show how this reduction can be modified to obtain the same hardness result for the optimization variant of the Dependent PTSPD.

The overall idea is to construct, for a given E-MINORITY-SAT instance, a highly restricted instance for the decision variant Dependent PTSPD. By highly restricted we mean that deadlines are used to allow only certain paths to appear in an optimal solution. We then simulate the NP decision process, that is, we simulate the assignment for the first set of variables, by using a gadget for each variable which allows an optimal solution to take one of two possible paths. By using customers which are present with a probability of $1/2$ and with properly defined dependencies, we create "copies" of these assignments. The second set of variables from the E-MINORITY-SAT instance is modeled by customers which are present with a probability of $1/2$. In this case proper dependencies can be used to directly create "copies" of these customers. We then build a new collection of gadgets for the clauses of the formula. The idea is that an assignment which satisfies the formula leads to a certain (controlled) delay. At the end, a special customer is added. The probability for a deadline violation at this customer in an optimal solution enables us to infer the minimum (over the assignments of the first set of variables) number of assignments for the second set of variables that satisfy the clauses. Finally, the cost of an optimal solution for the constructed instance allows us to infer the probability with which the deadline is actually violated in an optimal solution, and this solves the given E-MINORITY-SAT instance.

Simulating the Assignment of the First k Variables. For the simulation of the NP decision process, we use the gadget depicted in Fig. 1. The distance function used here (and also throughout the whole instance) is the Euclidean distance. The starting customer here is v_0. The presence probabilities are 1 for all customers except x_1 and $\bar{x}_1$. The presence probabilities for these two customers are $1/2$ and additionally x_1 is present if and only if $\bar{x}_1$ is absent[3]. By using large enough penalty values and appropriate deadlines, we can force

---

[3] We use the following convention for all the x customers in the reduction: A group of customers with the same label is either completely present or none of those customers is present. Customers with a label of x_i are present if and only if the customers with the label $\bar{x}_i$ are absent.

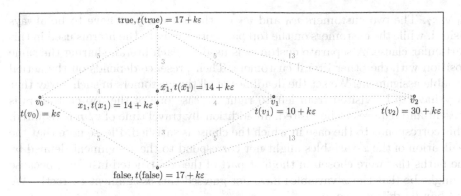

Fig. 1. The gadget used to simulate the variable assignment.

a solution to take one of the following two paths: $v_0, true, v_1, x_i, \bar{x}_i, false, v_2$ or $v_0, false, v_1, x_i, \bar{x}_i, true, v_2$ [4]. The first path corresponds to an assignment of *true* for the corresponding x_i from the E-MINORITY-SAT formula, while the second corresponds to an assignment of *false* for x_i. The goal of the stochastic customers is to guess such chosen assignment. In case the guess is wrong, the travel time within this gadget is by a value of $d(x_i, \bar{x}_i) = \varepsilon$ larger than if the guess would be correct. The reason why we are guessing the chosen assignment is that we can use dependencies between stochastic customers to create "copies" which we require at later stages of the construction to check the clauses. The drawback is that we have to handle the case in which the guess is wrong, but this is not an issue as we will see soon.

It is clear that we can put k of these gadgets in a row to simulate the assignment of the first k variables. Since we might have some additional delay of ε for each of the gadgets, the deadlines all contain some sort of slack. As long as this distance is small enough compared to the other distances in the gadget, this does not change the overall characteristic of possible optimal paths.[5]

The situation now is as follows. A solution in which the gadgets are visited one after another and in which one of the two paths is chosen for each of the gadgets is always better (of lower cost) than a solution which does not visit the customers in this order. This just follows from the fact that we can use sufficiently large penalty values for the involved customers. Additionally, a realization of the presence for the customers $x_1, x_2, \ldots, x_k$ (and therefore also for the customers $\bar{x}_1, \bar{x}_2, \ldots, \bar{x}_k$) might result in a delay. If the realization correctly guesses the assignment chosen by the path, the delay is 0, otherwise the delay is at least ε.

Verifying the Clauses. To verify the clauses' satisfiability, we build gadgets as depicted in Fig. 2. Here we illustrate the construction for the clause $x_1 \vee \bar{x}_3 \vee$

---

[4] In fact, there are four and not two possible paths, since the order of x_i and $\bar{x}_i$ can be changed without affecting the quality of the solution, but they are analogous.

[5] Additionally, we save some travel time going from v_1 over $x_1 / \bar{x}_1$ to true / false. Again, this does not incur in relevant changes, for the very same reasons.

$\bar{x}_4 \lor x_7$. The two customers w_0 and w_1 on the bottom part have to be always visited, while the customers on the top part correspond to the literals used in this particular clause. A separate customer is used for each literal, sharing the same position with the other literal customers. Their presence depends on the actual variable assignment. We set the deadlines of all the customers in such a way that they have to be visited from left to right. In case one of the top customers is present and requires to be visited, an additionally travel time of ε/m is required. This corresponds to the case in which the clause is satisfied. Please note that the realization of the x-variables might not correspond to the assignment defined by the paths that were chosen in the first part of the constructed instance, because it might be that the x-variables have not guessed this assignment correctly. We will handle this case soon.

It is clear that we can put m of these gadgets next to each other to check all the clauses given in the E-MINORITY-SAT instance. Here the deadlines of subsequent gadgets have to be adapted accordingly. If the whole formula is satisfied by the realization of the x-variables, we get a delay of ε/m for each of the clauses for a total delay of ε. This means that the total delay after all clause gadgets is at least ε if the realization of the x-variables is not correctly guessing the assignment defined by the chosen paths, or if the guess is correct and the x-variables satisfy the formula. Otherwise, the delay is at most $\varepsilon - \varepsilon/m$.

Putting Everything Together. We will now place the two sets of gadgets next to each other. Additionally, we will add a special customer z after the clause gadgets. Let T be the arrival time at z without any of the delays. We now use for z a deadline of $t(z) = T + \varepsilon - \varepsilon/m$. In an optimal tour, z is the last customer visited. After visiting z, the vehicle returns to the depot. z is also the only customer whose deadline is violated in an optimal solution. We set the penalty value for z to be $h(z) = 2^{2n}$.

Let us analyze the costs of a solution which does not violate any deadlines except at customer z and which therefore corresponds to an assignment of the first k variables of the original E-MINORITY-SAT instance. There is no need to care about travel times, since the costs are completely dominated by the penalty at customer z. We are late at customer z if the guess of the decisions that have been made in the first part of the constructed instance is wrong. This happens with a probability of $1 - 2^{-k}$. If our guess is correct, we arrive at

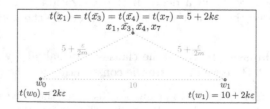

Fig. 2. The gadget used to verify if a clause is satisfied or not.

the clause part without any additional delay. The deadline at customer z is then violated if and only if all the clauses are satisfied by the correctly guessed first k variables and the randomly assigned remaining $n - k$ variables. Each assignment of the remaining $n - k$ variables occurs with the same probability of 2^{-n+k}. Let r denote the number of satisfying assignments of the remaining $n - k$ variables for the given assignment of the first k variables. We can now write the expected penalty as $(1 - 2^{-k} + 2^{-k}2^{-n+k}r)h(z) = 2^{2n} - 2^{2n-k} + 2^n r$.

If there exists an assignment of the first k variables of the original E-MINORITY-SAT instance such that the formula is satisfied for at most half of the assignments of the remaining $n - k$ variables, then there exists a solution for the Dependent PTSPD instance with costs of at most $2^{2n} - 2^{2n-k} + 2^n 2^{n-k}/2 = 2^{2n} - 2^{2n-k-1}$. On the other hand, if no such solution exists, then every solution for the Dependent PTSPD instance has higher cost. Considering the relatively small travel times, we can set the bound of the constructed instance to $2^{2n} - 2^{2n-k-1} + 2^n - 1$. In this way we are able to solve the original E-MINORITY-SAT problem with the decision version of the Dependent PTSPD.

Theorem 6. *The decision version of the Dependent Probabilistic Traveling Salesman Problem with Deadlines is $NP^{\#P[1]}$-hard, even for Euclidean instances.*

4.3   Hardness of the Optimization Variant

To obtain the same hardness result for the optimization variant of the Dependent PTSPD, we have to modify the construction slightly. The same technique has been used in [8] and therefore we only describe the main ideas at this point. Instead of a single final customer z, we add three customers z_1, z_2, z_3. z_1 is located at the position of z and is forced to be visited immediately after the clause part by imposing proper deadlines and penalties. The three new customers then form an equilateral triangle. This triangle is now placed in a way such that z_2 is closer to the depot than z_3. The deadline which was formerly imposed on z will now be prolonged by the sidelength of the triangle and imposed on z_2. No deadline is imposed on z_3. In this way we offer an optimal solution two choices: visiting at the end of the tour z_1, z_2, z_3 or z_1, z_3, z_2. In the first case the probability that the deadline at z_2 will be violated is lower than in the second case. On the other hand, the travel times are larger in the first case. Placing the triangle in a proper way and using adequate sidelengths, we are able to solve the original E-MINORITY-SAT instance by computing an optimal solution for this modified Dependent PTSPD instance. In case the optimal solution finishes with z_1, z_2, z_3 a solution for the E-MINORITY-SAT instance exists, if the optimal solution finishes with z_1, z_3, z_2, no such solution exists.

Theorem 7. *The optimization version of the Dependent Probabilistic Traveling Salesman Problem with Deadlines is $NP^{\#P[1]}$-hard, even for Euclidean instances.*

5 Discussion and Conclusions

In this paper we have investigated a very powerful scenario-based model of stochastic optimization problems. Using the framework of weighted counting we prove upper bounds for the complexity of various computational tasks related to these problems. Additionally, we show that these upper bounds are matched for a practical existing stochastic vehicle routing problem.

Many stochastic optimization problems inherit NP-hardness from their non-stochastic counterparts, which are usually contained as a special case. We believe that it is possible to obtain much stronger hardness results for a large number of these problems and we hope that our work can help in obtaining such stronger hardness results.

It would also be very interesting to better understand which properties of stochastic optimization problems are actually responsible for their hardness. In particular, it would be interesting to understand if the same hardness results can be obtained for (non artificial) stochastic optimization problems without any dependencies among the random variables.

Our work also motivates the usage of approximation algorithms (for both, the objective function and the optimization variant). Objective functions which are #P-hard might still allow for efficient approximations. In this case, the class NP would be an upper bound for the complexity of an approximative version of the decision variant (the upper bound for the optimization variant would change accordingly). On the other hand, inapproximability results for certain stochastic optimization problems could even further strengthen the already existing hardness results.

References

1. Birge, J.R., Louveaux, F.V.: Introduction to Stochastic Programming. Springer, Berlin (1997)
2. de Campos, C., Stamoulis, G., Weyland, D.: A structured view on weighted counting with relations to quantum computation and applications. Technical report, Electronic Colloquium on Computational Complexity, TR13-133 (2013)
3. Dyer, M., Stougie, L.: Computational complexity of stochastic programming problems. Math. Program. **106**(3), 423–432 (2006)
4. Fortnow, L., Reingold, N.: PP is closed under truth-table reductions. Inf. Comput. **124**(1), 1–6 (1996)
5. Jaillet, P.: A priori solution of a traveling salesman problem in which a random subset of the customers are visited. Oper. Res. **36**(6), 929–936 (1988)
6. Littman, M.L., Goldsmith, J., Mundhenk, M.: The computational complexity of probabilistic planning. J. Artif. Intell. Res. **9**(1), 36 (1998)
7. Valiant, L.G.: The complexity of computing the permanent. Theor. Comput. Sci. **8**, 189–201 (1979)

8. Weyland, D., Montemanni, R., Gambardella, L.M.: Hardness results for the probabilistic traveling salesman problem with deadlines. In: Mahjoub, A.R., Markakis, V., Milis, I., Paschos, V.T. (eds.) ISCO 2012. LNCS, vol. 7422, pp. 392–403. Springer, Heidelberg (2012)
9. Weyland, D., Montemanni, R., Gambardella, L.M.: An improved heuristic for the probabilistic traveling salesman problem with deadlines based on GPGPU. In: Moreno-Díaz, R., Pichler, F., Quesada-Arencibia, A. (eds.) EUROCAST 2013, Part I. LNCS, vol. 8111, pp. 332–339. Springer, Heidelberg (2013)

A Hybrid Heuristic Approach Based on a Quadratic Knapsack Formulation for the Max-Mean Dispersion Problem

Federico Della Croce, Michele Garraffa[✉], and Fabio Salassa

Politecnico di Torino, Corso Duca degli Abruzzi, 24, 10129 Torino, Italy
michele.garraffa@polito.it

Abstract. The paper deals with the Max-Mean Dispersion Problem ($Max - MeanDP$) belonging to the general category of clustering problems. The aim of such problems is to find a subset of a set which maximizes a measure of dispersion/similarity between elements. To tackle the problem a two phases hybrid heuristic combining a mixed integer non linear solver and a local branching procedure is developed. Computational results, performed on literature instances, show that the proposed procedure outperforms the state-of-the-art approaches.

Keywords: Max-Mean dispersion problem · Quadratic programming · Matheuristics

1 Introduction

In recent years, several mathematical programming based heuristics have been developed in order to solve complex combinatorial optimization problems. These heuristics rely on a general purpose solver for the solution of subproblems which can be more tractable. To this end, linear programming (LP) and mixed integer linear programming (MILP) models have been exploited so far, since they can take advantage of the impressive effort in the improvements of the related solvers. Moreover, there has been a significant evolution in the performances on solving non linear programming (NLP) models, in particular in specific case of the quadratic - quadratic integer programming (QP -QIP). In this paper, we consider the Max-Mean Dispersion Problem ($Max - MeanDP$), also called Equitable Dispersion Problem (EDP), and propose a hybrid heuristic approach based on the solution of a QIP formulation. $Max - MeanDP$ belongs to a general category of clustering problems whose aim is to find a subset M of a set N which maximizes a measure of dispersion/similarity of the elements in M. More formally, suppose N is a set of elements with cardinality n, and D a matrix whose components $d_{i,j}$ indicate distance/proximity between item $i \in N$ and $j \in N$. We assume that the matrix D is symmetric, namely $d_{i,j} = d_{j,i}$ $\forall i,j \in N$, where the values on the diagonal are equal to 0 ($d_{i,i} = 0$, $\forall i \in N$). When the measure of dispersion/similarity of the elements in M is the sum of

© Springer International Publishing Switzerland 2014
P. Fouilhoux et al. (Eds.): ISCO 2014, LNCS 8596, pp. 186–197, 2014.
DOI: 10.1007/978-3-319-09174-7_16

the $d_{i,j}$s between elements $i, j \in M$ (that is $\sum_{i,j \in M} d_{i,j}$) and the cardinality of subset M is given a priori ($|M| = m$ with m predefined), then we have the $Max - SumDP$ [3] which is known to be strongly NP-Hard. When the measure of dispersion/similarity of the elements in M is the minimum of the $d_{i,j}$s between elements $i, j \in M$ (that is $\min_{i,j \in M} d_{i,j}$) and the cardinality of subset M is given a priori, we have the $Max - MinDP$ [1,7], which is also known to be strongly NP-Hard. Finally, when the measure of dispersion/similarity of the elements in M is the average of the distances between elements $i, j \in M$ (that is $\frac{\sum_{i,j \in M} d_{i,j}}{|M|}$), but the cardinality of subset M is not given a priori, we have the $Max - MeanDP$. It is of interest to exploit the relationship between $Max - SumDP$ and $Max - MeanDP$. Let $OPT_{m=i}(Max - SumDP)$ denote the optimal solution value of $Max - SumDP$ for $m = i$. The optimal solution of $Max - MeanDP$ can be simply computed by iteratively solving $Max - SumDP$ for $m = 1, ..., n$ and subsequently taking the maximum $\frac{OPT_{m=i}(Max - SumDP)}{i}$ $\forall i \in \{1, ..., n\}$.

This problem has a real importance in fields like architectural space planning and analysis of social networks (as claimed in [4]), and in such domains $d_{i,j}$s can violate the triangular inequality $d_{i,j} \leq d_{i,k} + d_{k,j} \forall i, k, j \in N$ and the non negativity condition $d_{i,j} \geq 0 \forall i, j \in N$. To the authors' knowledge, the state of the art literature on $Max - MeanDP$ is quite limited. In [6], $Max - MeanDP$ is shown to be strongly NP-Hard whenever $d_{i,j}$s can take both positive and negative values. Those authors have also presented a mixed integer non linear programming (MINLP) formulation and an equivalent ILP formulation. In [4], a randomized GRASP with path relinking is proposed for $Max - MeanDP$. The presented computational experiments dealt with a set of real world instances from a social network application. The authors also noted that a specific ILP solver (CPLEX) iteratively applied to the ILP formulation of $Max - SumDP$ reached better performances than the same solver applied just once to the ILP formulation of $Max - MeanDP$. In this paper, we propose a hybrid heuristic procedure that exploits this feature and repeatedly solves a QIP formulation of $Max - SumDP$ within a matheuristic framework. The proposed approach proved to be computationally superior to the approach proposed in [4].

2   Mathematical Formulations

$Max - MeanDP$ has a straightforward non linear and fractional formulation, as introduced in [6]. Define vector $x \in \{0, 1\}^n$ where, for each component x_i, we have $x_i = 1$ if and only if element i is included in the subset M, otherwise 0. The formulation below follows directly from the definition of $Max - MeanDP$:

MaxMeanDP. **Trivial Formulation 1**

$$\max \frac{\sum_{i=1}^{n-1} \sum_{j=i+1}^{n} d_{i,j} x_i x_j}{\sum_{i=1}^{n} x_i} \tag{1}$$

subject to:

$$\sum_{i=1}^{n} x_i \geq 1 \tag{2}$$

$$x_i \in \{0,1\} \forall i \in N \tag{3}$$

Since D is symmetric, it can be written in the following vectorial form:

$$\min -\frac{\frac{1}{2}x^T D x}{u^T x} \tag{4}$$

subject to:

$$u^T x \geq 1 \tag{5}$$

$$x \in \{0,1\}^n \tag{6}$$

where $u^T = \overbrace{(1,1,...,1)}^{n}$, and, for convenience, the problem is converted into a minimization problem where the sign of the objective is changed.

The following proposition indicates that if the integrality constraints are relaxed, then the resulting mathematical program is not convex if no other assumptions are given on the matrix D.

Proposition 1. *The function $f(x) : \Gamma \to \mathbb{R}$:*

$$f(x) = -\frac{\frac{1}{2}x^T D x}{u^T x} \tag{7}$$

where $\Gamma = \{x \in [0,1]^n : u^T x \geq 1\}$, is convex if and only if $D \preceq 0$.

Proof. Set $u^T x = y$. The above function can be written as:

$$f(x,y) = -\frac{\frac{1}{2}x^T D x}{y}$$

over the domain $\Gamma' = \{x \in [0,1]^n, y \in \mathbb{R} : y = u^T x \geq 1\}$

The Hessian $\nabla^2 f(x,y)$ is equal to:

$$\nabla^2 f(x,y) = -\frac{1}{y^3}\begin{bmatrix} Dy^2 & -Dxy \\ -(Dx)^T y & x^t Dx \end{bmatrix} = -\frac{1}{y^3}D\begin{bmatrix} y \\ -x \end{bmatrix}\begin{bmatrix} y \\ -x \end{bmatrix}^T$$

Function $f(x,y)$ is defined on the convex set Γ' and is convex if and only if $\nabla^2 f(x,y) \succeq 0$, thus if $D \preceq 0$.

With real data, the condition $D \preceq 0$ typically does not hold, hence the problem given by the continuous relaxation of (4) – (6) is not convex in the general case. NLP solvers can clearly be applied to this formulation (here we used XPRESS-SLP by Fair-Isaac), as it is shown in Section 4, even though just local maxima can be guaranteed.

On the other hand, the following straightforward QIP model holds for $Max-SumDP$:

MaxMeanDP. **QIP Formulation 1**

$$\max \sum_{i=1}^{n-1} \sum_{j=i+1}^{n} d_{i,j} x_i x_j \qquad (8)$$

subject to:

$$\sum_{i=1}^{n} x_i = m \qquad (9)$$

$$x_i \in \{0,1\} \forall i \in N \qquad (10)$$

Even if general convexity results do not hold also for the continuous relaxation of this QIP, we note that for $Max - SumDP$, the formulation corresponds to a 0/1 quadratic knapsack problem (with equality constraint and unitary weights) that has been much more tackled in the literature (see, e.g. [5]) and can be efficiently tackled by means of QIP solvers, such as, for instance, CPLEX. In [4], computational experiments dealt only with ILP formulations of $Max-MeanDP$ and $Max - SumDP$ showing that the iterative solution of $Max - SumDP$ was superior to the one-shot solution of $Max - MeanDP$. In our preliminary tests on $Max - SumDP$, we determined that CPLEX 12.5 was more efficient when applied to QIP Formulation 2 rather than to its standard linearization indicated in [4]. In fact, the latter required a computational time which is higher by more than an order of magnitude on instances with $n = 35$. The purpose of this work is to embed the repeated solution of the QIP formulation of $Max - SumDP$ into a heuristic framework for $Max - MeanDP$.

3   A Hybrid Heuristic Approach

In [4], it was shown that in their real world small instances the value of m associated with the optimal solution lies in an interval, such that there are high quality solutions with values of m inside this interval. Although these promising intervals may be disjoint in large instances, the basic conclusion is very interesting: if a set C of good candidate values for m is selected, the use of the pseudo-polynomial models can be restricted only to values of m in C. In the following, an interval, i.e., the set of all possible integer values between two extreme points m_1 and m_2, where $m_1, m_2 \in \mathbb{N}$ and $1 \leq m_1 \leq m_2 \leq n$, is indicated with $[m_1, m_2]$.

The proposed hybrid algorithm for $Max - MeanDP$ is composed by two phases:

PHASE ONE. It consists in the selection of good candidate values of m to be included in set C, while also providing good feasible solutions;

PHASE TWO. Given C, a matheuristic procedure is used in order to enhance the quality of the solutions provided in PHASE ONE $\forall m \in C$.

Finally, the best solution is provided (whatever is the value of m).

PHASE ONE makes use of a decision tree with k branches (where k is a parameter to be defined experimentally) to seek promising intervals. Since the subproblems with $m = 1$ and $m = n$ can be solved to optimality in $O(1)$ and $O(n)$, the search is restricted to $[2, n-1]$. Such interval is split into k intervals whose width differs by one unit at most, and an evaluation value is computed for each of them. Then, the interval with the best evaluation is expanded at each step, resulting in a best first expansion of the decision tree. The evaluation of an interval $[m_1, m_2]$ is performed by considering the quality of the solution provided by the QIP solver after a fixed amount of time, using QIP Formulation 2 with $m = m_1$ and $m = m_2$. Thus, we evaluate a set of contiguous values of m by computing two solutions at its extreme points and summing their objectives. Call $h(m)$ the result provided by the solver after T_s seconds. The evaluation $\lambda([m_1, m_2]) \in \mathbb{R}$ of an interval $[m_1, m_2]$ can be computed as:

$$\lambda([m_1, m_2]) = h(m_1) + h(m_2) \tag{11}$$

A set S of feasible solutions for the $Max - MeanDP$ is computed in PHASE ONE, during the evaluations of intervals thanks to the computation of function h. In other words, the output of PHASE ONE is not limited to the definition of C, but for each value $m' \in C$ a solution s with $m = m'$ is computed, included in S and provided as output.

Figure 1 shows how the decision tree is expanded during PHASE ONE, under the assumption that gray nodes are associated to the intervals with the highest evaluation λ at each step. Only the initial three steps of the expansion are depicted. Table 1 enumerates solutions computed in these steps in order to expand nodes. These solutions are indicated with the notation $\widehat{x}_{m'}$ where m' is the value of m for the solution.

Table 1. Solutions computed in PHASE ONE

First step	Second step	Third step
$\widehat{x}_2$	$\widehat{x}_{120}$	$\widehat{x}_{144}$
$\widehat{x}_{100}$	$\widehat{x}_{140}$	$\widehat{x}_{148}$
$\widehat{x}_{200}$	$\widehat{x}_{160}$	$\widehat{x}_{152}$
$\widehat{x}_{300}$	$\widehat{x}_{180}$	$\widehat{x}_{156}$
$\widehat{x}_{400}$		
$\widehat{x}_{499}$		

In the example, PHASE ONE yields the following sets C and S after the initial three steps.

$$C = \{2, 100, 120, 140, 144, 148, 152, 156, 160, 180, 200, 300, 400, 499\} \tag{12}$$

$$S = \{\widehat{x}_2, \widehat{x}_{100}, \widehat{x}_{120}, \widehat{x}_{140}, \widehat{x}_{144}, \widehat{x}_{148}, \widehat{x}_{152}, \widehat{x}_{156}, \widehat{x}_{160}, \widehat{x}_{180}, \widehat{x}_{200}, \widehat{x}_{300}, \widehat{x}_{400}, \widehat{x}_{499}\} \tag{13}$$

In general, PHASE ONE ends as soon as the overall number of computed solutions is greater than or equal to a constant γ, thus at its end $|S| = \gamma$.

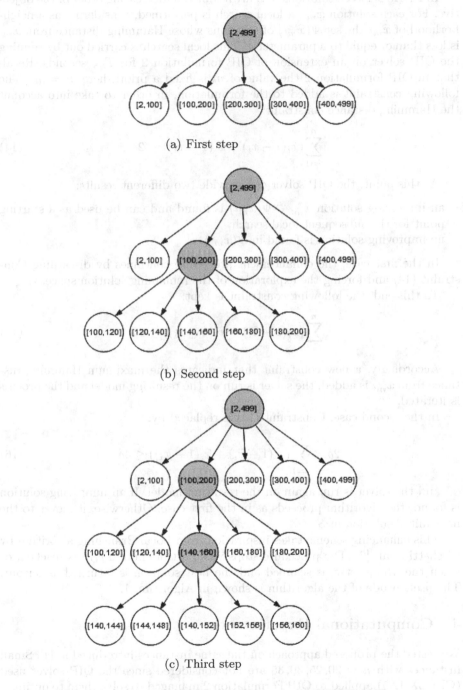

(a) First step

(b) Second step

(c) Third step

Fig. 1. An example which shows three steps of PHASE ONE

In PHASE TWO, solutions in S are scanned in decreasing order of the objective. For each solution $\widehat{\boldsymbol{x}}_{m'}$, a local search is performed, considering as a neighborhood of $\widehat{\boldsymbol{x}}_{m'}$ the set $\Psi(\widehat{\boldsymbol{x}}_{m'})$ of solutions whose Hamming distance from $\widehat{\boldsymbol{x}}_{m'}$ is less than or equal to a parameter δ. This local search is carried out by running the QIP solver on an extension of QIP formulation 2 for T_{LS} seconds. Recall that in QIP formulation 2 the value of m is fixed a priori, here $m = m'$. The following constraint is added to the formulation, in order to take into account the Hamming distance constraint:

$$\sum_{i=1}^{n}(x_i(1-\widehat{x}_i) + \widehat{x}_i(1-x_i)) \leq 2\delta \tag{14}$$

At this point, the QIP solver can provide two different results:

- an improving solution $\widehat{\boldsymbol{x}}_{m'}^* \in \Psi(\widehat{\boldsymbol{x}}_{m'})$ is found and can be used as a starting point for the subsequent local search;
- no improving solution is found in $\Psi(\widehat{\boldsymbol{x}}_{m'})$.

In the first case, the enhancement procedure continues by discarding Constraint (14) and forcing the exploration of the remaining solution space.

To this end, the following constraint is kept:

$$\sum_{i=1}^{n}(x_i(1-\widehat{x}_i) + \widehat{x}_i(1-x_i)) \geq 2\delta \tag{15}$$

Accordingly, a new constraint that indicates the maximum Hamming distance from $\widehat{\boldsymbol{x}}_{m'}^*$ is added, the solver is run on the resulting model and the process is iterated.

In the second case, Constraint (14) is replaced by:

$$2\delta < \sum_{i=1}^{n}(x_i(1-\widehat{x}_i) + \widehat{x}_i(1-x_i)) \leq 4\delta \tag{16}$$

and the solver is run again on the resulting model. If an improving solution is found, the algorithm proceeds as in the first case. Otherwise, it moves to the next initial solution in S.

This enhancing scheme takes inspiration from Local Branching algorithm by Fischetti et al. [2]. The procedure stops when a time limit T_{TOT} is reached, or when the whole set S is scanned and the best solution is returned as output. The pseudocode of the algorithm is shown in Algorithm 1.

4   Computational Experiments

We tested the proposed approach on the same instances introduced in [4]. Small instances with $n = 20, 25, 30, 35$ are not considered since the QIP solver used (CPLEX 12.5) applied to QIP Formulation 2 managed to solve them to optimality in few seconds. Thus, large instances (40 in total) where $n = 150$ and $n = 500$ are considered in the following. These instances belong to two categories:

Algorithm 1. QIP Matheuristic

Compute $\hat{x}_2$ and $\hat{x}_n$ by solving to optimality $Max - SumDP$ for $m = 1$ and $m = n$

$C \leftarrow \{2, n-1\}$ $S \leftarrow \{\hat{x}_2, \hat{x}_{n-1}\}$ $S' \leftarrow \emptyset$

Set $I \leftarrow \{[2, n-1]\}$

Set $counter \leftarrow 2$

while $counter < \gamma$ **do**

    Extract best evaluated interval from I

    Split the selected interval $[m_0, m_k]$ in k intervals $[m_0, m_1] \cdots [m_{k-1}, m_k]$

    Compute $k - 1$ solutions $\hat{x}_{m_1} \cdots \hat{x}_{m_{k-1}}$ with $m = m_1 \cdots m_{k-1}$ by running

    the QIP solver for T_s seconds

    $I \leftarrow I \cup \{[m_0, m_1] \cdots [m_{k-1}, m_k]\}$

    $C \leftarrow C \cup \{m_1 \cdots m_{k-1}\}$

    $S \leftarrow S \cup \{\hat{x}_{m_1} \cdots \hat{x}_{m_{k-1}}\}$

    $counter \leftarrow counter + k - 1$

end while

while S is not empty AND time limit T_{TOT} is not reached **do**

    Extract best solution $\hat{x}$ in S

    Compute $\hat{x}'$ with a Local Branching procedure starting from $\hat{x}$

    $S' \leftarrow S' \cup \{\hat{x}'\}$

end while

Output: Best solution in S'

Type I distances $d_{i,j}$s are uniformly distributed in $[-10, 10]$;
Type II distances $d_{i,j}$s are uniformly distributed in $[-10, -5] \cup [5, 10]$.

Three approaches have been used to solve these instances:

- GRASP with path relinking proposed in [4];
- the QIP hybrid heuristic presented in the previous section;
- XPRESS-SLP non linear solver applied to Trivial Formulation 2.

The first approach was tested on an Intel Core Solo 1.4GHz with 3GB of RAM, while the other two approaches were tested on an Intel Core i5-3550 3.30GHz with 4GB of RAM. The parameter settings of the hybrid heuristic discussed in the previous section have been established after some preliminary tests. In PHASE ONE, the number of subintervals k generated at each iteration is set to 5. The time limit T_S is set to $\lceil \frac{n}{200} \rceil$ seconds, as this was the time typically required by the QIP solver to find a good solution. The number of solutions (and promising values for m) γ generated in PHASE ONE is equal to $\lceil \frac{n}{10} \rceil$, such that all the most promising values for m can be selected. In PHASE TWO, parameters are inherent to the Local Branching procedure. The parameters δ is set to 3, while the time limit T_{LS} for the local search steps is set to 3 second. Finally, the overall time limit for the procedure is set to 60 seconds for instances with $n = 150$ and to 600 seconds for instances with $n = 500$.

Table 2 and Table 3 report the results of the two phases of the algorithm, in order to evaluate the effectiveness of the second phase. They show that PHASE TWO leads to a sharp increase of the solution quality with respect to PHASE

Table 2. Results on [4]'s instances

Instances	[4]GRASP			PHASE ONE			PHASE TWO			XPRESS-SLP		
	best value	m	time (s)	best value	m	time (s)	best value	m	time (s)	best value	m	time (s)
MDPI1_150	**45,92**	53	83	45,89	59	19	45,91	52	61	45,55	52	5
MDPI2_150	43,33	41	72	43,31	53	19	**43,39**	42	64	43,03	59	12
MDPI3_150	39,64	43	60	39,89	46	19	**39,98**	50	65	38,38	58	13
MDPI4_150	43,70	57	76	43,94	60	19	**44,04**	58	61	43,76	58	12
MDPI5_150	**42,48**	49	71	**42,48**	49	19	**42,48**	49	61	40,63	55	11
MDPI6_150	**43,67**	40	73	43,67	45	19	43,72	43	69	43,27	53	10
MDPI7_150	**46,08**	53	60	45,99	51	19	**46,08**	52	63	45,44	52	8
MDPI8_150	42,39	45	61	42,26	41	19	**42,44**	43	62	41,71	53	12
MDPI9_150	**42,14**	42	64	41,82	49	19	41,82	49	67	41,49	53	22
MDPI10_150	**41,80**	41	55	**41,80**	41	19	**41,80**	41	52	41,80	41	23
MDPII1_150	56,72	49	62	**57,48**	50	19	**57,48**	50	61	55,95	59	5
MDPII2_150	57,80	47	61	57,69	48	19	**57,82**	46	59	56,18	54	14
MDPII3_150	58,28	45	59	**58,42**	44	19	**58,42**	44	65	57,49	50	15
MDPII4_150	**57,38**	47	59	56,74	51	19	57,29	48	64	55,21	60	10
MDPII5_150	**54,23**	42	48	54,09	37	19	54,14	35	62	51,05	54	7
MDPII6_150	**56,44**	49	58	**56,44**	49	19	**56,44**	49	61	55,10	55	9
MDPII7_150	**58,89**	48	61	58,47	49	19	58,88	49	64	57,62	54	10
MDPII8_150	**57,97**	54	66	**57,97**	54	19	**57,97**	54	59	57,97	54	14
MDPII9_150	**58,30**	42	56	58,09	42	19	**58,30**	42	66	55,84	52	9
MDPII10_150	56,92	39	55	56,18	41	19	**57,18**	41	62	54,25	52	20

Table 2. (Continued).

Instances	[4]GRASP			PHASE ONE			PHASE TWO			XPRESS-SLP		
	best value	m	time (s)	best value	m	time (s)	best value	m	time (s)	best value	m	time (s)
MDPI1_500	78,60	152	716	79,92	155	171	**81,25**	160	611	77,23	175	135
MDPI2_500	76,87	150	682	76,61	176	171	**77,45**	154	638	75,17	177	262
MDPI3_500	**75,69**	128	668	74,68	165	171	75,31	145	602	73,86	172	182
MDPI4_500	81,81	166	647	81,33	163	171	**82,28**	146	604	79,89	185	277
MDPI5_500	78,57	140	683	79,30	143	171	**80,01**	147	608	78,75	170	339
MDPI6_500	79,64	156	732	80,46	152	170	**81,12**	149	624	77,40	169	246
MDPI7_500	75,50	146	607	76,86	159	171	**78,09**	143	608	74,18	182	221
MDPI8_500	76,98	151	666	78,60	179	171	**79,01**	168	611	78,22	176	294
MDPI9_500	75,72	128	635	76,51	159	171	**76,98**	164	603	75,45	175	317
MDPI10_500	80,38	137	849	80,61	143	171	**81,24**	137	617	80,09	159	370
MDPII1_500	108,15	165	766	108,49	162	171	**109,16**	170	612	107,78	174	234
MDPII2_500	103,29	121	656	103,45	112	171	**105,06**	158	608	103,46	169	275
MDPII3_500	106,30	140	710	106,24	159	171	**107,64**	158	613	101,74	206	267
MDPII4_500	104,62	154	725	103,83	169	171	**105,37**	163	611	101,01	197	111
MDPII5_500	103,61	149	707	105,91	164	171	**106,37**	161	614	103,85	182	253
MDPII6_500	104,81	158	713	104,09	163	171	**105,52**	172	623	102,57	169	229
MDPII7_500	104,50	148	626	104,94	151	171	**106,61**	142	607	103,26	181	335
MDPII8_500	100,02	135	609	103,28	156	171	**103,41**	154	617	99,28	178	245
MDPII9_500	104,93	130	636	105,33	138	171	**106,20**	142	601	104,58	177	239
MDPII10_500	103,50	144	649	102,88	150	171	**103,79**	150	623	102,13	176	428

Table 3. Average results on [4]'s instances grouped by size and type

Instances	[4] GRASP			PHASE ONE			PHASE TWO			XPRESS-SLP		
	best value	m	time (s)	best value	m	time (s)	best value	m	time (s)	best value	m	time (s)
Type I	60,55	95,90	378,10	60,80	104,40	94,95	**61,22**	99,60	337,55	59,77	113,70	138,45
Type II	80,83	95,30	369,08	81,00	99,45	95,00	**81,65**	101,40	337,60	79,32	117,65	136,43
n = 150	50,20	46,30	63,07	50,13	47,95	19,00	**50,28**	46,85	62,40	49,09	53,90	12,03
n = 500	91,18	144,90	684,11	91,66	155,90	170,95	**92,59**	154,15	612,75	90,00	177,45	262,86

ONE. As a matter of fact, PHASE ONE of our hybrid heuristic achieves the best heuristic solution 6 times out of 40 on this specific dataset. At the same time, the best result is achieved 33 times out of 40 when PHASE TWO is subsequently executed.

The solutions provided by XPRESS-SLP have a low quality, which indicates that the solver gets generally stuck in local maxima. The GRASP with path relinking approach proposed in [4] is dominated by our heuristic both for $n = 150$ and $n = 500$. However, Table 3 shows that the dominance increases as the instance size grows. In fact, PHASE TWO provides the best solution in 19 instances over 20 when $n = 500$. Table 3 also shows that the type of the instance does not affect the performances of the three approaches.

5   Conclusions

The paper proposed a two phase matheuristic for the $Max - MeanDP$. The first phase selects a set of promising values for the cardinality of unknown subset and generates a pool of initial solutions. The second phase enhances the quality of the solutions provided by the previous phase. As shown above, this hybrid heuristic globally outperforms the best heuristic in the state of the art. In our opinion, the quality of the results indicated that QIP models and related solvers can also be taken into consideration in the design of matheuristics for other combinatorial optimization problems with a QIP formulation. Since the QIP solver does not provide tight upper bounds, a future development of this work is to determine a bounding method that allows to prune suboptimal solutions, in order to improve the overall performances of the heuristic presented.

References

1. Della Croce, F., Grosso, A., Locatelli, M.: A heuristic approach for the max-min diversity problem based on max-clique. Comput. Oper. Res. **36**, 2429–2433 (2009)
2. Fischetti, M., Lodi, A.: Local branching. Math. Program. **98**, 23–47 (2003)
3. Ghosh, J.B.: Computational aspects of the maximum diversity problem. Oper. Res. Lett. **19**, 175–181 (1996)
4. Martí, R., Sandoya, F.: GRASP and path relinking for the equitable dispersion problem. Comput. Oper. Res. **40**, 3091–3099 (2013)
5. Pisinger, D.: The quadratic knapsack problem - a survey. Discrete Appl. Math. **155**, 623–648 (2007)
6. Prokopyev, O.A., Kong, N., Martinez-Torres, D.L.: The equitable dispersion problem. Eur. J. Oper. Res. **197**, 59–67 (2009)
7. Resende, M.G.C., Martí, R., Gallego, M., Duarte, A.: GRASP and path relinking for the maxmin diversity problem. Comput. Oper. Res. **37**, 498–508 (2010)

A Constraint Generation Approach
for the Two-Machine Flow Shop Problem
with Jobs Selection

Federico Della Croce[1,2]([✉]), Christos Koulamas[3], and Vincent T'kindt[4]

[1] D.A.I., Politecnico di Torino, Torino, Italy
[2] CNR, IEIIT, Torino, Italy
federico.dellacroce@polito.it
[3] College of Business, Florida International University,
Florida, USA
koulamas@fiu.edu
[4] Université Francois-Rabelais, CNRS, LI EA 6300,
OC ERL CNRS 6305, Tours, France
tkindt@univ-tours.fr

Abstract. We consider a job selection problem in a two-stage flow shop. The objective is to select the best job subset with a given cardinality to minimize the makespan. This problem is known to be ordinary NP-hard and the current state of the art algorithms can solve instances with up to 3000 jobs. We introduce a constraint generation approach to the integer linear programming (ILP) formulation of the problem according to which the constraints associated with nearly all potential critical paths are relaxed and then only the ones violated by the relaxed solution are sequentially reinstated. This approach yields a new solution algorithm capable of solving problems with up to 100000 jobs or more.

1 Introduction

We consider a job selection problem in a two-stage flow shop according to the following specifications. There is a set of n jobs j, $j = 1, ...n$ all of them available at time zero; each job j must be processed non-preemptively with known integer processing times a_j, b_j on two continuously available machines M_1, M_2 respectively where the order of processing is $M_1 \rightarrow M_2$ for all jobs. Each machine can process at most one job at a time and the two operations of each job cannot overlap. The flow shop is available for some amount of time d; the goal is to maximize the number of selected jobs to be completed while the flow shop is available and also to minimize the length of the total availability period.

When the availability period is preset, for instance when there are preset daily operating hours, a natural application of this problem is as follows. Since the jobs cannot be preempted, the shop manager should ensure that an appropriate subset of the available jobs is selected each day in order to maximize the number of jobs completed before the shop closes. The remaining jobs become available for

© Springer International Publishing Switzerland 2014
P. Fouilhoux et al. (Eds.): ISCO 2014, LNCS 8596, pp. 198–207, 2014.
DOI: 10.1007/978-3-319-09174-7_17

processing on the next day and so on. On the other hand, if the number of jobs to be processed is given, then the problem reduces to selecting the appropriate subset of jobs such that the length of the availability period d is minimized.

It is implicitly assumed that all jobs have comparable values. If this is not the case, then a job-specific rejection cost should be assigned to each job and the job selection problem becomes the corresponding scheduling problem with job rejection. The literature on shop scheduling problems with job rejection is reviewed by Shabtay et al. [9]. It is mentioned there that most two-stage shop problems with job rejection are NP-hard even with the equal job rejection cost assumption.

When the availability period d is preset, using the extended three-field notation [10], the considered flow shop problem is denoted as $F2|d_j = d|n_T$. When the number of jobs to be selected is given in advance and the aim is to find the minimum value of d, the considered flow shop problem is denoted $F2|d_j = d, unknown\ d|\epsilon(d/n_T)$. Finally, if both the number of selected jobs must be maximized and the length of the availability period must be minimized and the goal is to search for the non dominated solutions, then the considered flow shop problem is denoted as $F2|d_j = d, unknown\ d|d, n_T$. Jozefowska et al. [5] showed that the $F2|d_j = d|n_T$ problem is ordinary NP-hard and proposed an $O(nd^2)$ pseudo-polynomial dynamic programming (DP) algorithm for the $F2|d_j = d|n_T$ problem (and for a more general weighted version). Della Croce et al. [3] proposed a branch and bound algorithm for the $F2|d_j = d|n_T$ problem. They also observed that the problem is solvable in $O(n \log n)$ time when the jobs and the machines are both ordered because in that case the problem resembles its single-machine counterpart. The ordering of the machines implies that all jobs have their smallest (largest) processing time on the same machine; the ordering of the jobs implies that if $a_i < a_j$ for any two jobs i, j, then $b_i < b_j$ as well. Panwalkar and Koulamas [7] considered the less-restrictive case of ordered machines (without ordered jobs) and proposed an $O(n^2)$ algorithm for the $F2|d_j = d|n_T$ problem.

It is of interest to exploit the relationship between the bi-criterion $F2|d_j = d|d, n_T$ problem and the corresponding single-objective $F2|d_j = d|n_T$ problem in which the common due date d is given. A solution for the $F2|d_j = d|d, n_T$ problem can be used to solve the $F2|d_j = d|n_T$ problem for any value of d. Alternatively, if the $F2|d_j = d|n_T$ problem is NP-hard, then the $F2|d_j = d|d, n_T$ problem is also NP-hard as shown in [11]. Tkindt et al. [11] showed that the $F2|d_j = d|d, n_T$ problem is solvable in $O(nD^2)$ time by implementing the DP algorithm for the $F2|d_j = d|n_T$ problem where D denotes the makespan of the corresponding $F2||C_{max}$ problem.

Tkindt et al. [11] proposed an integer linear programming (ILP) formulation and a branch & bound (B&B) algorithm for the $F2|d_j = d, unknown\ d|\epsilon(d/n_T)$ problem capable of solving problems with up to 3000 jobs. Then, by applying an ϵ-constraint approach on the number of tardy jobs that repeatedly solved $O(n)$ instances of the $F2|d_j = d, unknown\ d|\epsilon(d/n_T)$ problem, they were able to solve to optimality the $F2|d_j = d|d, n_T$ problem with up to 500 jobs.

The objective of this paper is to show that the knapsack-like ILP formulation proposed in [11] for the $F2|d_j = d, unknown\ d|\epsilon(d/n_T)$ problem can be efficiently tackled by means of a constraint generation approach. As mentioned in [2], in order to solve linear programming models with a large number of constraints, constraint generation techniques (see also [6] and [8]) are often used where a relaxation of the formulation containing only a subset of the constraints is first solved. Then a separation procedure is applied which adds to the relaxation any inequality of the formulation that is violated by the current solution. The process is iterated until no violated inequality can be found. In this work, all linear programming formulations share the feature of having a linear number of constraints where only a small part of these constraints is necessary to reach the optimal solution of the original problem. The computational results reveal a dramatic improvement in the performances with respect to the literature.

Our approach yields a new solution algorithm capable of solving problems with up to 100000 jobs and more. It should be pointed out that, to the best of our knowledge, the largest size of an NP-hard scheduling problem solvable by an ILP formulation is the $1||n_{wT}$ problem. According to [1], problem instances with up to 50000 jobs can be handled by their algorithm. Our constraint generation approach doubles the problem size in a two-stage shop environment compared to a single machine. Correspondingly, by means of the ϵ-constraint approach the limit size of the $F2|d_j = d|d, n_T$ problems instances solvable to optimality is also strongly increased.

2   ILP Formulation and a Constraint Generation Approach

2.1   The ILP Formulation for the $F2|d_j = d, unknown\ d|\epsilon(d/n_T)$ Problem

Let a_i (b_i) denote the processing time of job i on machine M_1 (M_2). We recall here the integer programming formulation of the $F2|d_j = d, unknown\ d|\epsilon(d/n_T)$ problem (hereafter denoted $F2$ for conciseness) proposed in [11] where it is assumed that the jobs are indexed and ordered using Johnson's algorithm (schedule first the jobs with $a_i \le b_i$ in nondecreasing order of a_i and then the jobs with $a_i > b_i$ in nonincreasing order of b_i [4]). Notice that, if the set Ω of the $(n - \epsilon)$ early jobs is fixed, then the optimal value of the common due date d is given by $d = C_{max}(J(\Omega))$ where J refers to Johnson's algorithm. Let d denote the unknown common due date and ϵ denote the given number of late jobs. Let us associate to each job i a binary variable x_i.

The ILP model is as follows.

$$\min d \tag{1}$$

$$\sum_{i=1}^{n} x_i = n - \epsilon \tag{2}$$

$$a_1x_1 + \sum_{i=1}^{n} b_i x_i \le d \tag{3}$$

$$\sum_{i=1}^{n} a_i x_i + b_n x_n \le d \tag{4}$$

$$\sum_{i=1}^{j} a_i x_i + \sum_{i=j}^{n} b_i x_i \le d \quad \forall j = 2, ..., n-1 \tag{5}$$

$$x_i \in \{0, 1\} \quad \forall i \in 1, ..., n \tag{6}$$

Here, constraint (2) implies that there are exactly ϵ late jobs while constraints (3–5) are critical-path constraints which define the value of the common due date d. Notice that d is always determined by the sum of the processing times of jobs $1, .., k$ on the first machine plus the sum of the processing times of jobs $k, .., n$ on the second machine where k depends on the selected early jobs and therefore constraints (3–5) consider all possible values of k with $1 \le k \le n$. Notice that in the critical path constraints (3–5), we explicited constraint (3) corresponding to $k = 1$ and constraint (4) corresponding to $k = n$. Finally constraints (6) indicate that the x_i variables are binary. We point out that the capacity constraints on machines M_1 and M_2 ($\sum_{i=1}^{n} a_i x_i \le d$ and $\sum_{i=1}^{n} b_i x_i \le d$, respectively) are not included in the model as they are dominated by constraints (4) and (3), respectively.

In [11], the proposed exact procedure failed to solve large size problems mainly due to the presence of constraints (5) that generated $O(n^2)$ nonzeroes in the constraints matrix inducing an out-of-memory status of the procedure if problems with more than 3000 variables were considered.

In this paper, a constraint generation approach is considered where, initially, constraints (5) are discarded. To this extent, it is of interest to determine the relative error between the optimal solution of the above model with and without constraints (5). Let $OPT(F2)$ be the optimal solution of problem $F2$ and $OPT(F2_{rel})$ be the optimal solution of the relaxed problem $F2_{rel}$ where constraints (5) are omitted. The following property holds.

Proposition 1. $\frac{OPT(F2)}{OPT(F2_{rel})} \le 2$ and this ratio is asymptotically tight.

Proof. To see that the inequality holds, it is sufficient to sum up constraints (3–4). Then, we obtain $\sum_{i=1}^{n}(a_i + b_i)x_i \le 2d - a_1x_1 - b_nx_n$, that is the sum of the processing times on both machines of the early jobs is inferior to $2d$. But then for any feasible solution of problem $F2_{rel}$ with value d', there exists a corresponding feasible solution of problem $F2$ with value $< 2d'$, hence $\frac{OPT(F2)}{OPT(F2_{rel})} \le 2$. The following instance with three jobs and $\epsilon = 1$ provides an asymptotically tight bound. Let $a_1 = b_1 = \delta$, $a_2 = M$, $b_2 = M - \delta$, $a_3 = M^2$ and $b_3 = M - 2\delta$ with $\delta << M$. Then, in the optimal solution of problem $F2$, the selected jobs are $1, 2$ with job 1 preceding job 2, hence, $OPT(F2) = a_1 + a_2 + b_2 = \delta + M + (M - \delta) = 2M$. Also, for problem $F2_{rel}$, the variables set to 1 are x_1, x_2. Correspondingly, both constraints (3) and (4) are tight with $d = M + \delta$ and $OPT(F2_{rel}) = M + \delta$. Hence the ratio $\frac{OPT(F2)}{OPT(F2_{rel})} \to 2$ for very large M and small δ suitably chosen. □

2.2 A Constraint Generation Approach

The model of Subsect. 2.1 presents a number of constraints which is linear in the number of jobs. Particularly, constraints (5) induce both a linear number of constraints and a quadratic number of nonzero elements. The ILP solvers run out of memory whenever instance sizes exceeding 3000 jobs are considered [11]. On the other hand, we have seen that by removing constraints (5), just a constant number of constraints and a linear number of nonzeroes are present. Also, the optimal solution value of the original problem (in terms of worst case analysis) is at most twice the optimal solution value of the relaxed problem with the related constraints removed.

We propose a constraint generation approach solving initially problem $F2_{rel}$ and then considering a separation procedure adding to the relaxation any inequality of the original formulation that is violated by the current solution. Preliminary testing showed that it was more effective to add just one constraint (namely the most violated) at a time rather than the violated constraints all together. Notice that, by applying Johnson's algorithm to the set of jobs j such that $x_j^{rel} = 1$ where x_j^{rel} is the value of x_j in the optimal solution of $F2_{rel}$, we can derive in linear time the critical job and corrrespondingly the related constraint. Whenever the optimal solution of the relaxed problem is reached, the constraint generation procedure checks if there is any violated constraint and in that case adds the most violated one and iterates until no more violated constraints are present. A pseudo-code of the approach is depicted in Algorithm 1.

Algorithm 1. A Constraint Generation Algorithm

1: End=$False$
2: **while** !End **do**
3:     Solve the ILP of $F2_{rel}$: $\bar{x}$ is its solution and $OPT(F2_{rel})$ its value
4:     Compute $d(\bar{x})$ the optimal value of the ILP of $F2$ with added constraints $x = \bar{x}$
5:     **if** $(d(\bar{x}) = OPT(F2_{rel}))$ **then**
6:         End=$True$
7:     **else**
8:         Let $\mathcal{C}$ be the constraint giving $d(\bar{x})$ in the ILP of $F2$ for $\bar{x}$
        //   ($\mathcal{C}$ is the most violated constraint)
9:         Add $\mathcal{C}$ to $F2_{rel}$
10:    **end if**
11: **end while**
12: **return** $\bar{x}$ as the optimal solution of $F2$

---

3   Computational Experiments

3.1   Experimentation Plan

A set of experimentations have been conducted to evaluate the efficiency of the constraint generation approach for the considered flow shop problem with job selection. We have considered two kinds of experimentations. First, for a given number of late jobs ϵ we have compared the constraint generation approach, referred to as $IPCG$, with state-of-the-art algorithms, whenever existing, and the solution by a mathematical solver of the original IP models, referred to as IP. The values of ϵ have been taken in the set $\{n-5, n-10, \frac{n}{2}\}$.

Next, we focus on the enumeration of strict Pareto optima for the common due date d and the number of late jobs n_T. T'kindt et al. ([11]) showed that there are exactly $(n+1)$ strict Pareto optima. For the $F2|d_j = d, d\ unknown|d, n_T$ problem, we enumerate the set of strict Pareto optima by a simple algorithm which, for each value $\epsilon \in \{0, 1, ..., n\}$, solves the corresponding $F2|d_j = d, d\ unknown, n_T = \epsilon|d$ problem by the constraint generation approach. The optimal solutions of all these ϵ-constraint problems constitute the set of strict Pareto optima.

All the testings have been done on a PC Computer Intel i5 with 4 cores of 2.6 GHz and 8 GB of RAM. Whenever we solved an IP model we used CPLEX 12.2 mathematical solver.

The $F2|d_j = d, unknown\ d|\epsilon(d/n_T)$ problem has been solved by T'kindt et al. ([11]) who provided a dedicated branch-and-bound algorithm, denoted by BaB. They showed that BaB outperformed IP and that the hardest instances are obtained for $\epsilon = \frac{n}{2}$ for the ϵ-constraint problem.

A set of random instances have been generated as in [11]. All processing times a_i and b_i have been drawn at random between 10 and 100 using an uniform law. For each problem size $n \in \{10000; 20000; ...; 100000\}$, 30 instances have been generated.

First, let us compare the state-of-the-art solution approaches BaB and IP to $IPCG$ on the hardest instances. The results are provided in Table 1. For each algorithm, the average and maximum CPU times are given in seconds, referred to as t_{avg} and t_{max}, and the average and maximum number of nodes, referred to as n_{avg} and n_{max}, explored to solve the instances. Notice that we limit ourselves to instances with up to 3000 jobs in size since for higher instances, neither BaB nor IP were able to find the optimal solutions. Table 1 shows that, even with the improvement of the efficiency of mathematical solvers, BaB still outperforms IP. However, both algorithms are strongly outperformed by $IPCG$: for instances with 3000 jobs in size, BaB requires on average 42.46 seconds whilst $IPCG$ requires 0.1 second. Both BaB and IP failed to solve instances with 4000 jobs. As the latter solves the IP model (1–6), the number of non-zero coefficients in the constraints lead to a model too large in memory. For BaB, the problem is similar since it solves the LP relaxation of model (1–6) during its processing. In the remainder of this section we only focus on algorithm $IPCG$.

Now, let us turn to the evaluation of $IPCG$ on large instances. Tables 2, 3 and 4 present the results obtained for different values of ϵ. The same information is

Table 1. Solving the flowshop problem with $\epsilon = \frac{n}{2}$: comparison with state-of-the-art algorithms

n	IP				BaB			
	t_{avg}	t_{max}	n_{avg}	n_{max}	t_{avg}	t_{max}	n_{avg}	n_{max}
1000	4.70	10	128.26	512	1.90	4	112.53	347
2000	20.23	45	285.03	663	11.13	17	136.86	379
3000	54.06	151	439.60	978	42.46	189	403.80	4097
4000	Out of Memory				Out of Memory			

n	IPCG			
	t_{avg}	t_{max}	n_{avg}	n_{max}
1000	0.03	1	40.13	166
2000	0.06	1	23.96	220
3000	0.10	1	34.60	202
4000	0.10	1	34.30	174

Table 2. Solving the flowshop problem with $\epsilon = \frac{n}{2}$

n	IPCG					
	t_{avg}	t_{max}	n_{avg}	n_{max}	it_{avg}	it_{max}
10000	0.46	1	36.06	174	1.0	1
20000	0.76	2	33.96	228	1.0	1
30000	1.36	5	35.66	331	1.0	1
40000	3.03	8	75.26	330	1.0	1
50000	6.13	16	181.86	600	1.0	1
60000	5.96	12	141.03	450	1.0	1
70000	6.26	10	160.20	527	1.0	1
80000	8.26	16	161.80	442	1.0	1
90000	9.33	17	144.66	400	1.0	1
100000	11.00	27	198.46	934	1.0	1

provided as in Table 1 with the addition of columns it_{avg} and it_{max} which provide the average and maximum number of iterations performed, *i.e.* the number of IP models solved. Table 2 shows that $IPCG$ provides very good results for $\epsilon = \frac{n}{2}$: for the largest tested instances the average CPU time is 11 s and less than 200 nodes are explored. Besides, columns it_{avg} and it_{max} show that only 1 iteration was necessary to solve the instances which means that no constraint has been added: an optimal solution of the IP model with only the critical paths going through the first and last jobs was always an optimal solution of the full IP.

Tables 3 and 4 show the results when we impose 10 and 5 early jobs. Surprisingly, these instances are even harder, for $IPCG$, than the case $\epsilon = \frac{n}{2}$. However, they still remain solvable in less than 210 s on the average for 100000 jobs. The average number of iterations done is, in contrast to what happened for the case $\epsilon = \frac{n}{2}$, is relatively important: for $\epsilon = (n - 5)$ and instances with 100000 jobs, 230.90 iterations have to be done on the average. Equivalently saying, almost 230 critical path constraints have been added on the average.

Table 3. Solving the flowshop problem with $\epsilon = n - 10$

n	$IPCG$					
	t_{avg}	t_{max}	n_{avg}	n_{max}	it_{avg}	it_{max}
10000	0.40	1	117.30	696	14.76	27
20000	1.46	4	372.63	1478	29.90	43
30000	3.80	8	678.76	2119	47.30	70
40000	8.73	16	1320.23	5011	68.80	89
50000	16.83	26	1872.10	6793	87.46	115
60000	36.10	62	3639.26	22673	116.70	149
70000	69.06	105	4513.16	26700	138.50	176
80000	107.66	140	5398.03	30913	167.63	194
90000	140.23	264	6657.63	15460	194.66	227
100000	208.33	311	9512.46	15310	219.80	253

Table 4. Solving the flowshop problem with $\epsilon = n - 5$

n	$IPCG$					
	t_{avg}	t_{max}	n_{avg}	n_{max}	it_{avg}	it_{max}
10000	0.43	1	141.13	606	23.66	38
20000	1.23	2	383.13	842	41.43	55
30000	3.40	5	599.10	1150	63.26	84
40000	8.10	11	1108.20	2065	85.60	101
50000	16.36	23	1468.93	2842	110.66	130
60000	32.16	44	2021.63	3455	137.06	156
70000	54.73	73	2797.33	4621	160.63	182
80000	86.20	112	4168.40	7088	183.33	203
90000	133.43	190	6139.33	10678	206.76	232
100000	196.86	254	9663.96	16279	230.90	258

Regarding the enumeration of the set of strict Pareto optima, the results are provided in Table 5. Columns t_{avg} and t_{max} give the average and maximum

Table 5. Enumeration of the strict Pareto optima for the flowshop problem

n	IPCG							
	t_{avg}	t_{max}	n_{avg}	n_{max}	it_{avg}	it_{max}	Mit_{avg}	Mit_{max}
500	32.30	50	8475.16	10557	600.70	789	12.53	23
1000	62.86	115	22413.10	40688	1131.03	1697	19.66	39
1500	152.50	269	33811.70	39205	1647.36	1976	25.76	48

CPU time, in seconds, required to perform the enumeration. Columns n_{avg} and n_{max} give the average and maximum total number of nodes explored by CPLEX when solving all the ϵ-constraint problems. Similarly, columns it_{avg} and it_{max} provide the average and maximum total number of iterations done (number of added constraints). At last, columns Mit_{avg} and Mit_{max} present the average and maximum number of iterations in the worst case: for a given instance, the worst case is given by the value of ϵ leading to the maximum number of iterations among all ϵ values. Then, for instance, Mit_{avg} is the average value over all instances of that worst case.

The results presented in Table 5 show that the enumeration of the strict Pareto optima can be done very efficiently: for instances with 1500 jobs, about 150 seconds are required on the average. Besides, it is important to remember that there are exactly $(n + 1)$ strict Pareto optima. Therefore, the time spent to enumerate them is mainly influenced by the number of such optima. From the columns it_{avg} and it_{max} we can derive that few critical path constraints are added: for instances with 1500 jobs, in the worst case, 1976 iterations are done. However, as 1501 ϵ-constraint problems are solved, we can deduce that 475 constraints have been added along the enumeration process. Columns Mit_{avg} and Mit_{max} confirm that conclusion since, for instances with 1500 jobs, at most 47 constraints have been added when solving an ϵ-constraint problem.

4   Conclusions

We considered a job selection problem in a two-stage flow shop with the objective of selecting the best job subset with a given cardinality to minimize the makespan. We introduced a constraint generation approach to the ILP formulation of the problem according to which the constraints associated with nearly all potential critical paths are relaxed and then only the ones violated by the relaxed solution are sequentially reinstated. Our approach yielded a new generation of solution algorithms capable of solving problems with up to 100000 jobs and more. The approach seems to be extendable to the other two well known two-stage shops, namely job shop and open shop that share with the flow shop the characteristic of being modeled as a multidimensional knapsack-like problem.

References

1. Baptiste, P., Della Croce, F., Grosso, A., Tkindt, V.: Sequencing a single machine with due dates and deadlines: an ILP-based approach to solve very large instances. J. Sched. **13**, 39–47 (2010)
2. Ben-Ameur, W., Neto, J.: A constraint generation algorithm for large scale linear programs using multiple-points separation. Math. Program. A **107**, 517–537 (2006)
3. Della Croce, F., Gupta, J.N.D., Tadei, R.: Minimizing tardy jobs in a flowshop with common due date. Eur. J. Oper. Res. **120**, 375–381 (2000)
4. Johnson, S.: Optimal two and three stage production schedules with set-up time included. Nav. Res. Logist. Q. **1**, 61–68 (1954)
5. Jozefowska, J., Jurisch, B., Kubiak, W.: Scheduling shops to minimize the weighted number of late jobs. Oper. Res. Lett. **16**, 277–283 (1994)
6. Junger, M., Reinelt, G., Thienel, S.: Practical problem solving with cutting plane algorithms in combinatorial optimization. In: Cook, W., Lovasz, L., Seymour, P. (eds.) Combinatorial Optimization. DIMACS series in discrete mathematics and theoretical computer science, pp. 111–152. American Mathematical Society, Providence (1995)
7. Panwalkar, S.S., Koulamas, C.P.: An algorithm for the variable common due date, minimal tardy jobs bicriteria two-machine flow shop problem with ordered machines. Eur. J. Oper. Res. **211**, 7–13 (2012)
8. Nemhauser, G.L., Wolsey, L.A.: Integer and Combinatorial Optimization. Wiley, New York (1988)
9. Shabtay, D., Gaspar, N., Kaspi, M.: A survey on offline scheduling with rejection. J. Sched. **16**, 3–28 (2013)
10. Tkindt, V., Billaut, J.-C.: Multicriteria Scheduling: Theory, Models and Algorithms. Springer, Heidelberg (2002)
11. Tkindt, V., Della Croce, F., Bouquard, J.L.: Enumeration of Pareto optima for a flow shop scheduling problem with two criteria. INFORMS J. Comput. **19**, 64–72 (2007)

Rectilinear Shortest Path and Rectilinear Minimum Spanning Tree with Neighborhoods

Yann Disser[1], Matúš Mihalák[2], Sandro Montanari[2](✉), and Peter Widmayer[2]

[1] Department of Mathematics, TU Berlin, Berlin, Germany
[2] Department of Computer Science, ETH Zurich, Zurich, Switzerland
sandro.montanari@inf.ethz.ch

Abstract. We consider a setting where we are given a graph $\mathcal{G} = (\mathcal{R}, E)$, where $\mathcal{R} = \{R_1, \ldots, R_n\}$ is a set of polygonal regions in the plane. Placing a point p_i inside each region R_i turns G into an edge-weighted graph G_p, $p = \{p_1, \ldots, p_n\}$, where the cost of $(R_i, R_j) \in E$ is the distance between p_i and p_j. The *Shortest Path Problem with Neighborhoods* asks, for given R_s and R_t, to find a placement p such that the cost of a resulting shortest st-path in $\mathcal{G}_p$ is minimum among all graphs $\mathcal{G}_p$. The *Minimum Spanning Tree Problem with Neighborhoods* asks to find a placement p such that the cost of a resulting minimum spanning tree is minimum among all graphs $\mathcal{G}_p$. We study these problems in the L_1 metric, and show that the shortest path problem with neighborhoods is solvable in polynomial time, whereas the minimum spanning tree problem with neighborhoods is APX-hard, even if the neighborhood regions are segments.

Keywords: Neighborhoods · Minimum spanning tree · Shortest path

1 Introduction

In computational geometry we typically assume to be able to estimate locations of objects as exact points in the plane. In many real world applications, however, obtaining this information without uncertainty might be unrealistic because of noise or uncertain measurements; therefore, standard techniques and algorithms cannot be applied.

A more realistic assumption is to consider, instead of exact points, uncertainty (or *neighborhood*) regions in which we are assured the objects will lie. In this setting, shapes and properties of geometric structures induced by the points (i.e., convex hulls, minimum spanning trees, etc.) will vary with their placements inside the regions. It then becomes crucial to inspect the best and worst possible placements for the considered application. Typically, these are placements minimizing or maximizing a certain cost function of some geometric structure induced by the placement points. In this paper, we study minimum spanning trees and shortest paths under the assumption that the neighborhood regions are rectilinear polygons, and distances are measured in the L_1 metric.

© Springer International Publishing Switzerland 2014
P. Foulhoux et al. (Eds.): ISCO 2014, LNCS 8596, pp. 208–220, 2014.
DOI: 10.1007/978-3-319-09174-7_18

Several geometric optimization problems have been studied in the setting "with neighborhoods", such as the *Traveling Salesman Problem* [4,5] , the problems of finding a *convex hull* [8] or *enclosing circle* [9], or the *Minimum Spanning Tree Problem* [2,8,11]. In these variants, one searches for a placement of points inside the neighborhood regions such that the resulting cost of the geometric structure induced by the points of the placement is as big or as small as possible. For example, the *Minimum Spanning Tree Problem with Neighborhoods* (MSTN for short) asks for a placement of points, one inside each neighborhood region, such that the cost of a Euclidean minimum spanning tree of these points is smallest among all such placements. Löffler and van Kreveld showed [8] this problem to be NP-hard when the neighborhood regions are squares (not necessarily disjoint). Yang et al. [11] showed that when the neighborhood regions are disjoint unit disks, the problem admits a PTAS (i.e., it can be approximated arbitrarily well). Dorrigiv et al. [2] later proved APX-hardness of MSTN in L_2 metric when the neighborhood regions are disjoint disks. They propose to study the setting where the regions may consist of other shapes, such as line segments or rectangles. Our results show that even if the regions consist of vertically or horizontally aligned segments, the problem remains hard to approximate.

Another problem considered in the setting "with Neighborhoods" is the so-called TOURINGPOLYGONS. Given a sequence of simple polygons in the plane, a start point s, and a target point t, TOURINGPOLYGONS is the problem to find a shortest tour that starts at s, visits the polygons in the given order, and ends at t. This problem is solvable in polynomial time whenever the polygons are convex and disjoint [3]. If the polygons are allowed to be non-convex and intersecting, the problem is NP-hard for any metric L_p, $p \geq 1$, unless the polygons are rectilinear regions (not necessarily convex) and distances are measured with the L_1 metric [3]. For several years, the complexity for the case of general non-convex yet disjoint polygons has been open, which motivated the design of approximation algorithms [10]. Recently, Ahadi et al. [1] proved TOURINGPOLYGONS to be NP-hard for non-convex disjoint polygons for every metric L_p, $p \geq 1$, even for degenerate polygons composed of two line segments joint at a common endpoint whose angles with the x-axis are in $\{0, \pm\pi/4, \pi/2\}$.

In this paper we study a generalization of TOURINGPOLYGONS relaxing the requirement to visit all polygons in a natural way. In addition to the set of polygons, we are given a set of *allowed traversals* represented as a graph defined on the polygonal regions. Given a start and a target polygon, we search for a path of minimum length traversing the polygons accordingly with the edges of the underlying graph. We call this the *Shortest Path Problem with Neighborhoods*, SPN for short. In general the problem remains NP-hard, since TOURINGPOLYGONS is a special case where the graph is the path induced by the order in which the polygons need to be visited. Our results show that SPN is solvable in polynomial time for the L_1 metric in case the polygons are rectilinear regions not necessarily convex.

2   Shortest Path with Neighborhoods

Let $\mathcal{G} = (\mathcal{R}, E)$ be a directed graph defined over a set of non-overlapping recti-linear polygons $\mathcal{R} = \{R_1, \ldots, R_n\}$. Placing a point p_i inside each R_i turns $\mathcal{G}$ into an edge-weighted graph $\mathcal{G}_{\boldsymbol{p}}, \boldsymbol{p} = \{p_1, \ldots, p_n\}$, where the cost of $(R_i, R_j) \in E$ is the L_1 distance between p_i and p_j. Given a pair $s, t \in \{1, \ldots, n\}$, the Shortest Path Problem with Neighborhoods, or SPN, asks for a placement $\boldsymbol{p}$ such that the cost of a shortest path between R_s and R_t in $\mathcal{G}_{\boldsymbol{p}}$ is smallest among all possible placements. We call such a placement an *optimum SPN placement*.

The SPN problem can be solved trivially if $(R_s, R_t) \in E$. In this case, a shortest st-path is the edge (R_s, R_t), and an optimum placement minimizes the length of this edge. If $(R_s, R_t) \notin E$, it is not clear a priori what sequence of rectangles constitutes a shortest st-path, for an optimum placement $\boldsymbol{p}$. Note that, if we know which regions constitute a shortest st-path, and in which order, the problem becomes the Touring Polygons Problem.

Given a finite set of points $P \subset \mathbb{R}^2$, the *Hanan grid* of P is induced by imposing horizontal and vertical lines through the points in P. In the following, we show that an optimum SPN placement lies on the intersections of the lines of the Hanan grid induced by the corners of the regions in $\mathcal{R}$. Based on this, we provide an algorithm that computes an optimum SPN placement in time $\mathcal{O}(n^2 k^2 \log nk + mnk^3)$, where $n = |\mathcal{R}|$, $m = |E|$, and k is the maximum number of corners of a region of $\mathcal{R}$. This result is a generalization of the polynomial-time algorithm for TouringPolygons with rectilinear regions.

2.1   Properties of Optimum Solutions

Since distances between points are measured in L_1 metric and the neighborhood regions are axis-parallel, one of the most trivial approaches is to consider as possible placement points only the corners of regions in $\mathcal{R}$. It is however easy to construct instances where every optimum placement contains at least one point that is not a corner of a region in $\mathcal{R}$. We do not have to consider many more points other than the corners of the regions, though. Lemma 1 (below) shows that there always exists an optimum SPN placement where all points are points of the Hanan grid induced by the corners of the regions in $\mathcal{R}$, lying on the perimeters of those regions. To prove this, we use a property of the L_1 metric defined in terms of bounding boxes of points in $\mathbb{R}^2$. Given $x, y \in \mathbb{R}^2$, the *bounding box* B_{xy} is the smallest axis-parallel rectangle containing x and y.

Proposition 1. *For every $x, y, z \in \mathbb{R}^2$,*

$$z \in \mathsf{B}_{xy} \iff \|xy\| = \|xz\| + \|zy\|$$
$$z \notin \mathsf{B}_{xy} \iff \|xy\| < \|xz\| + \|zy\|.$$

Lemma 1. *There exists an optimum placement $\boldsymbol{p}$ such that every $p_i \in \boldsymbol{p}$ lies on the perimeter of R_i and is a grid point of the Hanan grid induced by the corners of the regions in $\mathcal{R}$.*

Proof. Let p be an optimum placement and P a shortest st-path in $\mathcal{G}_p$. We show how to move points in p not satisfying the lemma to points of the Hanan grid on the perimeter of the regions in a way such that the resulting placement is still optimum. We distinguish between regions on (visited by) P and not on P.

A point in p of a region not on P not satisfying the lemma can be trivially moved to an arbitrary corner of that region. Since the cost of P in the resulting placement is the same as in $\mathcal{G}_p$, the resulting placement is still optimum.

We first show how to move points of regions on P not satisfying the lemma to the perimeter in a way such that the resulting placement is still optimum. Then, we show that every remaining point still not satisfying the lemma can be moved to a Hanan grid point on the perimeter of its region.

Note that p_s of R_s lies on its perimeter, otherwise we can obtain a better placement by moving it to a point on the perimeter of R_s closest to the point in the successor of R_s on P. The same argument holds by simmetry for p_t.

Let $R_j \notin \{R_s, R_t\}$ be a region on P, and consider $p_i, p_k \in p$, where R_i is the predecessor of R_j on P and R_k is its successor. Consider the bounding box $\mathsf{B}_{p_i p_j}$, and let p_c be a point on the perimeter of R_j contained in B_{ij}. By Proposition 1 and triangle inequality, we have

$$\|p_i p_j\| + \|p_j p_k\| = \|p_i p_c\| + \|p_c p_j\| + \|p_j p_k\| \ge \|p_i p_c\| + \|p_c p_k\|.$$

Thus, moving p_j to p_c does not increase the cost of P. The resulting placement is still optimum, and p_j now lies on the perimeter of R_j. We can apply this operation to every point in the interior of its region.

We now show how to move points in p to Hanan grid points on the perimeters of their regions in such a way that the resulting placement is still optimum. By the above, we can assume each point of p to be lying on the perimeter its region, and that only points of regions on P may not be grid points.

Let $p_j = (x_j, y_j) \in p$ be a point on the perimeter of R_j not on the Hanan grid. Since R_j is axis-parallel, p_j lies on a line of the grid. Thus, either x_j is the x-coordinate of a grid point, or y_j is the y-coordinate of a grid point. We consider only the latter case; the former is symmetric.

Let x_l be the largest x-coordinate of a grid point lying to the left of p_j, and x_r be the smallest x-coordinate of a grid point lying to the right of p_j. We define the set $\{(x, y) \in \mathbb{R}^2 \mid x_l < x < x_r\}$ as the *vertical stripe* of p_j.

Consider a sequence $R_i, \ldots, R_k$ of consecutive regions on P of maximal length such that R_j is in the sequence, and every point in p of a region in the sequence lies in the vertical stripe of p_j. None of the points in the sequence is a grid point; however, the y-coordinate of all such points are y-coordinates of grid points. We first consider the case where $R_i \ne R_s$ and $R_k \ne R_t$.

Let $R_{i'}$ be the predecessor of R_i on P, and $R_{k'}$ be the successor of R_k on P. If $p_{i'}$ lies to the left of the vertical stripe of p_j, we move every point $p_i, \ldots, p_k$ horizontally to the x-coordinate x_l. Otherwise, we move them horizontally to x_r. Figure 1 illustrates an example of such a moving. The cost difference of P before and after moving the points can be expressed as

Fig. 1. Moving points in a vertical stripe.

$$\sum_{(R_a, R_b) \in P'} \|p_a, p_b\| - \|p'_a, p'_b\|, \tag{1}$$

where P' is the sub-path between $R_{i'}$ and $R_{k'}$, and p'_a (resp. p'_b) is the new location of p_a (p_b). Since points are only moved horizontally, their y-differences do not change. Thus, (1) can be rewritten as

$$\sum_{(R_a, R_b) \in P'} |x_a - x_b| - |x'_a - x'_b|. \tag{2}$$

Before moving them, all points $p_i, \ldots, p_k$ are contained in the vertical stripe of p_j; therefore the cost of P' before the moving is at least $|x_{i'} - x_i| + |x_{k'} - x_k|$. After the moving, the x-coordinates of all $p_i, \ldots, p_k$ become $x' \in \{x_l, x_r\}$. Thus, (2) is at least

$$|x_{i'} - x_i| + |x_{k'} - x_k| - |x_{i'} - x'| - |x_{k'} - x'|. \tag{3}$$

If $p_{i'}$ and $p_{k'}$ lie on the same side of the vertical stripe of p_j, the new coordinate x' is closer to both $x_{i'}$ and $x_{k'}$. If $p_{i'}$ and $p_{k'}$ lie on different sides of the vertical stripe, then $|x_{i'} - x'| + |x_{k'} - x'| = |x_{i'} - x_{k'}|$. In both cases, (3) is positive; that is, the cost of P does not increase and the new placement is still optimum.

The case where $R_i = R_s$ follows trivially from above, because we can define $p_{i'}$ to be the point p_i itself. In this way, the distance between p_i and $p_{i'}$ is always 0, and the direction of the moving depends only on the position of $p_{k'}$. The same holds also for the remaining cases. ☐

2.2 Algorithm

We now present an algorithm that computes an optimum SPN placement by exploiting the structural properties of optimal placements established in Lemma 1. To do so, we create an auxiliary graph from $\mathcal{R}$ and E with the property that a shortest path between two designated vertices of this graph yields a minimum SPN placement. Such a path can be found using standard shortest path techniques, such as Dijkstra's algorithm. The auxiliary graph $\mathcal{D} = (V_\mathcal{D}, E_\mathcal{D})$ is defined as follows. There is a vertex in $V_\mathcal{D}$ for every point on the perimeter of a region that is also a point of the Hanan grid induced by the corners of the regions in $\mathcal{R}$, and two additional vertices v_s and v_t. In the following, we say "a vertex v of region R_i" to indicate a vertex corresponding to a point of R_i. There is an edge in $E_\mathcal{D}$ from v_s to every vertex of R_s, and from every vertex of R_t to v_t. Also, let u be a vertex of R_i and R_j be a region such that $(R_i, R_j) \in E$. For every segment composing the perimeter of R_j, there is an edge in $E_\mathcal{D}$ from u to its closest vertex on that segment. Furthermore, there is an edge in $E_\mathcal{D}$ from u to the next vertex along the perimeter of R_i, in both directions. We assign a cost to an edge $(u, v) \in E_\mathcal{D}$ equal to 0 if either $u = v_s$ or $v = v_t$, and equal to $\|uv\|$ otherwise. The following theorem shows that a shortest path between v_s and v_t in $\mathcal{D}$ yields an optimum SPN placement.

Theorem 1. *Given a shortest path $P_\mathcal{D}$ from v_s to v_t in $\mathcal{D}$, let p be a placement as follows. For each region $R_i \in \mathcal{R}$, if R_i has vertices on $P_\mathcal{D}$, p contains the first of them. Otherwise, p contains one of its corners chosen arbitrarily. The placement p is an optimum SPN placement.*

Proof. Consider the vertices on $P_\mathcal{D}$ chosen as points of p in the order as they appear on $P_\mathcal{D}$. Since the regions of these points are connected in $\mathcal{G}$, $P_\mathcal{D}$ corresponds to an st-path P in $\mathcal{G}$. By triangle inequality, the cost of P in $\mathcal{G}_p$ is at most the cost of $P_\mathcal{D}$. For the sake of contradiction, suppose there exists an optimum placement q and a shortest st-path Q in $\mathcal{G}_q$ with cost smaller than the cost of P in $\mathcal{G}_p$. Without loss of generality, we can assume the points in q to satisfy Lemma 1. Thus, every point of q corresponds to a vertex of $\mathcal{D}$. We construct a path $Q_\mathcal{D}$ from v_s to v_t in $\mathcal{D}$ as follows. The first edge is (v_s, q_s); after that, for every edge (R_i, R_j) on Q, consider the points $q_i, q_j \in q$, the bounding box $\mathsf{B}_{q_i q_j}$, and the at most two segments on the perimeter of R_j on which q_j lies. By construction, q_i is connected in $\mathcal{D}$ to a vertex on both segments; let v be one of them chosen arbitrarily such that $v \in \mathsf{B}_{q_i q_j}$. We add to $Q_\mathcal{D}$ the path that from q_i goes to v, and follows the perimeter of R_j to q_j. By Proposition 1, the cost of this path is equal to $\|q_i q_j\|$. The last edge on $Q_\mathcal{D}$ is (q_t, v_t). To see that the cost of $Q_\mathcal{D}$ is equal to the cost of Q in $\mathcal{G}_q$ it is sufficient to notice that the first and the last edge of $Q_\mathcal{D}$ have cost 0 and, for every edge (R_i, R_j) on Q, the sub-path from q_i to q_j in $Q_\mathcal{D}$ has cost equal to $\|q_i q_j\|$. This results in a contradiction, because we have then found a path from v_s to v_t with cost smaller than $P_\mathcal{D}$. $\square$

The above theorem shows how to construct an optimum SPN placement once a shortest path between v_s and v_t in $\mathcal{D}$ is known. Since edge costs in $\mathcal{D}$ are greater or equal than 0, we can find such a path with Dijkstra's algorithm in time $\mathcal{O}(|V_\mathcal{D}| \log |V_\mathcal{D}| + |E_\mathcal{D}|)$. The sizes of $V_\mathcal{D}$ and $E_\mathcal{D}$ depend on the number of points on the perimeters of the regions that are the grid points of the Hanan grid induced by the corners of $\mathcal{R}$. To evaluate this number, consider a line of the Hanan grid. Each time this line intersects (cut in two nonempty parts) an orthogonal segment on the perimeter of a region, an additional vertex is introduced. Conversely, each segment of the perimeter of a region can in the worst case be intersected by every grid line orthogonal to it. If k is the maximum number of corners of a region in $\mathcal{R}$ (and therefore on the number of segments of its perimeter), and $|\mathcal{R}| = n$, the number of grid lines is $\mathcal{O}(nk)$. Thus, the number of grid points lying on the perimeter of one region is $\mathcal{O}(nk^2)$, and the size of $V_\mathcal{D}$ is $\mathcal{O}(n^2k^2)$. To evaluate the size of $E_\mathcal{D}$, consider an edge $(R_i, R_j) \in E$ and a vertex v of R_i. By construction, there is an edge from v to a vertex on each of the at most k segments on the perimeter of R_j. Furthermore, v is connected to at most two vertices on the perimeter of R_i. If we have $|E| = m$ edges and $\mathcal{O}(nk^2)$ vertices in each region, the size of $E_\mathcal{D}$ is $\mathcal{O}(mnk^3)$. Thus, computing a shortest path from v_s to v_t in $\mathcal{D}$ with Dijkstra's algorithm takes time $\mathcal{O}(n^2k^2 \log nk + mnk^3)$.

3   Minimum Spanning Tree with Neighborhoods

In the Minimum Spanning Tree Problem with Neighborhoods, or MSTN, we are given a set of regions $\mathcal{R} = \{R_1, \ldots, R_n\}$ and an underlying graph $\mathcal{G} = (\mathcal{R}, E)$. The problem asks for a placement $\boldsymbol{p}$ such that the cost of a minimum spanning tree in $\mathcal{G}_{\boldsymbol{p}}$ is smallest among all possible placements.

It is known [2] that, if distances are measured in L_2 norm and the neighborhood regions are disks, the MSTN problem does not admit an FPTAS unless P = NP. We will adapt their proof and show that MSTN does not admit an FPTAS for the L_1 metric even for non-overlapping axis-parallel segments.

The reduction is from the planar 3-SAT problem. Planar 3-SAT is a variant of 3-SAT where the graph associated with the formula is planar. The graph contains a vertex for each variable and each clause, and there is an edge from a variable to a clause if the clause contains a literal of that variable. Planar 3-SAT was shown to be NP-hard by a reduction from the standard 3-SAT problem [7]. Furthermore, it was shown that in the plane embedding used in the reduction there always exists a so-called *spinal path* passing through every vertex corresponding to a variable without crossing any edge of the graph. Knuth and Raghunathan [6] observed that there always is a simple embedding where the variables are arranged on a straight line (the spinal path), and the clauses are drawn as three legged segments completely above or below them, in a way such that none of the legs cross each other. Figure 2 shows an example of such an embedding.

The reduction starts from a plane embedding of an instance of planar 3-SAT and constructs an instance of MSTN such that a solution to the latter indicates whether the former is satisfiable. First, we define three types of gadgets: a gadget for each variable, a gadget for each clause, and a gadget for the spinal path. Then, we show how to replace each variable, clause and the spinal path with a

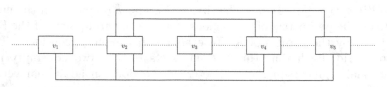

Fig. 2. A planar 3-SAT instance on 5 variables. Dashed lines are parts of the spinal path, solid lines are clauses.

corresponding gadget resulting in an instance of MSTN. From our construction, it will be easy to see that the size of the resulting MSTN instance is polynomially bounded. Finally, we provide two threshold values t_1 and t_2, with $t_1 < t_2$, and we prove that an optimum solution of the constructed MSTN instance has a cost smaller than t_1 if and only if the initial 3-SAT formula is satisfiable. If the formula is not satisfiable, the cost of an optimum solution of the MSTN instance is at least t_2. This proves that MSTN does not admit an FPTAS unless $P = NP$.

An important tool in the definitions of the gadgets is a so-called wire. A *wire* is a set of points (i.e., regions) placed in close succession, so that any minimum spanning tree (for any placement) will contain the edges connecting the points. To ensure this, it is sufficient to place two consecutive points in a wire at a suitably small distance. Since the edges between consecutive points in wires do not form a cycle, any minimum spanning tree in any placement will contain the edge connecting them. However, this suitably small distance must still be large enough to guarantee that a wire can be realized with a polynomial number of points. Since in the following construction the smallest non-zero distance between any two regions (other than those for the wires) is at least $d/2$, for a constant $d := 0.25$, a suitably small value for the points of a wire is, for example, $d/4$.

3.1 Reduction Gadgets

Variable Gadget. For each variable there are $k = 6c + 6$ segments of length α, where c is the maximum number of clauses in which the variable appears as a literal that are completely above or below the variable vertex in the embedding. Note that $k \geq 12$, because a variable appears at least once in a clause. In the following we specify the value of the parameter α more precisely, and we show it to be polynomial in the number of clauses and variables.

As illustrated in Fig. 3, the segments are placed along the perimeter of a rectangle with sides of length $3c\alpha + d$ and $3\alpha + d$. In its interior we place a wire for every two segments consecutive in clockwise order. Each of these wires ends on the line bisecting the angle formed by the corresponding segments; for parallel segments, the endpoint is at distance d from their common point. For perpendicular segments, the endpoint is at distance d from the intersection of the lines passing through the segments. We connect these wires in the bounded region in a tree-like structure as in Fig. 3. We call this arrangement of wires in the internal region a *k-tree*.

A placement of points inside a variable gadget is called a *configuration* if, for every two consecutive segments in clockwise order, the placement contains either their two closest points or their two farthest points. For a variable gadget there exist exactly two different configurations. To see this, consider two consecutive segments in a variable gadget and a configuration placement. If the placement contains their two closest points, we can place points in the remaining segments in exactly one way in order to obtain a configuration. Similarly, if the placement contains their two farthest points, we have exactly one way to place points in the remaining segments. We associate these two possible configurations with the two assignments to the variable.

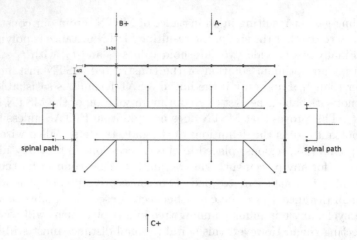

Fig. 3. A variable gadget with $k = 18$. The variable appears in A with negative sign and in B, C with positive sign. Thick lines are segments, the rest wires.

Clause Gadget. Clause gadgets are composed of at most three wires meeting at a single point following the embedding. As in Fig. 3, each wire of a clause gadget approaches the common point of two adjacent horizontal segments of a variable gadget. Clauses that are located above the spinal path in the rectilinear embedding approach variable gadgets from above, while clauses that in the rectilinear embedding are located below the spinal path approach variable gadgets from below. Furthermore, clause wires approach a variable gadget in the same clockwise order as the edges connecting the variable vertex to the corresponding clauses in the rectilinear embedding.

A clause wire terminates at distance $1 + 2d$ from the common point of the approached segments along the vertical line passing through it. The approached segments are chosen such that their common point is contained in a configuration satisfying the clause. That is, an edge with cost $1 + 2d$ connects the clause wire to the segments in a configuration placement satisfying the clause.

Spinal Path Gadget. The spinal path gadget consists of wires following the embedding of the planar 3-SAT instance. As in Fig. 3, the spinal path gadget approaches every variable gadget twice, once from the left and once from the right. For each side, the spinal path wire is split in two parts, each approaching two adjacent vertical segments. The point at which a part terminates is located at distance 1 from the common point of the approached segments along the horizontal line passing through it.

3.2   The Reduction

Given a rectilinear embedding of an instance of planar 3-SAT, we create an instance of MSTN and provide two threshold values t_1, t_2, with $t_1 < t_2$. We show that if the 3-SAT instance is satisfiable, then there is a placement with a minimum spanning tree of cost at most t_1, and if the 3-SAT instance is unsatisfiable, then the cost of a minimum spanning tree for any placement is at leat t_2.

Theorem 2. *MSTN with L_1 metric and axis-parallel segments is* APX-*hard.*

Proof. To create an instance of MSTN, replace in the given embedding every variable, clause, and the spinal path with a gadget as explained above. The wires forming the spinal path, the m clause gadgets and the k-trees in the internal region of each variable gadget have a fixed cost in every MST, denoted as c_{wires}. The remaining cost of the spanning tree is given by connecting the segments of the variable gadgets to the k-trees and the spinal path and clause wires.

Suppose there exists a satisfying assignment. Then, we place points in each variable gadget in a configuration according to its value in the assignment. We provide an upper bound t_1 on the cost of a minimum spanning tree in this placement by constructing a spanning tree and evaluating its cost. For each pair of consecutive segments having their closest points in the placement, the spanning tree connects them to the k-tree of the corresponding variable with cost d. If there is a total of K segments among all variable gadgets, the spanning tree requires a cost of $(K/2)d$ to connect all of them to the k-trees (note that K is even). For each clause gadget, consider a variable satisfying it in the assignment. We connect the corresponding endpoint of the clause wire to one of the segments it approaches with an edge with cost $1 + 2d$. Overall, the cost for connecting all the clause wires to the tree is $m(1 + 2d)$. For each part of the spinal path gadget approaching a variable gadget, exactly one of its endpoint approaches a point of the placement. The spanning tree contains the $2n$ edges of cost 1 connecting them. Overall, the cost of an optimum MSTN solution in case a satisfying assignment exists is therefore at most

$$t_1 := c_{wires} + (K/2)d + (1 + 2d)m + 2n.$$

If there is no satisfying assignment, we show that the cost of an optimum MSTN solution is at least $t_2 := t_1 + d$. To see this, consider an optimum placement where every point in a segment is one of its extreme points. The existence of such an optimum placement is guaranteed by the fact that wires approaching variable gadgets and wires of the k-trees terminate either to the left or to the right of horizontal segments, and above or below vertical segments.

We first provide an upper bound on the minimum spanning tree cost for such a placement. By constructing a spanning tree and evaluating its cost in the placement. Then, we use this upper bound to show that, in every minimum spanning tree, clause wires are connected to the tree either with an edge from one of the endpoints to one of the approached segments, or with an edge from one of its endpoints to the approached k-tree endpoint. Finally, we show that the cost of any optimum MSTN solution is at least t_2.

The spanning tree contains the wires composing the spinal path, the clauses, and all k-trees. Every segment in a variable gadget is connected to an endpoint of the k-tree with an edge of cost d. Every part of a wire of the spinal path approaching a variable gadget is connected to one of the approached segments by one of its endpoints with an edge with cost 1. Similarly, an endpoint of each clause wire chosen arbitrarily is connected to a k-tree of a variable appearing in that clause with an edge with cost $1 + 3d$. The cost of such a spanning tree is

$$c_{wires} + 2n + m(1 + 3d) + Kd. \tag{4}$$

We now prove that, in every minimum spanning tree, each clause is connected to it either with an edge from one of its endpoint to an approached segment, or with an edge from one of its endpoints to the corresponding k-tree endpoint. Suppose this is not the case, and there is a clause whose endpoint in the MST is connected neither to an approached segment, nor to the corresponding k-tree endpoint. By construction, the next closest object is located at distance at least α, where α is the above defined length of the segments of the variable gadgets. Since the spinal path, clauses and k-trees wires are part of every MST, setting

$$\alpha := 2n + m(1 + 3d) + Kd + 1$$

we get a contradiction, because the cost of a minimum spanning tree would then be greater than (4). Thus, in every minimum spanning tree every clause is connected via one of its endpoints to one of the approached regions.

Finally, we show that if the formula is not satisfiable, any optimum MSTN solution has cost greater than t_2. Clearly, we cannot provide a configuration for each variable gadget such that every clause where that variable appears can be connected to it with an edge with cost $1 + 2d$, otherwise the formula would be satisfiable. Therefore, in an optimum solution, either at least one variable gadget is not set in a configuration, or every variable gadget is in a configuration and for at least one clause no wire endpoint approaches a point of the placement.

In the former case, the cost of a minimum spanning tree is at least $c_{wires} + 2n + (K/2)d + j(1 + 2d) + (m - j)\delta$, where j is the number of clauses that can be satisfied by the assignment corresponding to the configuration and δ is the minimum cost necessary to connect a clause that is not satisfied by the assignment. By the above, we know that in any minimum spanning tree a clause wire is connected to one of the approached segments or the corresponding k-tree endpoint. Since every variable gadget is in a configuration, the smallest distance δ between a non satisfied clause wire and a point in the placement is at least $1 + 3d$. Thus, any optimum MSTN solution where every variable gadget is set in a configuration results in a minimum spanning tree with cost at least t_2.

In the latter case, there exists at least one variable gadget that is not in a configuration. Let then a be the overall number of segments for which the point in the placement is not the closest or the farthest to the point in one of the consecutive segments. Note that a is even, therefore $a \geq 2$, and the cost of a spanning tree is at least

$$c_{wires} + 2n + \frac{(K + a)}{2}d + m(1 + 2d) \geq t_2.$$

Suppose now that there exists an FPTAS for MSTN. Given an instance of planar 3-SAT, we construct the gadget presented above and calculate t_1.

We then set a parameter $\epsilon < d/t_1$, so a $(1 + \epsilon)$-approximate solution to the MSTN problem would tell us whether the cost of the corresponding optimum solution is smaller than t_1 or greater than t_2, and thus, whether there exists a satisfying assignment for the planar 3-SAT instance.   □

4 Conclusions

We considered the Shortest Path Problem and the Minimum Spanning Tree Problem with Neighborhoods in the L_1 metric and showed that the former can be solved efficiently if the neighborhood regions are rectilinear polygons not necessarily convex, while the latter does not admit a PTAS unless P = NP even if the regions are axis-parallel segments. An interesting open problem is to consider variants of SPN and MSTN where the goal is to find placements *maximizing* the cost of shortest paths and minimum spanning tree, respectively.

Acknowledgments. This work was supported by the EU FP7/2007-2013 (DG CONNECT.H5-Smart Cities and Sustainability), under grant agreement no. 288094 (project eCOMPASS) and by the Alexander von Humboldt-Foundation.

References

1. Ahadi, A., Mozafari, A., Zarei, A.: Touring disjoint polygons problem is NP-hard. In: Widmayer, P., Xu, Y., Zhu, B. (eds.) COCOA 2013. LNCS, vol. 8287, pp. 351–360. Springer, Heidelberg (2013)
2. Dorrigiv, R., Fraser, R., He, M., Kamali, S., Kawamura, A., López-Ortiz, A., Seco, D.: On minimum-and maximum-weight minimum spanning trees with neighborhoods. In: Erlebach, T., Persiano, G. (eds.) WAOA 2012. LNCS, vol. 7846, pp. 93–106. Springer, Heidelberg (2013)
3. Dror, M., Efrat, A., Lubiw, A., Mitchell, J.S.B.: Touring a sequence of polygons. In: Proceedings of the 35th Annual ACM Symposium on Theory of Computing (STOC), pp. 473–482 (2003)
4. Dumitrescu, A., Mitchell, J.S.B.: Approximation algorithms for TSP with neighborhoods in the plane. In: Proceedings of the Twelfth Annual ACM-SIAM Symposium on Discrete algorithms (SODA), pp. 38–46 (2001)
5. Elbassioni, K.M., Fishkin, A.V., Mustafa, N.H., Sitters, R.A.: Approximation algorithms for euclidean group TSP. In: Caires, L., Italiano, G.F., Monteiro, L., Palamidessi, C., Yung, M. (eds.) ICALP 2005. LNCS, vol. 3580, pp. 1115–1126. Springer, Heidelberg (2005)
6. Knuth, D., Raghunathan, A.: The problem of compatible representatives. SIAM J. Discrete Math. **5**(3), 422–427 (1992)
7. Lichtenstein, D.: Planar formulae and their uses. SIAM J. Comput. **11**(2), 329–343 (1982)
8. Löffler, M., Kreveld, M.: Largest and smallest convex hulls for imprecise points. Algorithmica **56**(2), 235–269 (2010)
9. Löffler, M., van Kreveld, M.J.: Largest bounding box, smallest diameter, and related problems on imprecise points. Comput. Geom. **43**(4), 419–433 (2010)

10. Pan, X., Li, F., Klette, R.: Approximate shortest path algorithms for sequences of pairwise disjoint simple polygons. In: Proceedings of the 22nd Canadian Conference on Computational Geometry (CCCG), pp. 175–178 (2010)
11. Yang, Y., Lin, M., Xu, J., Xie, Y.: minimum spanning tree with neighborhoods. In: Kao, M.-Y., Li, X.-Y. (eds.) AAIM 2007. LNCS, vol. 4508, pp. 306–316. Springer, Heidelberg (2007)

Lovász and Schrijver N_+-Relaxation on Web Graphs

Mariana Escalante$^{(\boxtimes)}$ and Graciela Nasini

CONICET and FCEIA - Universidad Nacional de Rosario, Rosario, Argentina
{mariana,nasini}@fceia.unr.edu.ar

Abstract. In this contribution we continue the study of the Lovász-Schrijver PSD-operator applied to the edge relaxation of the stable set polytope of a graph. The problem of obtaining a combinatorial characterization of graphs for which the PSD-operator generates the stable set polytope in one step has been open since 1990. In an earlier publication, we named these graphs N_+-perfect. In the current work, we prove that the only imperfect web graphs that are N_+-perfect are the odd-cycles and their complements. This result adds evidence for the validity of the conjecture stating that the only graphs which are N_+-perfect are those whose stable set polytope is described by inequalities with near-bipartite support. Finally, we make some progress on identifying some minimal forbidden structures on N_+-perfect graphs which are also rank-perfect.

1 Introduction

Perfect graphs were introduced by Berge in the early sixties [1]. A graph is *perfect* if each of its induced subgraphs has chromatic number equal to the cardinality of a maximum cardinality clique in the subgraph.

According to the results in [6] the family of perfect graphs constitute a class where the Maximum Weighted Stable Set Problem (MWSSP) can be solved in polynomial time. Some years later, the same authors proved a beautiful result [8]: for every graph G,

$$G \text{ is perfect} \Leftrightarrow \text{TH}(G) = \text{STAB}(G) \Leftrightarrow \text{TH}(G) = \text{CLIQUE}(G) \Leftrightarrow \\ \text{STAB}(G) = \text{CLIQUE}(G) \Leftrightarrow \text{TH}(G) \text{ is polyhedral,} \quad (1)$$

where $\text{STAB}(G)$ is the stable set polytope of G, $\text{CLIQUE}(G)$ is its clique relaxation and $\text{TH}(G)$ is the *theta body* of G defined by Lovász [10].

In the early nineties, Lovász and Schrijver introduced the PSD-operator N_+ which, applied over the edge relaxation of $\text{STAB}(G)$, generates the positive semi-definite relaxation $N_+(G)$ stronger than $\text{TH}(G)$ [11].

As it holds for perfect graphs, MWSSP can be solved in polynomial time for the class of graphs for which $N_+(G) = \text{STAB}(G)$. We will call these graphs N_+-*perfect*.

Partially supported by grants PIP-CONICET 241, PICT-ANPCyT 0361, PID UNR 415, PID UNR 416.

P. Fouilhoux et al. (Eds.): ISCO 2014, LNCS 8596, pp. 221–229, 2014.
DOI: 10.1007/978-3-319-09174-7_19

Our main goal is to obtain a characterization of N_+-perfect graphs similar to the one given in (1) for perfect graphs. More precisely, we would like to find an appropriate polyhedral relaxation of STAB(G) playing the role of CLIQUE(G) in (1). Following this line, in a recent publication [3], we proposed the following conjecture:

Conjecture 1. The stable set polytope of every N_+-perfect graph can be described by facet inducing inequalities with near-bipartite support.

In [2] the validity of this conjecture on near-perfect graphs is established. In fact, the following theorem is proved.

Theorem 1 [2]. *Let G be an N_+-perfect and a properly near-perfect graph. Then, either G or its complement is an odd cycle.*

Later, in [3] we extended its validity to fs-*perfect* graphs, a superclass of near-perfect graphs defined as those graphs for which the stable set polytope is completely described by clique constraints and a single full-support inequality.

The main contribution of this paper is to prove the validity of the conjecture on one more infinite family of graphs, the web graphs.

2   Preliminaries

Given a graph $G = (V, E)$ a *stable set* is a subset of mutually non-adjacent nodes in G. The maximum cardinality of a stable set is denoted by $\alpha(G)$, the *stability number of G*. The *stable set polytope* is the convex hull of the incidence vectors of the stable sets in the graph G and it is denoted by STAB(G).

The polyhedron

$$\text{FRAC}(G) = \{x \in [0,1]^V : x_i + x_j \leq 1 \text{ for every } ij \in E\}$$

is the *edge relaxation* of STAB(G).

A *clique* Q is a subset of pairwise adjacent nodes in G. Every incidence vector of a stable set must satisfy clique constraints, i.e., $\sum_{i \in Q} x_i \leq 1$. These constraints define the *clique relaxation* of the stable set polytope, CLIQUE(G). In general, STAB(G) $\subset$ CLIQUE(G). Chvátal [5] showed that perfect graphs are exactly those graphs for which equality holds.

Minimally imperfect graphs are those graphs that are not perfect but after deleting any node they become perfect. The Strong Perfect Graph Theorem states that the only minimally imperfect graphs are the odd cycles and their complements [4].

The *support* of a valid inequality for STAB(G) is the subgraph induced by the nodes having positive coefficient in it. We say that an inequality is a *full-support inequality* if its support is the whole graph.

In [14] Shepherd called a graph G *near-perfect* if its stable set polytope is defined only by non-negativity constraints, clique constraints and the *full-rank constraint*

$$\sum_{u \in V} x_u \leq \alpha(G).$$

Clearly, every node induced subgraph of a near-perfect graph is also near-perfect [14].

Due to results of Chvátal [5] near-perfect graphs constitute a superclass of perfect graphs. According to Padberg [13] minimally imperfect graphs are also near-perfect graphs.

Near-bipartite graphs, defined in [15], are those graphs such that removing all neighbours of an arbitrary node and the node itself, leaves the resulting graph bipartite.

Given integer numbers k and n such that $n \geq 2(k+1)$, the *web* graph, denoted by W_n^k, is the graph having node set $\{1, \ldots, n\}$ and such that ij is an edge if i and j differ by at most k (mod n) and $i \neq j$.

If $k = 1$, W_n^1 is a cycle. If $k \geq 2$ and $n \leq 2k + 2$ W_n^k is a perfect graph and W_{2k+3}^k is the complementary graph of the $(2k+3)$-cycle.

In [18] Wagler characterized all near-perfect web graphs:

Theorem 2 [18]. *A web graph is near-perfect if and only if it is perfect, an odd hole, the web W_{11}^2 or it has stability number 2.*

If $W_{n'}^{k'}$ is a node induced subgraph of W_n^k then it is a *subweb* of W_n^k. In [17] Trotter characterized for which values of n' and k', $W_{n'}^{k'}$ is a subweb of W_n^k.

Theorem 3 [17]. *If $k \geq 1$ and $n \geq 2(k+1)$ the graph $W_{n'}^{k'}$ is a subweb of W_n^k if and only if*

$$\frac{k'}{k} \leq \frac{n'}{n} \leq \frac{k'+1}{k+1}.$$

2.1   The N_+-Operator

As we have already mentioned, in this paper we focus on the behaviour of the N_+-operator defined by Lovász and Schrijver [11] on the edge relaxation of the stable set polytope.

We denote by $\mathbf{e}_0, \mathbf{e}_1, \ldots, \mathbf{e}_n$ the vectors of the canonical basis of $\mathbb{R}^{n+1}$ (where the first coordinate is indexed zero), $\mathbf{1}$ the vector with all components equal to 1 and $\mathbb{S}_+^n$ the space of n-by-n symmetric and positive semidefinite matrices with real entries.

Given a convex set K in $[0, 1]^n$, let

$$\text{cone}(K) = \left\{ \begin{pmatrix} x_0 \\ x \end{pmatrix} \in \mathbb{R}^{n+1} : x = x_0 y; \ y \in K \right\}.$$

Then, we define the polyhedral set

$$M(K) = \{ Y \in \mathbb{S}_+^{n+1} : \ Y\mathbf{e}_0 = \text{diag}(Y),$$
$$Y\mathbf{e}_i \in \text{cone}(K),$$
$$Y(\mathbf{e}_0 - \mathbf{e}_i) \in \text{cone}(K), \ i = 1, \ldots, n\},$$

where $\text{diag}(Y)$ denotes the vector whose i-th entry is Y_{ii}, for every $i = 0, \ldots, n$.

Projecting this polyhedral lifting back to the space $\mathbb{R}^n$ results in

$$N_+(K) = \left\{ x \in [0,1]^n : \begin{pmatrix} 1 \\ x \end{pmatrix} = Y\mathbf{e}_0, \text{ for some } Y \in M(K) \right\}.$$

In practice, we prove that a point $x \in [0,1]^n$ belongs to $N_+(K)$ by showing the existence of a symmetric PSD matrix Y of the form

$$Y = \left(\begin{array}{c|c} 1 & x^t \\ \hline x & \bar{Y} \end{array} \right) \tag{2}$$

where x^t stands for the transpose of column vector x and $\bar{Y}$ is an $n \times n$ matrix with columns $\bar{Y}_i$ for $i = 1, \ldots, n$, satisfying the following conditions:

1. $\bar{Y}_{ii} = x_i$,
2. If $x_i = 0$ then $\bar{Y}_i = \mathbf{0}$,
3. If $x_i = 1$ then $\bar{Y}_i = x$,
4. If $0 < x_i < 1$ then $\frac{1}{x_i}\bar{Y}_i \in K$ and $\frac{1}{1-x_i}(x - \bar{Y}_i) \in K$,

for every $i = 1, \ldots, n$.

In [11], Lovász and Schrijver proved that $N_+(K)$ is a relaxation of the convex hull of integer solutions in K.

If we let $N_+^0(K) = K$ then k-th application of the N_+-operator is $N_+^k(K) = N_+(N_+^{k-1}(K))$ for every $k \geq 1$. The authors in [11] showed that $N_+^n(K) = \text{conv}(K \cap \{0,1\}^n)$.

In this work we focus on the behaviour of a single application of the N_+-operator on the edge relaxation of the stable set polytope of a graph. Then, in order to simplify the notation we write $N_+(G) = N_+(\text{FRAC}(G))$.

In [11] it is shown that

$$\text{STAB}(G) \subset N_+(G) \subset \text{TH}(G) \subset \text{CLIQUE}(G).$$

Also from results in [11], we know that graphs for which every facet defining inequality of $\text{STAB}(G)$ has a near-bipartite support is N_+-perfect. Then, Conjecture 1 establishes that these graphs are the only N_+-perfect graphs.

In particular, perfect and near-bipartite graphs are N_+-perfect. In addition, it can be proved that every subgraph of an N_+-perfect graph is also N_+-perfect. A graph G that is not N_+-perfect is called N_+-imperfect.

Using the properties of the N_+-operator, if G' is an N_+-imperfect subgraph of G then G is also N_+-imperfect.

In [7] and [9] it was proved that all the imperfect graphs with at most 6 nodes are N_+-perfect graphs, except for the two imperfect near-perfect graphs depicted in Fig. 1. The graph on the left is denoted by G_{LT} and the other one is denoted by G_{EMN}.

A graph G' is an *odd subdivision* of a graph G if it is obtained by replacing an edge of G by a path of odd length.

As a consequence of the results in [9] we have the following:

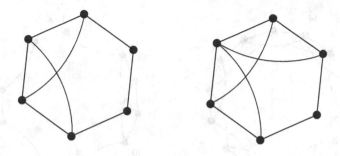

Fig. 1. The graphs G_{LT} and G_{EMN}.

Lemma 1. *If G is N_+-imperfect and G' is obtained after the odd subdivision of an edge in G, then G' is also N_+-imperfect.*

This result becomes relevant in the proof of the validity of the conjecture on web graphs since there we show that most of the web graphs have an odd subdivision of the graph G_{LT} as a node induced subgraph.

3   The Conjecture on Web Graphs

The fact that the conjecture holds on web graphs will follow after proving that the only N_+-perfect webs are either perfect or minimally imperfect webs.

Theorem 1 [2] asserts that every near-perfect graph satisfies the conjecture, therefore, from Theorem 2, we only need to consider web graphs with stability number at least three. It is known that the stability number of W_n^k is $\alpha(W_n^k) = \left\lfloor \frac{n}{k+1} \right\rfloor$. Then, if $n \leq 3k+2$, W_n^k is near-perfect and from Theorem 1 the conjecture holds on these web graphs. Therefore, from now on we can consider web graphs W_n^k with $n \geq 3k + 3$.

Now we are able to present the following result:

Theorem 4. *If $n \geq 9$ and $n \neq 10$, W_n^2 has an odd subdivision of G_{LT} as a node induced subgraph.*

Proof. Let $n \geq 9$ and $\{1, \ldots, n\}$ be the node set of W_n^2. Assume that we delete the six consecutive nodes in the set $\{n-5, n-4, \ldots, n\}$. Note that if we find a subset $T^s = \{v_1, \ldots, v_{2s}\}$ of $\{1, \ldots, n-6\}$, with $s \geq 1$, $v_1 = 1$, $v_{2s} = n - 6$ and such that T^s induces a path in W_n^2, then $T^s \cup \{n-4, n-3, n-2, n-1\}$ induces in W_n^2 an odd subdivision of G_{LT}.

For example, in the web W_{14}^2 the set $T^3 = \{1, 2, 4, 5, 7, 8\}$ induces a path and $T^3 \cup \{10, 11, 12, 13\}$ induces an odd subdivision of G_{LT}. See Fig. 2.

Then, in order to prove the result we show the existence of such a set T^s with the above required properties for every $n \geq 9$ and $n \neq 10$.

We divide the rest of the proof into four different cases according to the value of $n - 6$.

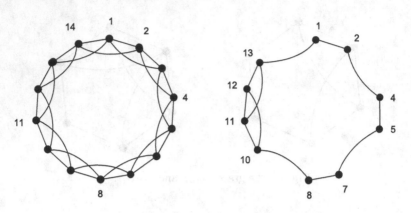

Fig. 2. The web graph W_{14}^2 and a node induced odd subdivision of G_{LT}.

– If $n - 6 = 4r + 3$ for some $r \geq 0$, then

$$T^{r+1} = \{2t - 1 : 1 \leq t \leq 2r + 2\}.$$

– If $n - 6 = 4r + 2$ then $r \geq 1$ since $n \geq 9$. In this case, we consider

$$T^{r+1} = \{2t - 1 : 1 \leq t \leq 2r + 1\} \cup \{n - 6\}.$$

– If $n - 6 = 4r + 1$ again $r \geq 1$. In this case, we find

$$T^{r+1} = \{2t : 1 \leq t \leq 2r\} \cup \{1, n - 6\}.$$

– If $n - 6 = 4r$ we have $r \geq 2$ since $n \geq 9$ and $n \neq 10$. In this case we consider

$$T^{r+1} = \{3 + 2t : 1 \leq t \leq 2r - 2\} \cup \{1, 2, 4, n - 6\}.$$

$\square$

Corollary 1. *If $n \geq 8$ and $n \neq 10$ then W_n^2 is N_+-imperfect.*

Proof. The web graph W_8^2 is an imperfect near-perfect graph and then it is N_+-imperfect.

Since every odd subdivision of G_{LT} is N_+-imperfect and the N_+-imperfection of a subgraph implies the N_+-imperfection of the graph itself, the result follows directly from the previous theorem. $\square$

In order to complete the analysis of the family of web graphs W_n^2 we need to prove that W_{10}^2 is N_+-imperfect. Note that it does not have an odd subdivision of G_{LT} as a node induced subgraph. Instead, we make use of the definition of $N_+(W_{10}^2)$.

Lemma 2. *The web graph W_{10}^2 is N_+-imperfect.*

Proof. The proof is based on finding a point $\bar{x} \in N_+(W_{10}^2) \setminus \text{STAB}(W_{10}^2)$.

Let us consider the point $\bar{x} = \lambda\mathbf{1} \in \text{FRAC}(W_{10}^2)$ for $\lambda = \frac{31}{100}$. Clearly it violates the full-rank constraint and therefore, $\bar{x} \notin \text{STAB}(W_{10}^2)$.

In order to prove that $\bar{x} \in N_+(W_{10}^2)$ we present a matrix Y as in (2) which represents the point in the higher dimensional space.

For this purpose we make use of the following definition.

Let $T : \mathbb{R}^n \to \mathbb{R}^n$ be such that $T(v_1, \dots, v_n) = (v_n, v_1, \dots, v_{n-1})$. The matrix $\text{circ}(u)$ is the $n \times n$-matrix whose first row is $T^0(u) = u$ and whose j-th row is given by $T^{j-1}(u) = T(T^{j-2}(u))$, for every $j \geq 2$.

Let $z = (\lambda, 0, 0, \beta, \gamma, \delta, \gamma, \beta, 0, 0)$ where

$$\gamma = \frac{853}{10000}, \quad \delta = \frac{336}{10000} \quad \text{and} \quad \beta = \frac{2234}{10000}.$$

If $\bar{Y} = \text{circ}(z)$ then it is not difficult to check that $Y \in \mathbb{R}^{11}$ defined as in (2) is PSD and it satisfies that

$$\frac{1}{\lambda}\bar{Y}_i \in \text{FRAC}(W_{10}^2) \quad \text{and} \quad \frac{1}{1-\lambda}(\lambda\mathbf{1} - \bar{Y}_i) \in \text{FRAC}(W_{10}^2).$$

$\square$

The following result implies that Conjecture 1 holds for web graphs.

Theorem 5. *If the web graph W_n^k is N_+-perfect then it is either a perfect or a minimally imperfect graph.*

Proof. Due to the fact that the conjecture is proved for near-perfect graphs (Theorem 1) we only need consider those webs which are not near-perfect and prove that none of them are N_+-perfect.

If the web W_n^k is not near-perfect then $k \geq 2$ and $n \geq 3k + 3$.

If $k = 2$ the result follows from Corollary 1 and Lemma 2.

Let $k \geq 3$ and $n \geq 3k + 3$. We will prove that every web W_n^k has a subweb of the form $W_{n'}^2$, for some $n' \geq 8$.

After Trotter's result (Theorem 3) $W_{n'}^2$ is a subweb of W_n^k if

$$\frac{2n}{k} \leq n' \leq \frac{3n}{k+1}.$$

Let $\Delta^k(n) = \frac{3n}{k+1} - \frac{2n}{k} = n\frac{k-2}{k(k+1)}$.

Observe that $\Delta^k(n)$ assumes its minimum value when n is minimum, i.e. when $n = 3k + 3$. In this case, $\Delta^k(n) = \frac{3(k-2)}{k} \geq 1$. Then, for every value of $n \geq 3k + 3$ we can find an integer n' satisfying

$$\frac{2n}{k} \leq n' \leq \frac{3n}{k+1}$$

and then $W_{n'}^2$ is a subweb of W_n^k.

Moreover, since $n \geq 3k + 3$ we have that $\left\lfloor \frac{3n}{k+1} \right\rfloor \geq 9$ and W_n^k has a subweb $W_{n'}^2$ for some $n' \geq 8$.

Finally, Corollary 1 and Lemma 2 prove the result. $\square$

4 On minimally N_+-imperfect subgraphs

In [2] we proved the validity of the conjecture on the family of near-perfect graphs. Rank-perfect graphs constitute a superclass of near-perfect graphs, then it seems natural to continue our work towards proving the conjecture on this family. In fact, while studying the N_+-perfect graphs which are also near-perfect graphs we could identify some minimal forbidden structures on this family.

We say that a graph is *minimally N_+-imperfect* if it is N_+-imperfect but deleting any node leaves an N_+-perfect graph. The results in [2] give the minimally N_+-imperfect graphs in the family of near-perfect graphs. In order to present them let us introduce some more definitions.

We denote by C_{2k+1} the cycle having node set $\{1, \ldots, 2k+1\}$ and edge set $\{i(i+1) : i \in \{1, \ldots, 2k\} \cup \{1(2k+1)\}$.

In [2] we consider two families of near-perfect graphs, named $\mathcal{W}^k$ and $\mathcal{H}^k$ for each $k \geq 2$.

Let $\mathcal{H}^2 = \mathcal{W}^2 = \{G_{LT}, G_{EMN}\}$. For $k \geq 3$, $\mathcal{W}^k$ is the family of graphs with node set $\{0, 1, \ldots, 2k+1\}$ such that:

– $G - 0 = C_{2k+1}$;
– there is no pair of consecutive nodes (in C_{2k+1}) with degree 2;
– the degree of node 0 is $k + 2$.

For $k \geq 3$, $\mathcal{H}^k$ is the family of graphs having node set $\{0, 1, \ldots, 2k+1\}$ such that:

– $G - 0$ is the complement of C_{2k+1};
– there is no pair of consecutive nodes (in C_{2k+1}) with degree $2k - 2$;
– the degree of node 0 is at most $2k$.

In Fig. 3 we have represented one of the graphs in the family $\mathcal{W}^9$ and one of the graphs in the family $\mathcal{H}^9$.

Using the results in [2] we have the following

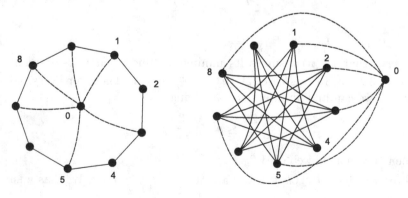

Fig. 3. A graph in the family $\mathcal{W}^9$ and $\mathcal{H}^9$.

Lemma 3. *Let G be a minimally N_+-imperfect graph. If G is a near-perfect graph then it is an odd subdivision of a graph in $\mathcal{H}^k \cup \mathcal{W}^k$ for some $k \geq 2$.*

In this contribution it was important to identify the odd subdivisions of G_{LT} as minimally N_+-imperfect structures in the webs, except for W_{10}^2.

In fact, we have proved that the web W_{10}^2 is a minimally N_+-imperfect rank-perfect graph which is not near-perfect. This shows some advance in order to characterize N_+-perfect rank-perfect graphs and also a line for our future research.

References

1. Berge, C.: Perfect graphs. In: Six Papers on Graph Theory, pp. 1–21. Indian Statistical Institute, Calcutta (1963)
2. Bianchi, S., Escalante, M., Nasini, G., Tunçel, L.: Near-perfect graphs with polyhedral $N_+(G)$. Electron. Notes Discrete Math. **37**, 393–398 (2011)
3. Bianchi, S., Escalante, M., Nasini, G., Tunçel, L.: Lovász-Schrijver PSD-operator and a superclass of near-perfect graphs. Electron. Notes Discrete Math. **44**, 339–344 (2013)
4. Chudnovsky, M., Robertson, N., Seymour, P., Thomas, R.: The strong perfect graph theorem. Ann. Math. **164**, 51–229 (2006)
5. Chvátal, V.: On certain polytopes associated with graphs. J. Comb. Theory B **18**, 138–154 (1975)
6. Grötschel, M., Lovász, L., Schrijver, A.: The ellipsoid method and its consequences in combinatorial optimization. Combinatorica **1**, 169–197 (1981)
7. Escalante, M., Montelar, M.S., Nasini, G.: Minimal N_+-rank graphs: progress on Lipták and Tunçel's conjecture. Oper. Res. Lett. **34**(6), 639–646 (2006)
8. Grötschel, M., Lovász, L., Schrijver, A.: Geometric Algorithms and Combinatorial Optimization. Springer, New York (1988)
9. Lipták, L., Tunçel, L.: Stable set problem and the lift-and-project ranks of graphs. Math. Program. **98**, 319–353 (2003)
10. Lovász, L.: On the Shannon capacity of a graph. IEEE Trans. **25**, 1–7 (1979)
11. Lovász, L., Schrijver, A.: Cones of matrices and set-functions and 0–1 optimization. SIAM J. Optim. **1**, 166–190 (1991)
12. Nasini, G.: Caracterización de los grafos fs-perfectos. Reunión Anual de Comunicaciones Científicas de la Unión Matemática Argentina, Tucumán, Argentina (2011)
13. Padberg, M.: On the facial structure of set packing polyhedra. Math. Program. **5**, 199–215 (1973)
14. Shepherd, F.B.: Near-Perfect matrices. Math. Program. **1**–3(64), 295–323 (1994)
15. Shepherd, F.B.: Applying Lehman's theorems to packing problems. Math. Program. **71**, 353–367 (1995)
16. Stauffer, G.: On the stable set polytope of claw-free graphs, Ph.D. thesis, EPF Lausanne (2005)
17. Trotter, L.: A class of facet producing graphs for vertex packing polyhedra. Discrete Math. **12**, 373–388 (1975)
18. Wagler, A.: Relaxing perfectness: which graphs are "almost perfect"? In: Groestchel, M.(ed.) The Sharpest Cut: The Impact of Manfred Padberg and His Work. SIAM, Philadelphia (2002)

The Envy-Free Pricing Problem
and Unit-Demand Markets

Cristina G. Fernandes, Carlos E. Ferreira, Álvaro J.P. Franco,
and Rafael C.S. Schouery [✉]

Department of Computer Science, University of São Paulo, São Paulo, Brazil
{cris,cef,alvaro,schouery}@ime.usp.br

Abstract. A common problem faced in economics is to decide the pricing of products of a company, since poorly chosen prices might lead to low profit. One important model for this is the unit-demand envy-free pricing problem, where one considers that every consumer buys the item that maximizes his own profit, and the goal is to find a pricing of the items that maximizes the expected profit of the seller. This problem is not in APX unless P = NP, but it is still interesting to be solved in practice. So, we present four new MIP formulations for it and experimentally compare them to a previous one from the literature. We describe three models to generate different random instances for general unit-demand auctions, that we designed for the computational experiments. Each model has a nice economic interpretation. Our results show that our MIP formulations are a great improvement both for solving the problem to optimality or in order to obtain solutions with small gap.

Keywords: Pricing problem · Envy-free allocations · Unit-demand auctions

1 Introduction

An interesting economical problem faced by companies that sell items or services is to maximize their profit by carefully choosing the pricing of their items. An item with low price can lead to a low profit but an item with high price can lead to fewer sells because consumers cannot (or do not want to) buy an expensive item. For that reason, it is necessary to account for consumers' preferences and affordability when deciding the pricing of the items. There is a variety of models that consider this problem [11,13,15] but, in particular, Rusmevichientong et al. [12] introduced non-parametrical models for this problem which rely on information about consumers' valuations for the items to decide on the pricing, and discussed the use of collected data (e.g., from a website) to sample consumers preferences. In those models, the general goal is to determine a price for each

Research partially supported by CAPES (Proc. 33002010176P0), CNPq (Proc. 302736/2010-7, 308523/2012-1, and 477203/2012-4), FAPESP (Proc. 2009/00387-7 and 2013/03447-6) and Project MaCLinC of NUMEC/USP.

© Springer International Publishing Switzerland 2014
P. Fouilhoux et al. (Eds.): ISCO 2014, LNCS 8596, pp. 230–241, 2014.
DOI: 10.1007/978-3-319-09174-7_20

item in a way that the overall profit of the seller is maximized. However, once the prices are defined, the consumers may behave in several ways. In the literature, there are several models (max-buying, min-buying, and rank-buying) [1,12] that try to capture the usual behavior of consumers in some situations.

Guruswami et al. [7] formalized one such model, suggested by Aggarwal et al. [1], defining the *envy-free pricing* for combinatorial auctions. In the more general setup, each consumer is willing to buy a bundle of items. Knowing the valuation of every bundle of items for each consumer, the seller has to decide on the pricing and on an allocation of the items to the consumers in an *envy-free manner*, that is, in a way that every consumer is at least as happy with the bundle of items assigned to him (which might even be empty) as with any other bundle of items, considering the prices. The seller's goal is to maximize his profit. Of course, in this general form, there is an issue on the amount of information involved, specifically for the valuations. So it is reasonable to consider, as Guruswami et al., particular cases that avoid this issue.

A well-studied such case is the *unit-demand* envy-free pricing problem, in which each consumer is willing (or is allowed) to buy at most one item, and there is an unlimited supply of each item. Guruswami et al. presented a $(2 \ln n)$-approximation for this problem, where n is the number of consumers, and proved that this problem is APX-hard (with each consumer valuation being 1 or 2 for the items of interest). Briest [3] further analyzed the hardness of approximation, considering the uniform budget case, where each consumer equally values all items he is interested in. Assuming specific hardness of the balanced bipartite independent set problem in constant degree graphs, or hardness of refuting random 3CNF formulas, there is no approximation for the uniform budget unit-demand envy-free pricing problem with ratio $O(\lg^\epsilon n)$ for some $\epsilon > 0$. Chen and Deng [4] showed that the unit-demand envy-free pricing problem can be solved in polynomial time if every consumer has a positive valuation for at most two items.

More recently, Shioda, Tunçel, and Myklebust [14] described a slightly more general model than the one of Guruswami et al. [7], and presented a mixed-integer programming (MIP) formulation for their model, along with heuristics and valid cuts for their formulation.

Regarding the generation of instances for auctions, Leyton-Brown et al. [9] proposed the well-know Combinatorial Auctions Test Suite (CATS), created to generate random instances for auctions with realistic bidding behavior for five real world situations. See also [5, Chap. 18]. Their generator produces instances for auctions where the consumers are interested in buying bundles of items.

1.1 Our Results

In this paper, we focus on solving the unit-demand envy-free pricing problem in practical cases. For this, we present new MIP formulations for this problem that can be adapted also to the model described by Shioda et al. [14]. We compare these new formulations to the one of Shioda et al. through computational experiments that indicate that our new formulations give better results in practice than the other one.

For the computational experiments, we used six different sets of instances, obtained from generators we designed. Since a simple adaptation of CATS for unit-demand auctions does not preserve the economic motivations, we designed three models to generate different random instances for unit-demand auctions, and we implemented the three corresponding generators. Each of the three models has a nice economic interpretation and can be used in other works on unit-demand auctions, such as [10,14]. To our knowledge these are the first unit-demand auctions instances generators proposed in the literature. Because of that, we believe these instances generators are a nice contribution by themselves, and we made them available as open-source software at https://github. com/schouery/unit-demand-market-models.

The paper is organized as follows. In the next section, we present some notation and describe formally the problem that we address. In Sect. 3, we present our new MIP formulations for the problem and revise the one by Shioda et al. [14]. In Sect. 4, we present the three instances generators and describe the economical motivation behind each one. In Sect. 5, we empirically compare our formulations for the unit-demand envy-free pricing problem against the one by Shioda et al. [14], using sets of instances produced by our generators. We conclude with some final remarks in Sect. 6.

2   Model and Notation

We denote by B the set of consumers and by I the set of items. A *valuation* is a non-negative rational matrix v indexed by $I \times B$ and v_{ib} represents the value of item i to consumer b. A *pricing* is a non-negative rational vector indexed by I. A (unit-demand) *allocation* is a binary matrix x indexed by $I \times B$, such that, for every consumer b, we have $x_{ib} = 1$ for at most one item i. Note that we consider that the items have unlimited supply, that is, the same item can be assigned to an arbitrary number of consumers (each one receives a copy of the item).

For a valuation v, a pricing p, and an allocation x, the *utility* of a consumer b, denoted by u_b, is $v_{ib} - p_i$ if consumer b receives item i, and 0 if consumer b receives no item. Notice that $v_{ib} - p_i$ is a measure of how cheap item i is compared with the valuation v_{ib}. An allocation x is *envy-free* if, for every consumer b, we have $u_b \geq 0$ and $u_b \geq v_{ib} - p_i$ for every item i. In other words, for every consumer b, there is no item that would give b a better utility, that is, b buys the "relatively cheapest" item.

Now we are ready to state the problem. The *unit-demand envy-free pricing problem* consists of, given a valuation v, finding a pricing p and an envy-free allocation x that maximize the seller's profit, which is the sum of the prices of items received by the consumers (considering multiplicities).

Shioda et al. [14] considered a problem a bit more general than the envy-free pricing problem. They considered that B represents consumers groups instead of single consumers and that every consumers group b has a value δ_b used as a tolerance when comparing the utilities provided by items, that is, if a consumers group b receives an item i, then $v_{ib} - p_i \geq v_{i'b} - p_{i'} + \delta_b$ for every item $i' \neq i$.

In our experiments, we noticed that the size of the consumers groups does not impact in the running time of the formulations. Also, in Shioda et al. experiments, they considered $\delta_b = 0$ for every consumers group b. With this in mind, and the fact that our formulation can be easy adapted to consider their model, we believe that this is a fair comparison with their work.

3 MIP Formulations

Before presenting the formulations, let us introduce two notations that will be used for "big-M" inequalities in the formulations presented in this section. For an item i, let $R_i = \max\{v_{ib} : b \in B\}$ and, for a consumer b, let $S_b = \max\{v_{ib} : i \in I\}$.

We start by presenting the formulation due to Shioda et al. [14] already considering $\delta_b = 0$ and one consumer per group ($N_b = 1$ for every consumers group b in their notation).

Consider a valuation v. We use the following variables: a binary matrix x indexed by $I \times B$ that represents an allocation, a rational vector p indexed by I that represents the pricing, and a rational matrix $\hat{p}$ indexed by $I \times B$ that represents the price paid for item i by each consumer b (that is, $\hat{p}_{ib} = 0$ if b does not receive i, and $\hat{p}_{ib} = p_i$ otherwise). The formulation, which we name (STM), consists of, finding x, p, and $\hat{p}$ that

$$
\begin{aligned}
\text{(STM)} \quad \max \quad & \sum_{b \in B} \sum_{i \in I} \hat{p}_{ib} \\
\text{s.t.} \quad & \sum_{i \in I} x_{ib} \leq 1, && \forall b \in B \\
& \sum_{i \in I \setminus \{k\}} (v_{ib} x_{ib} - \hat{p}_{ib}) \geq v_{kb} \sum_{i \in I \setminus \{k\}} x_{ib} - p_k, && \forall b \in B, \forall k \in I && (1) \\
& v_{ib} x_{ib} - \hat{p}_{ib} \geq 0, && \forall b \in B, \forall i \in I \\
& \hat{p}_{ib} \leq p_i, && \forall b \in B, \forall i \in I \\
& \hat{p}_{ib} \geq p_i - R_i(1 - x_{ib}), && \forall b \in B, \forall i \in I \\
& x_{ib} \in \{0,1\}, \quad p_i \geq 0, \quad \hat{p}_{ib} \geq 0, && \forall b \in B, \forall i \in I.
\end{aligned}
$$

From the formulation (STM), we developed a new and stronger formulation by changing inequalities (1). We call (I) this improved formulation, which we present below.

$$
\begin{aligned}
\text{(I)} \quad \max \quad & \sum_{b \in B} \sum_{i \in I} \hat{p}_{ib} \\
\text{s.t.} \quad & \sum_{i \in I} x_{ib} \leq 1, && \forall b \in B \\
& \sum_{i \in I} (v_{ib} x_{ib} - \hat{p}_{ib}) \geq v_{kb} - p_k, && \forall k \in I, \forall b \in B && (2) \\
& v_{ib} x_{ib} - \hat{p}_{ib} \geq 0, && \forall b \in B, \forall i \in I \\
& \hat{p}_{ib} \leq p_i, && \forall b \in B, \forall i \in I \\
& \hat{p}_{ib} \geq p_i - R_i(1 - x_{ib}), && \forall b \in B, \forall i \in I \\
& x_{ib} \in \{0,1\}, \quad p_i \geq 0, \quad \hat{p}_{ib} \geq 0, && \forall b \in B, \forall i \in I.
\end{aligned}
$$

Lemma 1. (I) *is a formulation for the envy-free pricing problem.*   □

Also, we noticed that inequalities (2) are unnecessary, which led us to another formulation that we call (L) (from the word *loose*), consisting of formulation (I) without inequalities (2).

Lemma 2. *Every integer solution of* (L) *is a solution of* (I).   □

Notice that it is possible that (L) is a weaker version of (I) in terms of relaxation guarantees.

Studying formulation (L), we developed another formulation with variables x and p as defined before, and a rational vector z indexed by B that represents the profit obtained from consumer b (that is, if consumer b receives item i then $z_b = p_i$, and $z_b = 0$ if b does not receive an item). One can view z as $z_b = \sum_{i \in I} \hat{p}_{ib}$. This formulation, which we name (P) (from the word *profit*), consists of finding x, p, and z that

$$
\begin{aligned}
\text{(P)}\quad \max\ & \sum_{b \in B} z_b \\
\text{s.t.}\quad & \sum_{i \in I} x_{ib} \leq 1, & \forall b \in B \\
& \sum_{i \in I} v_{ib} x_{ib} - z_b \geq v_{kb} - p_k, & \forall k \in I, \forall b \in B \\
& \sum_{i \in I} v_{ib} x_{ib} - z_b \geq 0, & \forall b \in B \\
& z_b \geq p_i - R_i(1 - x_{ib}), & \forall b \in B, \forall i \in I \\
& x_{ib} \in \{0,1\}, \quad p_i \geq 0, \quad z_b \geq 0, & \forall b \in B, \forall i \in I.
\end{aligned}
$$

Lemma 3. (P) *is a formulation for the envy-free pricing problem.*   □

As one would expect, (P) is a formulation weaker than (L).

Finally, we present our last formulation that was developed changing the focus from the price paid by a consumer to the utility that a consumer obtained. We use the following variables: x and p as defined before, and a rational vector u indexed by B that represents the utilities of the consumers. The formulation, which we name (U) (from the word *utility*), consists of finding x, p, and u that

$$
\begin{aligned}
\text{(U)}\quad \max\ & \sum_{i \in I} \sum_{b \in B} v_{ib} x_{ib} - \sum_{b \in B} u_b \\
\text{s.t.}\quad & \sum_{i \in I} x_{ib} \leq 1, & \forall b \in B \\
& u_b \geq v_{ib} - p_i, & \forall i \in I, \forall b \in B \\
& u_b \leq v_{ib} x_{ib} - p_i + (1 - x_{ib})(R_i + S_b), & \forall i \in I, \forall b \in B \\
& u_b \leq \sum_{i \in I} v_{ib} x_{ib}, & \forall b \in B \\
& x_{ib} \in \{0,1\}, \quad p_i \geq 0, \quad u_b \geq 0, & \forall b \in B, \forall i \in I.
\end{aligned}
$$

Lemma 4. (U) *is a formulation for the envy-free pricing problem.*   □

After our presentation at ISCO, we became aware of a formulation similar to (I) proposed in the literature for another problem, the so called Network Pricing Problem with Connected Toll Arcs [8].

For some MIP formulation (F) and a valuation v, we denote by $LR_F(v)$ the value of an optimal solution for the linear relaxation of (F) when the instance is v. Our next result formalizes the relaxation quality guarantees for the formulations presented in this section.

Theorem 5. *For every instance v of the unit-demand envy-free pricing problem, $LR_I(v) \leq LR_{STM}(v)$ and also $LR_I(v) \leq LR_L(v) \leq LR_P(v) \leq LR_U(v)$. Moreover, there exists an instance v where $LR_I(v) < LR_{STM}(v)$ and $LR_I(v) = LR_L(v) < LR_P(v) < LR_U(v)$.*   □

The instance mentioned above was generated in one of our empirical experiments. We did not find an instance v where $LR_I(v) < LR_L(v)$.

Before presenting and discussing the experimental comparison of these formulations, we describe the random instance generators that we used in the computational experiments.

4   Multi-Item Auctions Test Suites

In this section, we propose three new models to generate different random instances for unit-demand auctions, each of them with a nice economic interpretation.

In these three models, we start by creating a bipartite graph where one side of the bipartition represents the items and the other, the consumers. An edge of this graph represents that the consumer gives a positive value to the corresponding item.

4.1   Characteristics Model

In the Characteristics Model, we consider that every item has a set of characteristics that are relevant to determine if a consumer has interest in such item. For example, imagine that the items are cars and, for simplicity, that every car has two relevant characteristics: color and engine model. In this model, every consumer has interest in a set of colors and in a set of engine models and, potentially, would buy any car that has one of the colors and one of the engine models that such consumer is interested in.

The idea is that each item has a profile of characteristics and each consumer desires some of these characteristics. This way, two items that have the same profile are desired by the same consumers, and consumers desire items that do not differ much.

It is natural that an item has a market value, and the valuations for that item (by interested consumers) are concentrated around this value. That is, the valuations for an item should not differ too much, and valuations that are far away from this market value are rare.

An instance of our problem, which is a valuation, will be represented by a weighted bipartite graph as we describe in the following.

Let m be the number of items, n be the number of consumers, c be the number of the characteristics of each item, o be the number of options for any characteristic, p be the number of options preferred by a consumer for any characteristic, ℓ be the minimum market value of an item, h be the maximum market value of an item, and d be the percentage of deviation used. The set of vertices is $I \cup B$ and the set of edges E is described below.

For every item i, we have a vector of size c (the characteristics of that item) where every entry is in $\{1, \ldots, o\}$ chosen independent and uniformly at random. For every consumer b, we construct a matrix $A \in \{0,1\}^{c \times p}$ where, for every row, we choose independent and uniformly at random p positions to be set to 1 and $o - p$ positions to be set to 0. For a consumer b and an item i, we have that $\{i, b\} \in E$ if and only if the characteristics of item i coincide with the preferences of consumer b, that is, if the characteristic k of item i has value v then $A_{kv} = 1$ where A is the matrix of consumer b. See an example of such characteristics and preferences in Fig. 1.

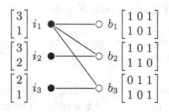

Fig. 1. Instance with two characteristics, each with three options. Each consumer wants items that have one of two options (given by their matrices) in every characteristic. Next to each item is the characteristics vector.

Finally, for every item i, we define $\bar{p}_i$ as the market price of item i, chosen independent and uniformly at random from the interval $[\ell, h]$. For every $\{i, b\}$ in E, we choose v_{ib} from $1.0 + \mathcal{N}(\bar{p}_i, (\bar{p}_i d)^2)$, where $\mathcal{N}(\mu, \sigma^2)$ denotes the Gaussian distribution with mean μ and standard deviation σ.

4.2   Neighborhood Model

In the Neighborhood Model, we consider that items and consumers are points in the plane and that every consumer has interest in items that are close to his location. Imagine, for example, that consumers are buying houses in a city and that each consumer has a favorite location to live in the city. In this model, a consumer will not buy a house too far away from its favorite location and will give more value for houses that are closer when comparing with farther houses.

This model selects the valuations based on a geometric relation. For this, we adapt a random process first introduced by Gilbert [6] for generating a Random Geometric Graph.

The idea is to distribute the items and the consumers in a 1×1 square W in $\mathbb{Q}^2$ and define the valuations according to the distance between a consumer and an item.

We are given the number of items m, the number of consumers n, and a radius r. First we assign a point in W independent and uniformly at random to each item and each consumer. In order to construct the valuation, we construct a graph $G = (I \cup B, E)$ such that $|I| = m$ and $|B| = n$ where there is an edge $\{i, b\}$ in E if the distance between i and b in W is at most r. Figure 2 represents one such graph.

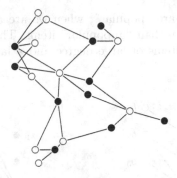

Fig. 2. The construction of the graph used in the Neighborhood Model with 10 items, 10 consumers and radius 0.33. Black vertices represent the items and white vertices represent the consumers.

Finally, to define the valuations, let $d(i, b)$ denote the (Euclidean) distance between i and b. For every consumer b, we choose a multiplier k_b in $[1, h]$, where h is given as input. If $\{i, b\} \in E$, then $v_{ib} = 1.0 + Mk_b/d(i, b)$, where M is a scaling factor also given as input.

This way, a consumer only gives a positive valuation to items sufficiently close to his location, and the valuation decreases as the distance increases. Two consumers that are at the same distance from an item still can evaluate it differently because they might have different k multipliers.

4.3 Popularity Model

In a market where the items are alike, the preferences of the consumers might be based on how popular a specific item is. In our case, we measure the popularity of an item by the number of consumers that are interested in it. For this, we consider a process where a popular item becomes more and more popular. We also consider that every item has a quality and that the market price of such item is directly proportional to its quality and inversely proportional to its popularity (because a company could sell this item cheaper if it is popular).

Let m be the number of items, n be the number of consumers, e be the number of edges, Q be the maximum quality of an item, and d be a percentage of deviation. We construct the bipartite graph as follows. We start with the empty bipartite graph with parts I and B such that $|I| = m$ and $|B| = n$. We add edges at random (as described below) until the graph has e edges.

To add an edge, we choose a consumer uniformly at random, and choose an item biased according to its degree. That is, an item i of current degree d_i has a weight of $d_i + 1$.

This process, called Preferential Attachment, was already used in the literature by Barabási and Albert [2] to generate graphs with properties that appear in the World Wide Web and in graphs that arise in biology, such as the interactions of the brain cells.

This way, items that are "popular", when we are adding an edge, have a higher chance to be chosen than "unpopular" items. The final result is a graph with a small number of items of large degree and a large number of items of small degree. See Fig. 3.

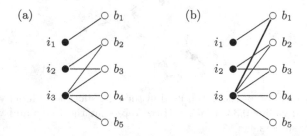

Fig. 3. From (a) to (b), the 8th edge (i_3, b_1) was added. In (a), the probabilities of choosing items i_1, i_2, and i_3 are 2/10, 3/10, and 5/10, respectively.

We also use the information of the (final) degree of an item to decide on the valuations. As in the Characteristic Model, an item i has a market price $\bar{p}_i$, and we choose the valuation v_{ib} according to $1.0 + \mathcal{N}(\bar{p}_i, (\bar{p}_i d)^2)$. But, in contrast to the Characteristic Model, we do not choose the market value uniformly at random in an interval. For each item i, we choose a random quality q_i in $(0, Q]$ and we define $\bar{p}_i$ as q_i/d_i. This way, the market price of an item increases with its quality and decreases with its degree. That is, an item has high desirability if it is "cheap" or has good quality.

5   Empirical Results

In this section, we present some empirical results involving the formulations. In our experiments, the new formulations produced better results than the previously known formulation from the literature, namely, (STM). In particular, formulation (L) was the best for small instances, and formulation (U) was the best for large instances (providing a small gap).

We used a computer with two Intel Xeon E5620 2.40 GHz processors, 64 GB of RAM running Gentoo Linux 64 bit. The MIPs were solved by CPLEX 12.1.0 in single-thread mode, with work memory limited to 4 GB. The tests were run using a time limit of one hour of CPU time.

For our tests, we generated six test sets, two for every one of our three models: one with 20 instances with 50, 100, 150, 200, 250, and 300 consumers and items, that we call "small instances", and other with 20 instances with 500, 1000, 1500, 2000, 2500 and 3000 consumers and items, that we call "large instances". Next we describe the parameters given to the models, and we denote by n the number of items (which is equal to the number of consumers).

For the Characteristic Model, we chose the number of options for a characteristic as $o = 8$, the number of options preferred by a consumer for any characteristic as $p = 7$, and the number of characteristics of each item as $\lceil \log(8/n)/\log(p/o) \rceil$ (this value was chosen so that the mean degree of the consumers is between 7 and 8), the minimum market value of an item as $\ell = 1$, the maximum market value of an item as $h = 100$, and the percentage of deviation as $d = 0.25$. For the Neighborhood Model, we chose the maximum multiplier of a consumer as $h = 3$, the radius as $\sqrt{8/n\pi}$ (in this way, the mean degree of the consumers is close to 8) and the scaling factor $M = 10$. Finally, for the Popularity Model, we chose the number of edges as $e = 8n$, the maximum quality of an item as $Q = 200$, and the percentage of deviation as $d = 0.25$. All instances, along with the source code for the generators, can be found at https://github.com/schouery/unit-demand-market-models.

Note that all of our instances are sparse, since we believe that in practice dense instances would be rare. That is, we expect that, in a market with many items, each consumer would have positive valuation usually only for a few items.

In our experiments, CPLEX had a better performance with formulation (L) than with any other formulation for small instances. First of all, with (L), CPLEX solved more small instances than with the other formulations. For example, with (STM), CPLEX could not solve (within one hour) any instance of the Popularity Model with 150 or more items. But, with (L), CPLEX solved 14 instances with 150 items and 7 instances with 200 items.

Also, for small instances that were not solved within one hour, (STM) and (P) provided a mean final gap much worse than the ones provided by (U), (I), and (L) (at least twice the gap for those formulations).

Even though formulation (L) was better than the others, we noticed that formulations (U) and (I) were not so far behind from formulation (L) as one can observe in Fig. 4.

When analyzing the mean final gap for large instances, formulation (U) seems better than the others. Actually, (U) was able to maintain a really small mean final gap even for instances with 3000 consumers and items (the mean final gap was 2 % for the Characteristics model, 3 % for the Neighborhood model, and 17 %

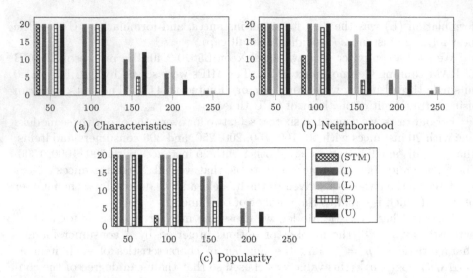

(a) Characteristics   (b) Neighborhood

(c) Popularity

Fig. 4. Number of solutions found by each MIP formulation.

for the Popularity model). Figure 5 presents this information for the Popularity model, where (U) proved to be much better than the other formulations.

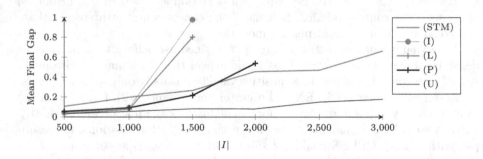

Fig. 5. Mean final gap for large instances for the Popularity Model not solved within the time limit.

We believe that this result is impressive because, theoretically (from Theorem 5), (U) is the weakest of our new formulations (we have no result that compares (U) and (STM)). Nonetheless, (U) is a small formulation when compared with (STM), (I) and (L) and, as a consequence, the linear relaxation can be solved more quickly for this formulation (and also for formulation (P)). Since (P) is a formulation stronger than (U) and they have the same size, we believe that formulation (U) outperformed formulation (P) because of the heuristics used by CPLEX to find good integer solutions.

6 Final Remarks

In this paper, we presented four new MIP formulations for the unit-demand envy-free pricing problem along with some empirical results comparing these formulations. Our results show that our formulations are better than the previous formulation from the literature.

We also presented three models for generating instances of unit-demand auctions. To our knowledge, these are the first tests suites for unit-demand auctions in the literature. The corresponding instances generators are available as open-source software.

References

1. Aggarwal, G., Feder, T., Motwani, R., Zhu, A.: Algorithms for multi-product pricing. In: Díaz, J., Karhumäki, J., Lepistö, A., Sannella, D. (eds.) ICALP 2004. LNCS, vol. 3142, pp. 72–83. Springer, Heidelberg (2004)
2. Barabási, A.L., Albert, R.: Emergence of scaling in random networks. Science **286**(5439), 509–512 (1999)
3. Briest, P.: Uniform budgets and the envy-free pricing problem. In: Aceto, L., Damgård, I., Goldberg, L.A., Halldórsson, M.M., Ingólfsdóttir, A., Walukiewicz, I. (eds.) ICALP 2008, Part I. LNCS, vol. 5125, pp. 808–819. Springer, Heidelberg (2008)
4. Chen, N., Deng, X.: Envy-free pricing in multi-item markets. In: Abramsky, S., Gavoille, C., Kirchner, C., Meyer auf der Heide, F., Spirakis, P.G. (eds.) ICALP 2010, Part II. LNCS, vol. 6199, pp. 418–429. Springer, Heidelberg (2010)
5. Cramton, P., Shoham, Y., Steinberg, R. (eds.): Combinatorial Auctions. MIT Press, Cambridge (2006)
6. Gilbert, E.N.: Random plane networks. J. Soc. Ind. Appl. Math. **9**(4), 533–543 (1961)
7. Guruswami, V., Hartline, J.D., Karlin, A.R., Kempe, D., Kenyon, C., McSherry, F.: On profit-maximizing envy-free pricing. In: Proceedings of the 16th Annual ACM-SIAM Symposium on Discrete Algorithms, pp. 1164–1173 (2005)
8. Heilporn, G., Labbé, M., Marcotte, P., Savard, G.: A polyhedral study of the network pricing problem with connected toll arcs. Networks **55**(3), 234–246 (2010)
9. Leyton-Brown, K., Pearson, M., Shoham, Y.: Towards a universal test suite for combinatorial auction algorithms. In: Proceedings of the 2nd ACM Conference on Electronic Commerce, pp. 66–76 (2000)
10. Myklebust, T.G.J., Sharpe, M.A., Tunçel, L.: Efficient heuristic algorithms for maximum utility product pricing problems. Research report, Department of Combinatorics and Optimization, University of Waterloo, November 2012
11. Oren, S., Smith, S., Wilson, R.: Product line pricing. J. Bus. **57**(1), S73–S79 (1984)
12. Rusmevichientong, P., Roy, B.V., Glynn, P.W.: A nonparametric approach to multiproduct pricing. Oper. Res. **54**(1), 82–98 (2006)
13. Sen, S.: Issues in optimal product design. In: Analytic Approaches to Product and Marketing Planning: The Second Conference, pp. 265–274 (1982)
14. Shioda, R., Tunçel, L., Myklebust, T.G.: Maximum utility product pricing models and algorithms based on reservation price. Comput. Optim. Appl. **48**(2), 157–198 (2011)
15. Smith, S.A.: New product pricing in quality sensitive markets. Mark. Sci. **5**(1), 70–87 (1986)

Mathematical Programming Models for Traffic Engineering in Ethernet Networks Implementing the Multiple Spanning Tree Protocol

Bernard Fortz[1], Luís Gouveia[2], and Martim Moniz[1(✉)]

[1] Départment d'Informatique, Université Libre de Bruxelles, Brussels, Belgium
martim.moniz@ulb.ac.be
[2] Centro de Investigação Operacional, Universidade de Lisboa, Lisboa, Portugal

Abstract. The Multiple Spanning Tree Protocol (MSTP), used in Ethernet networks, maintains a set of spanning trees that are used for routing the demands in the network. Each spanning tree is allocated to a pre-defined set of demands. In this paper we present two mixed integer programming models for the Traffic Engineering problem of optimally designing a network implementing MSTP, such that link utilization is minimized. This is the first approach that focuses on using exact methods to solve this problem. We present tests in order to compare the two formulations, in terms of formulation strength and computing time.

Keywords: Traffic engineering · Telecommunications · Multiple spanning tree protocol · Mixed-integer programming

1 Introduction

With the increasing demand for Internet and cloud computing services, the need for large scale data centers has become paramount. Nowadays, these data centers already hold a huge number of servers (up to 10 k) and the propensity is for that number to rise in the near future. This emphasizes the critical necessity of improving the performance of telecommunication networks in these data centers. Traffic Engineering is a wide array of optimization methods, that aims at finding the best network configuration, in order to improve different traffic-oriented performance measures, such as delay, delay variation, packet loss and throughput.

Most data centers are mainly used to either perform computation or to host Internet services. They support simultaneously multiple applications that run on a set of virtual machines, distributed on physical servers. In these data centers, switched Ethernet networks are becoming popular, as they are more effective in the management of traffic.

Loops in the network's topology can result in broadcast radiation, i.e., the repeatedly rebroadcasting of broadcast traffic and consequential network flooding. As such, Ethernet networks only activate, at a given time, a cycle-free subset

© Springer International Publishing Switzerland 2014
P. Fouilhoux et al. (Eds.): ISCO 2014, LNCS 8596, pp. 242–254, 2014.
DOI: 10.1007/978-3-319-09174-7_21

of the existing links. To ensure this, these networks implement the Institute of Electrical and Electronics Engineers (IEEE) 802.1d standard [1], also known as Spanning Tree Protocol. As the protocol's name indicates, the topology of a switched networks using 802.1d must be a spanning tree. In graph theory, a spanning tree of a graph G is a connected, acyclic subgraph that spans all the nodes of G. As a consequence, a spanning tree of G will only contain $n-1$ edges, where n is the number of nodes in G. Note, however, that it is important to keep redundant links to ensure automatic backup paths in case of link failure. Hence, one of the drawbacks of the 802.1d standard is that the Ethernet network ends up only using a small number of existing links.

The IEEE 802.1q standard [2] enables large Ethernet networks to be partitioned in multiple smaller Virtual Local Area Networks (VLANs), simplifying the Ethernet network design. This allows for isolation between different applications and/or data center costumers, as a server belonging to a given VLAN can only communicate with other servers established in the same VLAN. The Multiple Spanning Tree Protocol (MSTP), standardized as 802.1s [3], allows for service providers to install different spanning trees (one per VLAN) over a single physical topology. This is highly advantageous for the traffic performances of Ethernet networks, as the traffic can be spread throughout a bigger number of links.

Nevertheless, MSTP is hard to use effectively due to the huge complexity of the problem of optimizing over multiple spanning trees in one single network.

In this paper, we study the problem of finding optimal designs for Ethernet networks implementing MSTP. This problem consists in designing networks with multiple VLANs, such that each VLAN is defined by a spanning tree that meets the required traffic demand. Additionally, all the VLANs must, jointly, verify the bandwidth capacity of the Ethernet network. The traffic oriented performance measure that we consider as optimization objective in this paper is the minimization of the worst-case link utilization in the network. The utilization of a link is defined as the ratio between its load (i.e. the sum of traffic flowing on it) and its bandwidth. Therefore, if the utilization of a link exceeds 1, a link is considered to be overloaded. In this paper, two mathematical programming models are presented and compared.

2 State of the Art

Ho [4] wrote an extensive review on the literature regarding Traffic Engineering problems for Ethernet networks implemented MSTP. He divided the approaches used to solve these problems in four classes. In the first approach, [5–7] proposed optimization techniques that map a set of VLANs to a given number of spanning trees. Meddeb [7] developed an algorithm to generate a set of spanning trees with a small number of links in common, and then introduced another greedy algorithm to map each VLAN to those spanning trees, while attempting to minimize the number of used links. Lim et al. [6] proposed a QoS-aware mechanism that maps VLANs with the intent of minimizing network throughput and delay.

He *et al.* [5] use an admission control algorithm to assign a group of VLANs to each given spanning tree, and then map each service to a VLAN, such that it minimizes the link load.

The second approach was used by de Sousa *et al.* [8] and Santos *et al.* [9,10] in three different papers. They introduced heuristic schemes that aim to balance the load in networks using MSTP by mapping a set of traffic flows to a set of given spanning trees. Different criteria were taken into consideration, including service disruption and network load balancing.

Santos *et al.* appear to be the only ones to model the problems using Mixed Integer Programming (MIP) [9,10]. They extend the traditional multi-commodity flow formulation, commonly used in problems with a single spanning tree [11], to the design of networks using MSTP. However, the authors seem to use MIP mainly as a means of measuring the quality of proposed heuristics. As such, we believe there is plenty of room for improvement, and more efficient models for networks using MSTP can be developed.

Third, Chen *et al.* [12] proposed an algorithm that defines a spanning tree for each node sending traffic in the network, while trying to achieve a good trade off between load balance and average delay.

The last approach was suggested by Padmaraj *et al.* [13] and Mirjalily [14]. In the first paper, the proposed heuristic updates weights assigned to the links in the network, in order to find a set of spanning trees with a good load balancing. In the second one, the suggested algorithm tries to find the best set of edge-disjoint spanning trees, and the best mapping of VLANs to that set.

Ho [4] argued that all these approaches were not applicable for large networks. Hence, he proposes a local search based algorithm that aims at minimizing the maximal link utilization for data center networks.

Note however that all of these proposals, including [4], are only heuristic approaches, that do not aim at finding the optimal solution, but merely one that is good enough. Moreover, with the exception of [9,10] no lower bounds on the optimal value of instances solved are known, and, as such, it is difficult to evaluate the quality of the proposed algorithms.

3   Problem Modeling

Consider a network defined on a graph $G = (N, A)$, where N is the set of nodes, with size n, and A is the set of links between those nodes. Link $\{i, j\} \in A$ represents an undirected link between nodes $i \in N$ and $j \in N$. Arc (i, j) represents that same link, but directed from i to j. To each link is assigned a capacity, denoted by $C_{\{i,j\}}$. This capacity is considered to be symmetric, in the sense that $C_{\{i,j\}} = C_{ij} = C_{ji}$.

Also, consider S to be the set of VLANs. For each VLAN $s \in S$, $d_s\{u, v\}$ represents the traffic demand between nodes $u \in N$ and $v \in N$. Traffic demands are also considered to be symmetric.

The objective of the problem considered is to find the optimal design of all VLANs $s \in S$, such that the maximum link utilization is minimized. At the same time, the following constraints must be verified:

- the topology of each VLAN is a spanning tree;
- all given traffic demands in a VLAN are routed;
- the total traffic flowing through a link respects its given capacity.

3.1 Multi-commodity Flow Formulation

It is now possible to detail the first mathematical programming model, denoted by Multi-commodity Flow Formulation (MFF). The set of decision variables of this model is defined as follows:

- $x_{(i,j)}^{\{u,v\},s} = 1$ if node i is the predecessor of node j ($\{i,j\} \in E$) on the unique path from node u to node v, on VLAN $s \in S$; 0 otherwise;
- $w_{\{i,j\}}^{s} = 1$ if link $\{i,j\} \in E$ is used in VLAN $s \in S$; 0 otherwise;
- $U^{max} =$ maximum value of link utilization.

Given this set of variables, any feasible solution for the problem must verify the following set of constraints:

(MFF)

$$\min_{x,w} \quad U^{max} \tag{1a}$$

s.t

$$\sum_{j:(u,j)\in E} x_{(u,j)}^{\{u,v\},s} = 1, \qquad\qquad u,v \in V : u < v, s \in S, \tag{1b}$$

$$\sum_{j:(i,j)\in E} x_{(i,j)}^{\{u,v\},s} - \sum_{j:(j,i)\in E} x_{(j,i)}^{\{u,v\},s} = 0,$$

$$u,v,i \in V : u < v, i \neq \{u,v\}, s \in S, \tag{1c}$$

$$x_{(i,j)}^{\{u,v\},s} + x_{(j,i)}^{\{u,v\},s} \leq w_{\{i,j\}}^{s}, \qquad u,v \in V : u < v, \{i,j\} \in E, s \in S, \tag{1d}$$

$$\sum_{\{i,j\}\in E} w_{\{i,j\}}^{s} = n - 1, \qquad\qquad\qquad s \in S, \tag{1e}$$

$$\sum_{s\in S} \sum_{\{u,v\}:u<v} d_s\{u,v\}(x_{(i,j)}^{\{u,v\},s} + x_{(j,i)}^{\{u,v\},s}) \leq C_{\{i,j\}} U^{max}, \quad \{i,j\} \in E, \tag{1f}$$

$$x_{(i,j)}^{\{u,v\},s} \in \{0,1\}, \qquad\qquad (i,j) \in E, u,v \in V : u < v, j \neq u, s \in S, \tag{1g}$$

$$w_{\{i,j\}}^{s} \in \{0,1\}, \qquad\qquad\qquad\qquad \{i,j\} \in E, s \in S, \tag{1h}$$

$$0 \leq U^{max} \leq 1 \tag{1i}$$

The objective of the model (1a) is to minimize the value of the variable U^{max}, which measures the maximum link utilization observed in all the network. As stated in the first section of this paper, link utilization is the measure of the ratio between each link's load and its capacity.

The flow conservation constraints (1b) and (1c) define, for each VLAN $s \in S$, a flow path between every pair of nodes, $u, v \in V$, through which traffic between those two switches is routed. Constraints (1d) guarantee that, in each VLAN, the flow paths will only make use of the links used in the according VLAN.

In (1e) the number of links used in each VLAN is set to $n - 1$, so that they form a spanning tree.

This first group of constraints (1b–1e), along with (1g–1h), defines, for each VLAN, a spanning tree.

The set of constraints (1f) defines the variable U^{max}. At the same time, (1f) is responsible for bounding the traffic quantity flowing through each link to the given capacity, as $U^{max} \in [0, 1]$ (1i).

Constraints (1g) and (1h) define, respectively, the set of variables x and w as binary.

Balakrishnan et al. [15] have suggested a set of constraints for enhancing the linking constraints, that can be extended for the multiple spanning trees, in the following way:

$$x_{(i,j)}^{\{u,v\},s} + x_{(j,i)}^{\{u,v'\},s} \leq w_{\{i,j\}}^s, \qquad u, v, v' \in V : v \neq v', \{i,j\} \in E, s \in S, \tag{2a}$$

$$x_{(i,j)}^{\{u,v\},s} + x_{(j,i)}^{\{u',v\},s} \leq w_{\{i,j\}}^s, \qquad u, u', v \in V : u \neq u', \{i,j\} \in E, s \in S, \tag{2b}$$

The first set of constraints, (2a), states that when edge $\{i, j\}$ is used in a given VLAN s, then all commodities originated in a given node u, will flow either from i to j, or from j to i. Equation (2b) describes an equivalent situation, for all commodities flowing to a given node v.

In the following specific cases, (2a) and (2b) can be written as equalities, instead of inequalities:

$$x_{(i,j)}^{\{u,j\},s} + x_{(j,i)}^{\{u,i\},s} = w_{\{i,j\}}^s, \qquad u \in V, \{i,j\} \in E : u < j, u < i, s \in S, \tag{3a}$$

$$x_{(i,j)}^{\{i,v\},s} + x_{(j,i)}^{\{j,v\},s} = w_{\{i,j\}}^s, \qquad v \in V, \{i,j\} \in E : i < v, j < v, s \in S. \tag{3b}$$

Equation (3a) assert that if edge $\{i, j\}$ is used in VLAN s, this necessarily means that either the commodity flowing between a given node u and j travels from i to j, or the commodity flowing between u and i travels in the opposite direction. (3b) describes an equivalent situation, where the common node between the two commodities is not the origin, but the destination.

Empirical evidence have revealed that by using (2a) and (2b) along with (3a) and (3b), in the special cases previously mentioned, strengthens the linear programming (LP) relaxation of MFF. For future reference, the model using all the constraints in MFF plus these valid inequalities will be denoted as MFF+.

3.2   Rooted Directed Formulation

One alternative formulation can be obtained by extending the formulation suggested by Martin [16] to the modelling of multiple spanning trees (see also [17,18]). We denote this model by Rooted Directed Formulation (RDF). The set of decision variables is defined as follows:

- $z_{i,j}^{u,s} = 1$ if node i is the "father" of node j, in an arborescence rooted at node u, in VLAN $s \in S$; 0 otherwise;
- $f_{(i,j)}^{u,s}$ = traffic quantity, originated from node u, going through link $(i,j) \in E$, in VLAN $s \in S$;
- $w_{\{i,j\}}^{s} = 1$ if link $\{i,j\} \in E$ is used in VLAN $s \in S$; 0 otherwise;
- U^{max} = maximum value of link utilization.

Given this set of variables, any feasible solution for the problem must verify the following set of constraints:

(RDF)

$$\min_{z,f,w} \quad U^{max} \tag{4a}$$

s.t

$$\sum_{i:(i,j)\in E} z_{i,j}^{u,s} = 1, \qquad\qquad u, j \in V : u \neq j, s \in S, \tag{4b}$$

$$z_{i,j}^{u,s} + z_{j,i}^{u,s} = w_{\{i,j\}}^{s}, \qquad\qquad u \in V, \{i,j\} \in E, \forall s \in S, \tag{4c}$$

$$\sum_{\{i,j\}\in E} w_{\{i,j\}}^{s} = n - 1, \qquad\qquad \forall s \in S, \tag{4d}$$

$$\sum_{j:(j,i)\in E} f_{(j,i)}^{u,s} - \sum_{j:(i,j)\in E \wedge j\neq u} f_{(i,j)}^{u,s} = d_s\{u,i\}, \quad u, i \in V : u \neq i, s \in S, \tag{4e}$$

$$\sum_{i:(u,i)\in E} f_{(u,i)}^{u,s} = \sum_{v\in V:v\neq u} d_s\{u,v\}, \qquad\qquad u \in V : u, s \in S, \tag{4f}$$

$$f_{(i,j)}^{u,s} \leq \sum_{v\in V\setminus\{u,i\}} d_s\{u,v\} z_{i,j}^{u,s}, \quad i, j, u \in V : u \neq j, \{i,j\} \in E, s \in S, \tag{4g}$$

$$\sum_{s\in S}\sum_{u\in V}(f_{(i,j)}^{u,s} + f_{(j,i)}^{u,s}) \leq C_{\{i,j\}} U^{max}, \qquad\qquad \{i,j\} \in E, \tag{4h}$$

$$z_{i,j}^{u,s} \in \{0,1\}, \qquad\qquad u \in V, (i,j) \in E : j \neq u, s \in S, \tag{4i}$$

$$w_{\{i,j\}}^{s} \in \{0,1\}, \qquad\qquad \{i,j\} \in E, s \in S, \tag{4j}$$

$$f_{(i,j)}^{u,s} \geq 0, \qquad\qquad u \in V, (i,j) \in E : j \neq u, s \in S, \tag{4k}$$

$$0 \leq U^{max} \leq 1 \tag{4l}$$

Once again the objective of the formulation is to minimize U^{max} (4a).

In each VLAN $s \in S$, the z variables define an arborescence rooted in each node $u \in V$. To ensure that each one of these arborescence forms a spanning tree, constraints (4b) define, for every node on each arborescence, a "father" node. Node i is the "father" of j, in an arborescence rooted at node u if there is a path from u to j passing from link $\{i,j\}$, in the direction i to j. An exception is the root node that, evidently, cannot have a "father" node. Constraint set (4c) ensures that all arborescences, in each VLAN, use the same links. These two sets of constraints, in conjunction with the set on the number of links used in (4d) and (4i–4h), define the structure of each VLAN as a spanning tree.

Equations (4e) and (4f) are traffic flow conservation constraints, responsible for defining the f variables. Note that there is a set of f variables for each arborescence. Compared to MFF, the number of flow variables is n times smaller as flows are aggregated by origin nodes. Equation (4e) states that the difference between the traffic quantity originated from node u, in VLAN s, entering and exiting a certain node i (different from u) must match, exactly, the traffic demand between node u and i, on that same VLAN. Constraint set (4f) defines the traffic flow quantity exiting each root node.

By constraint set (4h), traffic flow is limited to the links used in the corresponding arborescence. In order to properly bound the f variables, having in account that z variables are binary, it is necessary to multiply the right hand side of the constraint by the maximum amount of traffic flow the corresponding link can be expected to carry. However, as flows are aggregated by origin, it is not possible to make this bound tight. The best estimation that can be achieved is that the traffic quantity flowing through a directed link (i, j), in any arborescence rooted in node u, and located in VLAN s, is at most equal to the sum of the traffic demands between node u, and every other node $k \in V$, with the exception of node i, as that commodity will evidently have been delivered just before link (i, j). Constraints (4h), as (1f) did in MFF, have the double task of defining variable U^{max} and bounding the flow quantity in each link to its capacity.

Finally, constraint sets (4i) to (4l) define the feasible space for each of the variables in the model.

Fernández and Luna-Mota [18] developed a set of valid inequalities to strengthen their arborescence-based model for a single spanning tree design. RDF was tested using an extension of these inequalities for the multiple spanning tree case. However, no empirical evidence was found that, in practice, they are useful for solving the considered problem.

3.3   Comparing the LP Relaxations

It is possible to prove that the LP relaxation of MFF+ is as strong as the LP relaxation of RDF. That is to say:

Theorem 1. *Let MFP and RDP be the polyhedra defined by the set of feasible solutions to the linear relaxations of, respectively, MFF+ and RDF. Then*

$$MFP \subseteq RDP.$$

Due to space restrictions, the formal proof of Theorem 1 won't be presented in this paper.

Computational results given in the next section show that for some instances, the LP relaxation of MFF+ is strictly stronger than the one of RDF.

4   Computational Experiments

In this section we present computational experiments that can help evaluate the quality of models RDF and MFF+ to solve instances of the problem of designing

Table 1. Description of each class of instances.

Class ID	#nodes	Density	#VLANs
1	4	1	1
2	4	1	5
3	6	0.8	2
4	6	0.8	4
5	8	0.4	3
6	8	0.4	6
7	8	0.6	2
8	8	0.8	2
9	8	0.8	3
10	10	0.3	2
11	10	0.3	4
12	10	0.5	2
13	10	0.5	3
14	12	0.3	4
15	14	0.3	2
16	14	0.5	3
17	16	0.4	4
18	16	0.4	5
19	20	0.2	1
20	20	0.2	2

Ethernet networks implementing MSTP. These experiments are divided in two parts. In the first experiment, we have measured the CPU time used by CPLEX to solve RDF and MFF+, in order to evaluate the effectiveness of the models in solving the problems to optimality. The second experiment compares both models' LP relaxation results. This way it is possible to assess the quality of each model in generating lower bounds.

These experiments were conducted using a set of randomly generated instances, with different characteristics: number of nodes, network density, and number of spanning trees. Respecting the first two parameters, it is possible to create the set of links, distributed randomly in the network. Each link will be randomly given a traffic capacity value of either 0.5, 0.75 or 1. Then, for each VLAN, traffic demands between nodes are calculated using the following formula, suggested by Fortz and Thorup [19].

$$d_s\{u, v\} = \alpha O_u D_v R_{(u,v)} e^{\frac{-L_2(u,v)}{2\Delta}} \qquad (5a)$$

For each node u, two random numbers, O_u and D_v are randomly generated in the interval $[0, 1]$. These values reflect, respectively, the activeness of each node as a sender and as a receiver. Another value, $R_{(u,v)}$, is generated in the same

Table 2. Time results for each class of instance.

Class ID	MFF+			RDF		
	#solved	Time avg	Time std	#solved	Time avg	Time std
1	5	0.06	0.05	5	0.04	0.05
2	5	0.07	0.01	5	0.02	0.01
3	5	2.74	1.32	5	0.43	0.16
4	5	10.84	7.10	5	0.85	0.42
5	5	3.70	2.08	5	0.27	0.17
6	5	8.33	7.05	5	0.25	0.25
7	5	498.75	368.59	5	6.26	3.80
8	5	979.84	736.79	5	48.20	53.21
9	3	307.40	173.07	5	50.61	31.36
10	5	41.41	53.71	5	0.72	1.00
11	5	126.62	138.78	5	2.65	3.09
12	2	478.17	579.58	5	140.5498	182.36
13	3	3733.48	3097.41	4	544.68	785.88
14	0	-	-	3	236.40	356.00
15	1	45.70	0.00	3	1957.82	1801.02
16	0	-	-	0	-	-
17	0	-	-	0	-	-
18	0	-	-	1	1307.49	0
19	0	-	-	0	-	-
20	0	-	-	1	227.30	0

interval, for each pair of nodes. α is another parameter given as input. In these tests, the Euclidian distance (L_2) was substituted by the length of the shortest path between each pair of nodes, with respect to the number of links. Δ is the largest distance verified in the network.

For each class of instances, five instances were generated and tested. Table 1 describes the classes of instances tested in these experiments in terms of number of nodes in the network (*#nodes*), network density, number of VLANs in the network (*#VLANs*). α was given a value of 0.1 for every instance.

Each instance was tested using both models, RDF and MFF+. A time limit of one hour was set for solving each instance's integer program. In this sense, Table 2 indicates the number of instances, per class, that CPLEX was able to solve in this time frame, for both formulations. Table 2 also presents the average CPU time (*Time avg*) it took for instances of each class to be solved to optimality, along with the standard deviation (*Time std*). Evidently, only instances solved in one hour or less are accounted when computing these statistics.

In these experiments we also computed, for each instance, the gap between the linear relaxation and the best known feasible solution. The gap value mea-

Table 3. Gap results, in percentage, for each class of instance.

Class ID	#ISF	MFF+		RDF	
		Gap avg	Gap std	Gap avg	Gap std
1	5	51.80	7.08	54.85	9.01
2	5	30.69	13.34	30.82	13.46
3	5	46.34	5.88	47.37	6.63
4	5	37.50	11.94	38.00	12.30
5	5	12.74	9.45	13.12	9.85
6	5	1.03	1.14	1.03	1.14
7	5	49.40	3.36	49.56	3.15
8	5	57.96	5.30	58.06	5.28
9	5	58.00	21.10	58.83	20.35
10	5	5.74	12.84	5.96	13.34
11	5	0.26	0.56	0.26	0.56
12	5	32.44	20.65	32.91	21.35
13	5	23.31	15.18	23.31	15.18
14	5	3.12	2.18	3.12	2.18
15	5	16.88	17.49	16.88	17.49
16	5	64.05	12.00	64.05	12.00
17	5	67.26	11.70	67.57	12.07
18	5	40.25	30.62	40.25	30.62
19	2	73.70	14.34	74.65	14.76
20	5	43.86	27.10	43.86	27.10

sures the ratio between the solution quality of the integer program and of its linear relaxation (see Formula 6a). In order to be able to make a valid comparison between the two models, for each instance, the gap was calculated taking into account the best integer solution found by any of the models. These results are presented in Table 3 (*Gap avg* and *Gap std*). Only instances where at least one integer solution was found by any of the models, within one hour, will be considered for these statistics. In this sense, for each class of instance we also indicate the number of instances where this condition was verified (*#ISF*).

$$Gap = \frac{IntegerSolution - RelaxationSolution}{IntegerSolution} \tag{6a}$$

All the tests were performed using ILOG CLEX 12.5, on a Intel Core i7 CPU 960 @ 3.20 GHz (x8) with 12 GB of memory with 64 bits, and running Ubuntu 12.04.1 LTS (GNU/Linux 3.2.0 − 26−generic x86_64).

Table 2 shows that RDF is much faster than MFF+. Consequently, RDF is able to go further, and solve bigger instances that MFF+ is unable to solve within one hour (Class IDs 9, 12–15, 18, 20).

In Table 3, however, we can observe that for some classes of instances (1–5, 7–10, 12, 17, 19), MFF+ was able to produce better lower bounds than RDF, evidenced by the smaller gap values. This way it is possible to conclude that for these instances the LP relaxation MFF+ is, in fact, strictly stronger than the LP relaxation of RDF.

Nevertheless, the differences between the gaps, when existing, are not all that substantial, with the average gain being less than 1 %.

5   Summary and Conclusions

In this paper we approached the problem of finding optimal designs for Ethernet networks implementing MSTP, with regards to link utilization. Slight variations of this problem have been widely treated in the literature. However, research has been focusing, almost exclusively, in heuristic methods. In fact, the only work [9, 10] that made use of exact methods, did so merely as a way of evaluating the quality of their proposed heuristic. Therefore, in our view, it is clear that there is still a lot of improvements that can be done using exact method approaches to these problems.

In this paper we propose two mixed-integer programming formulations for the considered problem. These formulations are based on models that have been previously used for the design of single spanning trees. The first formulation, MFF, makes use of multi-commodity flows to design the spanning trees. A set of valid inequalities were proposed, that were able to further strengthen this formulation. The combination of MFF and these valid inequalities was named MFF+. The second formulation, RDF, defines multiple arborescences rooted at each node of a VLAN, thusly ensuring the required spanning tree topology.

Various numerical experiments were made using both models implemented on CPLEX. By observing the results it is clear that RDF is much faster than MFF+. We were also able to conclude that the LP relaxation of MFF+ is a stronger formulation than the LP relaxation of RDF. Nevertheless, results suggest that MFF+ does not always produce better lower bounds than RDF. Furthermore, when MFF+ is indeed able to do so, the gain seems to be low.

Finally, the results also show huge gap fluctuations. Moreover, there is no evidence of a monotonous progression of the gap size. We believe that this unpredictableness is a direct result of the *max-min* structure of problem.

We are currently working on developing other formulations for this problem, such that they may be even faster than RDF and/or stronger than MFF+. In the future, in order to tackle larger and more realistic sized instances, we want to use Benders' decomposition in conjunction with the proposed models. This can be used in a branch-and-cut framework, where Benders' cuts are generated in each node of the branch-and-bound tree. While tests imply that RDF is a more appropriate formulation for the optimal solving of small size instances of this problem, MFF+ may yet prove itself useful for that framework, due to its propensity to produce better lower bounds.

References

1. IEEE Computer Society: Ieee standard for information technology- telecommuni-cations and information exchange between systems- local and metropolitan area networks- common specifications part 3: Media access control (mac) bridges. ANSI/IEEE Std 802.1D, 1998 Edition, i-355 (1998)
2. IEEE Computer Society: Ieee standard for local and metropolitan area networks virtual bridged local area networks. IEEE Std 802.1Q-2005 (Incorporates IEEE Std 802.1Q1998, IEEE Std 802.1u-2001, IEEE Std 802.1v-2001, and IEEE Std 802.1s-2002), 0_1-285 (2006)
3. IEEE Computer Society: Ieee standards for local and metropolitan area networks–virtual bridged local area networks– amendment 3: Multiple spanning trees. IEEE Std 802.1s-2002 (Amendment to IEEE Std 802.1Q, 1998 Edition), 0_1-211 (2002)
4. Ho, T.: Traffic engineering techniques for data center networks. Ph.D. thesis, Ecole polytechnique de Louvain, Université catholique de Louvain (2012)
5. He, X., Zhu, M., Chu, Q.: Traffic engineering for metro ethernet based on multiple spanning trees. In: International Conference on Networking, International Conference on Systems and International Conference on Mobile Communications and Learning Technologies (ICNICONSMCL'06), pp. 97–97. IEEE (2006)
6. Lim, Y., Yu, H., Das, S., Lee, S.S., Gerla, M.: QoS-aware multiple spanning tree mechanism over a bridged LAN environment. In: IEEE Global Telecommunications Conference (IEEE Cat. No. 03CH37489), GLOBECOM '03, vol. 6, pp. 3068–3072. IEEE (2003)
7. Meddeb, A.: Multiple spanning tree generation and mapping algorithms for carrier class ethernets. In: IEEE Globecom 2006, pp. 1–5. IEEE (2006)
8. de Sousa, A., Soares, G.: Improving load balance and minimizing service disruption on ethernet networks with IEEE 802.1S MSTP. In: Workshop on IP QoS and Traffic Control, pp. 25–35 (2007)
9. Santos, D., de Sousa, A., Alvelos, F., Dzida, M., Pióro, M., Zagozdzon, M.: Traffic engineering of multiple spanning tree routing networks: the load balancing case. In: Next Generation Internet Networks, NGI '09, pp. 1–8 (2009)
10. Santos, D., de Sousa, A., Alvelos, F., Dzida, M., Pióro, M.: Optimization of link load balancing in multiple spanning tree routing networks. Telecommun. Syst. 48(1–2), 109–124 (2010)
11. Magnanti, T.L., Wolsey, L.A.: Optimal trees. In: Ball, M.O., et al. (eds.) Handbooks in Operations Research and Management Science, vol. 7, pp. 503–615. Elsevier Science, Amsterdam (1995)
12. Chen, W., Jin, D., Zeng, L.: Design of multiple spanning trees for traffic engineering in metro ethernet. In: 2006 International Conference on Communication Technology, pp. 1–4. IEEE (2006)
13. Padmaraj, M., Nair, S., Marchetti, M., Chiruvolu, G., Ali, M., Ge, A.: Metro ethernet traffic engineering based on optimal multiple spanning trees . In: Second IFIP International Conference on Wireless and Optical Communications Networks, WOCN 2005, pp. 568–572. IEEE (2005)
14. Mirjalily, G., Sigari, F.A., Saadat, R.: Best multiple spanning tree in metro ethernet networks. In: 2009 Second International Conference on Computer and Electrical Engineering, pp. 117–121. IEEE (2009)
15. Balakrishnan, A., Magnanti, T.L., Wong, R.T.: A dual-ascent procedure for large-scale uncapacitated network design. Oper. Res. 37(5), 716–740 (1989)

16. Martin, R.K.: Using separation algorithms to generate mixed integer model reformulations. Oper. Res. Lett. **10**(3), 119–128 (1991)
17. Conforti, M., Cornuéjols, G., Zambelli, G.: Extended formulations in combinatorial optimization. 4OR **8**(1), 1–48 (2010)
18. Fernández, E., Luna-Mota, C., Hildenbrandt, A., Reinelt, G., Wiesberg, S.: A flow formulation for the optimum communication spanning tree. Electron. Notes Discrete Math. **41**, 85–92 (2013)
19. Fortz, B., Thorup, M.: Increasing internet capacity using local search. Comput. Optim. Appl. **29**(1), 13–48 (2004)

Graph Compact Orthogonal Layout Algorithm

Kārlis Freivalds$^{(\boxtimes)}$ and Jans Glagoļevs

Institute of Mathematics and Computer Science, University of Latvia,
Raina Bulvaris 29, Riga LV-1459, Latvia
karlis.freivalds@lumii.lv,
jansglagolevs@gmail.com

Abstract. There exist many orthogonal graph drawing algorithms that minimize edge crossings or edge bends, however they produce unsatisfactory drawings in many practical cases. In this paper we present a grid-based algorithm for drawing orthogonal graphs with nodes of prescribed size. It distinguishes by creating pleasant and compact drawings in relatively small running time. The main idea is to minimize the total edge length that implicitly minimizes crossings and makes the drawing easy to comprehend. The algorithm is based on combining local and global improvements. Local improvements are moving each node to a new place and swapping of nodes. Global improvement is based on constrained quadratic programming approach that minimizes the total edge length while keeping node relative positions.

1 Background

Graph drawing algorithms provide a visually appealing way to present the structure of a graph. Several graph drawing styles are commonly used, each underlining some property of the graph suitable for a particular application. We deal with the orthogonal drawing style where edges are represented by chains of horizontal and vertical line segments connecting the nodes. The goal is to obtain an aesthetically pleasing drawing of a given graph. Common aesthetic criteria include alignment of nodes, small area, few bends and crossings, short edge length. Overlaps of objects are not allowed.

Most of prior work on orthogonal drawing algorithms is dedicated to producing drawings of some provable quality aspect. This is the case of the popular topology-shape-metric approach [5,9,17] where the number of bends is minimized respecting some planar embedding. See [6] for an experimental evaluation of these algorithms. There are a number of works achieving proven area bounds, or bounds on the number of bends or both [1–3]. Unfortunately these are worst case bounds and often a given particular graph can be laid out much better as these algorithms produce. Most of the current orthogonal drawing algorithms perform poorly in a practical setting. Even simple heuristics often yield a significant improvement of the drawing quality [8,16].

Supported in part by project No. 2013/0033/2DP/2.1.1.1.0/13/APIA/VIAA/027.

© Springer International Publishing Switzerland 2014
P. Fouilhoux et al. (Eds.): ISCO 2014, LNCS 8596, pp. 255–266, 2014.
DOI: 10.1007/978-3-319-09174-7_22

We explore the orthogonal drawing problem from a practical point of view where the goal is to produce nice-looking layouts of typical graphs. To achieve this, crossings, bends, area, etc. should be minimized together in some proportion for each particular graph so that the user cannot spot obvious ways of improvement. Such goal is aimed in [3] where the layout process is divided in three phases – node placement, edge routing and port assignment. Unfortunately their node placement phase is weak – each node is placed in a new row and column thus producing large area and long edges. We generally follow this strategy, but implement each phase in a different way.

In this paper we present an orthogonal layout algorithm which produces good drawings for many practical graphs. At first, the algorithm assigns positions to nodes by putting them on a grid while minimizing the total edge length. Edges are routed afterwards by using standard techniques from integrated circuit layout [14] and minimally adjusting the node placement [10]. Minimizing the total edge length also helps to keep the number of crossings and bends low, although they are not directly minimized. Placing nodes on the grid is an essential ingredient of the algorithm that ensures non-overlapping and nice alignment of nodes characteristic to the orthogonal style.

2   Overview of the Algorithm

We consider the drawing model where nodes are represented by rectangles of a given minimum size, edges are represented with orthogonal polylines connecting the associated nodes. Overlaps between nodes or between nodes and edges are not allowed and some minimum distance δ between them has to be ensured. Only point-wise crossings of edges are allowed (no overlaps of segments of the same direction). We allow nodes to stretch to accommodate adjacent edges but excessive stretching should be minimized. We do not require strict grid placement of nodes and edges but include alignment as an aesthetic criterion to be maximized.

Similarly to [3], the layout process is divided into three phases - node placement, edge routing and normalization (see Fig. 1). Node placement is the main phase since it influences the drawing quality the most and other phases depend on it. Then comes edge routing which finds routes for edges that are short and with few bends [14]. Our employed routing algorithm does not minimize crossings, although some local crossing minimization heuristics can be easily incorporated in the routing algorithm. After routing, there can be overlaps of edge segments and the minimum distance requirement is violated. The third phase performs mental map preserving layout adjustment [7,10] to remove overlaps while minimizing the node movement.

The quality of the obtained drawings is mostly influenced by the node placement phase which is the main contribution of this paper. Edge routing and overlap removal phases will not be described any further since good solutions exist in the given references. In the node placement phase nodes are placed in a two-dimensional rectangular grid each node occupying one or more grid cells.

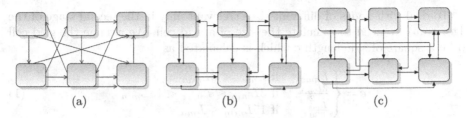

Fig. 1. Graph obtained after each of the 3 phases: **(a)** node placement; **(b)** edge routing; **(c)** normalization.

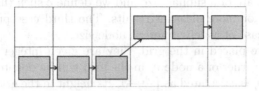

Fig. 2. Situation where compaction is needed to improve layout.

Node placement is divided into two stages. In the first stage all nodes are treated to be of the same size occupying exactly one grid cell. In the second stage a node can take several grid cells proportional to its given size. The second stage could be used alone, but obtaining an initial approximation with the unit size nodes often results in better layouts.

Similarly to [3], we formalize the node placement as an optimization problem to minimize the total edge length subject to constraints that no two nodes cover the same grid cell. The basis of our algorithm is inspired by the simulated annealing idea. We perform a greedy optimization that iteratively moves a nodes to a better place or swaps two nodes. This process is augmented with some random displacement. The idea of using a grid for node placement (although for straight-line drawings) together with simulated annealing is used in drawing of biochemical networks [12,13,15]. But in our algorithm we extend it with repeated global compaction steps that allow to escape from the many local minima of the optimization problem. Figure 2 shows a simple example where no node can be moved to improve the layout but the compaction step of our algorithm in vertical direction produces the optimal layout. In such small examples random displacements helps to find the optimum, but similar cases when groups of nodes have to be shifted often occur in larger graphs where randomization is too weak.

3   Detailed Description

In the input to the node placement algorithm we are given a graph $G = (V, E)$ with a node set V and edge set E to be laid out and the minimum width w_i and height h_i of each node. In output the algorithm gives the top-left corner (x_i, y_i) of each node.

To deal with nodes of different sizes (relevant only in the second stage of the algorithm) we need to calculate the size of a grid cell. We assume the grid cell to be a square of side length c which is calculated as

$$c = \begin{cases} L_{max} & \text{if } L_{max} < 3L_{min} \\ \frac{3L_{min}}{2} & \text{if } 3L_{min} \leq L_{max} < 15L_{min} \\ \frac{L_{max}}{30} & \text{if } 15L_{min} \leq L_{max} \end{cases} \tag{1}$$

where $L_{min} = \min(\min(w_i + \delta), \min(h_i + \delta))$ and $L_{max} = \max(\max(w_i + \delta), \max(h_i + \delta))$. The main case of c is the middle one. The first case is chosen when all nodes are of a similar size and we define c such that all boxes take only one grid cell for more pleasant results. The third case prevents excessive memory usage in case of widely different node sizes.

When nodes are placed in the grid, they are given integer coordinates and sizes. The top-left corner of a node v_i in the grid will be denoted by (x'_i, y'_i). Its width in the grid w'_i is calculated as $\lceil \frac{w_i + \delta}{c} \rceil$. Its height in the grid h'_i is calculated as $\lceil \frac{h_i + \delta}{c} \rceil$.

We can use different functions for the edge length to be minimized. Common examples include Euclidean or Manhattan distance. To deal with nodes of different sizes better, we use a distance function $d(v_i, v_j)$ between two nodes v_i and v_j defined as follows:

$$d(v_i, v_j) = d_e(v_i, v_j) + \frac{1}{20} \min \left(\frac{|x_i^c - x_j^c|}{w'_i + w'_j}, \frac{|y_i^c - y_j^c|}{h'_i + h'_j} \right), \tag{2}$$

where $d_e(v_i, v_j)$ is the Euclidean distance between the node rectangle borders and $x_i^c = x'_i + 1/2w'_i$ and $y_i^c = y'_i + 1/2h'_i$ are center coordinates of the nodes. The second addend helps to align node centers when the distance between their borders is approximately equal. The constant $1/20$ was chosen experimentally to balance the need for short edges with alignment of node centers.

3.1 Compaction

An essential step of the proposed layout algorithm is compaction, which performs global layout improvements and simultaneously creates new empty places in the grid. We use a quadratic programming approach [7, 10] where compaction in one dimension is expressed as minimization of a quadratic function subject to two-variable linear constraints. The function is constructed to minimize the total edge length but constraints keep the minimum distances between nodes and maintain their relative ordering.

Compaction is done separately in horizontal and vertical directions. Let us consider the horizontal direction; the vertical one is similar. The relative ordering is expressed as a visibility graph. A visibility graph is a directed graph with the same set of nodes V but with a different set of edges S. There is a directed edge $(i, j) \in S$ in the visibility graph if and only if $x'_j > x'_i$ and it is possible to connect nodes v_i and v_j with a horizontal line segment without overlapping any other node. The visibility graph can be constructed with a sweep-line algorithm in time $|V| \log |V|$ but in our case we can extract it directly from the grid in time proportional to the number of grid cells.

We construct the following optimization problem

$$\text{minimize} \sum_{(i,j)\in E} (z_i + \frac{1}{2}w'_i - (z_j + \frac{1}{2}w'_j))^2 \tag{3}$$

$$\text{subject to } z_j - z_i \geq d_{ij}, (i,j) \in S$$

where $d_{ij} = \gamma \cdot w'_i$ and $\gamma \geq 1$ is a coefficient that defines how much empty space will be left between nodes. To obtain the maximum compaction we should set $\gamma = 1$. Such setting is desirable at the final few iterations of the algorithm but otherwise using $\gamma > 1$ leaves some empty places between nodes giving additional freedom for node movement to find a better solution.

To perform compaction, the visibility graph is constructed from the current node positions, the optimization problem is constructed and its minimum is found by using the solver described in [10]. The node positions are calculated as $x'_i = \lfloor z_i \rfloor$. Since w'_i are integer and $\gamma \geq 1$, the rounded values satisfy $x'_j - x'_i \geq w'_i$ and non-overlapping of nodes is ensured. The rounded solution may not be optimal with respect to the integer variables x'_i but is good enough for our purposes. Note that compaction uses a quadratic edge length function but node swapping uses a linear distance defined by Eq. (2). In our case such mismatch does not create obvious bad effects since (2) is used when swapping nodes and compaction respects the obtained ordering via the constraint graph.

Compaction is used also to switch from the first stage of the algorithm where all nodes are of a unit size to the second stage with the real node sizes. The switch is done by simply compacting with d_{ij} calculated from the new node sizes. An example of horizontal compaction with different γ values is shown in Fig. 3.

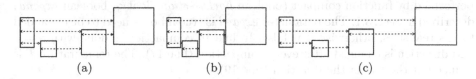

(a)   (b)   (c)

Fig. 3. An example graph of four nodes with their horizontal visibility graph (dashed). (a) before compaction; (b) after compaction with $\gamma = 1$; (c) after compaction with $\gamma = 2$.

3.2   Algorithm Pseudocode

The pseudocode for the node placement is shown in Algorithm 1. As the first steps, the grid of size $5\sqrt{|V|} \times 5\sqrt{|V|}$ is initialized and nodes are randomly placed in the grid. The first stage of the algorithm (lines 6–20) treats each node of size 1. The grid is dynamically expanded during layout, if required. The algorithm works in iterations and the number of iterations *iterationCount* is taken proportional to $\sqrt{|V|}$. At each iteration, local optimization is performed that decreases the total edge length. The optimization process is based on the simulated annealing idea. It requires the notion of temperature T which influences how much node positions are perturbed by randomness. The starting temperature T is set equal to $2\sqrt{|V|}$ to allow nodes to be placed almost everywhere initially. The temperature is smoothly reduced by a cooling coefficient k until it reaches the lowest temperature T_{min}; we take T_{min} equal to 0.2. The cooling coefficient k is calculated in line 5 such that T reaches T_{min} in *iterationCount* iterations.

To perform local optimization, every node is moved to a location that minimizes the total length of its adjacent edges. We use a heuristic to calculate this location approximately. Calculating the optimal location is expensive and actually is not needed since the added random displacement disturbs it anyway. We calculate an initial estimate to node's position (x,y) that minimizes the Manhattan distance to the adjacent nodes. Such point is found as a median of the neighbors' centers. A random displacement proportional to the temperature T is added to that point.

Then we search the closest place to (x,y), where v_j can be put (line 10). We calculate the Manhattan distance d of the closest free place to (x,y). Then we check all cells within Manhattan distance $d+1$ from (x,y) and choose the position with the least total edge length according to (2) to place v_j. If this place is different from the location of v_j from the previous iteration, we leave the node there. Otherwise, we try to swap it with the nodes nearby. We do this by checking the nodes residing in adjacent grid cells to v_j. For each of these we calculate the gain of the total edge length if we swap the adjacent node with v_j. If the gain is positive we swap the nodes.

Compaction is performed every 9-th iteration each time changing direction (lines 15–18). The variable *compactionDir* defines direction in which compaction is be performed, *true* for horizontal direction *false* for vertical. Compaction is performed by function compact (boolean *horDirection*, float γ, boolean *expand*) described in Sect. 3.1. The parameter *expand* is true if boxes need to be expanded to it's real sizes, otherwise it is false. In line 16 compaction is done with $\gamma = 3$ and direction is changed after every compaction (line 17). The temperature T is reduced at the end of the iteration (line 19).

In lines 21 and 22 switching from the first stage to the second is done by performing compaction with the new node sizes. The second stage (lines 23–37) is similar to the first one, only all boxes are treated with their prescribed sizes. Searching for a place for a node has to check if all grid cells under a larger node are free. This modification influences node swapping – there may be cases when adjacent nodes of different sizes cannot be swapped. This is the main motivation why the first stage with unit node sizes is beneficial. In line 33 compaction is done with gradually decreasing γ which becomes 1 in the last 3 compactions.

In this way the available free space for node movement is gradually reduced giving more emphasis to node swapping.

1   Initialize the grid of size $5\sqrt{|V|} \times 5\sqrt{|V|}$ Put nodes randomly into grid (treat them all 1×1 sized);

2   compactionDir=true;

3   iterationCount=$90\sqrt{|V|}$;

4   T=$2\sqrt{|V|}$;

5   k=$(0.2/\mathrm{T})^{1/\mathrm{iterationCount}}$;

6   **for** ($i=0$; $i<iterationCount/2$; $i++$) **do**

7   **for** ($j=1$; $j \leq |V|$; $j++$) **do**

8   x=neighboursMedianX(v_j) + random($-T, T$);

9   y=neighboursMedianY(v_j) + random($-T, T$);

10   Put v_j near (x,y);

11   **if** v_j *has not changed it's place from the previous iteration* **then**

12 | Try to swap v_j with nodes nearby;

13   **end**

14   **end**

15   **if** *iterationCount* mod *9* == *0* **then**

16   compact(compactionDir, 3, false);

17   compactionDir=!compactionDir;

18   **end**

19   T=T·k;

20   **end**

21   compact(true, 3, true);

22   compact(false, 3, true);

23   **for** ($i=iterationCount/2+1$; $i<iterationCount$; $i++$) **do**

24   **for** ($j=1$; $j \leq |V|$; $j++$) **do**

25   x=neighboursMedianX(v_j) + random($-T \cdot w'_j, T \cdot w'_j$);

26   y=neighboursMedianY(v_j) + random($-T \cdot h'_j, T \cdot h'_j$);

27   Put v_j near (x,y);

28   **if** v_j *has not changed it's place from the previous iteration* **then**

29 | Try to swap v_j with nodes nearby;

30   **end**

31   **end**

32   **if** *iterationCount* mod *9* == *0* **then**

33   compact$\left(\text{compactionDir, max} \left(1, 1 + \frac{2(\text{iterationCount}-i-30)}{0.5\text{iterationCount}} \right), \text{false} \right)$;

34   compactionDir=!compactionDir;

35   **end**

36   T=T·k;

37   **end**

Algorithm 1. Main algorithm

3.3   Other Starting Layouts

The algorithm described above starts with assigning random positions to the nodes and later it needs relatively many iterations to find a good layout. We consider two other initial layouts for increased speed or quality – force directed placement and arrangement by breadth first search (BFS). The constants of the algorithm have to be adjusted depending on the chosen initial layout.

One possibility is to use force directed placement for the initial positions of nodes. Fast methods [11] are known for the force directed placement. Node coordinates in the grid are initialized from the rounded results of the force directed placement. Since force-directed placement gives a good approximation to the minimum edge length, *iterationCount* can be set constant and we made it equal to 100. Compaction is done at every 3-rd iteration. The starting temperature should be small, we set $T = 3$.

Another possibility for the starting layout is to use incremental placement where nodes are added one by one. We execute a breadth-first search starting from some arbitrary chosen node and add nodes to the grid in this order. The position for each node is chosen as a free place that minimizes the total distance to already placed nodes. We found that BFS placement gives good results with small graphs or graphs with a small degree. For this starting layout we chose coefficients as follows: $iterationCount = 10\sqrt{|V|}$, $T = 0.2\sqrt{|V|}$. Compaction is done after every 3-rd iteration.

4   Results

We have tested our algorithm on many artificial and real world graphs. Figure 8 shows some examples. The algorithm produces pleasant drawings with small area and low number of crossings and edge bends. For small graphs like these, all three proposed initial placement methods produce similar results.

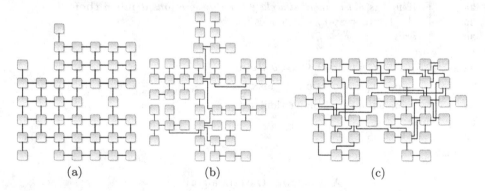

(a)   (b)   (c)

Fig. 4. Examples of the tested graphs. (a) partial grid graph; (b) tree graph; (c) random graph.

To test the quality and performance of our algorithm we run it on three automatically generated graph classes – partial grids, random trees and random graphs, see Fig. 4. Three modifications of the algorithm with different initial placements were tested for each class, the time for initialization is included in the measurement. For each class of graphs random instances were generated of progressively increasing size, 10 instances for each size. The running time and the average number of crossings per edge were calculated. To be independent of the routing algorithm, edges were treated as straight line segments connecting the node centers for the crossing calculation. The results were averaged over the 10 generated instances.

The running time measurements are similar for all three graph classes. The running time mostly depends on the number of iterations chosen in each case of initialization, random case being the slowest and force-directed case the fastest. The measurements indicate that the time for initial placement does not add much overhead.

The results for the partial grid graphs are shown in Fig. 5. A partial grid graph is a square grid with the specified number of nodes where 10 % of nodes are randomly removed. The results show that the algorithm with all the initial placement methods produce a planar layout of small instances (up to about 1000 nodes) but further only force-directed initialization is able to recover the graph structure correctly, BFS initialization being the worst. It has to be mentioned that, if we increase the number of iterations of the BFS case to match the random case, we obtain drawings of similar quality. But our intention for the BFS method was to check whether we can improve running time with a better initialization. Tests showed that BFS initialization does not give any advantage over the random one.

The quality on tree graphs is similar for all three modifications (Fig. 6). None is able to produce completely planar drawings of larger instances, although the crossing count is small. The BFS method has slightly more crossings than the other two.

Random graphs are generated by including randomly chosen node pairs as edges in the graph with density $|E| = 1.2|V|$. The quality on random graphs is similar for all three methods (Fig. 7). That is expected since random graphs cannot be drawn with significantly less crossings than any of these methods produce.

Overall, the best initialization method is force-directed, which produce the best drawing quality in the least running time. Of course, this option depends on the quality and performance of the available force-directed placement implementation.

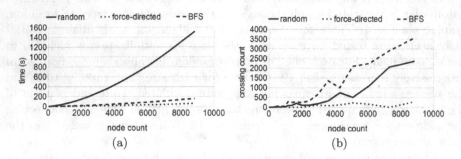

Fig. 5. The running time and crossing count depending on node count of partial grid graphs. (a) running time; (b) crossing count.

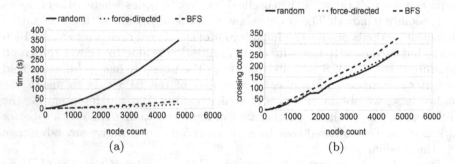

Fig. 6. The running time and crossing count depending on node count of tree graphs. (a) running time; (b) crossing count.

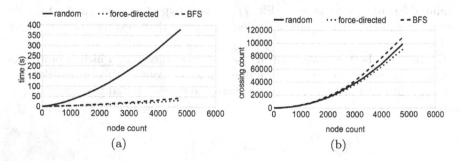

Fig. 7. The running time and crossing count depending on node count of random graphs. (a) running time; (b) crossing count.

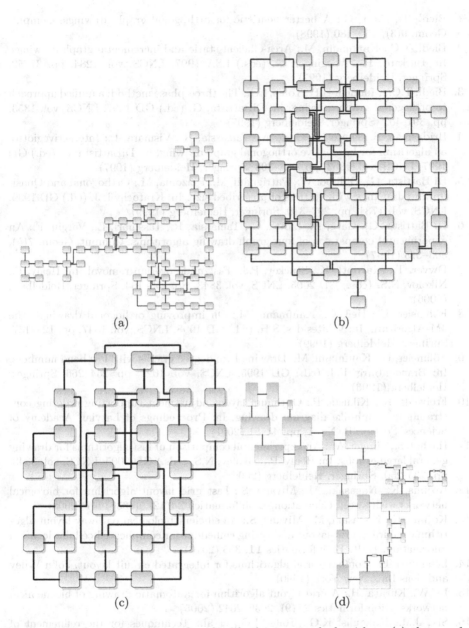

(a)

(b)

(c)

(d)

Fig. 8. Examples of layouts produced with the proposed algorithm. (a) the graph presented in [13]; (b) the graph presented in [6]; (c) the graph presented in [4]; (d) a graph with nodes of different sizes.

References

1. Biedl, T., Kant, G.: A better heuristic for orthogonal graph drawings. Comput. Geom. **9**(3), 159–180 (1998)
2. Biedl, T.C., Kaufmann, M.: Area-efficient static and incremental graph drawings. In: Burkard, R., Woeginger, G. (eds.) ESA 1997. LNCS, vol. 1284, pp. 37–52. Springer, Heidelberg (1997)
3. Biedl, T.C., Madden, B.P., Tollis, I.G.: The three-phase method: a unified approach to orthogonal graph drawing. In: DiBattista, G. (ed.) GD 1997. LNCS, vol. 1353, pp. 391–402. Springer, Heidelberg (1997)
4. Bridgeman, S., Fanto, J., Garg, A., Tamassia, R., Vismara, L.: Interactivegiotto: an algorithm for interactive orthogonal graph drawing. In: DiBattista, G. (ed.) GD 1997. LNCS, vol. 1353, pp. 303–308. Springer, Heidelberg (1997)
5. Di Battista, G., Didimo, W., Patrignani, M., Pizzonia, M.: Orthogonal and Quasi-upward drawings with vertices of prescribed size. In: Kratochvíl, J. (ed.) GD 1999. LNCS, vol. 1731, pp. 297–310. Springer, Heidelberg (1999)
6. Di Battista, G., Garg, A., Liotta, G., Tamassia, R., Tassinari, E., Vargiu, F.: An experimental comparison of four graph drawing algorithms. Comput. Geom. **7**(5), 303–325 (1997)
7. Dwyer, T., Marriott, K., Stuckey, P.J.: Fast node overlap removal. In: Healy, P., Nikolov, N.S. (eds.) GD 2005. LNCS, vol. 3843, pp. 153–164. Springer, Heidelberg (2006)
8. Fößmeier, U., Heß, C., Kaufmann, M.: On improving orthogonal drawings: the 4M-Algorithm. In: Whitesides, S.H. (ed.) GD 1998. LNCS, vol. 1547, pp. 125–137. Springer, Heidelberg (1999)
9. Fößmeier, U., Kaufmann, M.: Drawing high degree graphs with low bend numbers. In: Brandenburg, F.J. (ed.) GD 1995. LNCS, vol. 1027, pp. 254–266. Springer, Heidelberg (1996)
10. Freivalds, K., Kikusts, P.: Optimum layout adjustment supporting ordering constraints in graph-like diagram drawing. In: Proceedings of Latvian Academy of Sciences, Section B, No. 1, pp. 43–51 (2001)
11. Hachul, S., Jünger, M.: An experimental comparison of fast algorithms for drawing general large graphs. In: Healy, P., Nikolov, N.S. (eds.) GD 2005. LNCS, vol. 3843, pp. 235–250. Springer, Heidelberg (2006)
12. Kojima, K., Nagasaki, M., Miyano, S.: Fast grid layout algorithm for biological networks with sweep calculation. Bioinformatics **24**(12), 1433–1441 (2008)
13. Kojima, K., Nagasaki, M., Miyano, S.: An efficient biological pathway layout algorithm combining grid-layout and spring embedder for complicated cellular location information. BMC Bioinformatics **11**, 335 (2010)
14. Lengauer, T.: Combinatorial algorithms for integrated circuit layout. John Wiley and Sons Inc., New York (1990)
15. Li, W., Kurata, H.: A grid layout algorithm for automatic drawing of biochemical networks. Bioinformatics **21**(9), 2036–2042 (2005)
16. Six, J.M., Kakoulis, K.G., Tollis, I.G., et al.: Techniques for the refinement of orthogonal graph drawings. J. Graph Algorithms Appl. **4**(3), 75–103 (2000)
17. Tamassia, R.: On embedding a graph in the grid with the minimum number of bends. SIAM J. Comput. **16**(3), 421–444 (1987)

State Space Reduced Dynamic Programming for the Aircraft Sequencing Problem with Constrained Position Shifting

Fabio Furini[1], Martin Philip Kidd[2](✉),
Carlo Alfredo Persiani[3], and Paolo Toth[2]

[1] LAMSADE, Université Paris-Dauphine,
Place du Maréchal de Lattre de Tassigny, 75775 Paris, France
fabio.furini@dauphine.fr
[2] DEI, University of Bologna, Viale Risorgimento 2, 40136 Bologna, Italy
{martin.kidd,paolo.toth}@unibo.it
[3] ENAV S.p.A, Italian Agency for Air Navigation Services,
Via Salaria 716, Roma, Italy
carlo.persiani@enav.it

Abstract. In this paper we present state space reduction techniques for a dynamic programming algorithm applied to the Aircraft Sequencing Problem (ASP) with Constrained Position Shifting (CPS). We consider the classical version of the ASP, which calls for determining the order in which a given set of aircraft should be assigned to a runway at an airport, subject to minimum separations in time between consecutive aircraft, in order to minimize the sum of the weighted deviations from the scheduled arrival/departure times of the aircraft. The focus of the paper is on a number of ways of improving the computation times of the dynamic programming algorithm proposed. This is achieved by using heuristic upper bounds and a completion lower bound in order to reduce the state space in the dynamic programming algorithm. We compare our algorithm to an approach based on mixed integer linear programming, which was adapted from the literature for the case of CPS. We show using real-world air traffic instances from the Milan Linate Airport that the dynamic programming algorithm significantly outperforms the MILP. Furthermore, we show that the proposed algorithm is capable of solving very large instances in short computation times, and that it is suitable for use in a real-time setting.

Keywords: Aircraft sequencing problem · Dynamic programming · Integer programming · Completion bounds · Constrained position shifting

1 Introduction

In this paper we consider the *aircraft sequencing problem* (ASP) [1–4], which can be formally defined as follows. As input we are given the set A of aircraft

© Springer International Publishing Switzerland 2014
P. Fouilhoux et al. (Eds.): ISCO 2014, LNCS 8596, pp. 267–279, 2014.
DOI: 10.1007/978-3-319-09174-7_23

that are scheduled to use the runway at an airport (either arriving or departing) during a certain time-window, and as output a time $x_i \in T$ should be assigned to each aircraft $i \in A$ for using the runway, as defined by a discrete set T of time instants. The assigned time x_i is restricted to fall within a time window $[e_i, \ell_i]$ for all $i \in A$, where $e_i \in T$ denotes the earliest time and $\ell_i \in T$ the latest time for i using the runway. Aircraft i is also associated with a scheduled (target) time $t_i \in T$ for using the runway, and the cost incurred by i for using the runway at some time $t \in T$ is given by

$$c(i, t) = g_i \max(0, t_i - t) + h_i \max(0, t - t_i),$$

where g_i and h_i are weights associated with i using the runway before and after its target time, respectively. If an aircraft $j \in A$ is scheduled to use the runway directly after an aircraft $i \in A$, it cannot use the runway earlier than $x_i + s_{ij}$, where s_{ij} denotes a minimum separation in time that has to be maintained between i and j for safety reasons.

We also consider so-called *Constrained Position Shifting* (CPS) [2,5]. The motivation behind CPS is to ensure that a degree of fairness is maintained among the various airlines requesting usage of the runway, by restricting deviations from the first-come-first-serve (FCFS) sequence of the aircraft. Let f_i denote the position of aircraft $i \in A$ in the FCFS sequence of the aircraft (in which the elements of A are sorted according to increasing values of t_i for all $i \in A$). CPS restricts the position of aircraft $i \in A$ in the final sequence (in which the elements of A are sorted according to increasing values of x_i for all $i \in A$) to fall within the range $[f_i - p_i^-, f_i + p_i^+]$, where p_i^- denotes the *maximum position shift* of aircraft i to an earlier position in the sequence and p_i^+ to a later position in the sequence. CPS is often used in practice and the values of p_i^- and p_i^+ are usually equal to a small constant such as 1, 2 or 3 [3]. The objective of the ASP with CPS (ASP-CPS) is to determine a vector $\boldsymbol{x} = [x_i]$ of arrival/departure times for the aircraft so as to minimize the cost function

$$\sum_{i \in A} c(i, x_i),$$

subject to all the constraints mentioned above.

Preprocessing steps. Some preprocessing steps may be performed in order to reduce the complexity of the problem without affecting the optimal solution. The first step is *time window tightening* [4]. Given an upper bound z_{UB} to the optimal solution value of the ASP, the size of the time window $[e_i, \ell_i]$ for aircraft $i \in A$ may be decreased by updating the values of e_i and ℓ_i to

$$e_i = \max[e_i, t_i - z_{UB}/g_i]$$

and

$$\ell_i = \min[\ell_i, t_i + z_{UB}/h_i].$$

The second step, called *position shift tightening*, is to determine a reduced set of positions to which an aircraft may be assigned in a feasible solution.

Let P denote the set of ordered pairs of aircraft (i, j) where either $\ell_i < e_j + s_{ji}$ or $f_i + p_i^+ \leq f_j - p_j^-$. Hence P contains ordered pairs of aircraft (i, j) where i cannot use the runway after j without violating the maximum delay constraints, the maximum shift constraints or the minimum separation constraints. Furthermore, let $P_i^+ = |\{j : (i, j) \in P\}|$, i.e. the minimum number of aircraft using the runway after aircraft i in any feasible solution, and $P_i^- = |\{j : (j, i) \in P\}|$, the minimum number of aircraft using the runway before aircraft i in any feasible solution. A set of potential feasible positions for aircraft $i \in A$ may then be defined as

$$\mathcal{F}(i) = \{k : P_i^- < k < |A| - P_i^+ \text{ and } f_i - p_i^- \leq k \leq f_i + p_i^+\}.$$

In order to simplify notation in the discussions that follow, we further define

$$\mathcal{F}_{\min}(i) = \min[k : k \in \mathcal{F}(i)]$$

and

$$\mathcal{F}_{\max}(i) = \max[k : k \in \mathcal{F}(i)]$$

as the first and last possible positions, respectively, in which aircraft i may be scheduled in a feasible solution.

Literature review. In this paper the focus is on exact approaches to the ASP. A survey of approaches and problems in the context of landing and take-off scheduling at airports was recently presented by Bennel *et al.* [6]. These problems have been subdivided into static and dynamic cases. In the static case it is assumed that complete information on the set of aircraft is known, while in the dynamic case the solution is revised each time new aircraft are introduced into the system. In this paper we consider the static case. Mixed integer linear programming (MILP) approaches are presented by Beasley *et al.* [4] and Abela *et al.* [7]. In [4] the authors consider the ASP for landings only and without considering CPS, while in [7] the authors proposed a MILP formulation for the dynamic ASP, and developed a branch-and-bound algorithm as a solution approach. An alternative formulation of the ASP was presented by Beasley and Pinol [8], where two objective functions were considered, one linear and one nonlinear. Dynamic programming approaches to the ASP-CPS were proposed by Psaraftis [9], Trivizas [10] and Balakrishnan and Chandran [3]. The algorithms presented in [9] and [10] considered the ASP-CPS without time windows, and in [9] it is assumed that all aircraft of the same type are identical. Balakrishnan and Chandran [3] present dynamic programming approaches for a number of different variants of the ASP-CPS, also providing complexity analyses of their algorithms.

Paper contribution. We focus in particular on developing a fast dynamic programming algorithm for solving the ASP-CPS with average maximum position shift values. We propose a dynamic programming approach with a recursion which extends the one proposed in [3] for discrete time models with aircraft dependent cost functions (i.e. such as minimizing the weighted delay of the aircraft). However, in [3] the dynamic programming algorithm is solved over a so-called *CPS network*, whereas our dynamic programming algorithm is applied directly to the ASP. Furthermore, in [3] results are obtained for a different objective function and also for cases of the ASP with *precedence constraints* (pairs of aircraft with a fixed relative order), a constraint which is not strictly necessary in the real world cases considered in this paper. Furthermore, we propose a number of new ways by which the computation time required by the dynamic programming algorithm may be improved, such as using heuristic upper bounds and completion lower bounds in order to reduce the state space. These techniques enables us to consider larger values of the maximum shift compared to the ones considered in [3]. Finally we compare the improved dynamic programming approach to the MILP proposed in [4] by using instances from that paper as well as real-world instances from the Milan Linate Airport (proposed in [1]).

2 Mixed-Integer Programming Formulation

A MILP formulation for the ASP was presented by Beasley *et al.* [4]. This formulation is extended in Model (1)–(11) below for the ASP-CPS. Here δ_{ij} is a binary decision variable such that

$$\delta_{ij} = \begin{cases} 1 & \text{if aircraft } i \in A \text{ uses the runway before aircraft } j \in A, \\ 0 & \text{otherwise,} \end{cases}$$

and α_i and β_i are decisions variables such that $\alpha_i = \max(0, t_i - x_i)$ and $\beta_i = \max(0, x_i - t_i)$ for aircraft $i \in A$. As mentioned before x_i is a decision variable denoting the time assigned to aircraft i for using the runway.

$$\text{Min} \quad \sum_{i \in A} (g_i \alpha_i + h_i \beta_i) \tag{1}$$

$$\text{s. t.} \quad e_i \leq x_i \leq \ell_i \qquad i \in A \tag{2}$$

$$\delta_{ij} + \delta_{ji} = 1 \qquad i, j \in A, i \neq j \tag{3}$$

$$x_i + s_{ij} - M(1 - \delta_{ij}) \leq x_j \qquad i, j \in A, i \neq j \tag{4}$$

$$\alpha_i \geq t_i - x_i \qquad i \in A \tag{5}$$

$$0 \leq \alpha_i \leq t_i - e_i \qquad i \in A \tag{6}$$

$$\beta_i \geq x_i - t_i \qquad i \in A \tag{7}$$

$$0 \leq \beta_i \leq \ell_i - t_i \qquad i \in A \tag{8}$$

$$t_i - \alpha_i + \beta_i = x_i \qquad i \in A \tag{9}$$

$$\mathcal{F}_{\min}(i) \leq \sum_{j \in A \setminus \{i\}} \delta_{ji} + 1 \leq \mathcal{F}_{\max}(i) \qquad i \in A \tag{10}$$

$$\delta_{ij} \in \{0, 1\} \qquad i, j \in A, i \neq j \tag{11}$$

where M is a large positive constant.

The objective function (1) aims at minimizing the total cost due to the weighted delay of the aircraft in the sequence. Constraints (2) ensure that the delay of each aircraft i respects its lower and upper bounds, while Constraints (3) ensure that either aircraft i uses the runway before aircraft j, or vice versa, for any two aircraft i and j. Constraints (4) imposes the minimum separation between aircraft while Constraints (5)–(9) ensure that $\alpha_i = \max(0, t_i - x_i)$ and $\beta_i = \max(0, x_i - t_i)$ holds. Finally, Constraints (10) ensure that the maximum position shift constraint is satisfied for all aircraft.

The parameter M in Constraints (4) may be replaced by $\ell_i + s_{ij} - e_j$ (as proposed in [4]), simplifying these constraints to

$$x_i + s_{ij}\delta_{ij} - (\ell_i - e_j)\delta_{ji} \le x_j, \qquad i, j \in A, \ i \ne j. \tag{12}$$

Constraints (4) may be strengthened by defining the sets described below. These sets were proposed in [4] only taking into account the maximum delay constraints. Here, however, the sets also take into account the maximum shift constraints.

Let P' be a subset of the set P (introduced in the previous section) such that it contains the ordered pairs of aircraft (i, j) where j may be delayed by i due to the minimum separation constraints, in other words where $f_i + p_i^+ \ge f_j - p_j^- - 1$ and $\ell_i + s_{ij} > e_j$. Furthermore, let Q denote the set of all ordered pairs of aircraft (i, j) such that neither (i, j) nor (j, i) are in P.

Constraints (4) may then be replaced by Constraints (13)–(15) below.

$$\delta_{ij} = 1 \qquad\qquad (i, j) \in P \qquad (13)$$
$$x_i + s_{ij} \le x_j \qquad\qquad (i, j) \in P' \qquad (14)$$
$$x_i + s_{ij}\delta_{ij} - (\ell_i - e_j)\delta_{ji} \le x_j \qquad (i, j) \in Q \qquad (15)$$

A number of valid inequalities were also proposed in [4], given by Constraints (16)–(18) below. Here Q' is the subset of Q where $t_i \le t_j$ for any ordered pair $(i, j) \in Q$, and Q'' is the subset of Q' where $t_i + s_{ij} > t_j$ for any ordered pair $(i, j) \in Q'$.

$$\sum_{i \in A} \sum_{j \in A \setminus \{i\}} \delta_{ij} = |A|\,(|A| - 1)\,/2 \tag{16}$$

$$\delta_{ij} \ge 1 - \frac{\beta_i + \alpha_j}{t_j - t_i} \qquad (i, j) \in Q' \qquad (17)$$

$$(t_i + s_{ij} - t_j)\delta_{ij} + (t_j + s_{ji} - t_i)\delta_{ji} \le \alpha_i + \beta_i + \alpha_j + \beta_j \qquad (i, j) \in Q'' \qquad (18)$$

3 Dynamic Programming Approach

As mentioned before, the recursion of the dynamic programming algorithm presented below is similar to expression (2) by Balakrishnan and Chandran [3]. In the following sections the recursion, the pseudocode of the algorithm, and the computation of bounds used for state pruning are discussed in detail.

3.1 Recursion

A state in the dynamic programming algorithm is defined by three parameters, namely a subset $S \subset A$ of aircraft, an aircraft $i \in A \setminus S$ such that $|S| + 1 \in \mathcal{F}(i)$ and a time instant $t \in [e_i, \ell_i]$. The state (S, i, t) represents the partial solution to the ASP-CPS where aircraft i uses the runway at time t after all of the aircraft in S have used the runway in an optimal way. The value of the state (S, i, t), denoted by $\mathcal{D}(S, i, t)$, is therefore defined as the optimal solution value, if one exists, of the ASP-CPS when considering the set of aircraft $S \cup \{i\}$, and imposing the additional constraints $x_i = t$ and $x_j \leq x_i$ for all $j \in S$. If no such solution exists then $\mathcal{D}(S, i, t) = \infty$.

A recursion is easily established between the values of the states, namely

$$\mathcal{D}(S, i, t) = \min_{(i', t') \in S \times T \,:\, t \geq t' + s_{i', i}} \left[\mathcal{D}(S \setminus \{i'\}, i', t') + c(i, t) \right]. \tag{19}$$

The set of pairs (i', t') considered in the minimization is, however, quite large and its size can be reduced in the following way. First of all, it is easy to verify that if aircraft i directly precedes aircraft j in the optimal sequence, then $x_j = x_i + s_{i,j}$ if $x_j > t_j$ (where x_j is the optimal time assigned for j to use the runway). Hence it will not be fruitful to consider values of t' such that $t > t' + s_{i', i}$. Noting further that $t' \in [e_{i'}, \ell_{i'}]$ and $|S| \in \mathcal{F}(i')$ should hold, the set $S \times T$ can be substituted in (19) by the set $T(S, i, t) \subseteq S \times T$ which, for each $i' \in S$ such that $|S| \in \mathcal{F}(i')$ and each $t' \in [e_{i'}, \ell_{i'}]$, contains the pair (i', t') if and only if

$$t = t' + s_{i', i} \quad \text{if } t > t_i$$
$$t \geq t' + s_{i', i} \quad \text{otherwise.}$$

An improved recursion is therefore

$$\mathcal{D}(S, i, t) = \min_{(i', t') \in T(S, i, t)} \left[\mathcal{D}(S \setminus \{i'\}, i', t') + c(i, t) \right]. \tag{20}$$

As initial values to this relation we have $\mathcal{D}(\varnothing, i, t_i) = 0$ if $1 \in \mathcal{F}(i)$, and by introducing a dummy aircraft i^* where $t_{i^*} = \max_{i \in I} \ell_i$ and $s_{ii^*} = 0$ for all $i \in A$, the optimal solution to the ASP-CPS for the input set A of aircraft is given by $\mathcal{D}(A, i^*, t_{i^*})$.

3.2 State Space Reduction Techniques

By applying the following state space reduction techniques, it may be shown that some states in the set $T(S, i, t)$ do not have to be considered in the recursion. This may be achieved by means of an upper bound z_{UB} to the optimal solution (as found, for instance, by a heuristic) and a *completion bound* $\mathcal{C}(S, i, t)$ for a state (S, i, t). The completion bound is defined such that $\mathcal{D}(S, i, t) + \mathcal{C}(S, i, t)$ is a lower bound to the optimal solution value, if one exists, of the ASP-CPS when considering the set of aircraft I, and imposing the additional constraints $x_i = t$ and $x_j \leq x'_j$ for any $j \in S$ and $j' \in I \setminus S$. Hence a completion bound is a lower bound on the additional weighted delay which will be added to $\mathcal{D}(S, i, t)$ for any complete solution (S, i, t) may eventually lead to. Therefore, if $\mathcal{D}(S, i, t) + \mathcal{C}(S, i, t) > z_{UB}$, there is no need to consider the state (S, i, t) in the recursion.

Completion bound. Given that some set S of aircraft uses the runway before aircraft $i \in A \backslash S$ does, let $\sigma(i, S) = \max(f_i - p_i^- - |S|, 1)$, i.e. $\sigma(i, S)$ represents the earliest possible position in which aircraft i can use the runway when taking the maximum shift constraints into consideration. The completion bound is then defined, for any state (S, i, t), by

$$\mathcal{C}(S, i, t) = \sum_{j \in A \backslash (S \cup \{i\})} c(j, \max[t_j, t + s_{\min} \sigma(j, S \cup \{i\})]),$$

which is the sum of the weighted delays of the aircraft in $A \backslash S$ assuming that each one is assigned to its earliest possible position, where $s_{\min} = \min_{i,j \in A} s_{ij}$.

Heuristic upper bound. In order to obtain an upper bound z_{UB} as discussed above we propose a tabu search heuristic. A solution in the tabu search heuristic is defined by a permutation of the given set of aircraft, and the runway times are calculated by assigning the aircraft to runway time slots in this order while taking the minimum separation constraints into consideration. A neighbour solution of a given feasible solution is defined as the solution obtained by performing a *swap*, i.e. exchanging the positions of two aircraft in the given solution. A swap may only be performed if both aircraft still are in feasible positions in the neighbour solution (with respect to the maximum position shift p). During each iteration of the tabu search heuristic the tabu-list (updated in a first-in-first-out fashion) contains pairs of aircraft representing the swaps performed within a predefined number of previous iterations (the tabu *tenure*). The neighbour solutions of the current solution (initially the FCFS order) obtained by performing swaps not forbidden by the tabu list are evaluated in the order in which the aircraft appear in the current sequence. Then either the first neighbour that improves the current solution, or otherwise the overall best non-improving neighbour, replaces the current solution in the next iteration. The search continues until a predefined number of iterations is reached and the best overall solution found is given as output.

3.3   Dynamic Programming Algorithm

The pseudocode of the dynamic programming algorithm is given in Algorithms 1 (for initialization) and 2 (for the recursion). In Algorithm 1 preprocessing calculations are performed. This includes calculating the sets $\mathcal{F}(i)$ for each $i \in A$ in Step 5 and calculation of sets U_k and V_k for $2 \leq k \leq |A|$ in Steps 7–9. Here U_k is the set of all aircraft that can be assigned to position k and V_k is the set of aircraft that cannot be assigned to a position larger than k. These sets are later used in Steps 3–7 of Algorithm 2 in order to avoid states that cannot lead to any feasible solution. This is achieved by determining the set of all remaining aircraft which cannot be assigned to any position after the current one. If the cardinality of this set is larger than 1, then the current state cannot lead to a feasible solution and the state is pruned. If it is exactly equal to 1, then this aircraft has to be assigned the current position, and if the set is empty then it poses no additional restrictions. Finally in Steps 10–16 of Algorithm 1 the values of the initial states are set.

In Algorithm 2, a pre-existing state (S', i', t') is identified in Steps 1–2 and in Steps 3–11 all possible pairs (i, t) such that $(i', t') \in \mathcal{T}(S, i, t)$ where $S = S' \cup \{i'\}$ are determined. The algorithm then considers $\mathcal{D}(S', i', t') + c(i, t)$ as a new possible value for $\mathcal{D}(S, i, t)$ according to (20).

The completion bound is calculated in Step 15, and it is only necessary to do this once for each state independent of the value of the state. In order to achieve this the algorithm remembers whether a state has been considered or not. If $\mathcal{D}(S, i, t) + \mathcal{C}(S, i, t) \geq z_{UB}$ then this value of the state does not have to be considered in the algorithm, and the current value of the state is set to infinity instead in Step 19 of Algorithm 2 (yet allowing the state to attain a finite value again in future steps of the algorithm).

Algoritmo 1. Initialization

Input : An input set A of aircraft and an upper bound z_{UB}.
Output: The sets U_k, V_k and $\mathcal{F}(i)$, the values $\mathcal{D}(\varnothing, i, t_i)$ and updated values of e_i and ℓ_i.

1 $U_k \leftarrow \varnothing$, $V_k \leftarrow \varnothing$ for $k = 2 \ldots |A|$;
2 **for** *all $i \in A$* **do**
3      $e_i \leftarrow \max[e_i, t_i - z_{UB}/g_i]$;
4      $\ell_i \leftarrow \min[\ell_i, t_i + z_{UB}/h_i]$;
5   Determine $\mathcal{F}(i)$;
6   **for** *all $k \in \mathcal{F}(i)$* **do** $U_k \leftarrow U_k \cup \{i\}$;
7      $k \leftarrow$ the largest element of $\mathcal{F}(i)$;
8      $V_k \leftarrow V_k \cup \{i\}$;
9   **if** $1 \in \mathcal{F}(i)$ **then**
10   **for** *all $t \in [e_i, t_i]$* **do**
11             $\mathcal{D}(\varnothing, i, t) = c(i, t)$;
12   Calculate completion bound $\mathcal{C}(\varnothing, i, t)$;
13   **if** $\mathcal{D}(\varnothing, i, t) + \mathcal{C}(\varnothing, i, t) \geq z_{UB}$ **then** $\mathcal{D}(\varnothing, i, t) = \infty$
14   **end**
15   **end**
16 **end**

---

4 Computational Experiments

For the purpose of performing computational experiments we consider three different sets of instances. The first set (presented in [1]) consists of 12 real air traffic instances of 60 aircraft each from Milan Linate Airport, the largest single runway airport in Italy. The second set consists of 7 instances of 10–44 aircraft presented in [4]. These two sets are used to compare the performances

---

Algoritmo 2. Dynamic programming algorithm

Input : An input set A of aircraft with updated time windows, an upper
bound z_{UB}, the sets U_k, V_k and $\mathcal{F}(i)$, and the values $\mathcal{D}(\varnothing, i, t_i)$.

Output: The optimal arrival/departure time x_i of each aircraft $i \in A$.

```
 1 for k = 2 ... |A| do
 2 │   for all states (S', i', t') where |S'| = k - 2 and D(S', i', t') ≠ ∞ do
 3 │   │   W ← A\S' ∩ U_k;
 4 │   │   if |W ∩ V_k| > 1 then W ← ∅;
 5 │   │   else if |W ∩ V_k| = 1 then W ← W ∩ V_k;
 6 │   │   for all i ∈ W do
 7 │   │   │   t_min = max(t' + s_{i',i}, e_i);
 8 │   │   │   t_max = min(t' + s_{i',i}, t_i);
 9 │   │   │   for t_min ≤ t ≤ t_max do
10 │   │   │   │   d = D(S', i', t') + c(i, t);
11 │   │   │   │   S = S' ∪ {i'};
12 │   │   │   │   if (S, i, t) is considered for the first time then
13 │   │   │   │   │   Calculate completion bound C(S, i, t);
14 │   │   │   │   end
15 │   │   │   │   if d < D(S, i, t) then
16 │   │   │   │   │   D(S, i, t) = d;
17 │   │   │   │   │   if D(S, i, t) + C(S, i, t) ≥ z_UB then D(S, i, t) = ∞
18 │   │   │   │   end
19 │   │   │   end
20 │   │   end
21 │   end
22 end
```

---

of the dynamic programming algorithm and the MILP formulation. A direct comparison with the performance of the dynamic programming algorithm proposed in [3] for discrete time models is not possible, since in the computational section of that paper only results are presented on the version of the ASP which minimizes makespan. The third set consists of merging the 12 instances from Milan Linate Airport into a large sequence of 173 aircraft, and sampling from this sequence instances consisting of 100–160 aircraft. The third set is used to show the effects of the heuristic upper bounds and the completion bound on the performance of the dynamic programming algorithm, and also to show the capability of this algorithm to solve large instances in very short computation times. These instances are larger than the realistic cases, and we also use a larger maximum shift value than that used in practice in order to enable the effects of using the various bounds to be more clear (since the computation times are small for realistic instances). In what follows $p_i^- = p_i^+ = p$ for all $i \in A$, as is customary in practice. All code was written in C in Linux Ubuntu on executed on an Intel(R) Core(TM)2 Duo CPU clocked at 2.33 GHz with 2 GB of RAM, and version 12.5 of CPLEX was used to solve the MILP. All computation times are presented in seconds.

Table 1. Results on the average time required by the exact dynamic programming algorithm using various bounding procedures. In each row the left-most value is given in seconds, while the preceding values in that row is given as a percentage of the first value. For each value of p and $|A|$, five instances were considered, and in cases where the algorithm exceeded the memory limit no result is given.

p	$|A|$	No pruning	No completion bound		With completion bound	
			FCFS	Tabu search	FCFS	Tabu search
4	100	1.73	96 %	116 %	86 %	76 %
	110	2.06	96 %	116 %	89 %	76 %
	120	2.62	96 %	114 %	85 %	72 %
	130	3.00	95 %	115 %	87 %	74 %
	140	3.79	90 %	108 %	82 %	68 %
	150	4.36	90 %	109 %	82 %	69 %
	160	4.68	95 %	114 %	86 %	74 %
5	100	9.65	94 %	90 %	82 %	47 %
	110	11.49	94 %	91 %	85 %	48 %
	120	14.45	95 %	91 %	83 %	46 %
	130	16.60	95 %	91 %	84 %	48 %
	140	19.81	94 %	90 %	83 %	47 %
	150		21.41	96 %	88 %	50 %
	160			23.01	93 %	53 %
6	100				31.39	65 %
	110					25.96
	120					29.75

In Tables 1 and 2 a comparison is made of the computation times and the number of states required by the dynamic programming algorithm when using various bounding procedures compared to the basic configuration where no pruning of states is done. For a heuristic upper bound we used either the FCFS value, or the value of the solution found by the tabu search with a tabu tenure of 50 and 1000 iterations. Results are also given with and without use of the completion bound. Furthermore, a memory limit of 1 GB was imposed, and where this limit was exceeded no result is given. For $p = 6$ and $n \geq 120$ all the configurations of the dynamic programming algorithm exceeded the memory limit.

It can be seen from these results that in all cases the use of the completion bound improves the performance of the dynamic programming algorithm. Moreover, if the completion bound is not used, using the tabu search value as an upper bound slightly reduces the number of states, but does not reduce, on average, the computation time. However, using the tabu search and the completion bound together significantly outperforms the other configurations, especially for

Table 2. Results on the average number of states required by the exact dynamic programming algorithm using various bounding procedures. In each row the left-most value is given as a number of states, while the preceding values in that row is given as a percentage of the first value. For each value of p and $|A|$, five instances were considered, and in cases where the algorithm exceeded the memory limit no result is given.

p	$|A|$	No pruning	No completion bound		With completion bound	
			FCFS	Tabu search	FCFS	Tabu search
4	100	182795.8	96 %	89 %	75 %	41 %
	110	201753.2	97 %	90 %	78 %	43 %
	120	235135.4	96 %	90 %	75 %	40 %
	130	252977.6	96 %	90 %	78 %	42 %
	140	285062.6	96 %	89 %	77 %	41 %
	150	308930.4	96 %	90 %	77 %	41 %
	160	328962.8	96 %	89 %	77 %	42 %
5	100	867619.6	96 %	87 %	73 %	36 %
	110	948986.8	97 %	89 %	77 %	39 %
	120	1116559.8	96 %	88 %	74 %	36 %
	130	1200858.2	96 %	88 %	76 %	39 %
	140	1349463.6	96 %	88 %	75 %	37 %
	150		1399297.4	92 %	79 %	40 %
	160			1358824.0	87 %	51 %
6	100				2098792.0	66 %
	110					1656792.8
	120					1725284.0

the largest instances. In fact, this configuration avoids exceeding the memory limit in more cases that the other configurations.

As mentioned before, realistic values of p are in the range from 1 to 3, and so for the comparison of the two exact approaches these values are used. In Table 3 the required computation times for the dynamic programming algorithm and the MILP formulation are given for each of the real-world instances we consider.

For the instances from [4] (BKSA) both algorithms require low computation times. This behaviour may be attributed to the fact that these instances consist of relatively small numbers of aircraft. However, in the majority of these cases the dynamic programming algorithm is able to outperform the MILP formulation. For the real-world Milan Linate Airport instances from [1] (FPT), the dynamic programming algorithm outperforms the MILP formulation by a significant margin. The computation times vary much for the MILP formulation, however for the dynamic programming algorithm the times are consistent, and increases as the maximum shift value increases. This is to be expected as the

Table 3. The required computation times for the dynamic programming (DP) algorithm and the MILP formulation where $p = 1, 2, 3$, for the 7 instances from [4] (BKSA) and the 12 instances from [1] (FPT). For the MILP formulation a time limit of one hour was imposed, and the time limit was reached in these two cases indicated by 'TL.'

| Instance | $|A|$ | $p = 1$ | | $p = 2$ | | $p = 3$ | |
		DP	MILP	DP	MILP	DP	MILP
BKSA1	10	0.00	0.01	0.01	0.02	0.03	0.02
BKSA2	15	0.02	0.02	0.03	0.13	0.13	0.16
BKSA3	20	0.04	0.02	0.17	0.04	0.52	0.03
BKSA4	20	0.03	0.11	0.12	0.91	0.39	3.73
BKSA5	20	0.03	0.13	0.11	4.92	0.31	23.93
BKSA6	30	0.00	0.00	0.01	0.01	0.03	0.01
BKSA7	44	0.03	0.20	0.08	0.27	0.08	0.27
FPT1	60	0.00	0.85	0.08	257.44	0.17	27.59
FPT2	60	0.00	1.12	0.07	TL	0.16	TL
FPT3	60	0.00	0.74	0.07	866.89	0.17	1153.95
FPT4	60	0.00	5.57	0.08	2396.44	0.17	324.79
FPT5	60	0.00	1.90	0.08	175.61	0.18	496.18
FPT6	60	0.00	0.48	0.08	9.73	0.17	12.58
FPT7	60	0.00	0.55	0.08	2.33	0.17	12.80
FPT8	60	0.00	0.37	0.08	2.28	0.17	2.51
FPT9	60	0.00	0.26	0.08	1.53	0.18	3.25
FPT10	60	0.00	0.53	0.08	1.33	0.18	3.90
FPT11	60	0.00	0.64	0.08	11.60	0.18	107.23
FPT12	60	0.00	0.14	0.08	0.91	0.18	1.88

complexity of the dynamic programming approach increases as the maximum shift value increases.

5   Conclusion

In this paper we proposed a dynamic programming algorithm which is capable of solving real-world air traffic instances in very short computation times, making it suitable for use in a real-time setting. We were able to utilize heuristic upper bounds and a completion lower bound in order to speed up the performance of the dynamic programming algorithm, and showed that this enabled us to solve very large instances in short computation times. Moreover, we compared the performance of the algorithm to that of a MILP formulation and concluded that it is a more effective approach to use for the real-world instances we considered.

For future work it could be fruitful to investigate, both from a theoretical and computational point of view, the effects on the dynamic programming algorithm of larger time windows, larger maximum shift values, and also maximum shift values that vary among the aircraft.

References

1. Furini, F., Persiani, C.A., Toth, P.: Aircraft sequencing problems via a rolling horizon algorithm. In: Mahjoub, A.R., Markakis, V., Milis, I., Paschos, V.T. (eds.) ISCO 2012. LNCS, vol. 7422, pp. 273–284. Springer, Heidelberg (2012)
2. Dear, R.G., Sherif, Y.S.: The dynamic scheduling of aircraft in high density terminal areas. Microelectron. Reliab. **29**(5), 743–749 (1989)
3. Balakrishnan, H., Chandran, B.G.: Algorithms for scheduling runway operations under constrained position shifting. Oper. Res. **58**(6), 1650–1665 (2010)
4. Beasley, J., Krishnamoorthy, M., Sharaiha, Y., Abramson, D.: Scheduling aircraft landing – the static case. Transp. Sci. **34**, 180–197 (2000)
5. de Neufville, R., Odoni, A.R.: Airport Systems: Planning, Design and Management. McGraw-Hill, New York (2003)
6. Bennel, J., Mesgarpour, M., Potts, C.: Airport runway scheduling. 4OR **9**, 115–138 (2011)
7. Abela, J., Abramson, D., Krishnamoorthy, M., De Silva, A., Mills, G.: Computing optimal schedules for landing aircraft. In: Proceeding 12th National ASOR Conference, Adelaide, Australia, pp. 71–90 (1993)
8. Beasley, J., Pinol, H.: Scatter search and bionomic algorithms for the aircraft landing problem. Eur. J. Oper. Res. **127**(2), 439–462 (2006)
9. Psaraftis, H.N.: A dynamic programming approach for sequencing groups of identical jobs. Oper. Res. **28**(6), 1347–1359 (1980)
10. Trivizas, D.A.: Optimal scheduling with maximum position shift (MPS) constraints: a runway scheduling application. J. Navig. **51**(2), 250–266 (1998)

Decomposition Algorithm for the Single Machine Scheduling Polytope

Ruben Hoeksma, Bodo Manthey, and Marc Uetz[✉]

Department of Applied Mathematics, University of Twente,
Enschede, The Netherlands
{r.p.hoeksma,b.manthey,m.uetz}@utwente.nl

Abstract. Given an n-vector p of processing times of jobs, the single machine scheduling polytope C arises as the convex hull of completion times of jobs when these are scheduled without idle time on a single machine. Given a point $x \in C$, Carathéodory's theorem implies that x can be written as convex combination of at most n vertices of C. We show that this convex combination can be computed from x and p in time $O(n^2)$, which is linear in the naive encoding of the output. We obtain this result using essentially two ingredients. First, we build on the fact that the scheduling polytope is a zonotope. Therefore, all of its faces are centrally symmetric. Second, instead of C, we consider the polytope Q of half times and its barycentric subdivision. We show that the subpolytopes of this barycentric subdivison of Q have a simple, linear description. The final decomposition algorithm is in fact an implementation of an algorithm proposed by Grötschel, Lovász, and Schrijver applied to one of these subpolytopes.

1   Introduction and Contribution

Given any point x in a d-dimensional polytope P, Carathéodory's theorem implies that x can be written as convex combination of at most $d + 1$ vertices of P. We are interested in an algorithmic version of Carathéodory's theorem for a specific polytope, namely the polytope C that arises as the convex hull of completion times of n jobs when these are sequenced non-preemptively without idle time on a single machine. More specifically, we are given a vector of positive processing times $p \in \mathbb{R}_+^n$ and some $x \in C$, and our goal is to compute an explicit representation of x by at most n vertices v^i of C, such that $x = \sum_i \lambda_i v^i$ for $\lambda_i \geq 0$ for all i and $\sum_i \lambda_i = 1$. We refer to this problem as *decomposition problem*.

The polytope C, also known as the single machine scheduling polytope, is well understood [13]. In particular, it is known to be a polymatroid, and the separation problem for C can be solved in $O(n \log n)$ time. Therefore, the existence of a polynomial time decomposition algorithm follows from the ellipsoid method [5]. A generic approach to compute a decomposition has been described

---

For a full version of this extended abstract including all proofs, see Ref. [7].

© Springer International Publishing Switzerland 2014
P. Fouilhoux et al. (Eds.): ISCO 2014, LNCS 8596, pp. 280–291, 2014.
DOI: 10.1007/978-3-319-09174-7_24

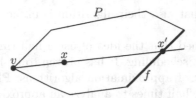

Fig. 1. Illustration of the decomposition algorithm by Grötschel, Lovász, and Schrijver. From some vertex $v \in P$, extend a half-line from v in direction $x - v$ until it intersects a lower dimensional face f of P in a point x'. The point x can be written as a convex combination of v and x'. Recurse with this face f and the intersection point x' to obtain a convex combination of vertices of f that yields x'.

by Grötschel, Lovász, and Schrijver [6]. We call this the *GLS method* in the following. Figure 1 depicts the idea behind the GLS method.

Yet, already Cunningham [3] remarked that it is interesting to find efficient combinatorial decomposition algorithms for specific polymatroids and that it is in general not straightforward to do so, even if the underlying optimization problem is well understood and can be solved efficiently. Decomposition of feasible points into vertices also plays an important role in algorithms for submodular function minimization, starting with work by Cunningham [2,3] and including the strongly polynomial time algorithms of Schrijver [15] and Iwata et al. [9].

Apart from this general motivation, the decomposition problem arises in the design of efficient mechanisms for optimization problems in private information settings [1,8]. In such problems, feasible points correspond to so-called interim (expected) allocations, which are computed as solutions to linear programming formulations. The decomposition is needed to translate these interim allocations back into actual implementations of the mechanism.

Let us sketch the state-of-the-art of the question to decompose a given point in the scheduling polytope C into vertices of C. An $O(n^9)$ algorithm follows directly from work by Fonlupt and Skoda [4] on the intersection of a line with a (general) polymatroid using the GLS method. However, a closer look reveals that an $O(n^3 \log n)$ implementation is also possible [8]. Still, this result is unsatisfactory in the following sense. For the permutahedron, Yasutake et al. suggested an $O(n^2)$ decomposition algorithm [16]. The permutahedron is precisely the single machine scheduling polytope for the special case where all processing times are 1. Hence, the natural question is if their $O(n^2)$ algorithm can be generalized to the scheduling polytope.

In this paper, we answer this question in the affirmative. Essentially, we show two things. First, we show that there is an $O(n^2)$ decomposition algorithm for the single machine scheduling polytope. The core of our algorithm remains the GLS method. However, we apply the algorithm to a specific subpolytope of a polyhedral subdivision of the polytope Q of *half times*, i.e., Q is obtained by shifting C by half the processing times of the jobs: $Q = C - p/2$. Second, we augment the algorithm by Yasutake et al. [16] by a simple, geometric interpretation. In particular, this shows that also their algorithm is in fact an implementation of the GLS method.

It should be mentioned that the idea of using half times, also referred to as midpoints, is not new in scheduling. It has proven to be helpful particularly for the design and analysis of approximation algorithms. Phillips et al. [12] were probably the first to use half times to analyze an approximation algorithm, and Munier et al. [11] were the first to use half times explicitly in the design of approximation algorithms.

The crucial ingredient to get our results is to exploit that the scheduling polytope is a zonotope. This means that all its faces are centrally symmetric. As each of the centers of a given face has a representation by two vertices, it suffices to decompose a given point into (certain) centers. To decompose a given point into centers, we consider the polyhedral subdivision of the scheduling polytope that is induced by these centers. This is also called a barycentric subdivision [10]. For the polytope of half times, we can show that this subdivision has a simple, linear description, which we can exploit algorithmically.

We believe that our results are interesting due to the following reasons. First, consider applying the GLS method directly to the scheduling polytope. In order to obtain an $O(n^2)$ implementation, one would have to compute a face f and the intersection point of the halfline through v and x with f in $O(n)$ time in each iteration. We do not see how to do this. Second, considering a naive, unit-cost encoding of the output, the $O(n^2)$ implementation is only linear in the output size. Third, our structural results shed new light on a well-studied object in polyhedral combinatorics, the single machine scheduling polytope.

2   The Single Machine Scheduling Polytope

Consider a set N of n jobs. Job $j \in N$ has processing time $p_j \in \mathbb{R}_+$. Non-preemptive schedules of jobs on a single machine are usually represented by vectors of either starting times s_j or completion times c_j. For any non-preemptive schedule without idle time, the starting time of job j is $s_j = \sum_{k<j} p_k$, where $k < j$ denotes that job k is scheduled before job j. Then the completion time of job j is $c_j = s_j + p_j$. For all sets $K \subseteq N$ of jobs, let

$$g(K) := \frac{1}{2} \left(\sum_{j \in K} p_j \right)^2 .$$

Queyranne [13] defined the single machine scheduling polytope using completion time vectors c and showed that it is described by the following system of inequalities:

$$\sum_{j \in K} c_j p_j \geq g(K) + \frac{1}{2} \sum_{j \in K} p_j^2 \qquad \qquad \text{for all } K \subset N \text{ and} \qquad (1)$$

$$\sum_{j \in N} c_j p_j = g(N) + \frac{1}{2} \sum_{j \in N} p_j^2 . \qquad (2)$$

If $p_j > 0$ for all $j \in N$, none of these inequalities is redundant, and the dimension is $n-1$ [13]. Note that, for the degenerate case where $p_k = 0$ for some jobs k, we would have to add constraints $0 \le c_k \le \sum_{j \in N} p_j$ in order to describe the convex hull of schedules. However, for all algorithmic purposes that we can think of, this degenerate case does not add anything interesting, since we can simply eliminate such jobs and reintroduce them afterwards. In particular, this is true for the problem we address here. Thus, we assume that $p_j > 0$ for all jobs $j \in N$ from now on.

In this paper, it is convenient to represent a schedule by h, the vector of half times, instead of by a vector of completion times. The *half time* of a job is the time at which the job has finished half of its processing. We have

$$h_j = s_j + \frac{1}{2}p_j = c_j - \frac{1}{2}p_j \ .$$

Equivalent to Queyranne's description, the single machine scheduling polytope in half times is completely described by

$$\sum_{j \in K} h_j p_j \ge g(K) \qquad \qquad \text{for all } K \subset N \qquad \qquad (3)$$

$$\sum_{j \in N} h_j p_j = g(N) \ , \qquad \qquad (4)$$

which is simply the scheduling polytope in completion times shifted by the vector $-p/2$. Let Q denote the single machine scheduling polytope in half times. The polytope Q is the set of all $h \in \mathbb{R}^n$ that fulfil (3) and (4).

The face lattice of the single machine scheduling polytope is well understood [13]. Every $(n-k)$-dimensional face f of Q corresponds one-to-one with an ordered partition of N into k sets. With an ordered partition, we mean a tuple $(S_1, \ldots, S_k)$ with $S_i \cap S_j = \emptyset$ for all $i \ne j$, $i, j \in \{1, \ldots, k\}$, and $\bigcup_{i=1}^{k} S_i = N$. The intended meaning is that inequalities (3) are tight for all $T_i := S_1 \cup \ldots \cup S_i$, $i \in \{1, \ldots, k\}$. This corresponds to convex combinations of all schedules where jobs in T_i are scheduled before jobs in $N \setminus T_i$, for all $i \in \{1, \ldots, k\}$. The schedules correspond to the ordered partitions $(\{\sigma(1)\}, \ldots, \{\sigma(n)\})$ for all permutations σ. Each such ordered partition corresponds to a vertex of Q as follows: let $(\{\sigma(1)\}, \ldots, \{\sigma(n)\})$ be an ordered partition and v the vertex it corresponds to, then

$$v_{\sigma(j)} = \frac{1}{2}p_{\sigma(j)} + \sum_{i=1}^{j-1} p_{\sigma(i)} \qquad \text{for all } j \in N \ . \qquad (5)$$

3   Zonotopes

In this paper, we make heavy use of the fact that the scheduling polytope is a zonotope.

Definition 1 (centrally symmetric polytope, zonotope). *Let $P \subseteq \mathbb{R}^n$ be a polytope. P is* centrally symmetric *if it has a center $c \in P$, such that $c + x \in P$ if and only if $c - x \in P$. If all faces of P are centrally symmetric, then P is called a* zonotope.

An equivalent definition of centrally symmetric is that there is a center $c \in P$ such that for all $x \in P$ also $2c - x \in P$.

Also zonotopes have alternative definitions. They are exactly the images of (higher-dimensional) hypercubes under affine projections, and they are exactly the Minkowski sum of line segments [17]. The standard textbook [17] example for zonotopes is the permutahedron, which is the scheduling polytope in completion times when all processing times are 1.

The scheduling polytope with arbitrary processing times is a zonotope, too. This can be seen in several ways. For example, the scheduling polytope can be obtained as affine transformation from a hypercube in dimension $\binom{n}{2}$ via linear ordering variables as follows [14, Theorem 4.1]: let the variable δ_{ij} for $i, j \in N$, $i < j$ be ordering variables. The intended meaning that $\delta_{ij} = 1$ if and only if job i is processed before job j. Then the vertices of this $\binom{n}{2}$-dimensional hypercube correspond one-to-one with all permutations, and the halftime h_j of any job j can be computed by

$$h_j = \frac{1}{2}p_j + \sum_{i<j} \delta_{ij}p_i + \sum_{i>j}(1 - \delta_{ji})p_i \ .$$

We summarize this brief discussion with the following Theorem.

Theorem 2 (Queyranne and Schulz [14, Theorem 4.1]). *The scheduling polytope is a zonotope.*

With respect to the centers of the faces of the scheduling polytope in halftimes, we have the following lemma.

Lemma 3. *Consider an arbitrary face f of Q, defined by the ordered partition $(S_1, \ldots, S_k)$, then the barycenter (or center of mass) $c(f)$ of f is given by*

$$c(f)_j = \sum_{\ell=1}^{i-1} \sum_{h \in S_\ell} p_h + \frac{1}{2} \sum_{h \in S_i} p_h \quad \text{for all} \quad j \in S_i \ . \tag{6}$$

Given that a face f of Q corresponds to some ordered partition $(S_1, \ldots, S_k)$, this is not difficult to verify. For a formal proof, we refer to the full version of this paper [7]. In particular, observe that all $j \in S_i$ have the same value, and the center of Q is the point c where all values c_i coincide, i.e., $c_1 = \ldots = c_n$. Note that this is no longer true if we consider the scheduling polytope in start or completion times. The property that all faces of a zonotope are centrally symmetric, as well as the simple description of these centers by Lemma 3, will be important for the design of the decomposition algorithm in Sect. 5.

4   Barycentric Subdivision

Consider the following, polyhedral subdivision of the scheduling polytope Q. For any vertex v of Q, define polytope Q_v^c as the convex hull of all barycenters $c(f)$ of faces f that contain v:

$$Q_v^c := \text{conv}\{c(f) \mid v \in f\} \ .$$

Then we have $Q = \bigcup_v Q_v^c$. By construction, v is the only vertex of Q that is also a vertex of Q_v^c. The subdivision thus obtained is also known as *barycentric subdivision* [10].

Another polyhedral subdivision of the scheduling polytope Q is obtained by subdividing the polytope according to orders as follows.

Definition 4. *Let $P \subseteq \mathbb{R}^n$ be a polytope. We define a relation $\sim$ on P as follows: for two points $x, y \in P$, we have $x \sim y$ if there exists a permutation $\sigma : \{1, \dots, n\} \to \{1, \dots, n\}$ such that both $x_{\sigma(1)} \leq \dots \leq x_{\sigma(n)}$ and $y_{\sigma(1)} \leq \dots \leq y_{\sigma(n)}$.*

Based on this definition, define for any vertex $v \in Q$ the polytope

$$Q_v^\sigma := \{x \in Q \mid x \sim v\} \ .$$

Because every permutation σ is represented by a vertex of Q, we have $Q = \bigcup_v Q_v^\sigma$, and v is the only vertex of Q that is also a vertex of Q_v^σ.

The following two lemmas encode the core and geometric intuition behind the decomposition algorithm that we develop in Sect. 5. They show that the two above polyhedral subdivisions are in fact equivalent. Thereby, we obtain an explicit description of the barycentric subdivision in terms of vertices and facets.

Lemma 5. *Let Q be the single machine scheduling polytope in half times, let v be an arbitrary vertex of Q and let σ denote a permutation such that $v_{\sigma(1)} \leq \dots \leq v_{\sigma(n)}$. Then Q_v^σ has the following, linear description:*

$$h_{\sigma(j)} \leq h_{\sigma(j+1)} \qquad \text{for all } j \in \{1, \dots, n-1\} \ , \qquad (7)$$

$$\sum_{j=1}^{k} h_{\sigma(j)} p_{\sigma(j)} \geq \frac{1}{2} \left(\sum_{j=1}^{k} p_{\sigma(j)} \right)^2 \qquad \text{for all } k \in \{1, \dots, n-1\} \ , \text{ and} \qquad (8)$$

$$\sum_{j \in N} h_j p_j = \frac{1}{2} \left(\sum_{j \in N} p_j \right)^2 \ . \qquad (9)$$

Proof. Since $Q_v^\sigma \subseteq Q$, (8) and (9) are satisfied for every point in Q_v^σ. Since σ is the only permutation with $v_{\sigma(1)} \leq \dots \leq v_{\sigma(n)}$, we have that h satisfies (7) if $h \sim v$. Therefore, (7) holds for any point in Q_v^σ.

It remains to be shown that (7), (8), and (9) imply $h \in Q_v^\sigma$. Let h satisfy (7), (8) and (9). For simplicity of notation and without loss of generality, let all vectors be sorted such that $h_i \leq h_j$ if and only if $i \leq j$. Then, for each j, we have

$$\left(\sum_{i=1}^{j} p_i\right) h_j \geq \sum_{i=1}^{j} p_i h_i \geq \frac{1}{2}\left(\sum_{i=1}^{j} p_i\right)^2 .$$

Thus, $h_j \geq \frac{1}{2}\sum_{i=1}^{j} p_i$ for all j. Now suppose h satisfies (7), (8), and (9), but $h \notin Q$. Then there is a set K of minimal cardinality, such that (3) is not satisfied. This means that

$$\sum_{i \in K} p_i h_i < \frac{1}{2}\left(\sum_{i \in K} p_i\right)^2 .$$

But then, for $j = \max_{k \in K} k$, we have

$$\sum_{i \in K \setminus \{j\}} p_i h_i = \sum_{i \in K} p_i h_i - p_j h_j < \frac{1}{2}\left(\sum_{i \in K} p_i\right)^2 - p_j h_j$$

$$\leq \frac{1}{2}\left(\sum_{i \in K} p_i\right)^2 - p_j \frac{1}{2}\left(\sum_{i=1}^{j} p_i\right)$$

$$\leq \frac{1}{2}\left(\sum_{i \in K} p_i\right)^2 - p_j \frac{1}{2}\left(\sum_{i \in K} p_i\right) = \frac{1}{2}\left(\sum_{i \in K \setminus \{j\}} p_i\right)^2 .$$

This contradicts that K is a set of minimal cardinality that does not satisfy (3). So (7), (8), and (9) imply $h \in Q$.

Now suppose $h \in Q \setminus Q_v^\sigma$, then $h \in Q_{v'}^\sigma$ for some other vertex $v' \in Q$, which would imply that (7) is not valid for h. Hence, $h \in Q_v^\sigma$.   □

Lemma 6. *Let Q be the single machine scheduling polytope in half times. Then, for all vertices v of Q, we have*

$$Q_v^c = Q_v^\sigma .$$

Proof. Lemma 3 implies that the vertices of Q_v^c are given by (6) for all $f \ni v$. From (6), we have $q \sim v$ for any vertex q of Q_v^c. It follows that $Q_v^c \subseteq Q_v^\sigma$.

Now, by Lemma 5, any vertex of Q_v^σ is obtained by having $n-1$ tight constraints among (7) and (8). Consider any such vertex q of Q_v^σ.

Let $\ell \in \{1, \ldots, n-1\}$. If (8) is tight for q for $k = \ell$, then (7) cannot be tight for q for $j = \ell$. This is because if (8) is tight for q and $k = \ell$, then jobs $1, \ldots, \ell$ are scheduled before jobs $\ell+1, \ldots, n$. Therefore,

$$q_{\ell+1} \geq \frac{1}{2}p_{\ell+1} + \sum_{j=1}^{\ell} p_j$$

and

$$q_\ell \leq \frac{1}{2}p_\ell + \sum_{j=1}^{\ell-1} p_j .$$

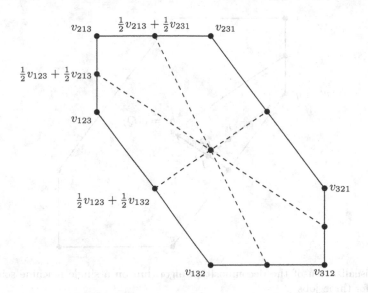

Fig. 2. Barycentric subdivision of a scheduling polytope with three jobs. v_{ijk} denotes the vertex corresponding to the order i, j, k

Thus, $q_\ell < q_{\ell+1}$ since all processing times are assumed to be positive. This implies that for any $\ell \in \{1, \ldots, n-1\}$, we have that q satisfies exactly one of the following: (8) is tight for $k = \ell$ or (7) tight for $j = \ell$. The inequalities (8) that are tight for q induce an ordered partition $(S_1, \ldots, S_k)$ that corresponds to a face $f \ni v$. The inequalities (7) that are tight for q ensure that $q_j = q_{j+1}$ for all $j \in S_i$ and any $i \in \{1, \ldots, k\}$.

It follows that $q = c(f)$ and, thus, q is a vertex of Q_v^c. Since this holds for any vertex of Q_v^σ, we have $Q_v^\sigma \subseteq Q_v^c$. Thus, $Q_v^\sigma = Q_v^c$.   □

For simplicity of notation, we define $Q_v := Q_v^c \ (= Q_v^\sigma)$.

Figure 2 illustrates the barycentric subdivision of the scheduling polytope. It shows the scheduling polytope for three jobs together with its barycentric subdivision (indicated by dashed lines). The subpolytope containing vertex v_{213} contains all vectors $h \in Q$ for which $h_2 \le h_1 \le h_3$. Its vertices are v_{213}, and all centers of faces on which v_{213} lies. Its facets are defined by $h_1 p_1 + h_2 p_2 + h_3 p_3 = (p_1 + p_2 + p_3)^2$ together with one of the following equalities:

$$h_1 p_1 + h_2 p_2 = (p_1 + p_2)^2 \ ,$$
$$h_2 p_2 = (p_2)^2 \ ,$$
$$h_2 = h_1 \ ,$$
$$h_3 = h_1 \ .$$

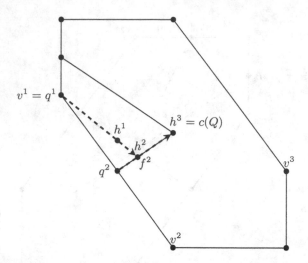

Fig. 3. Visualization of the decomposition algorithm on a single machine scheduling polytope for three jobs

5 Decomposition Algorithm for the Single Machine Scheduling Polytope

Based on Lemma 5, we next develop a decomposition algorithm for the schedul-ing polytope that runs in time $O(n^2)$. This algorithm can be seen as a gen-eralization of an algorithm recently proposed by Yasutake et al. [16] for the permutahedron. We argue here that this algorithm is in fact an application of the GLS method [6, Theorem 6.5.11]. Before giving the pseudo code for the decomposition algorithm, we describe the high level idea.

We know that any point $h \in Q$ lies in a subpolytope Q_v of the barycentric subdivision of Q, namely for a vertex v for which $v \sim h$ according to Definition 4.[1] Moreover, Q_v is described by inequalities (7) and (8), and the vertices of Q_v consist of the points $\frac{v+v'}{2}$ for all vertices v' of Q. This means that a decomposition of h into vertices of Q_v also yields a decomposition into vertices of Q.

The idea of the algorithm is as follows: We find a decomposition of h into vertices of Q_v by using the GLS method [6, Theorem 6.5.11]. The idea of this algorithm is illustrated in Fig. 3: Given $h = h^1 \in Q_v$ (we have $v = v^1$), we extend the difference vector $h^1 - v^1$ towards the intersection with a lower dimensional face of Q_v (this will be a facet of Q_v, unless we accidentally hit a face of even lower dimension). Then recurse with this intersection point and the face on which it lies. To arrive at the claimed computation time, it is crucial that both the intersection point and the face(t) on which it lies can be computed in time $O(n)$. This is indeed possible because of Lemma 5. As the number of iterations is bounded by the dimension of Q_v, which is equal to the dimension of Q, this

---

[1] In case of ties, h lies on the intersection of several of such subpolytopes, namely those corresponding to vertices v with $v \sim h$. We can break such ties arbitrarily.

gives an $O(n^2)$ implementation. Finally, by the fact that all vertices of Q_v can be written as $\frac{v+v'}{2}$ for vertices v' of Q, we obtain a decomposition of h into at most n vertices of Q.

In order to describe the technical details of the algorithm, we use the following notation.

v: vertex of Q corresponding to the permutation $1, 2, \ldots, n$; we have $v = v^1$;
J^t: set of indices associated with a face f^t of Q_v;
f^t: face of Q_v associated with J^t such that $x_j = x_{j+1}$ for all $x \in f^t$ and all $j \in \{1, \ldots, n-1\} \setminus J^t$;
q^t: vertex of f^t;
v^t: vertex of Q such that $q^t = \frac{1}{2}(v + v^t)$;
h^t: point in f^t;
$\tilde{\kappa}_t$: scalar such that $h^t = \tilde{\kappa}_t q^t + (1 - \tilde{\kappa}_t)h^{t+1}$;
κ_t: scalar corresponding to q^t in the convex combination $h = \sum_t \kappa_t q^t$.
λ_t: scalar corresponding to v^t in the convex combination $h = \sum_t \lambda_t v^t$.

Moreover, for ease of notation and without loss of generality, we assume that the given point $h \in Q$ satisfies $h_1 \leq \ldots \leq h_n$.[2]

Algorithm 1. Decomposition Algorithm

---

    input : processing times p, point $h \in Q$ with $h_1 \leq \ldots \leq h_n$
    output: at most n vertices v^t of Q and coefficients $\kappa_t \in [0,1]$
1 $t := 1, h^1 := h, J^1 := \{i \in \{1, \ldots, n-1\} \mid h_i^1 < h_{i+1}^1\}$;
2 let v be the vertex with $v_1 \leq \ldots \leq v_n$;
    while $J^t \neq \emptyset$ **do**
3       $q^t := \text{VERTEX}(J^t)$;
4       $v^t := 2q^t - v$;
5       $\tilde{\kappa}_t := \min_{j \in J^t}(h_{j+1}^t - h_j^t)/(q_{j+1}^t - q_j^t)$;
6       $h^{t+1} := (h^t - \tilde{\kappa}_t q^t)/(1 - \tilde{\kappa}_t)$;
7       $J^{t+1} := \{i \in J^t \mid h_i^{t+1} < h_{i+1}^{t+1}\}$;
8       $\kappa_t := (1 - \sum_{\tau=1}^{t-1} \kappa_\tau)\tilde{\kappa}_t$;
9       $t := t + 1$;
10 $q^t := h^t$;
11 $v^t := 2q^t - v$;
12 $\kappa_t := 1 - \sum_{\tau=1}^{t-1} \kappa_\tau$;
13 $\lambda_1 := \frac{1}{2} + \frac{1}{2}\kappa_1$;
    for $\tau \in \{2, \ldots, t\}$ **do**
14      $\lambda_\tau := \frac{1}{2}\kappa_\tau$;

The subroutine $\text{VERTEX}(J^t)$ computes the vertex corresponding to the face associated with J^t as follows: Let $J^t(i)$ denote the i-th element in J^t and define

---

[2] This comes at the expense of sorting, which costs $O(n \log n)$ time and falls within the $O(n^2)$ time complexity of the proposed algorithm.

$J^t(0) = 1$. Then, for $j \in \{J^t(i), \ldots, J^t(i+1) - 1\}$, we compute

$$q_j^t = \sum_{k=1}^{J^t(i)-1} p_k + \frac{1}{2} \sum_{k=J^t(i)}^{J^t(i+1)-1} p_k.$$

Note that vertex q^t can be computed in linear time per iteration by just computing $P_i^t := \sum_{k=J^t(i)}^{J^t(i+1)-1} p_k$ for all i, in time $O(n)$. Then, $q_1^t = \frac{1}{2}P_1^t$, and for $j \in \{J^t(i), \ldots, J^t(i+1) - 1\}$ and $k \in \{J^t(i+1), \ldots, J^t(i+2) - 1\}$, the values for q^t are computed iteratively as $q_k^t = q_j^t + \frac{1}{2}(P_i^t + P_{i+1}^t)$.

Theorem 7. *For any* $h \in Q$, *Algorithm 1 computes a convex combination of vertices of* Q *for* h *in* $O(n^2)$ *time.*

Proof (sketch). By lines 5 and 7 of the algorithm, the cardinality of J^t strictly decreases in each iteration. None of the steps within each of at most $n - 1$ iterations takes more than $O(n)$ time, so the total computation time of the algorithm is indeed $O(n^2)$. It remains to be shown that the pseudo code given in Algorithm 1 indeed computes a correct convex combination for h. For a formal proof of this claim, we refer to the full version of this paper [7].   □

6   Conclusions

The obvious question is if our algorithm can be generalized for zonotopes. In order to do that, we would have to find explicit expressions for the centers of symmetry, as well as the faces of the resulting barycentric subdivision that is induced by these centers.

Acknowledgements. We thank Maurice Queyranne for pointing us to the paper by Yasutake et al. [16], and Marc Pfetsch and Michael Joswig for helpful remarks concerning zonotopes.

References

1. Cai, Y., Daskalakis, C., Weinberg, S.M.: Optimal multi-dimensional mechanism design: reducing revenue to welfare maximization. In: Proceedings of 53rd Annual IEEE Symposium on Foundations of Computer Science (FOCS), pp. 130–139. IEEE (2012)
2. Cunningham, W.H.: Testing membership in matroid polyhedra. J. Comb. Theory B **36**, 161–188 (1984)
3. Cunningham, W.H.: On submodular function minimization. Combinatorica **5**, 186–192 (1985)
4. Fonlupt, J., Skoda, A.: Strongly polynomial algorithm for the intersection of a line with a polymatroid. In: Cook, W., Lovász, L., Vygen, J. (eds.) Research Trends in Combinatorial Optimization, pp. 69–85. Springer, Heidelberg (2009)
5. Grötschel, M., Lovász, L., Schrijver, A.: The ellipsoid method and its consequences in combinatorial optimization. Combinatorica **1**, 169–197 (1981)

6. Grötschel, M., Lovász, L., Schrijver, A.: Geometric Algorithms and Combinatorial Optimization: Algorithms and Combinatorics, vol. 2. Springer, Heidelberg (1988)
7. Hoeksma, R., Manthey, B., Uetz, M.: Decomposition algorithm for the single machine scheduling polytope. Technical Report TR-CTIT-13-25, CTIT, University of Twente. http://eprints.eemcs.utwente.nl/24630/
8. Hoeksma, R., Uetz, M.: Two dimensional optimal mechanism design for a sequencing problem. In: Goemans, M., Correa, J. (eds.) IPCO 2013. LNCS, vol. 7801, pp. 242–253. Springer, Heidelberg (2013)
9. Iwata, S., Fleischer, L., Fujishige, S.: A combinatorial strongly polynomial time algorithm for minimizing submodular functions. J. ACM 48(4), 761–777 (2001)
10. Lee, C.W.: Subdivisions and triangulations of polytopes. In: Handbook of Discrete and Computational Geometry, chapter 17, 2nd edn. Chapman & Hall/CRC, Beca Raton (2004)
11. Munier, A., Queyranne, M., Schulz, A.S.: Approximation bounds for a general class of precedence constrained parallel machine scheduling problems. In: Bixby, R.E., Boyd, E.A., Ríos-Mercado, R.Z. (eds.) IPCO 1998. LNCS, vol. 1412, pp. 367–382. Springer, Heidelberg (1998)
12. Phillips, C., Stein, C., Wein, J.: Scheduling jobs that arrive over time. In: Sack, J.-R., Akl, S.G., Dehne, F., Santoro, N. (eds.) WADS 1995. LNCS, vol. 955, pp. 86–97. Springer, Heidelberg (1995)
13. Queyranne, M.: Structure of a simple scheduling polyhedron. Math. Program. 58(1), 263–285 (1993)
14. Queyranne, M., Schulz, A.S.: Polyhedral approaches to machine scheduling. Preprint 408-1994, TU Berlin (1994)
15. Schrijver, A.: A combinatorial algorithm minimizing submodular functions in strongly polynomial time. J. Comb. Theory B 80, 346–355 (2000)
16. Yasutake, S., Hatano, K., Kijima, S., Takimoto, E., Takeda, M.: Online linear optimization over permutations. In: Asano, T., Nakano, S., Okamoto, Y., Watanabe, O. (eds.) ISAAC 2011. LNCS, vol. 7074, pp. 534–543. Springer, Heidelberg (2011)
17. Ziegler, G.M.: Lectures on Polytopes. Graduate Texts in Mathematic, vol. 152. Springer, New York (1995)

Subexponential Fixed-Parameter Algorithms for Partial Vector Domination

Toshimasa Ishii[1], Hirotaka Ono[2]([⊠]), and Yushi Uno[3]

[1] Graduate School of Economics and Business Administration, Hokkaido University,
Sapporo 060-0809, Japan
ishii@econ.hokudai.ac.jp

[2] Department of Economic Engineering, Faculty of Economics, Kyushu University,
Fukuoka 812-8581, Japan
hirotaka@econ.kyushu-u.ac.jp

[3] Department of Mathematics and Information Sciences, Graduate School of Science,
Osaka Prefecture University, Sakai 599-8531, Japan
uno@mi.s.osakafu-u.ac.jp

Abstract. Given a graph $G = (V, E)$ of order n and an n-dimensional non-negative vector $\mathbf{d} = (d(1), d(2), \ldots, d(n))$, called demand vector, the vector domination (resp., total vector domination) is the problem of finding a minimum $S \subseteq V$ such that every vertex v in $V \setminus S$ (resp., in V) has at least $d(v)$ neighbors in S. The (total) vector domination is a generalization of many dominating set type problems, e.g., the dominating set problem, the k-tuple dominating set problem (this k is different from the solution size), and so on, and subexponential fixed-parameter algorithms with respect to solution size for apex-minor-free graphs (so for planar graphs) are known. In this paper, we consider maximization versions of the problems; that is, for a given integer k, the goal is to find an $S \subseteq V$ with size k that maximizes the total sum of satisfied demands. For these problems, we design subexponential fixed-parameter algorithms with respect to k for apex-minor-free graphs.

1 Introduction

Given a graph $G = (V, E)$ of order n and an n-dimensional non-negative vector $\mathbf{d} = (d(1), d(2), \ldots, d(n))$, called *demand vector*, the *vector domination* (resp., *total vector domination*) is the problem of finding a minimum $S \subseteq V$ such that every vertex v in $V \setminus S$ (resp., in V) has at least $d(v)$ neighbors in S. These problems were introduced by [17], and they contain many existing problems, such as the minimum dominating set and the k-tuple dominating set problem (this k is different from the solution size) [18,19], and so on. Indeed, by setting $\mathbf{d} = (1, \ldots, 1)$, the vector domination becomes the minimum dominating set forms, and by setting $\mathbf{d} = (k, \ldots, k)$, the total vector dominating set becomes

This work is partially supported by KAKENHI No. 23500022, 24700001, 24106004, 25104521 and 25106508, the Kayamori Foundation of Informational Science Advancement and The Asahi Glass Foundation.

P. Fouilhoux et al. (Eds.): ISCO 2014, LNCS 8596, pp. 292–304, 2014.
DOI: 10.1007/978-3-319-09174-7_25

the k-tuple dominating set. If in the definition of total vector domination, we replace open neighborhoods with closed ones, we get the *multiple domination*. In this paper, we sometimes refer to these problems just as *domination problems*. Table 1 of [7] summarizes how related problems are represented in the scheme of domination problems. Many variants of the basic concepts of domination and their applications have appeared in [19, 20].

Since the vector or multiple domination includes the setting of the ordinary dominating set problem, it is obviously NP-hard, and the parameterized complexity is considered. It is well-known that the ordinary dominating set problem is $W[2]$-complete; it is unlikely FPT with respect to the solution size. For treewidth as another parameter, it is shown that the vector domination problem is $W[1]$-hard with respect to treewidth [2]. This result and Courcelle's meta-theorem about MSOL [9] imply that the MSOL-based algorithm is not applicable for the vector domination problem. Very recently, a polynomial-time algorithm for the vector domination of graphs with bounded clique-width has been proposed [5]. Since $cw(G) \leq 3 \cdot 2^{tw(G)-1}$ holds where $tw(G)$ and $cw(G)$ respectively denote the treewidth and clique-width of graph G [8], their polynomial-time algorithm implies the polynomial-time solvability of the vector domination problem for graphs of bounded treewidth. Also, polynomial-time algorithms for the domination problems of graphs with bounded branch-width are independently proposed in [21, 22]. It is known that $\max\{bw(G), 2\} \leq tw(G) + 1 \leq \max\{3bw(G)/2, 2\}$, where $bw(G)$ denotes the branchwidth of graph G [24]. Due to the linear relation of treewidth and branchwidth, the above results imply the polynomial-time solvability of all the domination problems (i.e., vector domination, total vector domination and multiple domination) for graphs of bounded treewidth. Furthermore, by extending the algorithms, we can show that these for apex-minor-free graphs are subexponential fixed-parameter tractable with respect to the size of the solution k^*, that is, there is an algorithm whose running time is $2^{O(\sqrt{k^*}\log k^*)}n^{O(1)}$. Note that the class of apex-minor-free graphs includes planar graphs.

In this paper, we consider the parameterized complexity of maximization variants of the domination problems for apex-minor-free graphs. For a given integer k, the goal is to find an $S \subseteq V$ with size k that maximizes the total sum of satisfied demands. We call the maximization problems *partial domination problems*, that is, *partial vector domination*, and so on. As the parameter, we adopt the given k itself. In the case of the ordinary dominating set, there is a $2^{O(\sqrt{k^*})}n^{O(1)}$-time algorithm for planar graphs (or apex-minor-free graphs), where k^* is the size of a dominating set of G. This subexponential running time is obtained by combining the following results [16]: (1) The branchwidth of planar graph G is $O(\sqrt{k^*})$ (this also holds for apex-minor-free graphs). (2) There is an algorithm whose running time is $O(3^{3bw(G)/2}|E|)$. This idea can be extended to the domination problems, and there are subexponential fixed-parameter algorithms with respect to solution size for the domination problems of apex-minor-free graphs, as mentioned above [21, 22].

In the case of partial domination problems, however, it might be difficult to bound the branchwidth of G itself by using k, because k could be much smaller than k^*. Instead, we try to choose a special S among all the optimal solutions. Roughly speaking, in this strategy, S and its neighbors are localized so that the branchwidth of the subgraph of G induced by S and its neighbors is bounded by $O(\sqrt{k})$. Then, we can expect a similar speeding-up effect; the points become (i) how we localize S, and (ii) the design of a fixed-parameter algorithm whose exponent is linear of $bw(G)$. This scheme is proposed by [15], and they have succeeded to design subexponential fixed-parameter algorithms with respect to k for the partial dominating set and the partial vertex cover of apex-minor-free graphs.

In this paper, we present subexponential fixed-parameter algorithms with respect to k for the partial domination problems, i.e., the partial vector domination, the partial total vector domination and the partial multiple domination, of apex-minor-free graphs. Their running times are all $2^{O(\sqrt{k}\log\min\{d^*+1,k+1\})}n^{O(1)}$, where $d^* = \max\{d(v) \mid v \in V\}$. Namely, we show that the results of [15] can be extended to vector versions. This subexponential fixed-parameter running time is obtained by a similar way as above; (i) the localization of S and (ii) the design of $2^{O(bw(G)\log\min\{d^*+1,k+1\})}n^{O(1)}$-time algorithms, whose parameters of are not only $bw(G)$ but also k and d^*. It is different from the case of the partial (ordinary) dominating set, but it eventually yields the subexponential fixed-parameter running time. It should be noted that the running time for the partial vector domination or the partial multiple domination generalizes the result of the partial (ordinary) dominating set, because the dominating set is included in the case of $d^* = 1$. In fact, the partial dominating set problem is equivalent to the partial vector dominating set or the partial multiple dominating set for $\mathbf{d} = (1, 1, \ldots, 1)$.

1.1   Related Work

The dominating set problem itself is one of the most fundamental graph optimization problems, and it has been intensively and extensively studied from many points of view. In the sense that the vector or multiple domination contains the setting of not only the ordinary dominating set problem but also many variants, there are an enormous number of related studies. Here we pick some representatives up.

As a research of the domination problems from the viewpoint of the algorithm design, Cicalese, Milanic and Vaccaro gave detailed analyses of the approximability and inapproximability [6,7]. They also provided some exact polynomial-time algorithms for special classes of graphs, such as complete graphs, trees, P_4-free graphs, and threshold graphs.

For graphs with bounded treewidth (or branchwidth), the ordinary domination problems can be solved in polynomial time. As for the fixed-parameter tractability, it is known that even the ordinary dominating set problem is W[2]-complete with respect to solution size k^*; it is unlikely to be fixed-parameter tractable [14]. In contrast, it can be solved in $O(2^{11.98\sqrt{k^*}}k^* + n^3)$ time for planar graphs, that is, it is subexponential fixed-parameter tractable [13]. The subexponent part comes from the inequality $bw(G) \le 12\sqrt{k^*} + 9$. Behind the inequality, there is a unified property of parameters, called *bidimensionality*. Namely, the subexponential fixed-parameter algorithm of the dominating set for planar graphs (more precisely, H-minor-free graphs) is based on the bidimensionality.

Partial Dominating Set is the problem of maximizing the number of vertices to be dominated by using a given number k of vertices, and our problems are considered as its generalizations. In [1], it was shown that partial dominating set problem is FPT with respect to k for H-minor-free graphs. Later, [15] gives a subexponential FPT with respect to k for apex-minor-free graphs. Although partial dominating set is an example of problems to which the bidimensionality theory cannot be applied, they develop a technique to reduce an input graph so that its treewidth becomes $O(\sqrt{k})$.

For the (not partial) vector domination, a polynomial-time algorithm for graphs of bounded treewidth has been proposed very recently [5]. In [23], it is shown that the vector domination for ρ-degenerated graphs can be solved in $k^{O(\rho k^{*2})}n^{O(1)}$ time, if $d(v) > 0$ holds for $\forall v \in V$ (positive constraint). Since any planar graph is 5-degenerated, the vector domination for planar graphs is fixed-parameter tractable with respect to solution size, under the positive constraint. Furthermore, the case where $d(v)$ could be 0 for some v can be easily reduced to the positive case by using the transformation discussed in [2], with increasing the degeneracy by at most 1. It follows that the vector domination for planar graphs is FPT with respect to solution size k^*. For the total vector domination and multiple vector domination, [22] presents first polynomial time algorithms for graphs of bounded treewidth (or bounded branchwidth). The same paper presents first subexponential fixed-parameter algorithms with respect to k^* for apex-minor-free graphs for all the domination problems (i.e., vector, total vector, and multiple domination problems). See also [21].

Other than these, several generalized versions of the dominating set problem are also studied. (k, r)-center is the problem that asks the existence of set S of k vertices satisfying that for every vertex $v \in V$ there exists a vertex $u \in S$ such that the distance between u and v is at most r; $(k, 1)$-center corresponds to the ordinary dominating set. The (k, r)-center for planar graphs is shown to be fixed-parameter tractable with respect to k and r [10]. For $\sigma, \rho \subseteq \{0, 1, 2, \dots\}$ and a positive integer k, $\exists[\sigma, \rho]$-dominating set is the problem that asks the existence of set S of k vertices satisfying that $|N(v) \cap S| \in \sigma$ holds for $\forall v \in S$ and $|N(v) \cap S| \in \rho$ for $\forall v \in V \setminus S$, where $N(v)$ denotes the open neighborhood of v. If $\sigma = \{0, 1, \dots\}$ and $\rho = \{1, 2, \dots\}$, $\exists[\sigma, \rho]$-dominating set is the ordinary dominating set problem, and if $\sigma = \{0\}$ and $\rho = \{0, 1, 2, \dots\}$, it is the independent set. In [4], the parameterized complexity of $\exists[\sigma, \rho]$-dominating set with respect to treewidth is also considered.

The remainder of the paper is organized as follows. In Sect. 2, we introduce some basic notations, problem definitions and then explain the branch decomposition. Section 3 is the main part of the paper. We present fixed-parameter algorithms with respect to the branchwidth and k and show how we can localize S. Then we obtain subexponential fixed-parameter algorithms with respect to k for apex-minor-free graphs.

2   Preliminaries

A graph G is an ordered pair of its vertex set $V(G)$ and edge set $E(G)$ and is denoted by $G = (V(G), E(G))$. We assume throughout this paper that all graphs are undirected, and simple, unless otherwise stated. Therefore, an edge $e \in E(G)$ is an unordered pair of vertices u and v, and we often denote it by $e = (u, v)$. Two vertices u and v are *adjacent* if $(u, v) \in E(G)$. For a graph G, the *(open) neighborhood* of a vertex $v \in V(G)$ is the set $N_G(v) = \{u \in V(G) \mid (u, v) \in E(G)\}$, and the *closed neighborhood* of v is the set $N_G[v] = N_G(v) \cup \{v\}$.

For a graph $G = (V, E)$, let $\mathbf{d} = (d(v) \mid v \in V)$ be an n-dimensional non-negative vector called a *demand vector*, where $n = |V(G)|$. Then, we call a set $S \subseteq V$ of vertices a $\mathbf{d}$-*vector dominating set* (resp., $\mathbf{d}$-*total vector dominating set*) if $|N_G(v) \cap S| \geq d(v)$ holds for every vertex $v \in V \setminus S$ (resp., $v \in V$). We call a set $S \subseteq V$ of vertices a $\mathbf{d}$-*multiple dominating set* if $|N_G[v] \cap S| \geq d(v)$ holds for every vertex $v \in V$. We may drop $\mathbf{d}$ in these notations if there are no confusions.

In this paper, we consider the *partial domination problems* defined as follows.

Partial domination problem: Given a graph $G = (V, E)$, a demand vector $\mathbf{d}$, and an integer $k \geq 0$, find a set S of vertices with cardinality at most k which maximizes the total sum $g(S)$ of demands satisfied by S.

In the setting of the *partial vector domination problem*, the function $g(S)$ is defined as
$$g(S) = \sum_{v \in S} d(v) + \sum_{v \in V \setminus S} \min\{d(v), |N_G(v) \cap S|\},$$
since each vertex $v \in S$ contributes not only 1 to the demand for each neighbor of v but also $d(v)$ to the demand for v. On the other hand, in the setting of the *partial total vector domination problem* (resp., the *partial multiple domination problem*), each vertex $v \in S$ contributes nothing (resp., only 1) to the demand for v, and hence the function $g(S)$ is defined as
$$g(S) = \sum_{v \in V} \min\{d(v), |N_G(v) \cap S|\} \quad (\text{resp.,} \sum_{v \in V} \min\{d(v), |N_G[v] \cap S|\}).$$

Notice that for each of these domination problems, if $d(v) = 1$ for all $v \in V$, then $g(S)$ is equal to the number of vertices to be dominated by S. Hence, the partial vector domination problem and partial multiple domination problem are both generalizations of the partial dominating set problem.

A graph is called *planar* if it can be drawn in the plane without generating a crossing by two edges. A graph G has a graph H as a *minor* if a graph isomorphic to H can be obtained from G by a sequence of deleting vertices, deleting edges, or contracting edges. A graph class C is *minor-closed* if for each graph $G \in C$, all minors of G belong to C. A minor-closed class C is *H-minor-free* for a fixed graph H if $H \notin C$. It is known that a graph is planar if and only if it has neither K_5 nor $K_{3,3}$ as a minor. An *apex graph* is a graph with a vertex v such that the removal of v leaves a planar graph. A graph class is *apex-minor-free* if it excludes some fixed apex graph. Notice that a planar graph is apex-minor-free and an apex-minor-free graph is H-minor-free.

2.1   Branch Decomposition

A *branch decomposition* of a graph $G = (V, E)$ is defined as a pair $(T = (V_T, E_T), \tau)$ such that (a) T is a tree with $|E|$ leaves in which every non-leaf node has degree 3, and (b) τ is a bijection from E to the set of leaves of T. Throughout the paper, we shall use the term *node* to denote an element in V_T for distinguishing it from an element in V.

For an edge f in T, let T_f and $T \setminus T_f$ be two trees obtained from T by removing f, and E_f and $E \setminus E_f$ be two sets of edges in E such that $e \in E_f$ if and only if $\tau(e)$ is included in T_f. The *order function* $w : E(T) \rightarrow 2^V$ is defined as follows: for an edge f in T, a vertex $v \in V$ belongs to $w(f)$ if and only if there exist an edge in E_f and an edge in $E \setminus E_f$ which are both incident to v. The *width* of a branch decomposition (T, τ) is $\max\{|w(f)| \mid f \in E_T\}$, and the *branchwidth* of G, denoted by $bw(G)$, is the minimum width over all branch decompositions of G.

In general, computing the branchwidth of a given graph is NP-hard [25]. On the other hand, Bodlaender and Thilikos [3] gave a linear time algorithm which checks whether the branchwidth of a given graph is at most k or not, and if so, outputs a branch decomposition with minimum width, for any fixed k. Also, as shown in the following lemma, for H-minor-free graphs, there exists a polynomial time algorithm for computing a branch-decomposition with width $O(bw(G))$ for any fixed H.

Lemma 1. *([11]) Let G be an H-minor-free graph. Then, a branch-decomposition of G with width $O(bw(G))$ can be computed in polynomial time for any fixed H.* $\square$

This lemma follows from the constant-factor approximation algorithm of [11] for computing a tree-decomposition of minimum width in H-minor-free graphs and the property that a tree-decomposition width w can be converted to a branch-decomposition with width at most $w + 1$ [24].

Here, we introduce the following basic properties about branch decompositions, which will be utilized in the subsequent sections (see e.g., [21, 22, Lemma 2] for its proof).

Lemma 2. *Let (T, τ) be a branch decomposition of G. For a tree T, let x be a non-leaf node and $f_i = (x, x_i)$, $i = 1, 2, 3$, be an edge incident to x (note that the degree of x is three). Then, $w(f_i) \setminus (w(f_j) \cup w(f_k)) = \emptyset$ for every $\{i, j, k\} = \{1, 2, 3\}$. Hence, $w(f_i) \subseteq w(f_j) \cup w(f_k)$.*

3   Subexponential Algorithm for Apex-Minor-Free Graphs

In this section, for apex-minor-free graphs, we give subexponential fixed-parameter algorithms, parameterized by k, for the partial vector domination problem, the partial total vector domination problem, and the partial multiple domination problem; namely, we show the following theorem.

Theorem 1. *Let $G = (V, E)$ be an apex-minor-free graph with $n = |V|$. The partial vector domination problem, the partial total vector domination problem, and the partial multiple domination problem can be solved in $2^{O(\sqrt{k}\log\min\{d^*+1,k+1\})}$ $n^{O(1)}$ time, where $d^* = \max\{d(v) \mid v \in V\}$.*

As mentioned in the following lemma, all of these three problems can be solved in $2^{O(w\log\min\{d^*+1,k+1\})}n^{O(1)}$ time, if a branch decomposition of G with width w is given (note that the problems maximizing g_1, g_2, and g_3, defined in Lemma 3, correspond to the partial vector domination problem, the partial total vector domination problem, and the partial multiple domination problem, respectively, as observed in Sect. 2). These properties follow from slightly modifying the algorithms of [21,22] based on a branch decomposition of a given graph for the vector domination problem, the total vector domination problem, and the multiple domination problem.

Lemma 3. *Let $G = (V, E)$ be a graph, $\mathbf{d}$ be a demand vector, and k be a nonnegative integer. If a branch decomposition of G with width w is given, then we can compute in $2^{O(w\log\min\{d^*+1,k+1\})}n^{O(1)}$ time a set S of vertices with cardinality at most k which maximizes a function $g_i(S;G)$ on S defined as follows for each $i = 1, 2, 3$.*

(i) $g_1(S; G) = \sum_{v \in S} d(v) + \sum_{v \in V \setminus S} \min\{d(v), |N_G(v) \cap S|\}$.
(ii) $g_2(S; G) = \sum_{v \in V} \min\{d(v), |N_G(v) \cap S|\}$.
(iii) $g_3(S; G) = \sum_{v \in V} \min\{d(v), |N_G[v] \cap S|\}$.

Proof. We consider only the case of the partial vector domination problem, i.e., the case of $g_1(S; G)$ (the partial total vector domination problem and the partial multiple domination problem can be treated similarly). We here show how to modify the algorithm of [21,22] based on a branch decomposition of a given graph for computing a minimum $\mathbf{d}$-vector dominating set in $2^{O(w\log(d^*+1))}n^{O(1)}$ time so that it can be applied to the partial vector domination.

We first sketch the algorithm of [21,22] based on a branch decomposition (T, τ) of $G = (V, E)$. Let $w : E(T) \to 2^V$ be the corresponding order function. We regard T with a rooted tree by choosing a non-leaf node as a root. For an edge $f = (y_1, y_2) \in E(T)$ such that y_1 is the parent of y_2, let $E_f = \{e \in E \mid \tau(e) \in V(T(y_2))\}$, and G_f be the subgraph of G induced by E_f, where $T(x)$ denotes the subtree of T rooted at $x \in V(T)$. The algorithm proceeds bottom-up in T, while computing $A_f(\mathbf{c})$ satisfying the following (*) for all vectors $\mathbf{c} \in \{\top, 0, 1, 2, \ldots, d^*\}^{|w(f)|}$ on $w(f)$ for each edge f in T.

(*) $A_f(\mathbf{c})$ is the cardinality of a minimum set $D_f(\mathbf{c}) \subseteq V(G_f)$ such that the demand of all $v \in V(G_f) \setminus w(f)$ is satisfied by $D_f(\mathbf{c})$(i.e., $v \in D_f(\mathbf{c})$ or $|N_{G_f}(v) \cap D_f(\mathbf{c})| \geq d(v)$), every vertex $v \in w(f)$ with $c(v) = \top$ belongs to $D_f(\mathbf{c})$, and every vertex $v \in w(f)$ with $c(v) = i \in \{0, 1, \ldots, d(v)\}$ satisfies $|N_{G_f}(v) \cap D_f(\mathbf{c})| \geq d(v) - i$ if $D_f(\mathbf{c})$ exists, and $A_f(\mathbf{c}) = \infty$ otherwise. (Intuitively, $D_f(\mathbf{c})$ is a minimum vector dominating set in G_f under the assumption that the demand of every vertex in $w(f)$ satisfied by $D_f(\mathbf{c})$ is restricted to $\mathbf{c}$.)

Let f be a non-leaf edge of T and f_1 and f_2 are two edges of T which are children of f, i.e., three edges f, f_1, and f_2 are incident to a common node y and f_1 and f_2 are contained in $T(y)$. By Lemma 2, we have $w(f) \subseteq w(f_1) \cup w(f_2)$ and $w(f_1) \setminus w(f) = w(f_2) \setminus w(f)$, and the value $A_f(\mathbf{c})$ is computed based on $A_{f_1}(\mathbf{c_1})$ and $A_{f_2}(\mathbf{c_2})$ for a pair of vectors $\mathbf{c_1} \in \{\top, 0, 1, 2, \ldots, d^*\}^{|w(f_1)|}$ on $w(f_1)$ and $\mathbf{c_2} \in \{\top, 0, 1, 2, \ldots, d^*\}^{|w(f_2)|}$ on $w(f_2)$ such that for each vertex $v \in w(f_1) \setminus w(f)(= w(f_2) \setminus w(f))$, we have $c_1(v) = c_2(v) = \top$ (i.e., $v \in D_{f_1}(\mathbf{c_1}) \cap (w(f_1) \setminus w(f))$ if and only if $v \in D_{f_2}(\mathbf{c_2}) \cap (w(f_1) \setminus w(f)))$ or $c_1(v) + c_2(v) = d(v)$. Roughly speaking, since the possible number of vectors $\mathbf{c} \in \{\top, 0, 1, 2, \ldots, d^*\}^{|w(f)|}$ is at most $(d^* + 2)^{|w(f)|} = 2^{|w(f)| \log (d^* + 2)}$ for each edge $f \in E(T)$, computing $A_f(\mathbf{c})$ for all vectors $\mathbf{c}$ on $w(f)$ takes $2^{O(w \log (d^* + 1))} n^{O(1)}$ time (notice that $\log (d^* + 2) = O(\log (d^* + 1))$ by $d^* \geq 1$); the total running time turns out to be $2^{O(w \log (d^* + 1))} n^{O(1)}$ by $|V(T)| = O(|E(G)|)$.

We modify this algorithm mainly in the following points (a) and (b) so that it can be applied to the partial vector domination.

(a) For each $f \in E(T)$, we compute $A'_f(\mathbf{c}, \ell)$ satisfying the following (**) for all vectors $\mathbf{c}$ on $w(f)$ such that $c(v) \in \{\top, \max\{0, d(v) - k\}, \max\{0, d(v) - k + 1\}, \ldots, d(v)\}$ for $v \in w(f)$ and all nonnegative integers $\ell \leq k$.

(**) Let $S_f(\mathbf{c}, \ell) \subseteq V(G_f)$ be a set of vertices with cardinality ℓ which maximizes $g_1(S_f(\mathbf{c}, \ell); G_f)$, under the assumption that every vertex $v \in w(f)$ with $c(v) = \top$ belongs to $S_f(\mathbf{c}, \ell)$, and every vertex $v \in w(f)$ with $c(v) = i \in \{\max\{0, d(v) - k\}, \max\{0, d(v) - k + 1\}, \ldots, d(v)\}$ satisfies $|N_{G_f}(v) \cap S_f(\mathbf{c}, \ell)| \geq d(v) - i$ (note that due to the constraint that $|S|$ is at most k, the satisfied demand for $v \notin S_f(\mathbf{c}, \ell)$ is at most k and we need not consider the case of $c(v) \leq d(v) - k - 1$). Let $A'_f(\mathbf{c}, \ell) = g_1(S_f(\mathbf{c}, \ell); G_f)$ if $S_f(\mathbf{c}, \ell)$ exists, and $A'_f(\mathbf{c}, \ell) = -\infty$ otherwise.

(b) Let f be a non-leaf edge of T and f_1 and f_2 are two edges of T which are children of f. The value $A'_f(\mathbf{c}, \ell)$ is computed based on $A'_{f_1}(\mathbf{c_1}, \ell_1)$ and $A'_{f_2}(\mathbf{c_2}, \ell_2)$ for a pair of vectors $\mathbf{c_1}$ on $w(f_1)$ and $\mathbf{c_2}$ on $w(f_2)$ such that for each vertex $v \in w(f_1) \setminus w(f)(= w(f_2) \setminus w(f))$, we have $c_1(v) = c_2(v) = \top$ or $c_1(v) + c_2(v) \in \{d(v), d(v)+1, \ldots, 2d(v)\}$, and $\ell_1 + \ell_2 - |w(f_1) \cap w(f_2) \cap S_{f_1}(\mathbf{c_1}, \ell_1) \cap S_{f_2}(\mathbf{c_2}, \ell_2)| = \ell$ (note that in the setting of the partial domination, all demands for a vertex are not satisfied and hence we need to consider the case of $c_1(v) + c_2(v) > d(v)$).

We analyze the time complexity of this modified algorithm. Let f be an edge in $E(T)$. The possible number of vectors $\mathbf{c}$ on $w(f)$ is at most $(\min\{d^*, k\} + 2)^{|w(f)|}$ since $c(v) \in \{\top, \max\{0, d(v) - k\}, \max\{0, d(v) - k + 1\}, \ldots, d(v)\}$ for $v \in$

$w(f)$; hence the term d^* in the time complexity of the original algorithm (for the vector domination) is replaced with $\min\{d^*, k\}$. For computing $A'_f(\mathbf{c}, \ell)$, we need to consider all possible cases of $(\ell_1, \ell_2, |w(f_1) \cap w(f_2) \cap S_{f_1}(\mathbf{c_1}, \ell_1) \cap S_{f_2}(\mathbf{c_2}, \ell_2)|)$ for a fixed pair $\mathbf{c_1}$ and $\mathbf{c_2}$; this part takes $O(k^3) = O(n^3)$ times the computation for the corresponding part in the original algorithm. Also, we need to consider all possible cases of $(c_1(v), c_2(v))$ with $d(v) \leq c_1(v) + c_2(v) \leq 2d(v)$ instead of $c_1(v) + c_2(v) = d(v)$ for $v \in (w(f_1) \setminus w(f)) \setminus S_{f_1}(\mathbf{c_1}, \ell_1)$; the number of such pairs $(c_1(v), c_2(v))$ is $O(\min\{d^*, k\}^2)$ because each of $c_1(v)$ and $c_2(v)$ takes a value between $\max\{0, d(v) - k\}$ and $d(v)$ as observed above. In the original algorithm, the number of such pairs $(c_1(v), c_2(v))$ is at most $d(v) + 1$, and hence in this part, the term d^* in the time complexity of the original algorithm is replaced with $O(\min\{d^*, k\}^2)$. Thus, we can observe that the modified algorithm can be implemented to run in $2^{O(w \log \min\{d^*+1, k+1\})} n^{O(1)}$ time. $\qquad\square$

Let G be an apex-minor-free graph. If $bw(G) = O(\sqrt{k})$ holds, then Lemma 3 proves Theorem 1, since a branch-decomposition with width $O(bw(G))$ can be computed in polynomial time by Lemma 1. Otherwise we will remove a set I of *irrelevant* vertices from G so that at least one optimal solution is a subset of $V \setminus I$ and optimal also for the problem in $G[V \setminus I]$, and we have $bw(G[V \setminus I]) = O(\sqrt{k})$; by applying Lemma 3 to $G[V \setminus I]$, we obtain Theorem 1. In order to identify a set of irrelevant vertices, we focus on a *lexicographically smallest solution*. These ideas follow from the ones given by Fomin et al. for solving the partial vertex cover problem or partial dominating set problem [15].

Definition 1. *Given an ordering* $\sigma = v_1, v_2, \ldots, v_n$ *of* V *and two subsets* X_1 *and* X_2 *of* V, *we say that* X_1 *is* lexicographically smaller than X_2, *denoted by* $X_1 \leq_\sigma X_2$, *if* $V_\sigma^i \cap X_1 = V_\sigma^i \cap X_2$ *and* $v_{i+1} \in X_1 \setminus X_2$ *for some* $i \in \{0, 1, \ldots, n\}$, *where* $V_\sigma^i = \{v_1, v_2, \ldots, v_i\}$ *for* $i \in \{1, 2, \ldots, n\}$ *and* $V_\sigma^0 = \emptyset$. *For a problem* P, *a set* $S \subseteq V$ *is called a* lexicographically smallest solution *for* P *if for any other solution* S' *for* P, *we have* $S \leq_\sigma S'$.

We will complete a proof of Theorem 1 by showing how to define an ordering σ of V and identify a set I of irrelevant vertices for the partial vector domination problem (resp., the partial total vector domination problem and the multiple domination problem) in Subsect. 3.1 (resp., 3.2).

3.1   Vector Domination

We consider the partial vector dominating set problem. Let $\sigma = v_1, v_2, \ldots, v_n$ of V be an ordering of V such that

$$d(v_1) + |N_G(v_1) \setminus V_0| \geq d(v_2) + |N_G(v_2) \setminus V_0| \geq \cdots \geq d(v_n) + |N_G(v_n) \setminus V_0|,$$

where $V_0 = \max\{v \in V \mid d(v) = 0\}$. Let $S_\sigma = \{v_{i_1}, v_{i_2}, \ldots, v_{i_k}\}$ be a lexicographically smallest solution for the problem where $i_1 < i_2 < \cdots < i_k$. Since each vertex $v \in S$ contributes to demands in $N_G[v]$, we have $g_1(S_\sigma; G[V_\sigma^{i_k} \cup N_G(V_\sigma^{i_k})]) = g_1(S_\sigma; G)$ (note that g_1 is defined in the statement of Lemma 3).

That is, S_σ is an optimal solution also for the partial vector dominating set in $G[V_\sigma^{i_k} \cup N_G(V_\sigma^{i_k})]$; S_σ is a set of vertices with cardinality at most k maximizing g_1 also in $G[V_\sigma^{i_k} \cup N_G(V_\sigma^{i_k})]$. Thus, we have only to treat a smaller instance, instead of the original instance. Below, we consider how to find such an instance.

Now we can observe that S_σ is a $(k,3)$-center in $G[V_\sigma^{i_k} \cup N_G(V_\sigma^{i_k})]$ (recall that a (k,r)-*center* of a graph H is a set W of vertices of H with size k such that any vertex in H is within distance r from a vertex of W).

Lemma 4. S_σ *is a* $(k,3)$-*center in* $G[V_\sigma^{i_k} \cup N_G(V_\sigma^{i_k})]$.

Proof. Let $G' = G[V_\sigma^{i_k} \cup N_G(V_\sigma^{i_k})]$. For proving the lemma, we will show that $N_{G'}^2[v] \cap S_\sigma \neq \emptyset$ holds for all $v \in V_\sigma^{i_k}$, where $N_H^2[v]$ denotes the set of vertices within distance 2 from a vertex v in a graph H.

Assume for contradiction that there exists a vertex v_j with $j < i_k$ such that $N_{G'}^2[v_j] \cap S_\sigma = \emptyset$. Since G has no path with length at most 2 connecting v_j and any vertex in S_σ which goes through a vertex in $V \setminus (V_\sigma^{i_k} \cup N_G(V_\sigma^{i_k}))$, it follows that $N_G^2[v_j] \cap S_\sigma = \emptyset$ also holds. Consider the set $S' = S_\sigma \cup \{v_j\} \setminus \{v_{i_k}\}$. Note that $|S'| = |S_\sigma|$. Then, we claim that $g_1(S';G) \geq g_1(S_\sigma;G)$ holds, i.e., S' is also optimal, which contradicts that S_σ is a lexicographically smallest solution.

This claim can be proved as follows. Observe that by deleting v_{i_k} from S_σ, the total sum $g_1(S_\sigma;G)$ of satisfied demands is decreased by at most $d(v_{i_k}) + |N_G(v_{i_k}) \setminus V_0|$. On the other hand, by $N_G^2[v_j] \cap S_\sigma = \emptyset$, an addition of v_j to $S_\sigma \setminus \{v_{i_k}\}$ increases $g_1(S_\sigma \setminus \{v_{i_k}\}; G)$ by exactly $d(v_j) + |N_G(v_j) \setminus V_0|$. It follows by the definition of σ that $g_1(S';G) - g_1(S_\sigma;G) \geq d(v_j) + |N_G(v_j) \setminus V_0| - (d(v_{i_k}) + |N_G(v_{i_k}) \setminus V_0|) \geq 0$. $\square$

Let G be apex-minor-free. Then, the following results about algorithms for computing (k,r)-centers are known; we denote the PTAS for the problem with $r = 3$ in the following lemma by algorithm A.

Lemma 5. *([12, Corollary 5.1]) Let G be an apex-minor-free graph. Then, there is a polynomial-time approximation scheme (PTAS) for the problem of finding a minimum set W of vertices such that W is a $(|W|, r)$-center in G; for each fixed constant $\epsilon > 0$, a $(1 + \epsilon)$-approximate solution for the problem can be obtained in polynomial time.*

By utilizing algorithm A in the following manner, we can find in polynomial time a set $V_\sigma^{i'}$ of vertices for some $i' \geq i_k$ such that $G[V_\sigma^{i'} \cup N_G(V_\sigma^{i'})]$ has a $((1 + \epsilon)k, 3)$-center for a positive constant ϵ:

Step 1: Let $i := n$.
Step 2: While $G[V_\sigma^i \cup N_G(V_\sigma^i)]$ does not have a $((1 + \epsilon)k, 3)$-center (this can be checked by algorithm A), let $i := i - 1$.

Let i' be the value of i when this procedure halts. By $i' \geq i_k$, we have $S_\sigma \subseteq V_\sigma^{i'}$ and $g_1(S_\sigma; G[V_\sigma^{i'} \cup N_G(V_\sigma^{i'})]) = g_1(S_\sigma; G)$, and S_σ is an optimal solution also for the problem in $G[V_\sigma^{i'} \cup N_G(V_\sigma^{i'})]$. Since $G[V_\sigma^{i'} \cup N_G(V_\sigma^{i'})]$ has a $((1 + \epsilon)k, 3)$-center, we have $bw(G[V_\sigma^{i'} \cup N_G(V_\sigma^{i'})]) = O(\sqrt{(1 + \epsilon)k}) = O(\sqrt{k})$ by the following Lemma 6; $V \setminus (V_\sigma^{i'} \cup N_G(V_\sigma^{i'}))$ can be regarded as a set I of irrelevant vertices mentioned above.

Lemma 6. *([15, Lemma 2]) Let G be an apex-minor-free graph. If G has a (k, r)-center, then the treewidth (branchwidth) of G is $O(r\sqrt{k})$.*

3.2   Total Vector Domination and Multiple Domination

We first consider the partial total vector dominating set problem. The difference between the partial total vector domination and the partial vector domination is that when a vertex v is selected as a member in a solution, v contributes nothing to the demand of v for the former problem, but the demand $d(v)$ is satisfied for the latter problem. For this problem, by defining an ordering $\sigma = v_1, v_2, \ldots, v_n$ of V as

$$|N_G(v_1) \setminus V_0| \geq |N_G(v_2) \setminus V_0| \geq \cdots \geq |N_G(v_n) \setminus V_0|,$$

we can obtain a counterpart of Lemma 4 for the partial vector dominating set problem; namely, a lexicographically smallest solution $S_\sigma = \{v_{i_1}, v_{i_2}, \ldots, v_{i_k}\}$ for the problem is a $(k, 3)$-center in $G[V_\sigma^{i_k} \cup N_G(V_\sigma^{i_k})]$, where $i_1 < i_2 < \cdots < i_k$.

Lemma 7. *S_σ is a $(k, 3)$-center in $G[V_\sigma^{i_k} \cup N_G(V_\sigma^{i_k})]$.*

Proof. Let $G' = G[V_\sigma^{i_k} \cup N_G(V_\sigma^{i_k})]$. Similarly to the proof of Lemma 4, it suffices to assume that there exists a vertex v_j with $j < i_k$ such that $N_G^2[v_j] \cap S_\sigma = \emptyset$, and derive a contradiction. Now, consider the set $S' = S_\sigma \cup \{v_j\} \setminus \{v_{i_k}\}$. Note that $|S'| = |S_\sigma|$. Then, we claim that $g_2(S'; G) \geq g_2(S_\sigma; G)$ holds, i.e., S' is also optimal, which contradicts that S_σ is a lexicographically smallest solution.

This claim can be proved as follows. Observe that by deleting v_{i_k} from S_σ, the total sum $g_2(S_\sigma; G)$ of satisfied demands is decreased by at most $|N_G(v_{i_k}) \setminus V_0|$. On the other hand, by $N_G^2[v_j] \cap S_\sigma = \emptyset$, an addition of v_j to $S_\sigma \setminus \{v_{i_k}\}$ increases $g_2(S_\sigma \setminus \{v_{i_k}\}; G)$ by exactly $|N_G(v_j) \setminus V_0|$. It follows by the definition of σ that $g_2(S'; G) - g_2(S_\sigma; G) \geq |N_G(v_j) \setminus V_0| - |N_G(v_{i_k}) \setminus V_0| \geq 0$.   □

The remaining parts can be treated in a similar way to the case of the partial vector dominating set. Also, we can treat the partial multiple vector dominating set, by replacing $N_G()$ with $N_G[]$ for the arguments for the partial total vector dominating set.

Summarizing the arguments given so far, we have shown Theorem 1.

References

1. Amini, O., Fomin, F.V., Saurabh, S.: Implicit branching and parameterized partial cover problems. J. Comput. Syst. Sci. **77**(6), 1159–1171 (2011)
2. Betzler, N., Bredereck, R., Niedermeier, R., Uhlmann, J.: On bounded-degree vertex deletion parameterized by treewidth. Discrete Appl. Math. **160**(1), 53–60 (2012)
3. Bodlaender, H.L., Thilikos, D.M.: Constructive linear time algorithms for branch-width. In: Degano, P., Gorrieri, R., Marchetti-Spaccamela, A. (eds.) ICALP 1997. LNCS, vol. 1256, pp. 627–637. Springer, Heidelberg (1997)

4. Chapelle, M.: Parameterized complexity of generalized domination problems on bounded tree-width graphs (2010). arXiv preprint arXiv:1004.2642
5. Cicalese, F., Cordasco, G., Gargano, L., Milanič, M., Vaccaro, U.: Latency-bounded target set selection in social networks (2013). arXiv preprint arXiv:1303.6785
6. Cicalese, F., Milanič, M., Vaccaro, U.: Hardness, approximability, and exact algorithms for vector domination and total vector domination in graphs. In: Owe, O., Steffen, M., Telle, J.A. (eds.) FCT 2011. LNCS, vol. 6914, pp. 288–297. Springer, Heidelberg (2011)
7. Cicalese, F., Milanic, M., Vaccaro, U.: On the approximability and exact algorithms for vector domination and related problems in graphs. Discrete Appl. Math. **161**(6), 750–767 (2013)
8. Corneil, D.G., Rotics, U.: On the relationship between clique-width and treewidth. SIAM J. Comput. **34**(4), 825–847 (2005)
9. Courcelle, B.: The monadic second-order logic of graphs. I. recognizable sets of finite graphs. Inf. Comput. **85**(1), 12–75 (1990)
10. Demaine, E.D., Fomin, F.V., Hajiaghayi, M., Thilikos, D.M.: Fixed-parameter algorithms for (k, r)-center in planar graphs and map graphs. ACM Trans. Algorithms (TALG) **1**(1), 33–47 (2005)
11. Demaine, E.D., Hajiaghayi, M.T., Kawarabayashi, K.i.: Algorithmic graph minor theory: decomposition, approximation, and coloring. In: 2005 46th Annual IEEE Symposium on Foundations of Computer Science, FOCS 2005, pp. 637–646. IEEE (2005)
12. Demaine, E.D., Hajiaghayi, M.: Bidimensionality: new connections between fpt algorithms and ptass. In: Proceedings of the 16th Annual ACM-SIAM Symposium on Discrete Algorithms. pp. 590–601. Society for Industrial and Applied Mathematics (2005)
13. Dorn, F.: Dynamic programming and fast matrix multiplication. In: Azar, Y., Erlebach, T. (eds.) ESA 2006. LNCS, vol. 4168, pp. 280–291. Springer, Heidelberg (2006)
14. Downey, R.G., Fellows, M.R.: Fixed-Parameter Tractability and Completeness. Cornell University, Mathematical Sciences Institute (1992)
15. Fomin, F.V., Lokshtanov, D., Raman, V., Saurabh, S.: Subexponential algorithms for partial cover problems. Inf. Process. Lett. **111**(16), 814–818 (2011)
16. Fomin, F.V., Thilikos, D.M.: Dominating sets in planar graphs: branch-width and exponential speed-up. SIAM J. Comput. **36**(2), 281–309 (2006)
17. Harant, J., Pruchnewski, A., Voigt, M.: On dominating sets and independent sets of graphs. Comb. Probab. Comput. **8**, 547–553 (1999)
18. Harary, F., Haynes, T.W.: Double domination in graphs. Ars Comb. **55**, 201–214 (2000)
19. Haynes, T.W., Hedetniemi, S.T., Slater, P.J.: Domination in Graphs: Advanced Topics, vol. 40. Marcel Dekker, New York (1998)
20. Haynes, T.W., Hedetniemi, S.T., Slater, P.J.: Fundamentals of Domination in Graphs. Marcel Dekker, New York (1998)
21. Ishii, T., Ono, H., Uno, Y.: (Total) vector domination for graphs with bounded branchwidth (2013). arXiv preprint arXiv:1306.5041
22. Ishii, T., Ono, H., Uno, Y.: (Total) vector domination for graphs with bounded branchwidth. In: Pardo, A., Viola, A. (eds.) LATIN 2014. LNCS, vol. 8392, pp. 238–249. Springer, Heidelberg (2014)
23. Raman, V., Saurabh, S., Srihari, S.: Parameterized algorithms for generalized domination. In: Yang, B., Du, D.-Z., Wang, C.A. (eds.) COCOA 2008. LNCS, vol. 5165, pp. 116–126. Springer, Heidelberg (2008)

24. Robertson, N., Seymour, P.D.: Graph minors. X. obstructions to tree-decomposition. J. Comb. Theor. Ser. B **52**(2), 153–190 (1991)
25. Seymour, P.D., Thomas, R.: Call routing and the ratcatcher. Combinatorica **14**(2), 217–241 (1994)

Efficient Approximation Schemes for the Maximum Lateness Minimization on a Single Machine with a Fixed Operator or Machine Non-Availability Interval

Imed Kacem[1](✉), Hans Kellerer[2], and Maryam Seifaddini[1]

[1] LCOMS EA 7306, Université de Lorraine, 57000 Metz, France
imed.kacem@univ-lorraine.fr
[2] ISOR, University of Graz, Graz, Austria

Abstract. In this paper we consider the single machine scheduling problem with one non-availability interval to minimize the maximum lateness where jobs have positive tails. Two cases are considered. In the first one, the non-availability interval is due to the machine maintenance. In the second case, the non-availability interval is related to the operator who is organizing the execution of jobs on the machine. The contribution of this paper consists in an improved FPTAS for the maintenance non-availability interval case and the elaboration of the first FPTAS for the operator non-availability interval case. The two FPTAS are strongly polynomial.

Keywords: Scheduling · Non-availability constraint · Lateness · FPTAS

1 Introduction

In this paper we consider the single machine scheduling problem with one non-availability interval to minimize the maximum lateness where jobs have positive tails. Two cases are considered. In the first one, the non-availability interval is due to the machine maintenance. In the second case, the non-availability interval is related to the operator who is organizing the execution of jobs on the machine. An operator non-availability period is a time interval in which no job can start, and neither can complete. The main difference between machine non-availability (MNA) and operator non-availability (ONA) consists in the fact that a job can be processed but cannot start neither finish during the ONA period. However, the machine non-availability interval is a completely forbidden period. Rapine et al. [14] have described the applications of this problem in the planning of a chemical experiments as follows: Each experiment is performed by an automatic system (a robot), during a specified amount of time, but a chemist is required to control its start and completion. At the beginning, the chemist launches the process (preparation step). The completion step corresponds to the experimental

© Springer International Publishing Switzerland 2014
P. Fouilhoux et al. (Eds.): ISCO 2014, LNCS 8596, pp. 305–314, 2014.
DOI: 10.1007/978-3-319-09174-7_26

analysis, which is to be done in a no-wait mode to stop chemical reactions. Here, the automatic system is available all the time, where the chemists may be unavailable due to planned vacations or activities. This induces operator (chemist) non-availability intervals when experiments (jobs) can be performed by the automatic system (machine), but cannot neither start nor complete.

The MNA case of this type of problems has been studied in the literature under various criteria (a sample of these works includes Lee [11], Kacem [8], Kubzin and Strusevich [10], Qi et al. [12,13], Schmidt [15], He et al. [6]). However, few papers studied the problem we consider in this paper. Lee [11] explored the Jackson's sequence JS and proved that its deviation to the optimal makespan cannot exceed the largest processing time, which is equivalent to state that JS is a 2-approximation. Recently, Yuan et al. developed an interesting PTAS for the studied problem [17]. Kacem [8] presented a Fully Polynomial Time Approximation Scheme (FPTAS) for the maximum lateness minimization. That is why this paper is a good attempt to design more efficient approximation heuristics and approximation schemes to solve the studied problem.

For the ONA case, few works have been published. Brauner et al. [1] considered the problem of single machine scheduling with ONA periods. They analyzed this problem on a single machine with the makespan as a minimization criterion and they showed that the problem is NP-hard with one ONA period. They also considered the problem with K ONA periods such that the length of each ONA period is no more than $\frac{1}{\lambda}$ times the total processing time of all jobs. They introduced a worst-case ratio smaller than $1 + \frac{2K}{\lambda}$ for algorithm LS (list scheduling). They presented an approximation algorithm with a worst-case ratio close to $2 + \frac{K-1}{\lambda}$. The natural case of periods where the duration of the periods is smaller than any processing time of any job, has been considered by Rapine et al [14]. They proved the problem can be solved in polynomial time, where there exists only one ONA period and they showed the problem is NP-hard if one has $K \geq 2$ small non-availability periods and the worst-case ratio of LS is no more than $\frac{K+1}{2}$ and the problem does not admit an FPTAS for $K \geq 3$ unless P = NP. Recently, Chen et al. [3] considered the single machine scheduling with one ONA period to minimize the total completion time. The problem is NP-hard even if the length of the ONA period is smaller than the processing time of any job. They have also presented an algorithm with a tight worst-case ratio of $\frac{20}{17}$. They showed that the worst-case ratio of SPT is at least $\frac{5}{3}$.

The contribution of this paper consists in an improved FPTAS for the maintenance non-availability interval case and the elaboration of the first FPTAS for the ONA interval case. The two FPTAS are strongly polynomial. These contributions can be summarized in Table 1 for the two cases.

The paper are organized as follows. Section 2 describes the exact formulation of the maintenance non-availability interval case and the improved FPTAS. Section 3 is devoted to the operator non-availability interval case and to the presentation of the proposed FPTAS. Finally, Sect. 4 gives some concluding remarks.

Table 1. Summary of results

	Result	Reference
MNA	2-approximation (Jackson's Rule)	Lee [11]
MNA	PTAS: $O(n \ln(n) + n.2^{1/\varepsilon})$	Yuan [17]
MNA	FPTAS: $O(n^3/\varepsilon^2)$	Kacem [8]
MNA	PTAS: $O(n \ln(n) + (1/\varepsilon)\, 2^{1/\varepsilon})$	This paper
MNA	FPTAS: $O(n \ln(n) + \min\{n, 1/\varepsilon\}^3/\varepsilon^2)$	This paper
ONA	PTAS: $O(n(\ln n)/\varepsilon + (n/\varepsilon^2)2^{1/\varepsilon})$	This paper
ONA	FPTAS: $O((n/\varepsilon)\ln(n) + n \min\{n, 1/\varepsilon\}^3/\varepsilon^3)$	This paper

2   Case Under MNA Interval

Here, the studied problem $(\mathcal{P})$ can be formulated as follows. We have to schedule a set J of n jobs on a single machine, where every job j has a processing time p_j and a tail q_j. The machine can process at most one job at a time and it is unavailable between T_1 and T_2 (i.e., $[T_1, T_2)$ is a forbidden interval). Preemption of jobs is not allowed (jobs have to be performed under the non-resumable scenario). All jobs are ready to be performed at time 0. With no loss of generality, we consider that all data are integers and that jobs are indexed according to Jackson's rule [2] (i.e., jobs are indexed in nonincreasing order of tails). Therefore, we assume that $q_1 \geq q_2 \geq ... \geq q_n$. The consideration of tails is motivated by the large set of scheduling problems such that jobs have delivery times after their processing [4]. Let $C_j(S)$ denote the completion time of job j in a feasible schedule S for the problem and let $\varphi_S(\mathcal{P})$ be the maximum lateness yielded by schedule S for instance $\mathcal{I}$ of $(\mathcal{P})$:

$$\varphi_S(\mathcal{I}) = \max_{1 \leq j \leq n} (C_j(S) + q_j) \tag{1}$$

The aim is to find a feasible schedule S by minimizing the maximum lateness. Due to the dominance of Jackson's order, an optimal schedule is composed of two sequences of jobs scheduled in nondecreasing order of their indexes.

If all the jobs can be inserted before T_1, the instance studied $(\mathcal{I})$ has obviously a trivial optimal solution obtained by Jackson's rule. We therefore consider only the problems in which all the jobs cannot be scheduled before T_1. Moreover, we consider that every job can be inserted before T_1 (i.e., $p_j \leq T_1$ for every $j \in J$).

In the remainder of this paper $\varphi^*(\mathcal{I})$ denotes the minimal maximum lateness for instance $\mathcal{I}$.

2.1   New Simplifications and PTAS

Now, let us describe our FPTAS. It uses a simplification technique based on merging small jobs [9].

1st STEP:

First, we simplify the instance $\mathcal{I}$ as follows. Given an arbitrary $\varepsilon > 0$. We assume that $1/\varepsilon$ is integer. We split the interval $[0, \max_{j \in J}\{q_j\}]$ in $1/\varepsilon$ equal length intervals and we round up every tail q_j to the next multiple of εq_{max} ($q_{max} = \max_{j \in J}\{q_j\}$). The new instance is denoted as $\mathcal{I}'$.

Proposition 1. *The obtained instance $\mathcal{I}'$ can be obtained in $O(n)$ time and it can be done with no $(1 + \varepsilon)$-loss.*

Proof. The modification can be done by setting $q_j := \lceil q_j/\varepsilon q_{max} \rceil \varepsilon q_{max}$ for every $j \in J$. Then, it can be done in $O(n)$ time. Moreover, since $\lceil q_j/\varepsilon q_{max} \rceil \varepsilon q_{max} \leq q_j + \varepsilon q_{max}$ then, $\varphi^*(\mathcal{I}') \leq \varphi^*(\mathcal{I}) + \varepsilon q_{max} \leq (1 + \varepsilon)\varphi^*(\mathcal{I})$ since q_{max} is a lower bound on the optimal maximum lateness.

2nd STEP:

J is divided into at most $1/\varepsilon$ subsets $J(k)$ $(1 \leq k \leq 1/\varepsilon)$ where jobs in $J(k)$ have identical tails of $k\varepsilon q_{max}$. The second modification consists in reducing the number of small jobs in every subset $J(k)$. Small jobs are those having processing times $< \varepsilon P/2$ where $P = p_1 + p_2 + ... + p_n$. The reduction is done by merging the small jobs in each $J(k)$ so that we obtain new greater jobs having processing times between $\varepsilon P/2$ and εP. The small jobs are taken in the order of their index in this merging procedure. At most, for every subset $J(k)$, a single small job remains. We re-index jobs according to nondecreasing order of their tails. The new instance we obtain is denoted as $\mathcal{I}''$. Clearly, the number of jobs remaining in the simplified instance $\mathcal{I}''$ is less than $3/\varepsilon$.

Proposition 2. *This reduction in the 2nd step cannot increase the optimal solution value of $\mathcal{I}'$ by more than $(1 + \varepsilon)$-factor. It can be done in $O(n)$ time.*

Proof. The proof is based on the comparison of a lower bound lb for $\varphi^*(\mathcal{I}')$ and a feasible solution σ for instance $\mathcal{I}''$. We will demonstrate that $\varphi_\sigma(\mathcal{I}'') \leq (1 + \varepsilon) lb$ which implies that $\varphi^*(\mathcal{I}'') \leq \varphi_\sigma(\mathcal{I}'') \leq (1 + \varepsilon) lb \leq (1 + \varepsilon) \varphi^*(\mathcal{I}')$.

The lower bound lb is the maximum lateness of the optimal solution σ_r of a special preemptive version of problem $\mathcal{I}'$ where the large jobs are supposed to be assigned (before or after the non-availability interval) as in the optimal solution of $\mathcal{I}'$. The principle of this lower bound is based on the splitting idea presented in [8]. Indeed, we split the small jobs so that the obtained pieces have an identical length of 1 and they keep their tails. It can be demonstrated that the pieces associated to the small jobs must be scheduled in σ_r according to the Jackson's order. At most, one piece of a certain job g will be preempted by the non-availability period. For more details on this lower bound, we refer to [8].

It is easy to transform such a relaxed solution σ_r to a close feasible solution σ for $\mathcal{I}''$. Indeed, we can remark that the small jobs of every subset $J(k)$ $(1 \leq k \leq 1/\varepsilon)$, except the subset $k(g)$ containing job g, are scheduled contiguously before of after the non-availability period. Hence, the associated merged jobs (in $\mathcal{I}''$) for these small ones (in $\mathcal{I}'$) will take the same order in sequence σ. The great jobs are common in the two instances and they will keep the same assignments in

σ. The only possible difference between σ and σ_r will consist in the positions of small jobs (from $\mathcal{I}'$) belonging to subset $k(g)$. For these small jobs, we construct the same associated merged jobs as in $\mathcal{I}''$ by scheduling them in σ as close in σ_r as possible. As a consequence, some small jobs (from $\mathcal{I}'$ and belonging to subset $k(g)$) will be moved after the non-availability period. Thus, it is easy to deduce that $\varphi_\sigma(\mathcal{I}'') \leq lb + \varepsilon P \leq (1 + \varepsilon) lb$.

Theorem 1. *Problem P has a Polynomial Time Approximation Scheme (PTAS) with a time complexity of $O(n\ln(n) + (1/\varepsilon)\, 2^{1/\varepsilon})$.*

Proof. The proof is based on the two previous propositions. The Jackson's order can be obtained in $n\ln(n)$. We construct the optimal solution of $\mathcal{I}''$ by an exhaustive search in $O((1/\varepsilon)\, 2^{1/\varepsilon})$. Then, we derive a feasible solution for $\mathcal{I}$, which can be done in $O(n)$.

Remark 1. The new PTAS has a lower time complexity compared to the one proposed by Yuan et al. [17] for which the time complexity is $O(n\ln(n) + n.2^{1/\varepsilon})$.

2.2   Improved FPTAS

Our FPTAS is similar to the one proposed by Kacem [8]. It uses the same technique but exploits also the modification of the input ($\mathcal{I} \rightarrow \mathcal{I}''$). First, we use the Jackson's sequence JS obtained for the modified instance $\mathcal{I}''$. Then, we apply the modified dynamic programming algorithm APS'_ε introduced in Kacem [8] on instance $\mathcal{I}''$.

The main idea of APS'_ε is to remove a special part of the states generated by a dynamic programming algorithm. Therefore, the modified algorithm becomes faster and yields an approximate solution instead of the optimal schedule (see Appendix 1). First, we define the following parameters:

$$\bar{n} = \min\{n, 3/\varepsilon\},$$

$$\omega_1 = \left\lceil \frac{2\bar{n}}{\varepsilon} \right\rceil,$$

$$\omega_2 = \left\lceil \frac{\bar{n}}{\varepsilon} \right\rceil,$$

$$\delta_1 = \frac{\varphi_{JS}(\mathcal{I}'')}{\omega_1}$$

and

$$\delta_2 = \frac{T_1}{\omega_2}.$$

We split $[0, \varphi_{JS}(\mathcal{I}''))$ into ω_1 equal subintervals $I_m^1 = [(m-1)\delta_1, m\delta_1)_{1 \leq m \leq \omega_1}$. We also split $[0, T_1)$ into ω_2 equal subintervals $I_s^2 = [(s-1)\delta_2, s\delta_2)_{1 \leq s \leq \omega_2}$ of length δ_2. Moreover, we define the two singletons $I_{\omega_1+1}^1 = \{\varphi_{JS}(\mathcal{I}'')\}$ and $I_{\omega_2+1}^2 = \{T_1\}$. Our algorithm APS'_ε generates reduced sets $\mathcal{X}_j^\#$ of states $[t, f]$ where t is the total processing time of jobs assigned before T_1 in the associated partial schedule and f is the maximum lateness of the same partial schedule. It can be described as follows:

Algorithm APS_ε'

(i). set $\mathcal{X}_1^\# = \{[0, T_2 + p_1 + q_1], [p_1, p_1 + q_1]\}$.

(ii). For $j \in \{2, 3, ..., \overline{n}\}$,

$\mathcal{X}_j^\# = \varnothing$.

For every state $[t, f]$ in $\mathcal{X}_{j-1}^\#$:

(1) Put $\left[t, \max\left\{f, T_2 + \sum_{i=1}^j p_i - t + q_j\right\}\right]$ in $\mathcal{X}_j^\#$

(2) Put $[t + p_j, \max\{f, t + p_j + q_j\}]$ in $\mathcal{X}_j^\#$ if $t + p_j \leq T_1$.

Remove $\mathcal{X}_{j-1}^\#$

Let $[t, f]_{m,s}$ be the state in $\mathcal{X}_j^\#$ such that $f \in I_m^1$ and $t \in I_s^2$ with the smallest possible t (ties are broken by choosing the state of the smallest f).

Set $\mathcal{X}_j^\# = \left\{[t, f]_{m,s} \,|\, 1 \leq m \leq \omega_1 + 1, 1 \leq s \leq \omega_2 + 1\right\}$.

(iii). $\varphi_{APS_\varepsilon'}(\mathcal{I}'') = \min_{[t,f] \in \mathcal{X}_{\overline{n}}^\#} \{f\}$.

Theorem 2. *Given an arbitrary $\varepsilon > 0$, Algorithm APS_ε' yields an output $\varphi_{APS_\varepsilon'}(\mathcal{I}'')$ such that:*

$$\varphi_{APS_\varepsilon'}(\mathcal{I}'') - \varphi^*(\mathcal{I}'') \leq \varepsilon\varphi^*(\mathcal{I}''). \tag{2}$$

Proof. The proof is similar to [8]. $\quad\blacksquare$

Lemma 1. *Given an arbitrary $\varepsilon > 0$, algorithm APS_ε' can be implemented in $O\left(\overline{n}\log\overline{n} + \overline{n}^3/\varepsilon^2\right)$ time.*

Proof. The proof is presented at the conference. $\quad\blacksquare$

The schedule obtained by APS_ε' for instance $\mathcal{I}''$ can be easily converted into a feasible one for instance $\mathcal{I}$. This can be done in $O(n)$ time. From the previous lemma and the proof of Theorem 2, the main result is proved and the following theorem holds.

Theorem 3. *APS_ε' is an FPTAS and it can be implemented in $O(n \log n + \min \{n, 1/\varepsilon\}^3/\varepsilon^2)$ time.*

3   Case Under Operator Non-availability Interval

Here, the studied problem (Π) can be formulated as follows. An operator has to schedule a set J of n jobs on a single machine, where every job j has a processing time p_j and a tail q_j. The machine can process at most one job at a time if the operator is available at the starting time and the completion time of such a job. The operator is unavailable during (T_1, T_2). Preemption of jobs is not allowed (jobs have to be performed under the non-resumable scenario). All jobs are ready to be performed at time 0. With no loss of generality, we consider that all data are integers and that jobs are indexed according to Jackson's rule and we assume that $q_1 \geq q_2 \geq \dots \geq q_n$. Let $C_j(S)$ denote the completion time of job j in a feasible schedule S ($C_j(S) \notin]T_1, T_2[$ and $C_j(S) - p_j \notin (T_1, T_2)$) and let $\varphi_S(\mathcal{I})$ be the maximum lateness yielded by schedule S for instance $\mathcal{I}$ of (Π):

$$\varphi_S(\mathcal{I}) = \max_{1 \leq j \leq n} (C_j(S) + q_j) \tag{3}$$

The aim is to find a feasible schedule S by minimizing the maximum lateness.

If all the jobs can be inserted before T_1, the instance studied ($\mathcal{I}$) has obviously a trivial optimal solution obtained by Jackson's rule. We therefore consider only the problems in which all the jobs cannot be scheduled before T_1. Moreover, we consider that every job can be inserted before T_1 (i.e., $p_j \leq T_1$ for every $j \in J$).

In the remainder of this paper $\varphi^*(\mathcal{I})$ denotes the minimal maximum lateness for instance $\mathcal{I}$.

Proposition 3. *If $p_j < T_2 - T_1$ for every $j \in J$, then problem (Π) has an FPTAS.*

Proof. In this case, it is easy to remark that Problem Π is equivalent to Problem $\mathcal{P}$ for which we can apply the FPTAS described in the previous section.

In the remainder, we consider the hard case where some jobs have processing times greater than $T_2 - T_1$. Let $\mathcal{K}$ be the subset of these jobs. In this case, two scenarios are possible:

- Scenario 1: there exists a job $s \in \mathcal{K}$ such that in the optimal solution it starts before T_1 and completes after T_2 (s is called the stradling job).
- Scenario 2: there is no stradling job in the optimal solution.

It is obvious that Scenario 2 is equivalent to Problem $\mathcal{P}$ for which we have an FPTAS. Thus, the last step necessary to prove the existence of an FPTAS for Problem Π is to construct a special scheme for Scenario 1. Without loss of generality, we assume that the stradling job s is known (indeed, it can be guessed among jobs of $\mathcal{K}$). The following proposition determines the time-window of the starting time of job s in the optimal solution.

Proposition 4. *Let t_s^* be the starting time of s in the optimal schedule. The following relation holds: $t_s^* \in [T_2 - p_s, T_1]$.*

Proof. Obvious since the stradling job s has to cover the operator non-availability period in the optimal schedule.

Proposition 5. *Scenario 1 has an FPTAS.*

Proof. The stradling job s is assumed to be known. Given an arbitrary $\varepsilon > 0$, we divide the interval $[T_2 - p_s, T_1]$ in $\lceil 1/\varepsilon \rceil$ equal-length sub-intervals $\bigcup_{h=1}^{\lceil 1/\varepsilon \rceil} D_h$ where

$$D_h = [T_2 - p_s + (h-1)\frac{T_1 - T_2 + p_s}{\lceil 1/\varepsilon \rceil}, T_2 - p_s + h\frac{T_1 - T_2 + p_s}{\lceil 1/\varepsilon \rceil}].$$

We consider a set of $\lceil 1/\varepsilon \rceil + 1$ instances $\{\mathcal{I}_1, \mathcal{I}_2, ..., \mathcal{I}_{\lceil 1/\varepsilon \rceil + 1}\}$ of Problem $\mathcal{P}$ where in $\mathcal{I}_h$ the stradling job starts at time

$$t_s^h = T_2 - p_s + (h-1)\frac{T_1 - T_2 + p_s}{\lceil 1/\varepsilon \rceil}$$

which is equivalent to an instance of Problem $\mathcal{P}$ with a set of jobs $J - \{s\}$ and a MNA period Δ_h:

$$\Delta_h = (T_2 - p_s + (h-1)\frac{T_1 - T_2 + p_s}{\lceil 1/\varepsilon \rceil}, T_2 + (h-1)\frac{T_1 - T_2 + p_s}{\lceil 1/\varepsilon \rceil}).$$

For every instance from $\{\mathcal{I}_1, \mathcal{I}_2, ..., \mathcal{I}_{\lceil 1/\varepsilon \rceil + 1}\}$, we apply the FPTAS described in the previous section for Problem $\mathcal{P}$ and we select the best solution among all the $\lceil 1/\varepsilon \rceil + 1$ instances. It is easy to see that if $t_s^* \in [t_s^h, t_s^{h+1})$ then, delaying s and the next jobs in the optimal schedule of $\mathcal{I}_{h+1}$ ($h = 1, 2, ..., \lceil 1/\varepsilon \rceil$) by setting $t_s^* = t_s^{h+1}$ will not cost more than

$$\frac{T_1 - T_2 + p_s}{\lceil 1/\varepsilon \rceil} \leq \varepsilon(T_1 - T_2 + p_s)$$

$$\leq \varepsilon p_s$$

Thus, the solution Ω_{h+1} obtained by APS'_ε for $\mathcal{I}_{h+1}(h = 1, 2, ..., \lceil 1/\varepsilon \rceil)$ is sufficiently close to optimal schedule for Scenario 1 if s is the stradling job and $t_s^* \in [t_s^h, t_s^{h+1})$. As a conclusion, Scenario 1 has an FPTAS.

Theorem 4. *Problem Π admits an FPTAS and this scheme can be implemented in $O\left(n\left(\ln n\right)/\varepsilon + n\min\{n, 3/\varepsilon\}^3/\varepsilon^3\right)$ time.*

Proof. The proof is a direct deduction from all the cases mentioned in this section.

Remark 2. By applying the same approach, a PTAS can be elaborated for Problem Π and it can be implemented in $O(n(\ln n)/\varepsilon + (n/\varepsilon^2)2^{1/\varepsilon})$ time.

4   Conclusion

In this paper, we considered the non-resumable case of the single machine scheduling problem with a non-availability interval. Our aim is to minimize the maximum lateness when every job has a positive tail. Two cases are considered. In the first one, the non-availability interval is due to the machine maintenance. In the second case, the non-availability interval is related to the operator who is organizing the execution of jobs on the machine. The contribution of this paper consists in an improved FPTAS for the MNA interval case and the elaboration of the first FPTAS for the operator non-availability interval case. The two FPTAS are strongly polynomial.

As future perspectives, we aim to consider other criteria for the single-machine problem as well as the study of multiple operator non-availability periods.

Acknowledgement. This work has been funded by the CONSEIL GENERAL DE LORRAINE: "opération réalisée avec le concours financier du Conseil Régional de Lorraine".

References

1. Brauner, N., Finke, G., Kellerer, H., Lebacque, V., Rapine, C., Potts, C., Strusevich, V.: Operator non-availability periods. 4 OR-Q. J. Oper. Res. **7**, 239–253 (2009)
2. Carlier, J.: The one-machine sequencing problem. Eur. J. Oper. Res. **11**, 42–47 (1982)
3. Chen, Y., Zhang, A., Tan, Z.: Complexity and approximation of single machine scheduling with an operator non-availability period to minimize total completion time. Inf. Sci. **251**, 150–163 (2013)
4. Dessouky, M.I., Margenthaler, C.R.: The one-machine sequencing problem with early starts and due dates. AIIE Trans. 4(3), 214–222 (1972)
5. Gens, G.V., Levner, E.V.: Fast approximation algorithms for job sequencing with deadlines. Discret. Appl. Math. **3**, 313–318 (1981)
6. He, Y., Zhong, W., Gu, H.: Improved algorithms for two single machine scheduling problems. Theor. Comput. Sci. **363**, 257–265 (2006)
7. Ibarra, O., Kim, C.E.: Fast approximation algorithms for the knapsack and sum of subset problems. J. ACM **22**, 463–468 (1975)
8. Kacem, I.: Approximation algorithms for the makespan minimization with positive tails on a single machine with a fixed non-availability interval. J. Comb. Optim. **17**(2), 117–133 (2009)
9. Kacem, I., Kellerer, H.: Approximation algorithms for no idle time scheduling on a single machine with release times and delivery times. Discret. Appl. Math. 2011 (2010). doi:10.1016/j.dam.2011.07.005
10. Kubzin, M.A., Strusevich, V.A.: Planning machine maintenance in two machine shop scheduling. Oper. Res. **54**, 789–800 (2006)
11. Lee, C.Y.: Machine scheduling with an availability constraints. J. Global Optim. **9**, 363–384 (1996)
12. Qi, X.: A note on worst-case performance of heuristics for maintenance scheduling problems. Discret. Appl. Math. **155**, 416–422 (2007)
13. Qi, X., Chen, T., Tu, F.: Scheduling the maintenance on a single machine. J. Oper. Res. Soc. **50**, 1071–1078 (1999)
14. Rapine, C., Brauner, N., Finke, G., Lebacque, V.: Single machine scheduling with small operator-non-availability periods. J. Sched. **15**, 127–139 (2012)

15. Schmidt, G.: Scheduling with limited machine availability. Eur. J. Oper. Res. **121**, 1–15 (2000)
16. Sahni, S.: Algorithms for scheduling independent tasks. J. ACM **23**, 116–127 (1976)
17. Yuan, J.J., Shi, L., Ou, J.W.: Single machine scheduling with forbidden intervals and job delivery times. Asia Pac. J. Oper. Res. **25**(3), 317–325 (2008)

A Multi-period Bi-level Stochastic Programming with Decision Dependent Uncertainty in Supply Chains

Yohanes Kristianto[⊠]

Department of Production, University of Vaasa, 65101 Vaasa, Finland
ykristiantonugroho@gmail.com

Abstract. The closed loop supply chain faces some challenges related to the complexity of setting production capacity, maximizing the product architecture modularity and operations scheduling when remanufacturing is included in the supply chain networks. A multi-period bi-level stochastic programming framework is used by setting product architecture modularity design is integrated with supply chain networks design at the upper level and multi-period operations scheduling at the lower level. The result show that supply chain tends to postpone the product architecture modularization until the end of product life is imminent. The bi-level optimization is proven to be good approach to get global optimum of the closed loop supply chain.

Keywords: Bi-level optimization · Operations scheduling · Product architecture modularity · Supply chain networks

1 Introduction

The implementation of product architecture modularity has its own key challenges in forward-to reverse logistics as follows.

1.1 Time to Introduce Product Architecture Modularization for Easy Recyclability

Product architecture modularity benefits to improve product recyclability. The level of recyclability can be improved by changing materials or physical layout without changing components interfaces. A contribution of [1] insists on the benefit of modularity to increase demands by allowing return policy. The more modular a product, the lower degree of interdependency among components [2] and the more components can be recycled, dis-assembled, recovered, and finally reused at the next cycle of product life. While modularity allows manufacturing and remanufacturing to be done without the explosion in costs from design, processing, inventory, and production that have been common [3, 4], it is necessary to decide on when the modularization is most beneficial to be launched.

© Springer International Publishing Switzerland 2014
P. Fouilhoux et al. (Eds.): ISCO 2014, LNCS 8596, pp. 315–324, 2014.
DOI: 10.1007/978-3-319-09174-7_27

1.2   The Implication of Modularization for Closed Loop Supply Chain Total Costs

While forward and reverse logistics promises the economics of production and the economics of recycling, it has challenges for instances (1) the conflicts of operational goals among supply chain members, (2) there is a lack of appropriate models for use as tools to manage the corresponding logistics flows associated with forward and reverse logistics, and (3) there is a lack of interest to return cores to their original equipment manufacturer (OEM) [5]. The implications for effectively closing the supply chain loop must be assessed toward economic benefit [6] and the increased customer wants for after life service and recyclable products [7]. Therefore the well function of reverse logistic networks aims to calculate the total collecting costs [8], minimize the remanufacturing cost [9] and to estimate the effect of return complexity for capacity planning in the remanufacturing operations [10].

1.3   Operations Scheduling for Forward–Reverse Manufacturing Systems in the Supply Chain

The hybrid manufacturing-remanufacturing supply chain considers recovery rates and part uniqueness as two factors that leads to stochastic routings and processing times. Furthermore the MRP would not effectively implemented in this kind of situation. The supply chain has to be designed by considering the process yields uncertainty [11, 12]. One reason is that component yield affects on the prices of new components [13]. Therefore, it is important to consider the re-scheduling and production capacity planning for remanufacturing at each time period and each processing facilities that are capable of damping the stochasticity of production rates due to remanufacturing process.

2   Research Questions (RQs)

Considering the mentioned challenges, the following research questions are raised. (RQ1) How do product architecture modularity and capacity planning are linked to operations scheduling? (RQ2) What solution method can efficiently optimize the both objectives?

3   Solution Methodology

We consider the following simple example of six sites supply chain networks. For each period t, product development investment for site j, $M_j(t)$ is calculated based on product architecture modularity m_{it} and its unit cost θ_{jt} according to the following equation:

$$M_j(t) = \theta_{jt}m_{jt} \quad j \in J, t \in T \tag{1}$$

3.1   GDP Formulation for Product Modularity-Supply Chain Integration (RQ1)

Considering the superstructure of supply chain networks (Fig. 1), the MILP is better presented in disjunctive formulation of generalized disjunctive programs (GDP), which consists of a linear program with disjunctions, but no Boolean variables. A particular logic operation in disjunctive programming that helps to generate a tighter formulation than other methods (i.e., hierarchical relaxations (HR) and big M) of the problem is the so-called Basic Step. Basic step symbolizes the interconnection between two sites of supply chain by logical relations for the Boolean variables $\Omega(Y)$. The $\Omega(Y)$ is expressed as propositional logic in terms of the different operators, as a set of linear equality and inequality constraints. The GDP can be used to solve the both of product modularity-supply chain integration and operations scheduling.

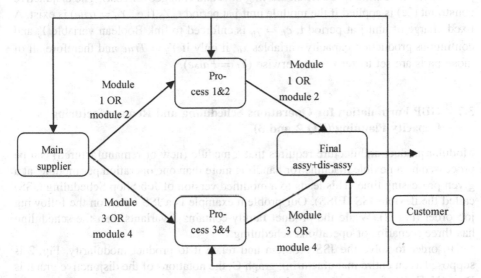

Fig. 1. Multi-period product architecture modularity and supply chain networks planning with remanufacturing capacity for period t

One block in Fig. 1 stands for one site with several number of production units that process several modules or components. For each period t, each integer variable $Y_{jt} \in (y_{11t}, y_{12t})$ for y_{11t}, y_{12t} signifies active/inactive state for module 1, $Y_{jt} \in (0, 1)$. If it is active, a set of continuous decision variables such as production capacity (x_{it}) and therefore direct costs of production and transportation are covered in it, indirect production cost c_{it} (i.e., administration, paper works, import duty, etc.) and the level of product modularity $M_i(t)$ are optimized. The GDP formulation of the integration is presented below:

$$\min Z = \sum_i c_{it} + f(m_{jt}, x_{it}) \tag{2a}$$

$$\text{s.t. } g(x) \leq 0 \tag{2b}$$

$$\begin{bmatrix} Y_{jt} \\ k_{jt}(m,x) \le 0 \\ c_{jt} = \gamma_{jt} \\ Scheduling\,equation\,(4) \end{bmatrix} \bigvee \begin{bmatrix} \neg Y_{jt} \\ B_{jt}(m,x) = 0 \\ c_{it} = 0 \end{bmatrix} i \in D \quad i \in I, j \in J, t \in T \qquad (2c)$$

$$\Omega(Y) = True, \quad x, m \in R^n, \quad t, c \ge 0, \quad Y \in \{True, False\}^n \qquad (2d)$$

Equation (2a) optimizes a Boolean Y_{jt} variable (whether the module j is being operated at period t or not) according to scenario s and continuous variable x_{it} of production capacity at stream i and time t. The global constraint $g(x) \le 0$ (2b) is employed for component transfer from one site to the next site, irrespective of the discrete choice of site allocation Y_{jt} (whether the site is operated or not). Y_{jt} may involve linear and nonlinear terms (i.e., mass balance and product specifications) and production scheduling (production starting time and changeover cost). The disjunctive constraint (2c) is applied if the module unit j at period t, Y_{jt} (i.e., $Y_{jt} = true$) is exist. A fixed charge of unit j at period t, $c_{jt} = \gamma_{jt}$ is enforced to link Boolean variable Y_j and continuous production capacity variables x_{it}, if only if $Y_{jt} = True$ and therefore all of those parts are set to zero for otherwise ($Y_{jt} = False$).

3.2 GDP Formulation for Operations Scheduling and Remanufacturing Capacity Planning (RQ 2 and 3)

Modular product architecture requires that a module (new or remanufactured) can be processed by a flexible machine for handling more than one operation per module, at a given processing time. This leads to a modified version of Job Shop Scheduling (JSS) called the flexible JSS (FJSS). Our problem example is a JSS based on the following job order (Fig. 2). While the product family contains 3 variants then the scheduling has three scenarios of operations scheduling.

In order to solve the JSS problem and relate it to product modularity, Fig. 2 is supposed as an entire manufacturing graph G, the notations of the disjunctive graph is D = (N; Z; W) for $D \in G$ where N is the set of nodes that represents machines (the machines (i,j) and jobs sequences (i,k) are prescribed), Z is the set of conjunctive arcs (for instance arcs $P1 \rightarrow P3$) and W is the set of disjunctive arcs (for instance arcs $P1 \leftrightarrow P2$). We use properties of disjunctive graph as defined by [14] as follows: (1) there must be a path from 1 to j (predecessor node) and from j to other node n within graph G, (2) there is no other path to connect i to j for $(i,j) \in W$, (3) The objective is to find any selection S_k (any disjunctive arc) in clique $Z_k, Z = \cup(Z_k : k \in M)$ for M is a set of machine for instance $M = \cup(P6, P7, P8)$ in such a way that the total make span (MS) is at minimum.

3.3 Multi-period Stochastic Operations Scheduling and Remanufacturing Capacity Planning

For multi-stage stochastic programming (SP) Models, it is necessary to track decisions at each node of the scenario tree. So, for Multi-stage SP, we will track decisions by

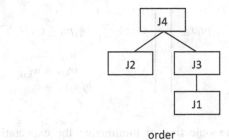

Fig. 2. An example of the relationship among order, jobs (J), operations (P), and tool group (TG)

node number and formulate the stochastic programming for T time periods and scenario tree S as follows [15]:

$$\min \Sigma_{s\in S} p_s \sum_i c_{it} + f\left(m_{jt}, w_{i,ts}, x_{i,t}\right) \tag{3a}$$

$$\text{s.t.} \geq C_{it} + pt_i x_{its} \quad i \in I, \left(s, s'\right) \in S, t \in T \tag{3b}$$

$$w_{i,ts} \in W_{i,s} \quad i \in I, \left(s, s'\right) \in S, t \in T \tag{3c}$$

$$w_{i,ts} = w_{i,ts'} \quad i \in I, \left(s, s'\right) \in S, t \in T \tag{3d}$$

$$\xi_{i,ts} w_{i,ts} \geq m_{j,t} x_{i,t} \quad i \in I, j \in J, t \in T, s \in S \tag{3e}$$

$$m_{j,t} \geq \xi_{i,ts} \quad i \in I, j \in J, t \in T, s \in S \tag{3f}$$

$$m_{j,ts}^{LB} \geq m_{j,ts} \quad j \in J, s \in S, t \in T \tag{3g}$$

$$
\begin{bmatrix}
C_{its} + \sum_{\substack{p\in J(i) \\ p\le j}} pt_i \le C_{kts} + \sum_{\substack{p\in J(k) \\ p\le j}} pt_k x_{kts} \\[4pt]
w_{k,ts} = w_{k,ts'} \\
w_{i,ts} = w_{i,ts'} \\
b_{j,(t+1)s} = b_{j,(t+1)s'}
\end{bmatrix}
\; \forall \;
\begin{bmatrix}
C_{kts} + \sum_{\substack{p\in J(k) \\ p\le j}} pt_k \le C_{its} + \sum_{\substack{p\in J(i) \\ p\le j}} pt_i x_{its} \\[4pt]
w_{i,ts} = w_{i,ts'} \\
w_{k,ts} = w_{k,ts'} \\
b_{j,(t+1)s} = b_{j,(t+1)s'}
\end{bmatrix}
$$

Over the left bracket is $P_{i,kts,s'}$ and over the right bracket is $P_{k,its,s'}$.

$$(3h)$$

Equation (3a) represents the objective of minimizing the expectation of some economic criterions, i.e., the probability of scenario s, p_s, production rate for manufacturing and remanufacturing of component i, $x_{it}, w_{i,ts}$ for time period t in scenario s. Following the upper level objective function, Constraint (3b) is a period-linking constraint for a particular scenario. Constraint (3c) represents integrality and bound restrictions on variables x_{its}. The non-anticipatory constraint (NAC) (3d) link decisions for different scenarios at certain period t. Constraint (3e) states that the production capacity of module i is bounded and has to anticipate endogenous production yield uncertainty, $\xi_{i,ts}$ at time t. Constraint (3e) consists of bilinear terms $m_{j,t} x_{i,t}$ and therefore the constraint is non-convex. Constraint (3f) states that the product modularity cannot less than $\xi_{i,ts}$. The $m_{j,ts}^{LB}$ is a lower bound (LB) for product architecture modularity of module j and is used as cut variables of the lower level (Constraint 3g). The disjunctive constraint (3h) states that after the indistinguishable scenarios s, s' has been made for two consecutive jobs i and k, in period t, $P_{i,kts,s'}$, the both scenarios must exist for the resolution of production scheduling just after revealing the exogenous uncertainty $\xi_{i,ts}$ at time t + 1 on facility j, $Y_{jt} \in \{0,1\}$.

While we have two levels (upper and lower) levels optimization, the linking between those two levels requires integer cuts for choosing supply chain configuration and continuous cuts for constraining the level of production capacities and product architecture modularity. Integer cuts are used to exclude subsets and supersets of previously obtained feasible configurations and to exclude infeasible configurations from future calculations. Continuous cuts are used to force values of state variables in upper level $(m_{jt}^{UL}, x_{it}^{UL})$ to be greater than or equal to their values in the lower level $(m_{jt}^{LL}, x_{it}^{LL})$ and thus the lower bound is improved at each iterations until upper and lower bounds are equal or their gaps are at allowable tolerance (Fig. 3).

4   Results

(RQ1) How do product architecture modularity and capacity planning are linked to operations scheduling?

Product modularity and capacity planning decision follows the upper level decision. The results show that the decision of opening or closing a facility does not depend solely on the product development or production cost (Table 1). Indeed, plant capacity and the level of product modularity are two decision variables for making capacity investment decision.

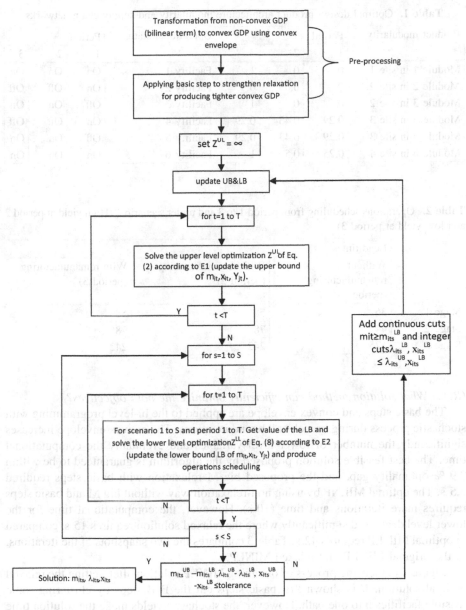

Fig. 3. The solution algorithm of the bi-level multi-period stochastic programming

The operations scheduling follows the lower level decision as follows:

An analysis of the scheduling results in Table 2 unveils that the stochastic solutions reported for Period 2 and 3 are not better than the one found for period 1. The discrepancy may be due to an additional in the number of time slots predefined for every processing unit in order to adjust the make span according to the remanufacturing capacity.

Table 1. Optimal design product architecture modularity and supply chain networks

Product modularity	Period			Operational status	Period		
	1	2	3		1	2	3
Module 1 in site 1	0	0	1	Facility 1	On	On	On
Module 2 in site 2	0	0,09	0,9	Facility 2	On	Off	Off
Module 3 in site 2	0	0	1	Facility 3	Off	On	On
Module 4 in site 3	0,29	0,43	0,29	Facility 4	On	Off	Off
Module 5 in site 3	0,29	0,43	0,29	Facility 5	Off	On	On
Module 6 in site 4	0,25	0,5	1	Facility 6	On	On	On

Table 2. Operations scheduling from period 1 to 3 and under scenario 1 (High yield at period 2 and low yield at period 3)

	Lead times		
	Without remanufacturing (period 1)	With remanufacturing (period 2)	With remanufacturing (period 3)
Stage 1	36	58	58
Stage 2	44	78	78
Stage 3	175	242	242

(RQ2) What solution method can efficiently optimize the both objectives?

The basic steps and convex envelope are applied to the bi-level programming with stochastic process during branch and bound iterations. The convex envelope increases significantly the number of lower level variables and eventually the computational time. The best feasible solution proposed by the algorithm is guaranteed to be within 2.9 % optimality gap, and the proposed big M relaxation with basic steps required 3,5 s. The optimal MILNP by using non-relaxation way without big M and basic steps requires more iterations and time (10 s). However, the computational time for the lower level decreased significantly where the relaxed solution requires 15 s, compared to optimal MINLP requires 42 s. Table 3 compares the two snapshots of the iterations, of the original MINLP and relaxed MINLP.

Figure 4 depicts the process of obtaining MINLP solution after getting the relaxed optimal solution. It is shown that basic steps cut the B-B steps by clustering some feasible facilities into one path. However, the stochastic yields make the solution time of lower level becomes longer and the gap between optimal and relaxed becomes wider (Table 3). Further investigation on the use of stronger relaxation of hierarchical relaxation (HR) is necessary by considering the reality that HR provides stronger relaxation than big M for nonlinear problems.

Table 3. Results from the six facilities supply chains

	Big-M after basic steps	
	Upper level	Lower level
Optimal solution	408,6	488,6
Relaxation	354,8	381,76
Number of constraints	253	723
Number of variables	115	483
Number of binaries	72	135
Number of nodes	14	14
Solution time (s)	3	9

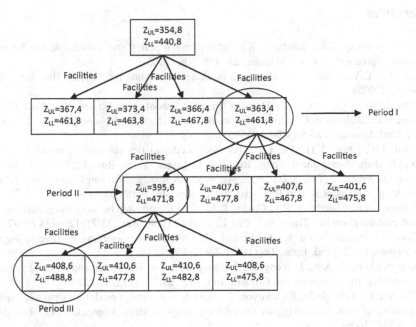

Fig. 4. Branch and Bound (B-B) method for the relaxation of Big M with basic steps

5 Concluding Remarks

This paper has given a general overview of bi-level stochastic programming application for optimizing the integration of product architecture modularity and supply chain networks planning by considering the operations constraints in terms of process scheduling and the opportunity of expanding the plant capacity. An example in supply chains networks planning with four locations and ten processing facilities has been presented to illustrate how GDP greatly facilitates the modeling of these problems. Furthermore, big-M relaxation and basic steps reformulations have been presented as two major mixed-integer algebraic models that can be systematically derived from a GDP model. The combination results in stronger continuous relaxations, and hence stronger lower bounds, which may potentially translate into a more efficient solution times, although this is not still satisfactory since the big M relaxation is weaker than Hierarchical Relaxation (HR) in handling non-convex and nonlinear problems. The application of basic steps following some basic rules strengthen the B-B method could be reduced in terms of its nodes and computational time.

Acknowledgments. The authors are most grateful to the two anonymous reviewers, who provided helpful comments on the presentation of this paper. This research is supported by the postdoctoral research funding from the Academy of Finland under decision number 269693 and project number 2700041211.

References

1. Mukhopadhyay, S.K., Setoputro, R.: Optimal return policy and modular design for build-to-order products. J. Oper. Manage. **23**, 496–506 (2005)
2. Baldwin, C.Y., Clark, K.B.: Managing in an age of modularity. Harvard Bus. Rev. **75**(5), 84–93 (2005)
3. Pine, J.: Mass Customization. Harvard Business School Press Boston, New York (1993)
4. Lee, H.L., Billington, C.: The evolution of supply chain models and practice at Hewlett-Packard. Interfaces **25**(5), 42–63 (1995)
5. Sheu, J.B., Chou, Y.H., Hu, C.C.: An integrated logistics operational model for green-supply chain management. Trans. Res. Part E: Logist. Trans. Rev. **41**, 287–313 (2005)
6. Zhu, Q., Sarkis, J., Lai, K.L.: Green supply chain management implications for "closing the loop". Trans. Res. Part E: Logist. Trans. Rev. **44**, 1–18 (2008)
7. Ellinger, A.E., Daugherty, P.J., Gustin, C.M.: The relationship between integrated logistics and customer service. Trans. Res. Part E: Logist. Trans. Rev. **33**(2), 129–138 (1997)
8. Kara, S., Rugrungruang, F., Kaebernick, H.: Simulation modelling of reverse logistics networks. Int. J. Prod. Econ. **106**(1), 61–69 (2007)
9. Kim, K., Song, I., Kim, J., Jeong, B.: Supply planning model for remanufacturing system in reverse logistics environment. Comput. Ind. Eng. **51**(2), 279–287 (2006)
10. Vlachos, D., Georgiadis, P., Iakovou, E.: A system dynamics model for dynamic capacity planning of remanufacturing in closed-loop supply chains. Comput. Oper. Res. **34**(2), 367–394 (2006)
11. Chiu, M.C., Okudan, G.: An integrative methodology for product and supply chain design decisions at the product design stage. J. Mech. Des. **133**, 1–15 (2011)
12. Chiu, M.C., Okudan, G.: An investigation on the impact of product modularity level on supply chain performance metrics: an industrial case study. J. Intell. Manuf. (2012). doi:10.1007/s10845-012-0680-3
13. Bakal, I., Ekcali, E.: Effects of random yield in remanufacturing with price-sensitive supply and demand. Prod. Oper. Manage. **15**(3), 407–420 (2006)
14. Balas, E.: Disjunctive Programming and a hierarchy of relaxations for discrete optimization problems. SIAM J. Alg. Disc. Math. **6**, 466–486 (1985)
15. Tarhan, B., Grossmann, I.E.: A multistage stochastic programming approach with strategies for uncertainty reduction in the synthesis of process networks with uncertain yields. Comput. Chem. Eng. **32**, 766–788 (2008)

{k}-Packing Functions of Graphs

Valeria Alejandra Leoni[1,2](✉) and Erica G. Hinrichsen[1]

[1] Depto. de Matemática, Escuela de Formación Básica, Facultad de Ciencias Exactas,
Ingeniería y Agrimensura, Universidad Nacional de Rosario, Rosario, Argentina
{ericah,valeoni}@fceia.unr.edu.ar
[2] CONICET, Buenos Aires, Argentina

Abstract. Given a positive integer k and a graph G, a k-limited packing in G (2010) is a subset B of its vertex set such that each closed neighborhood has at most k vertices of B. As a variation, we introduce the notion of a {k}-*packing function* f of G which assigns a non-negative integer to the vertices of G in such a way that the sum of $f(v)$ over each closed neighborhood is at most k. For fixed k, we prove that the problem of finding a {k}-packing function of maximum weight ({k}PF) can be reduced linearly to the problem of finding a k-limited packing of maximum cardinality (kLP). We present an $O(|V(G)| + |E(G)|)$ time algorithm to solve {k}PF on strongly chordal graphs. We also use monadic second-order logic to prove that both problems are linear time solvable for graphs with clique-width bounded by a constant.

1 Introduction

The notion of a k-limited packing in a graph was introduced by Gallant et al. in 2010 as a generalization of a 2-packing in a graph [10]. Given a simple undirected graph G with vertex set $V(G)$ and a positive integer k, a set $B \subseteq V(G)$ is a k-*limited packing* in G if each closed neighborhood has at most k vertices of B. Observe that a k-limited packing in G can be considered as a function $f : V(G) \rightarrow \{0,1\}$ such that $\sum_{w \in N_G[v]} f(w) \leq k$ for all $v \in V(G)$, where $N_G[v]$ denotes the closed neighborhood of vertex v. The maximum possible size of a k-limited packing in G is denoted by $L_k(G)$.

When $k = 1$, a k-limited packing in G is a *2-packing* in G and $L_k(G)$ is the well-known number $\rho(G)$.

This concept is a good model for many utility location problems in operations research. In most of them, the utilities are necessary but probably obnoxious. That is why it is of interest to place the maximum number of utilities in such a way that no more than a given number of them is near each agent in a given scenario.

The above definition induces the study of the following optimization problem, already introduced (as a decision problem) in [5]:

Partially supported by grantsPICT ANPCyT 0482 (2011–2013) and 1ING 391 (2012–2014).

P. Fouilhoux et al. (Eds.): ISCO 2014, LNCS 8596, pp. 325–335, 2014.
DOI: 10.1007/978-3-319-09174-7_28

k-**LIMITED PACKING** (kLP), for fixed positive integer k.
Given a graph G, find a k-limited packing in G of size $L_k(G)$.

The study of the computational complexity of kLP was started in [5]. Regarding NP-complete results, on the one hand, it is known that kLP is NP-complete even for instances given by split graphs [5], and also for bipartite graphs [6]. On the other hand, kLP is polynomial time solvable for strongly chordal graphs [5]. Regarding graph classes defined by forbidding subgraphs with "too many" induced path with four vertices, kLP is known to be polynomial time solvable for P_4-tidy graphs (that include cographs and P_4-sparse graphs) [5].

In view of other applications, in this paper we consider that in each vertex of the graph we are allowed to locate more than one utility. We introduce the notion of a $\{k\}$-packing function (defined in Sect. 3) and consider for a fixed positive integer k, the optimization problem concerning the existence of a $\{k\}$-packing function of maximum weight in a given graph ($\{k\}$PF). We introduce the study of the computational complexity of $\{k\}$PF. Specifically, in Sect. 4 we show that $\{k\}$PF can be reduced in linear time to kLP. The reduction is based on the strong product of graphs. Using this reduction, we derive that $\{k\}$PF is polynomial time solvable on strongly chordal graphs. Based on this result, in Sect. 5 we present an $O(|V(G)| + |E(G)|)$ time algorithm for solving $\{k\}$PF on strongly chordal graphs. Using the same reduction and monadic second-order logic (MSOL), in Sect. 6 we prove that both problems can be solved in linear time for graphs that have clique-width bounded by a constant. As a by-product, we generalize existing results concerning polynomial time solvable instances of kLP.

2   Notation and Background

2.1   Graphs, p-Graphs and Clique-Width

In this paper we use the term *graph* for a simple (without self loops or multiple edges) undirected graph. We use the term *labeled graph* for a graph having labels which are associated with its vertices, such that each vertex has exactly one label. A *p-graph* is a labeled graph with vertex labels in $\{1, \ldots, p\}$. An unlabeled graph is considered as a 1-graph such that all its vertices are labeled by 1.

For a graph G, $V(G)$ and $E(G)$ denote respectively its vertex and edge sets, and for $v \in V(G)$, $N_G[v]$ the *closed neighborhood* of v in G.

K_n denotes the complete graph on n vertices, i.e. the graph with $|V(K_n)| = n$ and $E(G)$ containing all the edges between two distinct vertices from $V(K_n)$.

A *chord* of a cycle C in a graph G is an edge (u, v) not in C such that u and v lie in C. A graph G is *chordal* if it does not contain an induced chordless cycle on n vertices for any $n \geq 4$. A graph G is *strongly chordal* if it is chordal and every cycle of even length in G has an odd chord, i.e. a chord that connects two vertices that are at odd distance apart from each other in the cycle.

Given G_1 and G_2 two graphs, the *strong product* $G_1 \otimes G_2$ is defined on the vertex set $V(G_1) \times V(G_2)$, where two vertices u_1v_1 and u_2v_2 are adjacent if and only if $u_1 = u_2$ and $(v_1, v_2) \in E(G_2)$, or $v_1 = v_2$ and $(u_1, u_2) \in E(G_1)$, or $(v_1, v_2) \in E(G_2)$ and $(u_1, u_2) \in E(G_1)$.

Given G_1 and G_2 two disjoint graphs and $v \in V(G_1)$, we denote $G_1[G_2/v]$ the graph H obtained by the substitution in G_1 of G_2 for v, i.e. $V(H) = V(G_1) \cup V(G_2) - \{v\}$ and

$$E(H) = E(G_2) \cup \{e : e \in E(G_1), \ e \text{ is not incident to } v\}$$

$$\cup \{(u, w) : u \in V(G_2), w \in V(G_1) \text{ and } w \text{ is adjacent to } v \text{ in } G_1\}.$$

With every p-graph G, an algebraic expression built using the following operations can be associated [3]:

- creation of a vertex with label i,
- disjoint union,
- renaming label i to label j,
- connecting all vertices with label i to all vertices with label j, for $i \neq j$.

If all the labels in the expression of G are in $\{1, \ldots, q\}$ for positive integer q, the expression is called a q-expression of G. It is clear to see that there is a $|V(G)|$-expression which defines G, for every graph G. For a positive integer q, $\mathcal{C}(q)$ denotes the class of p-graphs which can be defined by q-expressions. The clique-width of a p-graph G, denoted by $cwd(G)$, is defined by $cwd(G) = \min\{q : G \in \mathcal{C}(q)\}$.

Some important classes of graphs have clique-width bounded by a constant. For example, $\mathcal{C}(1)$ is the class of edgeless graphs, $\mathcal{C}(2)$ the class of cographs (cf. [4]), trees have clique-width at most 3 and P_4-tidy graphs have clique-width at most 4 [4].

A useful property concerning the clique-width of a graph is the following:

Lemma 1 [4]. *Given two disjoint graphs G_1 and G_2 and $v \in V(G_1)$, it follows that $cwd(G_1[G_2/v]) = \max\{cwd(G_1), cwd(G_2)\}$.*

Clique-width is a measure of the difficulty of decomposing a graph in a kind of tree-structure that yields efficient graph algorithms.

2.2 Graphs as Logical Structures and Monadic Second-Order Logic

The vocabulary $\{E\}$ consisting of one binary relation symbol E is denoted by τ_1. For a graph G, $G(\tau_1)$ denotes the presentation of G as a τ_1-structure $<V, E>$, where V is the domain of the logical structure $(V(G))$ and E is the binary relation corresponding to the adjacency matrix of G.

Regarding graph properties, if a formula can be defined using

- vertices and sets of vertices of a graph,
- the logical operators OR, AND, NOT, (denoted by $\vee$, $\wedge$, $\neg$),
- the logical quantifiers $\forall$ and $\exists$ over vertices and sets of vertices,
- the membership relation $\in$, to check whether an element belongs to a set,

- the equality operator $=$ for vertices and
- the binary adjacency relation *adj*, where $adj(u, v)$ holds if and only if vertices u and v are adjacent,

the formula can be said to be expressible in τ_1-monadic second-order logic, $\mathrm{MSOL}(\tau_1)$ for short.

An optimization problem P is a $\mathrm{LinEMSOL}(\tau_1)$ optimization problem over graphs, if it can be defined in the following form: Given a graph G presented as a τ_1-structure and functions $f_1, \ldots, f_m$ associating integer values to the vertices and edges of G, find an assignment z to the free variables in θ such that

$$\sum_{\substack{1 \le i \le l \\ 1 \le j \le m}} a_{ij} |z(X_i)|_j =$$

$$= \max \left\{ \sum_{\substack{1 \le i \le l \\ 1 \le j \le m}} a_{ij} |z'(X_i)|_j : \theta(X_1, \ldots, X_l) \text{ is true for } G \text{ and } z' \right\},$$

where θ is an $\mathrm{MSOL}(\tau_1)$ formula having free set variables $X_1, \ldots, X_l$, $a_{ij} : 1 \le i \le l$, $1 \le j \le m$ are integer numbers and $|z(X_i)|_j := \sum_{a \in z(X_i)} f_j(a)$. More details on the definition of $\mathrm{MSOL}(\tau_1)$ and $\mathrm{LinEMSOL}(\tau_1)$ can be found for example in [1,2,7,13].

It has been shown that $\mathrm{MSOL}(\tau_1)$ is particularly useful when combined with the concept of the graph parameter clique-width as the following theorem —first stated in [1] and then reinforced in [12]— shows:

Theorem 1 [1,12]. *Let q be a constant and $\mathcal{C}$ be a class of graphs of clique-width at most q (i.e. $\mathcal{C} \subseteq \mathcal{C}(q)$). Then every $\mathrm{LINMSOL}(\tau_1)$ problem on $\mathcal{C}$ can be solved in linear time.*

3 $\{k\}$-Packing Functions

In this section we introduce formally the notion of a $\{k\}$-packing function of a graph.

For a given graph G and a function $f : V(G) \to \mathbb{R}$ we denote $f(A) = \sum_{v \in A} f(v)$, where $A \subseteq V(G)$. The *weight* of f is $f(V(G))$.

Definition 1. *Given a graph G and a positive integer k, a $\{k\}$-packing function of G is a function $f : V(G) \to \mathbb{Z}_0^+$ such that for all $v \in V(G)$,*

$$f(N_G[v]) \le k.$$

The maximum possible weight of a $\{k\}$-packing function of G is denoted by $L_{\{k\}}(G)$.

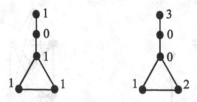

Fig. 1. A graph G with $L_3(G) = 4$ and $L_{\{3\}}(G) = 6$.

Fig. 2. $L_3(K_3) = L_{\{3\}}(K_3) = 3$.

Since any k-limited packing in G can be seen as a $\{k\}$-packing function of G (Figs. 1, 2), it is clear to see that

$$L_k(G) \leq L_{\{k\}}(G).$$

In this paper we introduce the study of the computational complexity of the following optimization problem:

$\{k\}$-**PACKING FUNCTION** ($\{k\}$PF), for fixed positive integer k.
Given a graph G, find a $\{k\}$-packing function of G of weight $L_{\{k\}}(G)$.

4   Linear Time Reduction of $\{k\}$PF to kLP

We begin this section by considering the graph transformation defined in the following way. Given a graph G and a positive integer k, $T^k(G)$ denotes the graph obtained from G by performing the strong product of G with the complete graph on k vertices K_k, i.e. $T^k(G) = G \otimes K_k$ (Fig. 3).

We can prove:

Theorem 2. *$\{k\}$PF can be reduced in linear time to kLP.*

Proof. Given a graph G and a positive integer k, consider the transformation $T^k(G)$ defined above and let $V(G) = \{u_1, \ldots, u_n\}$ and $V(K_k) = \{v_1, \ldots, v_k\}$.

Let f be a $\{k\}$-packing function of G and define the set $B \subseteq V(T^k(G))$ in the following way:

$$B := \{u_j v_r : 1 \leq r \leq f(u_j), j = 1, \ldots, n\}.$$

From the way B was built, it is clear to see that B is a k-limited packing in $T^k(G)$.

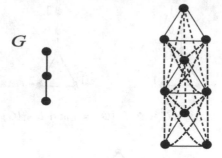

G

Fig. 3. Transformation $T^k(G)$ for the graph G in the left and $k = 3$.

Conversely, let $\widetilde{B}$ be a k-limited packing in $T^k(G)$. This implies

$$|N_{T^k(G)}[w] \cap \widetilde{B}| \leq k, \quad \text{for each } w \in V(T^k(G)). \tag{1}$$

Notice that if $w = u_i v_j$ is any vertex in $V(T^k(G))$, then

$$N_{T^k(G)}[w] = \{u_i v_r : r = 1, \ldots, k\} \cup \{u_s v_r : (u_i, u_s) \in E(G), \ r = 1, \ldots, k\}. \tag{2}$$

Let us define f over $V(G)$ in the following way:

$$f(u_i) := |\{u_i v_j : j = 1, \ldots, k\} \cap \widetilde{B}|, \quad \text{for each } i = 1, \ldots, n.$$

From (1) and (2) follows that f is a well-defined $\{k\}$-packing function of G.   □

Remark 1. $L_{\{k\}}(G) = L_k(T^k(G))$, for every graph G and positive integer k.

On the one hand, from Theorem 2 it is clear that finding NP-complete instances of $\{k\}$PF is a way to determine NP-complete instances of kLP.

On the other hand, given a positive integer k and two graph classes $\mathcal{R}$ and $\mathcal{T}$ such that if $G \in \mathcal{R}$ then $T^k(G) \in \mathcal{T}$, it is clear to see that provided that kLP is polynomial time solvable for $\mathcal{T}$, then $\{k\}$PF is polynomial time solvable for $\mathcal{R}$. For instance, if we consider strongly chordal graphs, it is not difficult to prove that if G is a strongly chordal graph, then $T^k(G)$ also is, for every fixed positive integer k. Therefore, as kLP is polynomial time solvable on strongly chordal graphs [5], then we derive that $\{k\}$PF also is. In the next section we present an $O(|V(G)| + |E(G)|)$ time algorithm to solve $\{k\}$PF on strongly chordal graphs.

5   An $O(|V(G)| + |E(G)|)$ Time Algorithm for $\{k\}$PF on Strongly Chordal Graphs

The algorithm that we present in this section follows some ideas given in [11] in the context of Y-dominating functions.

We first recall a known characterization of strongly chordal graphs [9]. A vertex v of a graph G is called *simplicial* in G if $N_G[v]$ is a clique in G, i.e. a subset of pairwise adjacent vertices in G. A *strong elimination ordering* of a graph G is an ordering $(v_1, \ldots, v_n)$ of its vertices such that:

1. v_i is simplicial in the subgraph G_i induced by $\{v_i, v_{i+1} \ldots, v_n\}$, and
2. for $i \leq j \leq k$, if v_j and v_k belong to $N_{G_i}[v_i]$ then $N_{G_i}[v_j] \subseteq N_{G_i}[v_k]$.

A graph is *strongly chordal* if and only if it has a strong elimination ordering.

In the remainder, we present a simple algorithm that takes G and a positive integer k as inputs, where G represents a strongly chordal graph with strong elimination ordering $(v_1, v_2, \ldots, v_n)$, and finds in $O(|V(G)| + |E(G)|)$ time, a maximum $\{k\}$-packing function of G.

Algorithm $\mathrm{MP}(G, k)$

Initialization

1. for $i = 1$ to n do
2. $f(v_i) = 0$
3. endfor

Increasing the weight of the function

4. for $i = 1$ to n do
5. $M = \min\{k - f(N_G[v]) : v \in N_G[v_i]\}$
6. $f(v_i) = M$
7. endfor
8. return function f.

Lemma 2. *Let G and k be respectively a strongly chordal graph and a positive integer k. Algorithm $\mathrm{MP}(G, k)$ solves $\{k\}PF$ for G in $O(|V(G)| + |E(G)|)$-time, if the strong elimination ordering of G is given.*

Proof. Consider the function returned by Algorithm $\mathrm{MP}(G, k)$. Clearly, the function f at the beginning of the first iteration of Steps 4–7 is a $\{k\}$-packing function of G. We assume that the function f of i-th iteration of Steps 4–7 is a $\{k\}$-packing function of G for $1 \leq i \leq n$. For each $v \in N_G[v_i]$, since $M \leq k - f(N_G[v])$, then $f(N_G[v]) + M \leq k$, which implies that the new function f obtained by changing the value of $f(v_i)$ in Step 6 is still a $\{k\}$-packing function of G, for $1 \leq i \leq n$. Therefore, the function returned by the Algorithm at Step 8 is a $\{k\}$-packing function of G.

In the following, we show that the function f returned from Algorithm $\mathrm{MP}(G, k)$ is a maximum $\{k\}$-packing function of G.

Let h be a maximum $\{k\}$-packing function of G such that the cardinality of $W = \{v \in V : f(v) \neq h(v)\}$ is minimum. We claim that $W = \emptyset$. Suppose that $W \neq \emptyset$ and let t be the smallest index such that $v_t \in W$. Notice that $h(v_x) = f(v_x)$ if $x < t$.

Case 1. $h(v_t) > f(v_t)$. At t-th iteration of Steps 4–7, $h(v_x) \geq 0 = f(v_x)$ for $x > t$. At Step 5, let $v_c \in N_G[v_t]$ such that $M = k - f(N_G[v_c])$. We have

$$h(N_G[v_c]) = h(N_G[v_c] - \{v_t\}) + h(v_t) \geq f(N_G[v_c] - \{v_t\}) + h(v_t) >$$
$$f(N_G[v_c] - \{v_t\}) + f(v_t) \geq k - M + M = k,$$

which contradicts the assumption that h is a $\{k\}$-packing function of G.

Case 2. $h(v_t) < f(v_t)$. Let $P = \{v \in N_G[v_t] : h(N_G[v]) + f(v_t) - h(v_t) > k\}$. We will prove that $P \neq \emptyset$. If not, $h(N_G[v]) + f(v_t) - h(v_t) \leq k$ for every $v \in N_G[v_t]$ and the function g defined by $g(v_t) = f(v_t)$ and $g(v) = h(v)$ otherwise, is a $\{k\}$-packing function of G with $g(V(G)) > h(V(G))$, a contradiction to the assumption that h is a maximum $\{k\}$-packing function of G.

Since for every vertex $v \in P$, $h(N_G[v]) + f(v_t) - h(v_t) > k \geq f(N_G[v])$, then $h(N_G[v]) - h(v_t) > f(N_G[v]) - f(v_t)$. Thus, $N_G[v] \cap \{v_x \in W : t < x, h(v_x) > f(v_x)\} \neq \emptyset$, for every $v \in P$.

Let s be the smallest index of vertices in P and b be the smallest index of vertices in $N_G[v_s] \cap \{v_x \in W : t < x, h(v_x) > f(v_x)\}$. There exist two positive integers c_1 and c_2 such that $f(v_t) = h(v_t) + c_1$ and $h(v_b) = f(v_b) + c_2$. Let $c = \min\{c_1, c_2\}$. We define h' as follows: $h'(v_t) = h(v_t) + c$, $h'(v_b) = h(v_b) - c$ and $h'(v) = h(v)$ otherwise. Clearly $h(V(G)) = h'(V(G))$ and $|\{v \in V(G) : f(v) = h'(v)\}| \geq |\{v \in V(G) : f(v) = h(v)\}| + 1$.

Notice that $P \subseteq N_G[v_t]$ and let us prove that $P \subseteq N_G[v_b]$. For this purpose, we analyze the two possible cases:

– $s \leq t$. Then $s \leq t < b$. Since v_t and v_b belong to $N_{G_s}[v_s]$ then, by definition of the strong elimination ordering, we have $N_{G_s}[v_t] \subseteq N_{G_s}[v_b]$. Since $P \subseteq N_{G_s}[v_t]$ then $P \subseteq N_{G_s}[v_b]$.
– $t < s$. Take any $v_k \in P$, thus $t < s \leq k$. Since v_s and v_k belong to $N_{G_t}[v_t]$, by definition of the strong elimination ordering, $N_{G_t}[v_s] \subseteq N_{G_t}[v_k]$ for every vertex $v \in P$. Since $v_b \in N_{G_t}[v_s]$, then $v_b \in N_{G_t}[v_k]$. In other words, we have $P \subseteq N_G[v_b]$.

In order to prove that h' is a $\{k\}$-packing function of G, it is enough to prove that $h'(N_G[v]) \leq k$ for each $v \in N_G[v_t]$.

If $v \in N_G[v_t]$ and $v \notin N_G[v_b]$, we have $h'(N_G[v]) = h'(N_G[v] - \{v_t\}) + h'(v_t) = h(N_G[v] - \{v_t\}) + h(v_t) + c = h(N_G[v]) + c \leq h(N_G[v]) + c_1 \leq k$ since $h(N_G[v]) + f(v_t) - h(v_t) \leq k$.

If $v \in N_G[v_t] \cap N_G[v_b]$, then $h'(N_G[v]) = h'(N_G[v] - \{v_t, v_b\}) + h'(v_t) + h'(v_b) = h(N_G[v] - \{v_t, v_b\}) + h(v_t) + c + h(v_b) - c = h(N_G[v]) \leq k$.

We conclude that h' is a maximum $\{k\}$-packing function of G such that $|\{v \in V : f(v) \neq h'(v)\}| < |\{v \in V : f(v) \neq h(v)\}|$, a contradiction to the assumption that $|W|$ is minimum.

The initialization in Steps 1–4 can be done in $O(|V(G)|)$ time. At i-th iteration of Steps 4–7, M can be computed in $O(|N_G[v_i]|)$ time by verifying $k - f(N_G[v])$ for every vertex $v \in N_G[v_i]$. Hence, the running time of Algorithm $MP(G, k)$ is $O(\sum_{v \in V} |N_G[v]|) = O(|V(G)| + |E(G)|)$.   □

6 {k}PF and kLP for Bounded Clique-Width Graphs

In this section, we combine monadic second-order logic in the context of k-limited packings with the reduction performed in Theorem 2 to find linear time solvable instances for both problems. We first have:

Proposition 1. *For each fixed positive integer k, kLP is a LinEMSOL(τ_1) optimization problem.*

Proof. Let k be a fixed positive integer. Following the notation of Sect. 2.2, kLP can be expressed as follows: Given a graph G presented as a τ_1-structure $G(\tau_1)$ and one evaluation function (the constant function that associates 1's to the vertices of G) and denoting by $X(v)$ the atomic formula indicating that $v \in X$, find an assignment z to the free set variable X in θ such that

$$|z(X)|_1 = \max\{|z'(X)|_1 : \theta(X) \text{ is true for } G \text{ and } z'\},$$

where $\theta(X)$ is following the MSOL(τ_1) formula

$$\theta(X) = \forall v \neg \left(\bigwedge_{1 \leq r \leq k+1} A_r(X, v, u_1, \ldots, u_r) \right),$$

with

$$A_1(X, v, u_1) := \exists u_1 \left[X(u_1) \wedge (adj(v, u_1) \vee v = u_1) \right]$$

and for each $r > 1$

$$A_r(X, v, u_1, \ldots, u_r) := \exists u_r \left[X(u_r) \wedge (adj(v, u_r) \vee v = u_r) \wedge \bigwedge_{1 \leq i \leq r-1} \neg(u_r = u_i) \right].$$

□

Now, by Theorem 1 and Proposition 1, we obtain the following result:

Theorem 3. *Let q be a constant and $\mathcal{C}$ be a class of graphs of clique-width at most q (i.e. $\mathcal{C} \subseteq \mathcal{C}(q)$). Then kLP can be solved in linear time on $\mathcal{C}$.*

Theorem 3 applies for instance to P_4-tidy graphs, a graph class already considered in [5] when studying the complexity of kLP. Hence, that result in [5] follows now from Theorem 3.

Next, notice that given a graph G and a positive integer k, we can obtain equivalently $T^k(G)$ by performing a substitution in G on each of its vertices by K_k. Taking into account Lemma 1 together with the fact that $cwd(K_k) \leq 2$, we remark that, for every positive integer k and any constant $q \geq 2$, if G is a graph in $\mathcal{C}(q)$, then $T^k(G)$ is in $\mathcal{C}(q)$. Therefore, if we recall the observations given below Theorem 2, we have proved:

Theorem 4. *Let q be a constant and $\mathcal{C}$ be a class of graphs of clique-width at most q (i.e. $\mathcal{C} \subseteq \mathcal{C}(q)$). Then $\{k\}PF$ can be solved in linear time on $\mathcal{C}$.*

7   Final Remarks

In this paper, we have introduced a new concept in graphs ($\{k\}$-packing function) as a variation of a k-limited packing, for each positive integer k. In this way, we have expanded the set of utility location problems that can be modeled using packings in graphs.

By relating the problems of finding a maximum $\{k\}$-packing function with the one of finding a maximum k-limited packing, we have opened another way of research that can be used to continue with the study of the computational complexity of kLP.

As already mentioned in Sect. 1, up to our knowledge, the only other known graph class in which kLP is polynomial time solvable is the class of P_4-tidy graphs [5]. We remark that it is not true that for every P_4-tidy graph G, $T^k(G)$ is a P_4-tidy graph. Actually, we solved in Sect. 6 the complexity of $\{k\}$PF in a superclass of P_4-tidy graphs. Besides, concerning kLP, the results in Sect. 6 improve and generalize the result in [5] related to P_4-tidy graphs.

We close this paper with the question of deciding if the algorithm in Sect. 5 can be modified to answer the complexity of $\{k\}$PF for other proper subclasses of chordal graphs that are characterized by certain elimination ordering.

Acknowledgments. We are grateful to H. Freytes for the discussions held with us around first- and second-order logics.

References

1. Courcelle, B.: The monadic second-order logic of graphs. I. Recognizable sets of nite graphs. Inf. Comput. **85**(1), 12–75 (1990)
2. Courcelle, B., Mosbah, M.: Monadic second-order evaluation on tree-decomposable graphs. Theoret. Comput. Sci. **109**, 49–82 (1993)
3. Courcelle, B., Engelfriet, J., Rozenberg, G.: Handle-rewriting hypergraph grammars. J. Comput. Syst. Sci. **46**, 218–270 (1993)
4. Courcelle, B., Makowsky, J.A., Rotics, U.: Linear time solvable optimization problems on graphs of bounded clique width. Theory Comput. Syst. **33**, 125–150 (2000)
5. Dobson, M.P., Leoni, V., Nasini, G.: The k-limited packing and k-tuple domination problems in strongly chordal, P_4-tidy and split graphs. Electron. Notes Discrete Math. **36**, 559–566 (2010)
6. Dobson, M.P., Leoni, V., Nasini, G.: The multiple domination and limited packing problems in graphs. Inf. Process. Lett. **111**, 1108–1113 (2011)
7. Ebbinghaus, H., Flum, J.: Finite Model Theory. Perspectives in Mathematical Logic. Springer, Berlin (1995)
8. Fagin, R.: Generalized RST-order spectra and polynomial-time recognizable sets. In: Karp, R. (ed.) Complexity of Computation. SIAM-AMS Proceedings, vol. 7, pp. 27–41. SIAM, Philadelphia (1974)
9. Farber, M.: Characterizations of strongly chordal graphs. Discrete Math. **43**, 173–189 (1983)
10. Gallant, R., Gunther, G., Hartnell, B., Rall, D.: Limited packings in graphs. Discrete Appl. Math. **158**(12), 1357–1364 (2010)

11. Lee, C.M., Chang, M.S.: Variations of Y-dominating functions on graphs. Discrete Math. **308**, 4185–4204 (2008)
12. Oum, S., Seymour, P.: Approximating clique-width and branch-width. J. Combinat. Theory Ser. B **96**, 514–528 (2006)
13. Papadimitriou, C.H.: Computational Complexity. Addison Wesley Longman, Reading (1994)

Robust Shift Scheduling in Call Centers

Sara Mattia[1], Fabrizio Rossi[2], Mara Servilio[2]([⊠]), and Stefano Smriglio[2]

[1] Istituto di Analisi dei Sistemi ed Informatica,
Consiglio Nazionale delle Ricerche, Rome, Italy
sara.mattia@cnr.iasi.it
[2] Dipartimento di Ingegneria e Scienze dell'Informazione e Matematica,
Università di L'Aquila, L'Aquila, Italy
{fabrizio.rossi,mara.servilio,stefano.smriglio}@univaq.it

Abstract. We propose a robust optimization model for shift scheduling in call centers. The model is designed to react to the deviations that often occur between the planned staffing levels and the actual number of employees that would be necessary to guarantee the desired level of service. Different perturbation patterns are considered giving rise to different uncertainty sets, and the corresponding algorithmic implications are discussed. A case study from an Italian Public Agency is finally presented, which shows how the proposed methodology improves the quality of the schedules. Interestingly, although the methodology is fairly sophisticated, it perfectly fits in a quite common managers current practice.

Keywords: Integer programming · Shift scheduling · Robust optimization

1 Introduction

Workforce management (WFM) in call centers is a complex process which receives historic data on call volumes as input and returns agent schedules as output. Traditionally, WFM is split into a sequence of almost separate steps: forecasting call volumes, determining the required *staffing levels* (that is, the number of agents required on duty in each time slot of a day in order to guarantee the desired level of service), translating them into agents work shifts (*shift scheduling*) and, finally, monitoring schedule inadherence at operational level and reacting accordingly (we refer to [1,7] for a comprehensive description of WFM). This decomposition has some drawbacks, as the actual deviations from the planned operations are only revealed and managed at short term level. This typically overcharges real-time operations managers, who exploit all available options to keep safe the required level of service. A recent stream of research in this context is devoted to overcome the problems caused by the separation between the *staffing stage* and the *shift scheduling stage*. The first receives in input the call volume estimates and translates them into staffing levels; the second determines an allocation of work shifts which guarantees a satisfactory trade-off between coverage of the staffing levels and personnel costs. Investigating options to smooth the separation between this

© Springer International Publishing Switzerland 2014
P. Fouilhoux et al. (Eds.): ISCO 2014, LNCS 8596, pp. 336–346, 2014.
DOI: 10.1007/978-3-319-09174-7_29

two stages is well-known to be relevant in practice, and it is interesting from a mathematical viewpoint as well. In fact, it basically amounts to integrate queueing theory techniques with integer programming models. A survey of papers following such a direction can be found in [1]. More recent studies introduce stochastic programming and robust optimization models [2,9,10].

A new *robust optimization* model is here presented, defined and validated along several joint projects with large Italian Telecommunications Companies and Public Agencies. In these experiences, call center managers reported to us the fact that too often the computed staffing levels do not cope with the actual demand. Our idea is then to design a robust optimization model for the shift scheduling stage which reacts to deviations in the staffing levels. It is important to remark that approximations in the staffing levels can derive from errors in the call volume estimates (often a quite difficult task) as well as from approximations made at the staffing stage. The LATTERS typically arise when applying queueing models to multi-queue/multi-skill systems, in which several assumptions of the standard models are violated (see, e.g. [8]).

We look at two possible patterns of staffing levels deviations, giving rise to different definitions of the *uncertainty set*, and discuss the related algorithmic issues. We finally present a case study showing that the resulting methodology provides high quality schedules which protect the system against unplanned deviations at quite reasonable cost. Interestingly, it also perfectly fits in a quite common practice, based on moving personnel from back to front office in order to manage schedule inadherence.

2   The Deterministic Model

We consider a discrete planning horizon $T = \{1, \ldots, m\}$, where time intervals typically correspond to 15 or 30 min. For each $t \in T$, let b_t the staffing level at period t, i.e. a positive integer representing the required number of agents on duty in period t. Work shifts can be *part-time* with duration 4 or 6 h or *full-time* with duration 8 h, and do not include breaks (which are assumed to be managed at real-time level). Let us denote by J the set of all possible shifts, and by c_j a cost associated to shift j.

The shift scheduling problem consists in determining the number of agents to assign to each shift in order to satisfy the staffing levels at minimum cost.

A basic integer programming model for this problem was introduced by Segal [16]. Let us define the shift matrix $S \in \{0,1\}^{m \times n}$ with $s_{tj} = 1$ if shift j covers period t and 0 otherwise, and denote by x_j the number of agents assigned to shift $j \in J$. The problem writes

$$\min \sum_{j \in J} c_j x_j \tag{1}$$

$$\sum_{j \in J} s_{tj} x_j \geq b_t \qquad\qquad t \in T$$

$$x_j \geq 0, \text{integer} \qquad\qquad j \in J$$

Observe that, since shifts do not include breaks, each column has consecutive ones. This property allows to reformulate the problem as a minimum cost flow problem [16].

One major criticism to this model, that we experienced in practical contexts, is that it enforces a full coverage of the staffing levels. On the other hand, in some cases allowing *understaffing* at some periods may result in a significant saving. Furthermore, an understaffing cost can be included in the model, to represent alternative agents recruitment options at operational level, such as taking them from the back office or asking for some overtime. A more "hidden" cost is also taken in great care by the managers, the one associated to *overstaffing*, arising when an excess of employees occurs (see, e.g., [19] for further insights).

Let us now introduce these amendments in the Segal model. Denote by $o_t, u_t \in \mathbb{R}_+$ the number of employees in excess (overstaffing) resp. defect (understaffing) at period t; let w_o, w_u the corresponding costs (assumed time-independent). Formulation (1) can be revised by introducing such variables acting as surplus and slack variables:

$$\min \sum_{j \in J} c_j x_j + \sum_{t \in T} (w_o o_t + w_u u_t) \tag{2}$$

$$\sum_{j \in J} s_{tj} x_j + u_t - o_t = b_t \qquad t \in T$$

$$o_t, u_t \geq 0 \qquad t \in T$$

$$x_j \geq 0, \text{integer} \qquad j \in J$$

3   The Robust Model

Recall that the staffing levels, i.e., the demand vector b in (1) and (2), result from a complex process involving two major steps: (i) the call volumes estimation from historic call center data and (ii) the translation into agents requirements, often realized by exploiting Erlang-like queueing models (we refer the reader to [1, 7] as starting points for the huge literature concerning this topic). Call volume estimation is intrinsically hard in practice. Furthermore, most of the analytical staffing methods are based on assumptions that are often invalid in practice (see, e.g. [8]), and require the additional use of simulation, especially in complex multi-queue/multi-skill systems (see again [7]). Among these assumptions, it is worth to recall that the staffing level expected for period t is typically considered independent from the staffing levels of the other time periods.

For these reasons, let us now suppose that the staffing levels b_t for $t \in T$ are *uncertain* and let us denote by U the uncertainty set including all possible realizations. We will refer to the problem with uncertainty as the *robust problem* and to the problem where b is fixed as the *nominal* (deterministic) *problem*.

Robust solutions are mostly computed using either *stochastic programming* or *robust optimization*. A stochastic programming approach [14] assumes that probability distributions for the uncertain parameters are given and it computes a solution with a given probability to be feasible for all the realizations. A traditional robust optimization method [3,5,17] finds a solution that is feasible for all possible realizations of the parameters in the given uncertainty set. In both cases the same solution is used for all the realizations (*static solution*). Usually, if the probability distributions for the parameters are reliable, stochastic programming produces less conservative solutions, but robust optimization models are computationally easier to solve. In particular, when the uncertainty affects only the right-hand-sides, a robust solution can be computed solving a deterministic problem with suitable values for the right-hand-sides [12]. To reduce the conservatism, the solution can be computed in stages (*dynamic solution*), that is, it can be computed in two or more steps using the partial information on the uncertainty that becomes available at every step. This approach is also called *optimization with recourse* and it makes the problem more difficult to solve. The complexity of multi-stage stochastic programming is investigated in [15], where it is discussed why a multi-stage linear problem with recourse is computationally intractable. *Two-stage robust optimization* (a.k.a. adjustable robust optimization) was introduced in [4], where it is proved that two-stage robust programming with unrestricted second stage (for each realization of the uncertainty it is possible to choose second stage variables arbitrarily) is computationally intractable. For this reason, the second stage IS often limited using so-called affine policies [6]. In [13] it is proved that, even with uncertainty on right-hand-sides only and with an LP nominal problem, the two-stage robust problem is NP-hard if the uncertainty set is a polyhedron, even for very simple polyhedra (e.g. a single knapsack constraint). Applications of two-stage optimization (both restricted and unrestricted) can be found in network design [11], production planning and inventory management [13].

A static approach is not applicable for our problem. Model (2) includes only equality constraints, therefore if we suppose to have uncertain right-hand-sides and to compute a static robust solution, this would lead to an infeasible robust problem. Moreover, a static solution does not correspond to what is done in practice, where it happens that the staffing levels computed before the realization of the uncertainty are adjusted by the operators when the real demand b becomes available. This naturally leads to a two-stage problem, where part of the solution (the staffing levels x) is computed before the realization of the uncertainty and the rest (overstaffing/understaffing $[u, o]$) is computed after. In this way we obtain a dynamic solution, i.e., a solution that can be (partially) changed according to the realization of the uncertainty.

The two-stage formulation for the robust shift scheduling problem with understaffing/overstaffing is the following:

$$\min_{x \in X} \left\{ \sum_{j \in J} c_j x_j + \max_{b \in U} \min_{[u,o] \in R(x,b)} \sum_{t \in T} (w_o o_t + w_u u_t) \right\} \qquad (3)$$

The set X contains the feasible x vectors that can be chosen in the first stage. To control the price of robustness, we require that the staffing cost of the computed robust solution (i.e., the cost related to the x variables) cannot exceed a given percent increase τ of the optimal planning cost $C = c^T x^*$ of the nominal problem:

$$X = \{x \in \mathbb{Z}_+^{|J|} : \sum_{j \in J} c_j x_j \leq (1 + \tau)C\} \tag{4}$$

Now, if we do not restrict in any way the value of the second stage (unrestricted second stage), the problem of computing the worst-case realization and the corresponding overstaffing/understaffing values reads:

[**Primal Second Stage problem**]

$$\max_{b \in U} \min_{[u,o] \in R(x,b)} \sum_{t \in T} (w_o o_t + w_u u_t) \tag{5}$$

where $R(x, b)$ is the set of the feasible $[o, u]$ pairs, given realization b and staff allocation x.

3.1 Solving the Robust Problem

Given a staff allocation $\bar{x}$ computed in the first stage and a realization $\bar{b}$, the corresponding understaffing/overstaffing levels are $o_t = \max\{0, -d_t\}$ and $u_t = \max\{0, d_t\}$, where $d_t = \bar{b}_t - \sum_{j \in J} s_{jt} \bar{x}_j$, for a resulting cost of $\sum_{t \in T}(w_o o_t + w_u u_t)$. This can be expressed by the following LP:

$$\min \sum_{t \in T} (w_o o_t + w_u u_t)$$
$$(y_t) \quad u_t - o_t = \bar{b}_t - \sum_{j \in J} s_{tj} \bar{x}_j \quad t \in T \tag{6}$$
$$o, u \geq 0$$

where, for every slot t, o_t and u_t cannot be both positive in the optimal solution due to the objective costs. The dual of (6) is:

$$\max \sum_{t \in T} (\bar{b}_t - \sum_{j \in J} s_{jt} \bar{x}_j) y_t$$
$$(o_t) \quad -y_t \leq w_o \quad t \in T \tag{7}$$
$$(u_t) \quad y_t \leq w_u \quad t \in T$$

Also (7) can be easily solved by inspection. Hence the second stage problem (5) can be alternatively formulated as:

[**Dual Second Stage problem**]

$$\max_{b \in B, y \in Y} \sum_{t \in T} (b_t - \sum_{j \in J} s_{tj} x_j) y_t \tag{8}$$

where Y is the feasible region of (7). This allows to reformulate the robust problem (3) as:

$$\min_{x \in X} \left\{ \sum_{j \in J} c_j x_j + \max_{b \in U, y \in Y} \sum_{t \in T} (b_t - \sum_{j \in J} s_{tj} x_j) y_t \right\} \qquad (9)$$

that can be rewritten as:

$$\min \sum_{j \in J} c_j x_j + \gamma$$

$$\gamma \geq \sum_{t \in T} (b_t - \sum_{j \in J} s_{tj} x_j) y_t \quad b \in U, y \in Y \qquad (10)$$

$$x \in X$$

Problem (10) can be solved by dynamically generating the constraints:

$$\gamma \geq \sum_{t \in T} (b_t - \sum_{j \in J} s_{tj} x_j) y_t \quad b \in U, y \in Y.$$

Namely, given a solution $[\bar{x}, \bar{\gamma}]$ of (10), the associated separation problem is the Dual Second Stage problem (8). Let $[b^*, y^*]$ be the corresponding optimal solution: if $\bar{\gamma} \leq \sum_{t \in T} (b_t^* - \sum_{j \in J} s_{tj} \bar{x}_j) y_t^*$, then no violated inequality exists, otherwise $[b^*, y^*]$ provides a violated inequality that can be added to the current formulation.

In general, the Dual Second Stage problem (8) is difficult to solve. A way to compute a solution is to use the Primal Second Stage problem (5) instead, that can be formulated as follows:

$$\max \sum_{t \in T} (w_o o_t + w_u u_t)$$

$$u_t - o_t = b_t - \sum_{j \in J} s_{tj} \bar{x}_j \quad t \in T \qquad (11)$$

$$o_t \leq M \alpha_t \quad t \in T$$

$$u_t \leq M(1 - \alpha_t) \quad t \in T$$

$$b \in U, o, u \geq 0, \alpha \in \{0, 1\}^{|T|}$$

Binary variables α are required to ensure that, for each t, at most one between o_t and u_t is positive. In fact, differently from (6) this is necessary as the problem is in maximization form. Once the solution $[u^*, o^*, \alpha^*, b^*]$ of (11) is known, we can use values $[u^*, o^*]$, to compute the corresponding optimal solution y^* of (7) that will provide a violated inequality (if any).

Solving (10) to optimality requires the embedding of the cutting plane algorithm into a branch-and-cut framework. However, in Sect. 4 we show that rounding down the optimal solution of the linear relaxation of (10) returns good quality solutions at cheap computational cost.

3.2   Uncertainty Set Structure

A key point in the above procedure is the solution of problem (11), that is in general NP-hard. In this context, we compare two different structures of the uncertainty set. Let b_t be the true value of the demand at period t. Uncertainty can be defined as:

$$U = \{\mathbf{b} : b_t = \tilde{b}_t + D_t z_t, \; |z_t| \leq 1, \; t \in T; z \in \mathbb{R}^T\} \tag{12}$$

where $\tilde{b}_t$ represents the nominal value of the staffing level at period t, z_t are variables indicating the percentage (positive or negative) of perturbation of the demand due to the uncertainty, $D_t = (b_t^{\max} - b_t^{\min})/2$ and $b_t^{\max}, b_t^{\min}$ are the maximum and minimum values that the variation can take for $t \in T$.

A classical restriction that is added to the uncertainty set to avoid system overprotection, consists in limiting the number of variations from the nominal values that can happen at the same time. If Γ is the number of time periods whose demand can deviate, binary variables ζ can be used to compute the number of demands with deviation, i.e.:

$$U_\Gamma = \{\mathbf{b} : b_t = \tilde{b}_t + D_t z_t, |z_t| \leq \zeta_t, \; t \in T; \sum_{t \in T} \zeta_t \leq \Gamma; z \in \mathbb{R}^T, \zeta \in \{0,1\}^T\} \tag{13}$$

It is shown in [18] that U_Γ can lead to an easy solvable problem (11). However, in our case, it is unlikely that variations in consecutive time periods are completely unrelated. Correlation between consecutive time periods can be modeled by adding to the uncertainty set the following constraints:

$$|D_t z_t - D_{t-1} z_{t-1}| \leq \Delta(t) \qquad t \in T \tag{14}$$

where $\Delta(t)$ indicates the maximum possible difference between the perturbations of two consecutive time periods. The uncertainty set U_Δ reads:

$$\begin{aligned} U_\Delta = \{\mathbf{b} : b_t = \tilde{b}_t + D_t z_t, \\ |D_t z_t - D_{t-1} z_{t-1}| \leq \Delta(t), |z_t| \leq \zeta_t, \; t \in T; \\ \sum_{t \in T} \zeta_t \leq \Gamma; z \in \mathbb{R}^T, \zeta \in \{0,1\}^T\} \end{aligned} \tag{15}$$

Unfortunately, U_Δ complicates the solution of (11), but in the next Section will show that the computational burden is limited.

4   Case Study

4.1   Instances Description

We deal with data extracted from a call center of a Public Italian Agency. Call center operating hours are from 7:00 a.m. to 11:00 p.m. The working period is divided into 64 time slots of 15 min length. Agents may have a full-time shift,

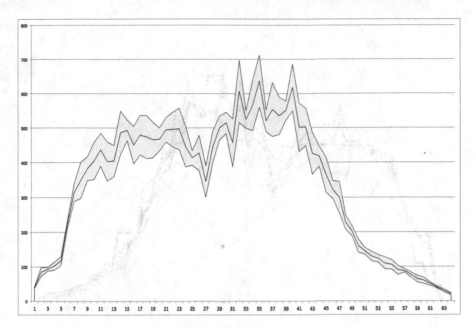

Fig. 1. Space U of possible realizations of b

made by 8 (32) consecutive working hours (slots), or a part-time shift, consisting of 4 (16) or 6 (24) consecutive working hours (slots). A workshift can start at any time period in the day but must terminate before 11.00 p.m. The shift cost depends on the contract type: a 4-h part-time shift has a cost of 72 €, a 6-h part-time shift has a cost of 96 € and a 8-h full-time shift has a cost of 112 €. The understaffing cost w_u is fixed at 120 €, while the overstaffing cost is $w_o = 20$ €.

We discuss two possible scenarios: the first considers only part-time agents (instance I-4-6) while in the second one all possible shifts are considered (instance I-4-6-8). Figure 1 represents the demand variation, i.e., the continuous line represents the nominal values $\tilde{b}_t$ and the dotted lines represent respectively $b^{\max}$ and $b^{\min}$. One can observe that the (worst) realized staffing values either increase or decrease of an amount varying from the 5 % to the 15 % of the nominal value.

In defining U_Γ and U_Δ we arranged with call center managers $\Gamma = 16$ and $\Delta(t) = 20$ for all $t \in T$. Finally, we fix the value τ in (4) to 0.1 (10 % planning budget increase allowed).

4.2   Results

Tables 1 and 2 report the costs obtained by the nominal vs the robust model evaluated with the two different uncertainty sets U_Γ and U_Δ. From the figures one can observe that the increase in workshifts costs request by robust models never exceeds 9 % of the workshifts cost of the nominal model. On the contrary,

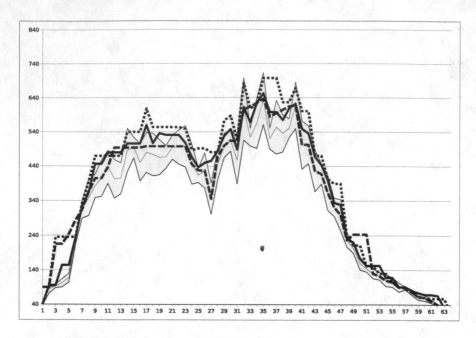

Fig. 2. Instance I-4-6: comparison among planned levels

robust models are obviously able to considerably decrease the cost needed to cope with the worst realization.

To better understand the added values of robust models, one can observe the slot-by-slot planned levels reported in Figs. 2 and 3. The planned level by the nominal solution is represented by the large dashed line, while the continuous (small dashed) one is associated with the U_Δ-robust (U_Γ-robust) planning. Both robust models are able to follow the demand variations, and the model U_Δ

Table 1. Uncertainty set U_Γ: nominal model vs robust model

Instance	Workshifts costs		Worst-case costs		Cutting plane	CPU
	Nominal	Robust U_Γ	Nominal	Robust U_Γ	Iterations	Time (s)
I-4-6	96,168	104,424	198,960	101,822	634	491
I-4-6-8	88,992	93,362	202,700	96,691	642	544

Table 2. Uncertainty set U_Δ: nominal model vs robust model

Instance	Workshifts costs		Worst-case costs		Cutting plane	CPU
	Nominal	Robust U_Δ	Nominal	Robust U_Δ	Iterations	Time (s)
I-4-6	96,168	98,859	93,360	50,845	468	597
I-4-6-8	88,992	89,437	97,760	51,718	333	675

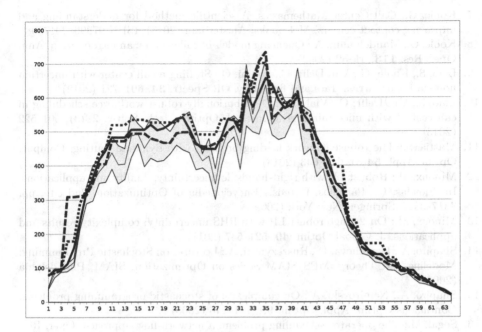

Fig. 3. Instance I-4-6-8: comparison among planned levels

appears very promising because smoothly follows the demand curve at a very low additional planning cost (about 2 % increase). It is also worth to mention that the computational burden of the robust model U_Δ appears negligible.

In conclusion, the computational experience shows that using a robust approach can lead to a relevant improvement with respect to the approach used in practice, consisting in computing a static solution and then in adapting it to the real value of the demand on the fly.

References

1. Aksin, Z., Armony, M., Mehrotra, V.: The modern call center: a multi-disciplinary perspective on operations management research. Prod. Oper. Manage. **16**(6), 665–688 (2007)
2. Atlason, J., Epelman, M.A., Henderson, S.G.: Optimizing call center staffing using simulation and analytic center cutting plane methods. Manage. Sci. **54**(2), 295–309 (2008)
3. Ben-Tal, A., Nemirowski, A.: Robust convex optimization. Math. Oper. Res. **23**, 769–805 (1998)
4. Ben-Tal, A., Goryashko, A., Guslitzer, E., Nemirovski, A.: Adjustable robust solutions of uncertain linear programs. Math. Program. **99**(2), 351–376 (2004)
5. Bertsimas, D., Sim, M.: Robust discrete optimization and network flows. Math. Program. Ser. B **98**, 49–71 (2003)
6. Chen, X., Zhang, Y.: Uncertain linear programs: extended affinely adjustable robust counterparts. Oper. Res. **57**(6), 1469–1482 (2009)

7. Koole, G.: Call Center Mathematics. A scientific method for understanding and improving contact centers. MG books, Amsterdam (2013). ISBN 9789082017908

8. Koole, G., Mandelbaum, A.: Queueing models of call centers: an introduction. Ann. Oper. Res. **113**, 41–59 (2002)

9. Liao, S., Koole, G., Van Delft, C., Jouini, O.: Staffing a call center with uncertain non-stationary arrival rate and flexibility. OR Spectr. **34**, 691–721 (2012)

10. Liao, S., Van Delft, C., Vial, J.P.: Distributionally robust workforce scheduling in call centres with uncertain arrival rates. J. Optim. Meth. Softw. **28**(3), 501–522 (2013)

11. Mattia, S.: The robust network loading problem with dynamic routing. Comput. Optim. Appl. **54**(3), 619–643 (2013)

12. Minoux, M.: Robust LP with right-handside uncertainty, duality and applications. In: Floudas, C., Pardalos, P. (eds.) Encyclopedia of Optimization, 2nd edn, pp. 3317–3327. Springer, New York (2009)

13. Minoux, M.: On 2-stage robust LP with RHS uncertainty: complexity results and applications. J. Global Optim. **49**, 521–537 (2011)

14. Shapiro, A., Dentcheva, D., Ruszczyński, A.: Lectures on Stochastic Programming: Modeling and Theory. MPS/SIAM Series on Optimization. SIAM, Philadelphia (2009)

15. Shapiro, A., Nemirovski, A.: On complexity of stochastic programming problems. Continuous Optim. Appl. Optim. **99**, 111–146 (2005)

16. Segal, M.: The operator-scheduling problem: a network-flow approach. Oper. Res. **22**(4), 808–823 (1974)

17. Soyster, A.: Convex programming with set-inclusive constraints and applications to inexact linear programming. Oper. Res. **21**, 1154–1157 (1973)

18. Thiele, A., Terry, T., Epelman, M.: Robust Linear Optimization With Recourse, Optimization Online (2010). http://www.optimization-online.org/DB_FILE/2009/03/2263.pdf

19. NextStep Workforce Planning. http://www.nextstepwfp.com/2012/01/how-much-is-over-staffing-costing-your-contact-centre

A Tabu Search Heuristic
for the Equitable Coloring Problem

Isabel Méndez Díaz[3], Graciela Nasini[1,2], and Daniel Severín[1,2]($\boxtimes$)

[1] Facultad de Ciencias Exactas, Ingeniería y Agrimensura,
Universidad Nacional de Rosario, Rosario, Argentina
{nasini,daniel}@fceia.unr.edu.ar
[2] CONICET, Rosario, Argentina
[3] Facultad de Ciencias Exactas y Naturales,
Universidad de Buenos Aires, Buenos Aires, Argentina
imendez@dc.uba.ar

Abstract. The *Equitable Coloring Problem* is a variant of the Graph
Coloring Problem where the sizes of two arbitrary color classes differ in
at most one unit. This additional condition, called equity constraints,
arises naturally in several applications. Due to the hardness of the problem, current exact algorithms can not solve large-sized instances. Such
instances must be addressed only via heuristic methods.

In this paper we present a tabu search heuristic for the Equitable Coloring Problem. This algorithm is an adaptation of the dynamic TABUCOL
version of Galinier and Hao. In order to satisfy equity constraints, new
local search criteria are given.

Computational experiments are carried out in order to find the best
combination of parameters involved in the dynamic tenure of the heuristic. Finally, we show the good performance of our heuristic over known
benchmark instances.

Keywords: Equitable coloring · Tabu search · Combinatorial
optimization

1 Introduction

The *Graph Coloring Problem* (GCP) is a very well-studied $\mathcal{NP}$-Hard problem
since it models many applications such as scheduling, timetabling, electronic
bandwidth allocation and sequencing problems.

Given a simple graph $G = (V, E)$, where V is the set of vertices and E is
the set of edges, a *k-coloring of* G is a partition of V into k sets $V_1, V_2, \ldots, V_k$,
called *color classes*, such that the endpoints of any edge lie in different color
classes. The GCP consists of finding the minimum number k such that G admits
a k-coloring, called the *chromatic number* of G and denoted by $\chi(G)$.

Some applications impose additional restrictions. For instance, in scheduling
problems, it may be required to ensure the uniformity of the distribution of

© Springer International Publishing Switzerland 2014
P. Fouilhoux et al. (Eds.): ISCO 2014, LNCS 8596, pp. 347–358, 2014.
DOI: 10.1007/978-3-319-09174-7_30

workload employees. Suppose that a set of tasks must be assigned to a set of workers so that pairs of tasks may conflict each other, meaning that they should not be assigned to the same worker. The problem is modeled by building a graph containing a vertex for every task and an edge for every conflicting pair of tasks. Workers are represented by colors. Then, in order for a coloring of this graph to represent a valid assignment of tasks to workers, the same number of tasks must be assigned to each worker. Since this is impossible when the number of tasks is not divisible by the number of workers, one can ask for the number of tasks assigned to two arbitrary workers can not differ by more than one. It is called *equity constraint* and the resulting problem is called *Equitable Coloring Problem* (ECP).

ECP was introduced in [1], motivated by an application concerning *garbage collection* [2]. Other applications of the ECP concern *load balancing problems* in multiprocessor machines [3] and results in *probability theory* [4]. An introduction to ECP and some basic results are provided in [5].

Formally, an *equitable k-coloring* (or just *k*-eqcol) of a graph G is a *k*-coloring satisfying the *equity constraint*, i.e. the size of two color classes can not differ by more than one unit. The *equitable chromatic number* of G, $\chi_{eq}(G)$, is the minimum k for which G admits a *k*-eqcol. The ECP consists of finding $\chi_{eq}(G)$ which is an $\mathcal{NP}$-Hard problem [5].

There exist some differences between GCP and ECP that make the latter harder to solve. It is known that the chromatic number of a graph is greater than or equal to the chromatic number of any of its induced subgraphs. Unfortunately, in the case of ECP, this property does not hold. For instance, if G is the graph shown in Fig. 1, by deleting v_5 from G, $\chi_{eq}(G)$ increases from 2 to 3.

As far as we know, there are few approximate and exact algorithms available in the literature related to ECP.

It was proved that, for any graph G, $\Delta(G) + 1$ is an upper bound of $\chi_{eq}(G)$ [6], where $\Delta(G)$ is the maximum degree of vertices in G. Based on this fact, a polynomial time algorithm for obtaining a *k*-eqcol of a graph G with $k \geq \Delta(G)+1$ is described in [7].

Two constructive heuristics called NAIVE and SUBGRAPH are given in [5] to generate greedily an equitable coloring of a graph. There also exist heuristic algorithms for constructing colorings that are "nearly" equitable [8,9], making

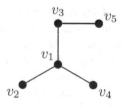

Fig. 1. An example.

emphasis on achieving a small difference between the sizes of the biggest class and the smallest one, although the equity constraint still might be violated.

The authors of [10] propose a tabu search heuristic to initialize an exact algorithm that solves ECP via Integer Linear Programming (ILP) techniques. Other exact algorithms for solving ECP are given in [11] and [12]. The first one also uses IPL techniques and the second one is based on a DSATUR enumeration scheme.

In this work, we propose a new heuristic based on the dynamic TABUCOL version of Galinier and Hao [13], one of the best tabu search algorithms for GCP [14]. Then, computational experiments are carried out in order to find the best combination of parameters involved in the dynamic tenure of our heuristic and to show the good performance of it over known benchmark instances.

The paper is organized as follows. In Sect. 2, we present TABUCOL and the dynamic variant of Galinier and Hao. In Sect. 3, we give our variant for ECP which we call TABUEQCOL. Finally, in Sect. 4 we report computational experiences and conclusions.

2   TABUCOL and its Variants

Tabu search is a metaheuristic method proposed by Glover [15] that guides a local search algorithm equipped with additional mechanisms that prevent from visiting a solution twice and getting stuck in a local optimum.

Let S be the solution space of the problem and $f : S \to \mathbb{R}$ be the objective function. The goal is to obtain a solution $s \in S$ such that $f(s)$ is minimum.

For each solution $s \in S$, consider a *neighborhood* $N(s) \subset S$ with two desirable (but not exclusionary) properties: (1) two solutions s and s' are neighbors when it is easy (from the computational point of view) to obtain s' from s, and to obtain $f(s')$ from $f(s)$ (for instance, in constant time), and (2) for any $s, s' \in S$, there exists a path $s = s_1, s_2, \ldots, s_m = s'$ such that $s_{i+1} \in N(s_i)$ for $i = 1, \ldots, m - 1$.

In general, neighbor solutions are very similar in some sense, and the difference between them can be seen as *features* that both solutions do not share. Consider a set of *features* P and a set $R \subset S \times P$ such that $(s, p) \in R$ if solution s presents a feature p.

Starting from an initial solution $s_0 \in S$, tabu search consists of generating a sequence of solutions $s_1, s_2, \ldots$ such that $s_{i+1} = \arg\min_{s \in N'(s_i)} f(s)$, where $N'(s_i)$ is a subset of $N(s_i)$ described below. In each iteration of this algorithm, a *movement* from s_i to s_{i+1} is performed and some feature of s_i is stored in a *tabu list* $L \subset P$. This list indicates whether a movement is allowed or forbidden: a solution s can be reached in the future only if s does not present any feature from L (this rule avoids from visiting a solution previously visited), except when s is better than the best solution found so far. This exception is called *aspiration* and the aspiration criterion is usually to check if the objective value of s is less than the value of currently-known best solution. Now, the set of allowed movements from s_i, $N'(s_i)$, is defined as

$$N'(s) = \{s' \in N(s) : f(s') < f(s^*) \ \lor \ (s',p) \notin R \ \forall \, p \in L\},$$

where s^* is the best solution found so far.

However, after several iterations, old features are no longer needed and it is better to remove them from the tabu list. This mechanism is usually implemented by assigning a "time of live" to each feature of the tabu list. Consider $live : L \to \mathbb{Z}$ and let $live(p)$ be the number of remaining iterations that p belongs to L. When a new feature p is inserted into L, $live(p)$ is assigned a value referred to as *tabu tenure* t. Then, in each iteration, the value of $live(p)$ is decreased by one unit until it reaches zero and p is removed from L. Above, we sketch a generic tabu search algorithm.

Data: initial solution s_0
Result: best solution found s^*
begin
    $L \leftarrow \varnothing$
    $s, s^* \leftarrow s_0$
    while *stopping criterion is not met* **do**
        for $p \in L$ **do**
            $live(p) \leftarrow live(p) - 1$
            if $live(p) = 0$ **then** $L \leftarrow L\backslash\{p\}$
        end
        $N'(s) \leftarrow \{s' \in N(s) : f(s') < f(s^*) \ \lor \ (s',p) \notin R \ \forall \, p \in L\}$
        choose a feature $p \in P$ such that $(s,p) \in R$
        $L \leftarrow L \cup \{p\}$
        $live(p) \leftarrow t$
        $s \leftarrow \arg\min_{s' \in N'(s)} f(s')$
        if $f(s) < f(s^*)$ **then** $s^* \leftarrow s$
    end
end

Algoritmo 1. TabuSearch

In order to implement a tabu search algorithm, some decisions must be taken: neighborhood of a solution, features of a solution, stopping criterion, how to choose the feature p to be stored in the tabu list and how to compute the tabu tenure t. In particular, the value of tabu tenure directly impacts *diversification* of the algorithm. A tabu search with low tenures behaves as a standard local search, where it frequently get trapped in local minima. On the other hand, a tabu search with high tenures tends to wander across solution space without converging towards the optimal solution.

TabuCol, the first tabu search algorithm designed for solving GCP, was proposed by Hertz and de Werra [16]. For a given graph $G = (V, E)$ and number $k \in \{1, \ldots, n\}$, where $n = |V|$, the goal of this algorithm is to find a k-coloring of G. In order to obtain a coloring that uses as few colors as possible, it is usual to embed TabuCol in a routine that, once a k-coloring is found, the algorithm can be restarted with $k \leftarrow k - 1$ and so on, until some criterion is met. Details of TabuCol are given below:

– *Search space and objective function.* A solution s is a partition $(V_1, V_2, \ldots, V_k)$ of the set of vertices. Let $E(V_i)$ be the set of edges of G with both endpoints in V_i. The objective function is defined as

$$f(s) = \sum_{i=1}^{k} |E(V_i)|.$$

Clearly, s is a k-coloring if and only if $f(s) = 0$.
– *Stopping criterion.* The algorithm stops when $f(s) = 0$ or when a maximum number of iterations is reached. Sometimes, a time limit is imposed.
– *Initial solution.* It is generated randomly. A suitable procedure given in [17] is the following. Start with empty sets $V_1, V_2, \ldots, V_k$ and, at each step, choose a non-considered vertex v randomly and put it into V_i with the smallest possible i such that $E(V_i)$ is not incremented. If it is not possible, choose a random number $j \in \{1, \ldots, k\}$ and put v into V_j.
– *Set of features.* It is $P = V \times \{1, \ldots, k\}$. A solution s presents a feature (v, i) if and only if $v \in V_i$, i.e. if v is assigned color i.
– *Neighborhood of a solution.* Let $C(s)$ be the *set of conflicting vertices* of a solution s, i.e.

$C(s) = \{v \in V : v \text{ is incident in some edge of } E(V_1) \cup E(V_2) \cup \ldots \cup E(V_k)\}$.

From a solution $s = (V_1, V_2, \ldots, V_k)$, a neighbor $s' = (V_1', V_2', \ldots, V_k')$ is generated as follows. Choose a conflicting vertex $v \in C(s)$. Let i be the color of v in s. Next, choose a color $j \in \{1, \ldots, k\} \setminus \{i\}$ and set

$$V_j' = V_j \cup \{v\}, \quad V_i' = V_i \setminus \{v\}, \quad V_l' = V_l \ \forall\, l \in \{1, \ldots, k\} \setminus \{i, j\}.$$

In other words, s' is a copy of s except that v is moved from class color V_i to V_j. We denote such operation with $s' = s(i \xrightarrow{v} j)$. Note that objective value can be computed in linear time from $f(s)$:

$$f(s') = f(s) + |\{vw \in E : w \in V_j\}| - |\{vw \in E : w \in V_i\}|.$$

Note also that searching all the neighbors of s requires exploring $(k-1)|C(s)|$ solutions. Original TABUCOL only explores a random subset of $N(s)$ while newer versions explore $N(s)$ completely.
– *Selection of feature to add in the tabu list.* Once a movement from s to $s(i \xrightarrow{v} j)$ is performed, $p = (v, i)$ is stored on tabu list and $live(p)$ is set to a fixed tabu tenure $t = 7$.

Later, Galinier and Hao [13] improved TABUCOL by using a dynamic tabu tenure that depends on the quality of the current solution, encouraging diversification of the search when solution is far from optimal. They proposed to assign a tenure of $t = \alpha|C(s)| + Random(\beta)$ where $Random(\beta)$ returns an integer randomly chosen from $\{0, \ldots, \beta - 1\}$ with uniform distribution. Based on experimentation, they suggest to use $\alpha = 0.6$ and $\beta = 10$. Other variants of TABUCOL are discussed in [14, 17].

3 TABUEQCOL: A Tabu Search for ECP

In this section, we present a new tabu search algorithm for ECP based on TABUCOL with dynamic tabu tenure, which we call TABUEQCOL.

Given a graph $G = (V, E)$ and a number $k \in \{1, \dots, n\}$, where $n = |V|$, the goal of TABUEQCOL is to find a k-eqcol of G.

Solution space consists of partitions of V into k sets $V_1, V_2, \dots, V_k$ such that they satisfy the equity constraint, i.e. for any pair of classes V_i and V_j, $||V_i| - |V_j|| \leq 1$. Objective function f is the same as in TABUCOL, so any solution s such that $f(s) = 0$ is indeed an equitable coloring. Also, set of features P is the same as in TABUCOL.

Stopping criterion depends on the experiment carried out. Usually, a time limit is imposed.

Let $s \in S$. Denote $W^+(s) = \{i : |V_i| = \lfloor n/k \rfloor + 1\}$ and $W^-(s) = \{i : |V_i| = \lfloor n/k \rfloor\}$, where V_i are the color classes of s. Since s satisfies the equity constraint, we have that $W^+(s)$ and $W^-(s)$ determine a partition of $\{1, \dots, k\}$ and, in particular, $|W^+(s)| = r$ where $r = n - k\lfloor n/k \rfloor$. From now on, we just write W^+ and W^-. These sets will be useful in the development of the algorithm.

We propose two greedy procedures for generating initial solution s_0.

Procedure 1. Start with empty sets $V_1, V_2, \dots, V_k$ and an integer $\tilde{r} \leftarrow 0$ (this value will have the cardinal of W^+). At each step, define set $I = \{i : |V_i| \leq M - 1\}$, where M is the maximum allowable size of a class:

$$M = \begin{cases} \lfloor n/k \rfloor + 1, & \text{if } \tilde{r} < r \\ \lfloor n/k \rfloor, & \text{if } \tilde{r} = r \end{cases}$$

(once we already have r class of size $\lfloor n/k \rfloor + 1$, the size of the remaining classes must not exceed $\lfloor n/k \rfloor$). Then, choose a non-considered vertex v randomly and put it into a class V_i such that $i \in I$ is the smallest possible and $E(V_i)$ is not incremented. If it is not possible, i is chosen ramdomly from I. To keep $\tilde{r}$ up to date, each time a vertex is added to a set V_i such that $|V_i| = \lfloor n/k \rfloor$, $\tilde{r}$ is incremented by one unit.

The previous procedure works fine for generating initial solutions from scratch. However, at this point it is common to know a $(k + 1)$-eqcol (i.e. in the cases where we previously ran tabu search with $k + 1$ and reached an equitable coloring) and we can exploit this coloring in order to improve the quality of the initial solution as follows.

Procedure 2. Let $\mathfrak{p} : \{1, \dots, k+1\} \to \{1, \dots, k+1\}$ be a bijective function (i.e. a random permutation) and let $V_1^*, V_2^*, \dots, V_k^*, V_{k+1}^*$ be the color classes of the known $(k + 1)$-eqcol. Set $V_i = V_{\mathfrak{p}(i)}^*$ for all $i \in \{1, \dots, k\}$, and $\tilde{r} = |W^+|$. Then, run Procedure 1 to assign a color to the remaining vertices which are those belonging to $V_{\mathfrak{p}(k+1)}^*$.

Regarding neighborhood of a solution $s \in S$ notice that, if n does not divide k, $W^+ \neq \emptyset$ and it is possible to move a vertex from a class of W^+ to W^-, keeping equity. That is, for all $v \in \cup_{i \in W^+} V_i$ and all $j \in W^-$, we have $s(i \xrightarrow{v} j) \in S$.

However, the number of allowed movements is rather limited when r is very low (for instance, $r = 1$) or very high ($r = k - 1$), so we need to add supplementary movements. Swapping the colors of two vertices simultaneously seems to work fine and as well can be used when n divides k.

From a solution $s = (V_1, V_2, \ldots, V_k)$, a neighbor $s' = (V_1', V_2', \ldots, V_k')$ is generated with two schemes:

- *1-move* (only applicable when n does not divide k). Choose a conflicting vertex $v \in C(s) \cap (\cup_{i \in W^+} V_i)$. Let i be the color of v in s. Next, choose a color $j \in W^-$. We have $s' = s(i \overset{v}{\to} j)$. Searching all the neighbors of s with this scheme requires exploring $(k - r)|C(s) \cap (\cup_{i \in W^+} V_i)|$ solutions.
- *2-exchange*. Choose a conflicting vertex $v \in C(s)$. Let i be the color of v in s. Next, choose another vertex u such that either $i < j$ or $u \notin C(s)$, where j is the color of u in s (the condition imposed to u prevents from evaluating 2-exchange on u and v twice). Then, set

$$V_j' = (V_j \setminus \{u\}) \cup \{v\}, \quad V_i' = (V_i \setminus \{v\}) \cup \{u\}, \quad V_l' = V_l \ \forall \, l \in \{1, \ldots, k\} \setminus \{i, j\}.$$

Note that objective value can be computed in linear time from $f(s)$:

$$f(s') = f(s) + |\{uw \in E : w \in V_i \setminus \{v\}\}| - |\{uw \in E : w \in V_j\}|$$
$$+ |\{vw \in E : w \in V_j \setminus \{u\}\}| - |\{vw \in E : w \in V_i\}|.$$

Searching all the neighbors of s with this scheme requires exploring a quadratic number of solutions.

Now, let s' be the next solution in the sucession; s' is obtained by applying either 1-move or 2-exchange to s, where vertex $v \in V_i$ in s and $v \notin V_i'$ in s'. In both schemes, $p = (v, i)$ is stored on tabu list and $live(p)$ is set to a dynamic tabu tenure $t = \alpha|C(s)| + Random(\beta)$ where α and β are parameters to be determined empirically. This is one of the purposes of the next section.

4 Computational Experiments and Conclusions

This section is devoted to perform and analyze computational experiments. They were carried out on an Intel i5 CPU 750@2.67Ghz with Ubuntu Linux O.S. and Intel C++ Compiler. We considered graphs from [18], which are benchmark instances difficult to color.

First, we test different combinations of values for parameters α and β from the dynamic tabu tenure in order to determine the combination that makes TABUEQCOL perform better. Then, we report the behaviour of TABUEQCOL over known instances by using the best combination previously found. We also compare its performance against tabu search algorithm given in [10].

Tuning parameters

We run TABUEQCOL over 16 instances with a predetermined value of k and an initial solution s_0 generated with Procedure 1. The same initial solution is used in all executions of TABUEQCOL for the same instance.

Results are reported in Table 1. First column is the name of the graph G. Second and third columns are the number of vertices and edges of G. Fourth and fifth columns are known lower and upper bound of $\chi_{eq}(G)$ (obtained by other means). The remaining columns are the time elapsed in seconds by the execution of TABUEQCOL when a k-eqcol is found within the term of 1 hour, for each combination. In the case TABUEQCOL is not able to find a k-eqcol, $f(s^*)$ is displayed between braces where s^* is the best solution found. Three last rows indicate the sum of objective function $f(s^*)$ over non-solved instances, percentage of instances TABUEQCOL solved successfully and the average time elapsed for these instances to be solved.

For the sake of simplicity, we refer to each combination with a capital letter.

Note that combination D has the least average time, however it has solved less instances than other combinations and the sum of objective values is also worse. We discard A, B, C, D, E and H with this criterion. By comparing the three remaining combinations, we have that G is faster than the other two. Even if we restrict the comparison to those 11 instances the 3 combinations solve simultaneously, we have 807 s for F, 562 s for G and 730 s for I, so G is still better.

We consider combination G ($\alpha = 0.9$ and $\beta = 5$) for TABUEQCOL.

Testing tabu search heuristic

For each instance, the following process is performed. First, execute NAIVE algorithm (described in [5]) in order to find an initial equitable coloring c of the current instance. Suppose that $k + 1$ is the number of colors of c. Then, obtain an initial solution s_0 of k color classes generated from c with Procedure 2, and run TABUEQCOL with parameters $\alpha = 0.9$ and $\beta = 5$. If a k-eqcol is found, start over the process with $k - 1$ color classes by running Procedure 2 and TABUEQCOL again. This process is repeated until 1 hour is elapsed or a $\underline{\chi_{eq}}$-eqcol is reached, and the best coloring found so far is returned.

In Table 2 we report results over 76 benchmark instances with at least 50 vertices (75 from [18] and one Kneser graph used in [10]). First 5 columns have the name of the graph G, number of vertices and edges, and best known lower and upper bound of $\chi_{eq}(G)$. Sixth column displays the number of colors of the initial equitable coloring c. Seventh and eighth columns display the value k of the best k-eqcol found after 30 s of execution of our algorithm and the time elapsed in seconds until such k-eqcol is reached. If the coloring is optimal, k is displayed in boldface. Next two columns show the same information after 1 h of execution, but if the best coloring is found within the first 30 s, these columns are left empty.

Time spent by NAIVE is not considered in the computation. However, NAIVE rarely spent more than 1 s (and never more than 4 s).

Last two columns show the same information for the tabu search described in [10]. If such information is not available, these columns are left empty. We recall that the values provided in [10] were computed on a different platform (1.8 Ghz AMD-Athlon with Linux and GNU C++ compiler).

Table 1. Execution of TabuEqCol with different combination of values

Instance	$|V|$	$|E|$	χ_{eq}	k	$\alpha = 0.3$			$\alpha = 0.6$			$\alpha = 0.9$		
					$\beta = 5$ A	$\beta = 10$ B	$\beta = 15$ C	$\beta = 5$ D	$\beta = 10$ E	$\beta = 15$ F	$\beta = 5$ G	$\beta = 10$ H	$\beta = 15$ I
DSJR500.1	500	3555	12	12	{3}	1	1	1	1	1	1	1	1
DSJR500.5	500	58862	120	131	{14}	{3}	{1}	{8}	{3}	3242	{5}	{3}	{1}
DSJR500.1c	500	121275	126	195	{4}	{1}	427	{3}	78	747	66	8	11
DSJC500.1	500	12458	5	13	{2}	55	41	38	63	47	39	83	57
DSJC500.5	500	62624	13	62	61	530	{1}	{1}	{1}	{2}	{1}	{2}	{1}
DSJC500.9	500	112437	101	148	{1}	106	104	94	91	80	100	90	121
DSJC1000.1	1000	49629	5	22	767	411	509	551	423	858	710	691	1059
DSJC1000.5	1000	249826	15	112	543	968	623	518	999	{2}	1853	{2}	{1}
DSJC1000.9	1000	449449	126	268	1850	1751	1822	1926	1725	1250	1808	1723	983
inithx.i.1	864	18707	54	54	{8}	{8}	{8}	{8}	{7}	{7}	{8}	{8}	{7}
latin_square_10	900	307350	90	131	1182	1080	1013	796	782	946	895	1298	778
flat300_28.0	300	21695	11	37	238	{1}	{1}	143	{1}	{1}	{1}	{2}	{2}
flat1000_76.0	1000	246708	14	112	228	548	1255	154	600	1681	245	780	3298
abb313GPIA	1557	53356	8	9	{27}	{44}	{15}	{2}	{10}	2801	1796	{1}	1304
qg.order40	1600	62400	40	40	26	31	17	25	26	20	24	25	26
wap01a	2368	110871	41	47	{21}	477	501	{6}	451	446	499	744	397
Sum of objective values					80	57	26	28	22	12	15	18	12
Success					50 %	69 %	69 %	63 %	69 %	75 %	75 %	63 %	69 %
Average Time					612	542	574	425	476	1010	670	544	730

Table 2. Execution of TABUEQCOL over benchmark instances

Instance	$|V|$	$|E|$	χ_{eq}	$\overline{\chi}_{eq}$	NAIVE	$\leq$ 30 sec. k	Time	$\leq$ 1 hour k	Time	[10] k	Time
miles750	128	2113	31	31	33	**31**	0.0			35	13
miles1000	128	3216	42	42	47	43	0.1			49	13
miles1500	128	5198	73	73	74	**73**	0.0			77	13
zeroin.i.1	211	4100	49	49	51	51	0.0			74	22
zeroin.i.2	211	3541	36	36	51	51	0.0			95	22
zeroin.i.3	206	3540	36	36	49	49	0.0			97	21
queen8_8	64	728	9	9	18	**9**	1.2			10	7
jean	80	254	10	10	10	**10**	0.0			**10**	3
anna	138	493	11	11	11	**11**	0.0			13	14
david	87	406	30	30	40	**30**	0.0			**30**	9
games120	120	638	9	9	9	**9**	0.0			11	6
kneser9_4	126	315	3	3	4	**3**	0.0			6	2
2-FullIns_3	52	201	5	5	9	**5**	0.0			8	1
3-FullIns_3	80	346	6	6	7	**6**	0.0			9	2
4-FullIns_3	114	541	7	7	12	**7**	0.1			11	5
5-FullIns_3	154	792	8	8	9	**8**	0.0			13	8
2-FullIns_5	852	12201	4	7	15	7	2.5				
3-FullIns_5	2030	33751	5	8	13	8	25				
4-FullIns_4	690	6650	6	8	14	8	0.4				
4-FullIns_5	4146	77305	6	9	21	14	20	9	254		
1-Insertions_6	607	6337	3	7	14	7	0.2				
2-Insertions_5	597	3936	3	6	6	**6**	0.0				
3-Insertions_5	1406	9695	3	6	8	6	1.2				
homer	561	1628	13	13	13	**13**	0.0				
huck	74	301	11	11	11	**11**	0.0				
latin_square_10	900	307350	90	130	460	169	30	130	1301		
DSJC125.1	125	736	5	5	8	**5**	0.8				
DSJC125.5	125	3891	9	18	27	19	0.1	18	788		
DSJC125.9	125	6961	42	45	66	45	0.4				
DSJC250.1	250	3218	4	8	13	9	0.1	8	32		
DSJC250.5	250	15668	11	32	65	33	7.2	32	69		
DSJC250.9	250	27897	63	83	136	83	1.2				
DSJR500.1	500	3555	12	12	12	**12**	0.0				
DSJR500.5	500	58862	120	131	135	133	0.1				
DSJR500.1c	500	121275	126	195	349	257	0.3				
DSJC500.1	500	12458	5	13	23	14	3.5	13	33		
DSJC500.5	500	62624	13	62	128	63	11				
DSJC500.9	500	112437	101	148	284	182	0.7				
DSJC1000.1	1000	49629	5	22	38	26	26	22	500		
DSJC1000.5	1000	249826	15	112	265	128	27	112	2261		
DSJC1000.9	1000	449449	126	268	575	329	20				
flat300_20_0	300	21375	11	34	81	38	9.2	34	463		
flat300_28_0	300	21695	11	36	65	39	3.3	36	3222		
flat1000_76_0	1000	246708	14	112	223	127	24	112	1572		
fpsol2.i.1	496	11654	65	65	85	78	0.1				
fpsol2.i.2	451	8691	47	47	62	60	0.0				
fpsol2.i.3	425	8688	55	55	80	79	0.0				
inithx.i.1	864	18707	54	54	70	66	0.1				
inithx.i.2	645	13979	30	93	158	93	7.2				
le450_15b	450	8169	15	15	17	16	0.3	**15**	107		
le450_15d	450	16750	15	16	30	22	9.6	16	599		
le450_25b	450	8263	25	25	25	**25**	0.0				
le450_25d	450	17425	25	27	31	27	29				
le450_5b	450	5734	5	5	12	7	7.2				
le450_5d	450	9757	5	8	18	8	15				
mug100_25	100	166	4	4	4	**4**	0.0				
mug88_25	88	146	4	4	4	**4**	0.0				
mulsol.i.1	197	3925	49	49	63	50	0.0				
mulsol.i.2	188	3885	31	48	58	48	0.1				
myciel6	95	755	7	7	11	**7**	0.0				
myciel7	191	2360	8	8	12	**8**	0.1				
qg.order40	1600	62400	40	40	64	42	22	**40**	47		
qg.order60	3600	212400	60	60	64	64	0.0	**60**	267		
queen8_12	96	1368	12	12	20	**12**	0.1				
queen9_9	81	1056	10	10	15	**10**	9.2				
queen10_10	100	1470	10	11	18	12	0.1	11	143		
school1	385	19095	15	15	49	**15**	12				
school1_nsh	352	14612	14	14	40	**14**	14				
wap01a	2368	110871	41	46	48	46	15				
wap02a	2464	111742	40	44	49	47	18	44	83		
wap03a	4730	286722	40	50	58	57	18	50	464		
abb313GPIA	1557	53356	8	9	17	13	28	10	283		
ash331GPIA	662	4181	3	4	8	4	2				
ash608GPIA	1216	7844	3	4	10	4	12				
ash958GPIA	1916	12506	3	4	10	5	11	4	41		
will199GPIA	701	6772	7	7	9	**7**	2.2				

Note that our approach reaches optimality in 29 instances and a gap of one unit between χ_{eq} and the best solution in 7 instances. In other words, it reaches a gap of at most one unit in roughly a half of the evaluated instances. Note also that TABUEQCOL improves the initial solution given by NAIVE in most cases (precisely, 63 instances).

On those instances the value of the best solution given by tabu search of [10] is known, our algorithm gives the same value or a better one. Despite the difference between platforms, it seems that our approach also runs faster.

An interesting fact is that each execution of TABUEQCOL needs no more than 500000 iterations to reach the best value since the largest number of iterations performed was 493204 and took place when TABUEQCOL found a 18-eqcol of DSJC125.5.

In the same sense, TABUEQCOL needs no more than 30000 iterations in each execution and the overall process needs no more than 30 s to reach the best value on 56 instances; justly those ones such that columns 9 and 10 are empty. On these instances, the largest number of iterations performed was 28791 and took place when TABUEQCOL found a 10-eqcol of queen9_9.

Conclusion

The Equitable Coloring Problem is a variation of the Graph Coloring Problem that naturally arises from several applications where the cardinality of color classes must be balanced. Just like Graph Coloring, the need to solve applications associated to this new $\mathcal{NP}$-Hard problem justifies the development of exact algorithms and heuristics. On large instances, known exact algorithms are unable to address them and heuristics such as NAIVE delivers poor solutions. Our tabu search heuristic based on TABUCOL has shown to improve these solutions and presented a fairly good performance, even if a limit of 30 s is imposed.

Acknowledgements. This work is partially supported by grants UBACYT 20020100100666, PICT 2010-304, PICT 2011-817, PID-UNR ING416 and PIP-CONICET 241.

References

1. Meyer, W.: Equitable coloring. Amer. Math. Mon. **80**, 920–922 (1973)
2. Tucker, A.: Perfect graphs and an application to optimizing municipal services. SIAM Rev. **15**, 585–590 (1973)
3. Das, S.K., Finocchi, I., Petreschi, R.: Conflict-free star-access in parallel memory systems. J. Parallel Distrib. Comput. **66**, 1431–1441 (2006)
4. Pemmaraju, S.V.: Equitable coloring extends Chernoff-Hoeffding bounds. In: Goemans, M.X., Jansen, K., Rolim, J.D.P., Trevisan, L. (eds.) APPROX-RANDOM 2001. LNCS, vol. 2129, p. 285. Springer, Heidelberg (2001)
5. Furmanczyk, H., Kubale, M.: Equitable coloring of graphs. In: Kubale, M. (ed.) Graph Colorings, pp. 35–53. American Mathematical Society, Providence (2004)
6. Hajnal, A., Szemerédi, E.: Proof of a conjecture of P. Erdös. Combin. theory and its app. **2**, 601–623 (1970)

7. Kierstead, H., Kostochka, A., Mydlarz, M., Szemerédi, E.: A fast algorithm for equitable coloring. Combinatorica **30**, 217–224 (2010)
8. Brélaz, D., Nicolier, Y., de Werra, D.: Compactness and balancing in scheduling. Math. Methods Oper. Res. **21**, 65–73 (1977)
9. Sulong, G.B.: Some balanced colouring algorithms for examination timetabling. Jurnal Teknologi **19**, 57–63 (1992)
10. Bahiense, L., Frota, Y., Noronha, T.F., Ribeiro, C.: A branch-and-cut algorithm for the equitable coloring problem using a formulation by representatives. Discrete Appl. Math. **164**(1), 34–46 (2014)
11. Méndez Díaz, I., Nasini, G., Severín, D.: A polyhedral approach for the equitable coloring problem. Discrete Appl. Math. **164**(2), 413–426 (2014)
12. Méndez Díaz, I., Nasini, G., Severín, D.: An exact DSatur-based algorithm for the equitable coloring problem. Electron. Notes Discrete Math. **44**, 281–286 (2013)
13. Galinier, P., Hao, J.-K.: Hybrid evolutionary algorithms for graph coloring. J. Comb. Optim. **3**(4), 379–397 (1999)
14. Galinier, P., Hertz, A.: A survey of local search methods for graph coloring. Comput. Oper. Res. **33**(9), 2547–2562 (2006)
15. Glover, F., McMillan, C., Novick, B.: Interactive decision software and computer graphics for architectural and space planning. Ann. Oper. Res. **5**(3), 557–573 (1985)
16. Hertz, A., de Werra, D.: Using tabu search techniques for graph coloring. Computing **39**(4), 345–351 (1987)
17. Blöchliger, I., Zufferey, N.: A graph coloring heuristic using partial solutions and a reactive tabu scheme. Comput. Oper. Res. **35**(3), 960–975 (2008)
18. Graph Coloring Benchmark Instances. http://www.cs.hbg.psu.edu/txn131/graphcoloring.html

Linear Arrangement Problems
and Interval Graphs

Alain Quilliot[1]($\boxtimes$) and Djamal Rebaine[2]

[1] LIMOS, UMR CNRS, Université Blaise Pascal,
6158 Clermont Ferrant, France
alain.quilliot@isima.fr
[2] Département d'Informatique et Mathématique,
Université du Québec à Chicoutimi, Saguenay, QC G7H 2B1, Canada
Djamal_Rebaine@uqac.ca

Abstract. We present in this paper new results on the Linear Arrangement Problem (LAP) for interval graphs. We first propose a new lower bounding scheme, which links LAP with the Max-Cut Problem, and then show that this lower bound is tight for unit interval graphs. Next, we focus on arbitrary interval graphs, and derive a polynomial time approximation algorithm.

Keywords: Interval graphs · Linear ordering · Linear arrangement

1 Introduction

Let $G = (X, E)$ be a non-oriented graph where X and E respectively denote the vertices and edges sets of G. The *Linear Arrangement* Problem (LAP) consists in finding a one-to-one mapping ϕ from X to $\{1, ..., |X|\}$ that minimizes

$$f(G, \phi) = \sum_{(x,y) \in E} |\phi(y) - \phi(x)|. \qquad (1)$$

The corresponding decision LAP was first shown to be *NP*-complete for arbitrary graphs [8, 9], and, next, for restricted graphs such as interval graphs [5] and bipartite graphs [9]. However, polynomial time algorithms were also designed for other special graphs such as trees [4]), unit interval graphs [6], paths, cycles, complete bipartite graphs, grid graphs [11], and restricted series-parallel graphs [1]. A survey is presented in [3].

Since LAP is NP-hard, the heuristic approach is therefore well justified to solve this problem. This paper mainly aims at providing tools for the design of approximation algorithms while focusing on interval graphs. It is organized as follows. Section 2 introduces notations and a linear ordering based reformulation of LAP. In Sect. 3, we propose a general lower bound, which links LAP with the well-known Max-Cut Problem. In the remaining sections, we first derive a very simple linear algorithm for the unit interval graph, and then we undertake the worst-case analysis of a heuristic algorithm that we propose. To do so, we proceed as follows. In Sects. 4 and 5, we introduce a restricted version CLAP of LAP and solve it in an exact way,

© Springer International Publishing Switzerland 2014
P. Fouilhoux et al. (Eds.): ISCO 2014, LNCS 8596, pp. 359–370, 2014.
DOI: 10.1007/978-3-319-09174-7_31

through local resolutions of its related *Max-Cut* problem. In Sect. 6, we derive an upper bound on the gap between the optimal value of LAP and CLAP. Finally, in Sect. 7, we present our concluding remarks. An appendix is included for those results left without a proof in the main text.

2   Preliminaries

A simple (non-oriented) graph with no loop is denoted by $G = (X, E)$: X (E) is the node (edge) set of G. We denote by (x, y) an edge with end-nodes x and y in X. If $A \subseteq X$, then G_A is the sub-graph induced by A from G. If $x \in X$, then $\Gamma_G(x) = \{y \in X$ such that $(x, y) \in E\}$ is the *neighbourhood* set. The complementary graph $G^c = (X, E^c)$ of G is defined by $E^c = \{(x, y)$ such that $(x, y) \notin E$ and $x \neq y\}$. A *triangle* of G is a clique with 3 nodes. An *anti-edge* is a pair $e = (x, y) = (y, x)$, with $x \neq y$ and $(x, y) \notin E$. A *fork* with *root* x is a (non-oriented) triple $f = \{x, y, z\}$ with (x, y), $(x, z) \in E$, and $(y, z) \notin E$. An *anti-fork* with *root* z is any triple $f = \{x, y, z\}$ such that $(x, y) \in E$ and $(x, z), (y, z) \notin E$. (Fig. 1)

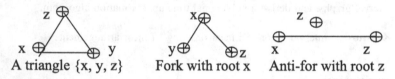

A triangle $\{x, y, z\}$ Fork with root x   Anti-for with root z

Fig. 1. Triangles, Forks and Anti-Forks

LAP Reformulation: A *linear ordering* of a set X is a binary order relation σ such that, for any pair x, y in X, $x \neq y$, we either have $(x \ \sigma \ y)$ or $(y \ \sigma \ x)$. Given are a graph $G = (X, E)$ and a linear ordering σ of X. For any edge $e = (x, y) \in E$, we denote by $BE(e, z, \sigma)$ the *elementary break* of e by z according to σ. We set $BE(e, z, \sigma) = 1$ if $(x \ \sigma \ z \ \sigma \ y)$ or $(y \ \sigma \ z \ \sigma \ x)$, and 0 otherwise. We Denote by $BG(G, \sigma)$ the *Global Break* of G according to σ. We set $BE(e, z, \sigma) = \Sigma_{e, z} BE(e, z)$. If $\phi(\sigma)$ is the one-to-one mapping from X into $\{1, ..., |X|\}$ which derives from σ. It then follows

$$f(G, \phi(\sigma)) = \sum_{(x,y) \in E} |\phi(y) - \phi(x)| = BG(G, \sigma) + |E| = \Sigma_{e,z} BE(e, z, \sigma) + |E|.$$

Therefore, solving LAP turns out to be the problem of seeking a linear ordering σ that minimizes the *Global Break* $BG(G, \sigma)$. We denote by $LAP(G)$ the optimal *Global Break* defined as $LAP(G) = \text{Inf}_\sigma B(G, \sigma)$.

3   A General Lower Bound

The problem of seeking a good linear ordering σ of the vertices of a given graph $G = (X, E)$ is the same as efficiently deciding, for any x in X, which vertices of the *neighbourhood* set $\Gamma_G(x)$ are located before x according to σ. This local decision process is linked to the well-known Max-Cut Problem [2, 7, 10] defined as follows:

Max-Cut Problem: Let $H = (Z, F)$ be a simple graph, and $\omega \geq 0$ an F-indexed vector. We denote by $Z = A \cup^{Ex} B$ a partition of Z into two disjoint subsets. The Max-Cut Problem is to find a partition $Z = A \cup^{Ex} B$ that maximizes $\Sigma_{x \in A, y \in B, (x,y) \in F} \omega_{(x,y)}$. We denote by Max-Cut$(H, \omega)$ the corresponding optimal value.

Let us now consider graph $G = (X, E)$. We denote by $Tr(G)$ the number of triangles of G. For any vertex x in X, we define the following:

– $H(x)$ is the complementary graph of the sub-graph of G which is induced by $\Gamma_G(x)$. Note that x is not a node of the graph $H(x)$, since G has no loop;
– $m(x)$ is the number of edges of $H(x)$; $V(x) = m(x) - Max\text{-}Cut(H(x), 1)$.

Theorem 1. *For any graph $G = (X, E)$, we have $LAP(G) \geq Tr(G) + \Sigma_x V(x)$.*

Proof. Let us consider a linear ordering σ of G, and introduce the following notation:

– Fk(G, σ) denotes number of *forks* $f = \{x, y, z\}$ of G, f with root x, such that

$$((x \, \sigma \, y) \wedge (x \, \sigma \, z)) \vee ((y \, \sigma \, x) \wedge (z \, \sigma \, x)).$$

– AFk(G, σ) is the number of *anti-forks* $f = \{x, y, z\}$ of G, f with root z, such that

$$(x \, \sigma \, z \, \sigma \, y) \vee (y \, \sigma \, z \, \sigma \, x).$$

Let us first check that

$$BG(G, \sigma) = Tr(G) + Fk(G, \sigma) + AFk(G, \sigma). \tag{2}$$

In order to do so, let us consider an edge $e = (x, y)$, and a node z, different from x and y. While counting BE(e, z, σ), we distinguish three cases:

Case 1: x, y and z define a triangle. Then, BE$(e, z, \sigma) = 1$ if either $(x \, \sigma \, z \, \sigma \, y)$ or $(y \, \sigma \, z \, \sigma \, x)$. In such a case no quantity BE$((x, z), y, \sigma)$, BE$((z, y), x, \sigma)$ is equal to 1. So, if x, y, z define a triangle, there exists exactly one node t in $\{x, y, z\}$ such that BE$(e(t), t, \sigma) = 1$, where $e(t)$ is the edge defined by $\{x, y, z\} - t$. We then get

$$\Sigma_{e = (x,y),\, z \text{ such that } (x,y,z) \text{ is a triangle}} BE(e, z, \sigma) = Tr(G).$$

Case 2: $f = \{x, y, z\}$ is a *fork* with root x. It follows BE$((x, y), z, \sigma) = 1$ if either $(x \, \sigma \, z \, \sigma \, y)$ or $(y \, \sigma \, z \, \sigma \, x)$, and then BE$((x, z), y, \sigma) = 0$. Conversely, BE$((x, z), y, \sigma) = 1$ if either $(x \, \sigma \, y \, \sigma \, z)$ or $(z \, \sigma \, y \, \sigma \, x)$, and then BE$((x, y), z, \sigma) = 0$. So, (x, y, z) yields an elementary break iff y and z are located on the same side of x according to σ. Hence,

$$\Sigma_{e = (x,y),\, z \text{ adjacent to exactly 1 extremity of } e} BE(e, z, \sigma) =$$
$$\Sigma_x \Sigma_{y, z \in \Gamma G(x),\, (y,x) \notin E, y, z \text{ located the same way with respect to } x, \sigma} 1 = Fk(G, \sigma).$$

Case 3: $f = \{x, y, z\}$ is an *Anti-Fork* with root z. Therefore, we have

$$\Sigma_{e = (x,y),\, z \text{ such that } (x,z) \notin E \text{ and } (y,z) \notin E} BE(e, z, \sigma) = AFk(G, \sigma).$$

We then derive (2) from the following

$$\Sigma_{e,z} BE(e, z, \sigma) = \Sigma_{e=(x,y),z \text{ such that } (x,y,z) \text{ is a triangle}} BE(e, z, \sigma) +$$

$$\Sigma_{e=(x,y),z \text{ adjacent to 1 extremity of } e} BE(e, z, \sigma) + \Sigma_{e=(x,y),z \text{ adjacent to no extremity of } e} BE(e, z, \sigma).$$

For any $x \in X$, a feasible solution $A(x, \sigma) \cup^{Ex} B(x, \sigma)$ of Max-Cut is defined on $H(x)$, with $\omega = 1$, by setting $A(x, \sigma) = \{y \in \Gamma_G(x), \text{ such that } y \ \sigma \ x\}$; $B(x, \sigma) = \{y \in \Gamma_G(x) \text{ such that } x \ \sigma \ y\}$. Its value, in the Max-Cut context, is

$$m(x) - \Sigma_{y,z \in \Gamma_G(x), (y,x) \notin E, y, z \text{ located the same way with respect to } x, \sigma} 1 \leq Max - Cut(H(x), \mathbf{1}).$$

It then follows that, for any $x \in X$, we have

$$\Sigma_{y, z \in \Gamma_G(x), (y,x) \notin E, y \text{ and } z, \text{ located the same way with respect to } x, \sigma} 1 \geq V(x).$$

Then, we derive

$$Fk(G, \sigma) = \Sigma_x \Sigma_{y, z \in \Gamma_G(x), (y,x) \notin E, y, z \text{ located the same side with respect to } x, \sigma} 1 \geq \Sigma_x V(x).$$

The result is thus established. □

Let us mention that we performed experiments which showed that this bound is often tight, specifically in the case of chordal graphs. Still, the Max-Cut Problem, which been extensively studied, is *NP*-Hard [9], even with $\omega = \mathbf{1}$. So, one may ask the question about the practical use of the above lower bound. The answer is two-fold:

1. Even though Max-Cut is NP-Hard, it may easier to handle than LAP. Indeed, it admits a natural quadratic $\{0, 1\}$ formulation, see for e.g. [4, 11]. Furthermore, instances related to $H(x)$, $x \in X$, are smaller than that of the original LAP.
2. We shall see in Sect. 4 that, in the case of interval graphs, the above bounding scheme leads to an efficient polynomial time approximation scheme.

4   The Case of Interval Graphs

We first introduce additional definitions related to interval graphs:

– A simple graph with no loop $G = (X, E)$ is an *interval graph* if it is the intersection graph of a set $[o(x), d(x)]$, $x \in X$, of closed intervals in the real line. Those intervals may be chosen such that points $o(x)$, $d(x)$, $x \in X$, are distinct. We assume this hypothesis to be always satisfied. Let us set:

  – $x \subset y$ if $o(x) < o(y)$ and $d(y) < d(x)$;
  – $x << y$ if $d(x) < o(x)$;
  – $x \, Ov \, y$ if $o(x) < o(y) < d(x) < d(x)$.

– In case X is an interval family with distinct endpoints, we say that a linear ordering σ of X is $(Ov, <<)$-*consistent* if it is consistent with both orderings Ov and $<<$. We denote by σ-*can* the *canonical linear ordering,* defined as follows:

x σ-*can* y if, and only if, o(x) < o(y).

– We say that a *fork* f = {x, y, z} with root x of such an interval graph G = (X, E) is a *strong fork* if there exists t ∈ {y, z} such that t ⊂ x, and that a triangle (x, y, z) is a *strong triangle* if at least some node is contained in another one (for instance z ⊂ x), as illustrated by Figs. 2, and Fig. 3, respectively.

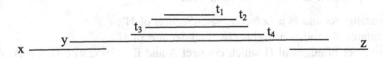

Fig. 2. *Strong fork* f = {x, y, z}   **Fig. 3.** *Strong triangle* = {x, y, z}

– We say that G is a *Unit Interval graph* if intervals [o(x), d(x)], x ∈ X, may be chosen in such way that no pair (x, y) exists such that x ⊂ y.
– Finally, a subset Y of X is a *Left-(Ov, <<)-Section* (*Right-(Ov, <<)-Section*) if, for any x, y ∈ X such that x ∈ Y and (y Ov x) ∨ (y << x), we have y ∈ Y (x ∈ Y).

4.1   A Direct Application on Unit Interval Graphs

In the case of unit interval graphs, from Theorem 1, we derive the following result:

Theorem 2. *If G = (X, E) is a unit interval graph, then σ-can is an optimal solution for LAP.*

Proof. Let us suppose that an *elementary break* (e = (x, y), z, σ-*can*) exists, and that x *Ov* y. It then follows that x σ-*can* y. If x << z then y σ-*can* z and z does not break e. Similarly, if z << x then z << x and z does not break e. It follows that x ∩ z is not empty. By the same way, y ∩ z is not empty and {x, y, z} forms a *triangle*. So, there is a one-to-one correspondence between *triangles* and *elementary breaks*. Thus, BG(G, σ-*can*) = *Tr*(G).

4.2   A Restricted Version of LAP

In the case of general interval graphs, σ-*can* may not be optimal. Moreover, optimal solutions may not even be (*Ov*, <<)-*consistent*: This is illustrated by Fig. 4.

LAP(G) = 11, optimal σ-*opt* such that y σ-*opt* z σ-*opt* x, while BG(G, σ-*can*) = 14.

Fig. 4. Non consistency of (*Ov*, <<)

Still, what can be easily checked is that σ-*can* produces a 2-approximation [5] if we refer to the standard definition of LAP.

Proposition 1. *If $G = (X, E)$ is an interval graph with m edges, then $BG(G, \sigma\text{-}can) \leq 2LOP(G) + m$.*

Proof: A *Global Break* oriented proof comes by induction on the cardinality of X. Let x_0 be the first (smallest) element of X according to σ-*can*, and σ-*opt* some optimal solution of LAP. By induction, we have

$$\Sigma_{e,z \neq x_0} BE(e, z, \sigma\text{-}can) \leq m - |\Gamma_G(x_0)| + 2.\Sigma_{e,z \neq x_0} BE(e, z, \sigma\text{-}opt). \tag{3}$$

Since all vertices of $\Gamma_G(x_0) \cup \{x_0\}$ are consecutive according to σ-*can*, we get that

$$\Sigma_{x \in \Gamma_G(x_0), z \in X} BE((x_0, x), z, \sigma\text{-}can) = \Sigma_{x, z \in \Gamma_G(x_0)} BE((x_0, x), z, \sigma\text{-}can)$$
$$= |\Gamma_G(x_0)| \times (|\Gamma_G(x_0)| - 1)/2.$$

On the other hand, if we refer to σ-*opt*, we get

$$\Sigma_{x \in \Gamma_G(x_0), z \in X} BE((x_0, x), z, \sigma\text{-}opt) \geq \Sigma_{x, z \in \Gamma_G(x_0)} BE((x_0, x), z, \sigma\text{-}opt) \geq$$
$$\lfloor |\Gamma_G(x_0)|/2 \rfloor (\lfloor |\Gamma_G(x_0)| \rfloor - 1)/2 + \lceil |\Gamma_G(x_0)|/2 \rceil (\lceil |\Gamma_G(x_0)|/2 \rceil - 1)/2. \tag{4}$$

The result follows by combining (3) and (4).   □

Still, experiments show that best linear orderings are more often $(Ov, <<)$-*consistent*. So, we are going to study the following restriction CLAP of LAP:

$(Ov, <<)$-*Consistent Linear Arrangement* Problem (*CLAP*): Compute a $(Ov, <<)$-consistent linear ordering σ which minimizes $BG(G, \sigma)$.

The following result bridges CLAP with Theorem 1.

Lemma 1. *If σ is $(Ov, <<)$-consistent, then we have $BG(G, \sigma) = Tr(G) + SFk(G, \sigma)$, where $SFk(G, \sigma)$ is the number of strong forks $f = (x, y, z)$, $x = Root(f)$ such that $((x \sigma y) \wedge (x \sigma z)) \vee (y \sigma x) \wedge (z \sigma x))$.*

Extending Theorem 1 to CLAP leads to introduce a specific version of Max-Cut: **$(Ov, <<)$-*Consistent Unit Cost Max-Cut* Problem (*C-Max-Cut*):** Given a graph $H = (Z, F)$, which is the complementary graph of an interval graph $H^c = (Z, F^c)$, and two disjoint subsets A_0 and B_0 of Z, such that:

– A_0 (B_0) is a *Left-$(Ov, <<)$-Section Right-$(Ov, <<)$-Section* of H^c;
– Both A_0 and B_0 define complete sub-graphs of $H^c = (Z, F^c)$.

Compute a partition $Z = A \cup^{Ex} B$ such that:

1. A contains A_0 and is a *Left-$(Ov, <<)$-Section* of H^c;
2. B contains B_0 and is a *Right-$(Ov, <<)$-Section* of H^c;
3. the number of edges of H which connect A and B $= |\{(x, y) \in E, x \in A, y \in B\}|$ is the largest possible;
4. A is maximal for the set inclusion order, provided points 1, 2, 3 are satisfied.

We denote by C-Max-Cut(H, A_0, B_0) the corresponding optimal value. We then define for the interval graph G = (X, E) and for any vertex x in X the following:

- $\Gamma_G^{Ov,\mathsf{C}}(x)$ is the (Ov, C)-*neighbourhood* set of x = {y ∈ $\Gamma_G(x)$, y ≠ x, such that (y ⊂ x) or (y Ov x) or (x Ov y)};
- H(x) is the complementary graph of the sub-graph induced by $\Gamma_G^{Ov,\mathsf{C}}(x)$;
- $A_0(x)$ = {y ∈ Z such that y Ov x}; $B_0(x)$ = {y ∈ Z such that x Ov y};
- m(x) is number of edges of H(x); CV(x) = m(x) − C-Max-Cut(H(x), $A_0(x)$, $B_0(x)$).

Then, we derive the following result.

Theorem 3. *CLAP(G) ≥ Tr(G) + Σ_x CV(x).*

Proof. For every x ∈ X, we set E*(x) = {non oriented pairs (y, z) such that:

- y ∈ $\Gamma_G^{Ov,\mathsf{C}}(x)$, z ∈ $\Gamma_G^{Ov,\mathsf{C}}(x)$; (y, z) ∉ E;
- at least one of both relations y ⊂ x or z ⊂ x holds;
- relation ((x σ y) ∧ (x σ z)) ∨ ((y σ x) ∧ (z σ x)) holds.

SFk(G, σ) may be written $\Sigma_{x \in X}|E*(x)|$. Since σ is $(Ov, <<)$-consistent, we may relax the "*at least ... y ⊂ x or z ⊂ x holds*" constraint which characterizes E*(x). So, for any x ∈ X : |E * (x)| ≥ $\Sigma_{y,z \in \Gamma G(x),\,(y,x) \notin E, y, z}$ are located the same way with respect to x,σ 1. For any x ∈ X we get a feasible solution A(x, σ) $\cup^{Ex}$ B(x, σ) of the *C-Max-Cut* instance defined by H(x), $\Lambda_0(x)$, $B_0(x)$, by setting A(x, σ) = {y ∈ $\Gamma_G^{Ov,\mathsf{C}}(x)$such that yσx} and B(x, σ) = {y ∈ $\Gamma_G^{Ov,\mathsf{C}}(x)$such that xσy}. Its value is m(x) − $\Sigma_{x,y \in \Gamma^{Ov,\mathsf{C}}G(x),\,(y,x) \notin E,y,}$ z located the same way with respect to x,σ 1 ≤ C-Max-Cut(H(x), $A_0(x)$, $B_0(x)$). It follows that, for any x ∈ X, we have |E * (x)| ≥ $\Sigma_{x,y \in \Gamma^{Ov,\mathsf{C}}G(x),(y,x) \notin E, y, z}$ located the same way with respect to $_{x,σ}$1 ≥ CV(x). Then, we get that SFk(G, σ) = $\Sigma_x |E * (x)| \geq \Sigma_x CV(x)$. Therefore, the result is established.   □

4.3   Solving C-Max-Cut and Evaluating CV(x)

The complexity of the Max-Cut problem in the case of the complementary graph of an interval graph is still an open question. However, things are easier with C-Max-Cut.

Theorem 4. *Given = (Z, F), A_0 and B_0 as in the definition of C-Max-Cut. Let us define, for every vertex z ∈ Z − $(A_0 \cup B_0)$, the following:*

- $d_H^-(A_0, z)$ = |{t ∈ Z − A_0 such that t << z }| + |{t ∈ A_0 such that t << z}|;
- $d_H^+(B_0, z)$ = |{t ∈ Z − B_0 such that z << t }| + |{t ∈ B_0 such that z << t }|.

Then we solve C-Max-Cut by setting:

- A = {z ∈ Z − $(A_0 \cup B_0)$ such that $d_H^-(A_0, z) \geq d_H^+(B_0, z)$} ∪ A_0;
- B = {z ∈ Z − $(A_0 \cup B_0)$ such that $d_H^-(A_0, z) < d_H^+(B_0, z)$} ∪ B_0.

Proof. For the sake of simplicity, we only consider here the case where $A_0 = B_0 =$ Nil. Clearly, (A^*, B^*) defines a feasible solution for *C-Max-Cut*. In addition, any feasible solution (A, B) defines an oriented graph structure $(Z, K_{A,B})$ on node set Z, for which arc set $K_{A,B}$ is as follows:

$(x, y) \in K_{A,B}$ if, and only if, $(x \in A, y \in A, y << x)$ or $(x \in B, y \in B, x << y)$. A node z is the origin of at least $\mathrm{Inf}(d_H{}^-(z), d^+(z))$. arcs of this digraph

$$(Z, K_{A,B}). \tag{5}$$

On the other hand, solving C-Max-Cut means seeking a partition $Z = A \cup^{Ex} B$ such that the number of edges of H which are not in the bipartite partial graph of H defined by A and B, i.e. for which both extremities belong to the same set either A or B, is the smallest possible. But, this number is equal to the number of arcs of the oriented graph $(Z, K_{A,B})$, and, from (5), we derive that it is at least equal to $\Sigma_{z \in Z}$ $\mathrm{Inf}(d_H{}^-(z), d^+(z)).(d_{\bar{H}}(z), d^{\pm}_H(z))$. However, this value is achieved in the case of (A^*, B^*). The result is thus established. $\qquad\square$

Corollary 1. *Let us denote by F_0 the edge subset of F defined by $F_0 = \{(x, y) \in F,$ such that $x \in A_0$ and $y \in B_0\}$. Then, C-Max-Cut(H, A_0, B_0) $\leq |(F - F_0|/2$.*

Proof. The proof follows from the fact that C-Max-Cut(H, A_0, B_0) $= \Sigma_{z \in Z}$ $\mathrm{Inf}(d_H{}^-(A_0, z), d^+(B_0, z))$.

Corollary 2. In Theorem 3, $CV(x) = \Sigma_{z \in Z} Inf (d_{H(x)}{}^-(z), d_{H(x)}{}^+(z))$, with

- $d_{H(x)}{}^-(z) = |\{t \in \Gamma_G^{Ov,C}(x),$ such that $t << z$ and $t \subset x \}|;$
- $d_{H(x)}{}^+(z) = |\{t \in \Gamma_G^{Ov,C}(x),$ such that $z << t$ and $t \subset x \}|.$

5   An Exact Solution σ-*bal* for CLAP

The objective of this section is to solve here CLAP in an exact way through local resolutions of *C-Max-Cut*.

Lemma 2. *C-Max-Cut(H, A_0, B_0) = C-Max-Cut(H, Nil, Nil).*

Next, we construct our solution σ-*bal*, by setting, for any pair x, y in X, x σ-*bal* y if, and only if, one among the following options holds:

$$
\begin{aligned}
&- \quad (x << y) \text{ or } (x\, Ov\, y); \\
&- \quad (x \subset y) \text{ and } d_{H(y)}{}^-(A_0(y), x) \leq d_{H(y)}{}^+(B_0(y), x); \\
&- \quad (y \subset x) \text{ and } d_{H(y)}{}^-(B_0(x), y) < d_{H(y)}{}^+(A_0(x), y).
\end{aligned}
\tag{6}
$$

Example 2. Let $G = (X, F)$ as depicted by Fig. 5.

$CV(A) = 0$; $CV(B) = 0$; $CV(C) = 1$; $CV(D) = 0$; $CV(E) = 0$; $Tr(G) = 2$; σ-*bal*: B σ-*bal* A σ-*bal* C σ-*bal* D σ-*bal* E; BG(G, σ-*bal*) = 3.

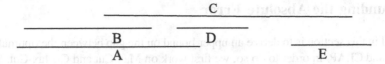

Fig. 5. A σ-bal construction

Lemma 3. *The σ-bal relation is transitive.*

We are now ready to state the optimality of σ-bal.

Theorem 5. *The relation σ-bal is an optimal solution of CLAP, which satisfies:*

1. $BG(G, \sigma\text{-}bal) \le BG(G, \sigma\text{-}can)$.
2. $Tr(G) \le BG(G, \sigma\text{-}bal) = Tr(G) + \Sigma_x CV(x) \le Tr(G) + Strong\text{-}Fork/2$,
   where Strong-Fork is the number of strong forks of the interval graph G.

Proof. From Lemma 1, we have that σ-*bal* is a $(<<, Ov)$-consistent linear ordering. The optimality of σ-*bal* (and so, the fact that $BG(G, \sigma\text{-}bal) \le BG(G, \sigma\text{-}can)$) with regards to CLAP derives from Lemma 1, 2, 3, and Theorem 4. Indeed, from Lemma 1, we have that $BG(G, \sigma\text{-}bal) = Tr(G) + SFk(G, \sigma\text{-}bal)$, with $SFk(G, \sigma\text{-}bal) = \Sigma_{x \in X}$ $|E^*(x)|$, where $E^*(x)$ is defined, for any $x \in X$, as the set of non-oriented pairs (y, z) such that $y \in \Gamma_G^{Ov,C}(x)$, $z \in \Gamma_G^{Ov,C}(x)$; $(y, z) \notin E$; at least one of both relations $y \subset x$ or $z \subset x$ holds; the relation $((x \ \sigma\text{-}bal \ y) \wedge (x \ \sigma\text{-}bal \ z)) \vee ((y \ \sigma\text{-}bal \ x) \wedge (z \ \sigma\text{-}bal \ x))$ holds. However, from Lemma 2 and 3, (5) and (6), we deduce that, for any $x \in X$, $|E^*(x)| = m(x) - C\text{-}Max\text{-}Cut(H(x))$, where $m(x)$ and $C\text{-}Max\text{-}Cut(H(x))$ respectively denote the number of edges of graph $H(x)$ and the optimal value of the $C\text{-}Max\text{-}Cut$ instance defined by graph $H(x)$. It then follows that $BG(G, \sigma\text{-}bal) = Tr(G) + SFk(G, \sigma\text{-}bal) = Tr(G) + \Sigma_x CV(x)$. Therefore, σ-*bal* achieves the lower bound of Theorem 4, and its optimality thus follows.

In order to get inequality (2) of the above statement, we first set, for any $x \in X$:

- $F(x)$ is the edge set of the graph $H(x)$ which is induced by the (Ov, C)-neighbourhood set $\Gamma_G^{Ov,C}(x)$;
- $F_0(x) = \{(y, z) \in F(x) \text{ such that } y \in A_0(x), z \in B_0(x)\}$.

Corollary 2 of Theorem 4 shows that $CV(x) \le |F(x) - F_0(x)|/2$. In addition, we see that $F(x) - F_0(x) = \{(y, z) \text{ such that } y << z, (x, y) \in E, (x, z) \in E, \text{ at least one}$ node t among $\{y, z\}$ is such that $t \subset x\}$. Then, we get a one-to-one correspondence between $F(x) - F_0(x)$ and the set of *strong forks* with root x. We then establish the result by using the fact that a *fork* has only one root. $\square$

We easily deduce that this result has an algorithmic interpretation:

Corollary 3. *Computing σ-bal may be done in $O(Arc\text{-}C)$ time, where Arc-C is the number of arcs of the digraph induced on X by the $\subset$ ordering.*

6 Bounding the Absolute Error

Our goal in this section is to derive an upper bound on the gap between the optimal value of LAP and CLAP. In order to do so, we first work on Max-Cut and C-Max-Cut. So, we consider $H = (Z, F)$, the complementary graph of an interval graph $H = (Z, F^c)$, together with two disjoint subsets A_0 and B_0 of Z. Max-Cut(H, **1**) and C-Max-Cut(H, A_0, B_0) = C-Max-Cut(H, Nil, Nil) respectively denote the optimal values of the corresponding *Max-Cut* and *C-Max-Cut* problems with unit costs. We also define the following:

- $X = Z - (A_0 \cup B_0)$;
- $F_0^c = \{(x, y) \in F^c,$ such that $x \in A_0$ and $y \in B_0\}$;
- $F_1^c = \{(x, y) \in F^c,$ such that $x \in X$ and $y \in X\}$;
- $F_2^c = \{(x, y) \in F^c,$ such that $x \in A_0 \cup B_0$ and $y \in X\}$.

Then, we derive the following Lemmas.

Lemma 4. *Given two graphs $H_1 = (Z, F_1)$ and $H_2 = (Z, F_1 \cup K)$. Then we have that Max-Cut(H_1, 1) $\leq$ Max-Cut(H_1, 1) $\leq$ Max-Cut(H_1, 1) + $|K|$.*

Lemma 5. *It is possible to assign, to every interval $x \in Z$, some interval $u(x)$ in such a way that*

- *function u is one-to-one: $x \neq y => u(x) \neq u(y)$;*
- *u maintains the Ov and $<<$ relations: x Ov y or $x << y => u(x)$ Ov*
- *$u(y)$ or $u(x) << u(y)$;*
- *both families $u(A_0) = \{u(x), x \in A_0\}$ and $u(B_0) = \{u(x), x \in B_0\}$ define complete sub-graphs in the sense of interval graphs;*
- *for any $x \in Z - (A_0 \cup B_0)$, $y \in Z$, we have $u(x) \cap u(y) = Nil$.*

We denote by H-*Nil* = $(u(Z),$ F-*Nil*$)$ the graph which results from this construction. Observe that H-Nil = $(u(Z),$ F-Nil$)$ may be viewed as been derived from H through withdrawal of the edges of $F_1^c \cup F_2^c \cup F_0^c$ of H^c.

Lemma 6. *It is possible to assign to every interval $x \in Z$ an interval $v(x)$ in such a way:*

- *function v is one-to-one: $x \neq y = > v(x) \neq v(y)$;*
- *v maintains the Ov and $\ll$ relations on $X = Z-(A_0 \cup B_0)$: for any $x, y \in X$, the following holds: x Ov y or $x < <y = > v(x) < <v(y)$;*
- *v maintains both families A_0 and B_0: $\forall x \in A_0 \cup B_0$, $v(x) = x$;*
- *for any $x, y \in X = Z-(A_0 \cup B_0)$, we have $v(x) \cap v(y) = Nil$;*
- *for any $x \in X = Z - (A_0 \cup B_0)$, $|\{y \in A_0 \cup B_0,$ such that $v(x) \cap y \neq Nil\}| \geq |\{y \in A_0 \cup B_0,$ such that $x \cap y \neq Nil\}|/2$.*

We denote by H - *Reduced* = $(v(Z),$ F - *Reduced*$)$ the graph produced from this construction. H-*Reduced* is the complementary graph of the interval graph defined by the family $v(x)$, $x \in Z$.

Lemma 7. *The following inequalities hold:*

$$C\text{-}Max\text{-}Cut(H, A_0, B_0) \leq C\text{-}Max\text{-}Cut(H\text{-}Reduced, A_0, B_0) \leq C\text{-}Max\text{-}Cut(H\text{-}Nil, u(A_0), u(B_0)).$$

Lemma 8. *The following holds:* $C\text{-}Max\text{-}Cut(H\text{-}Nil, u(A_0), u(B_0)) = Max\text{-}Cut$ *(H-Nil, 1).*

Lemma 9. *The following holds:* $C\text{-}Max\text{-}Cut(H\text{-}Reduced, A_0 B_0) = Max\text{-}Cut$ *(H-Reduced, 1).* We are now ready to evaluate from above the gap induced by imposing the *(Ov, <<)-consistency* on *Max-Cut* solutions:

Theorem 7. *The following inequalities hold:*

$$0 \leq Max\text{-}Cut(H, 1) - C\text{-}Max\text{-}Cut(H) \leq |F_1^c| + |F_2^c|/2.$$

Proof. Let us observe that, if we refer to the related edge subsets $F\text{-}Reduced_1^c$, $F\text{-}Reduced_2^c$ and $F\text{-}Reduced_0^c$ induced by the graph H-*Reduced* according to the notations which were introduced at the beginning of Sect. 6, then we have that:

$$F\text{-}Reduced_0^c = F_0^c; \quad F\text{-}Reduced_1^c = Nil; \quad F\text{-}Reduced_2^c \subseteq F_2^c.$$

First, we may deduce from the last condition imposed in Lemma 6, to the construction of H-*Reduced*, that $|F\text{-}Reduced_2^c| \geq |F_2^c/2|$. From Lemma 9 we have that $Max\text{-}Cut(H\text{-}Reduced, 1) = C\text{-}Max\text{-}Cut(H\text{-}Reduced, A_0, B_0)$, and from Lemma 8 we also have that $C\text{-}Max\text{-}Cut(H\text{-}Reduced, A_0, B_0) \geq C\text{-}Max\text{-}Cut(H, A_0, B_0)$. Now, from Lemma 4, we have that $Max\text{-}Cut(H, 1) - Max\text{-}Cut(H\text{-}Reduced, 1)$ is non negative and bounded from above by $|F_1^c| + |F_2^c\text{-}F\text{-}Reduced_2^c| = |F_1^c| + |F_2^c| - |F\text{-}Reduced_2^c| \leq |F_1^c| + |F_2^c|/2$. The result is thus established. $\square$

From Theorem 7 we derive a bound for the gap between the optimal values of LAP and CLAP which improves the result of Proposition 1.

Theorem 8. *Let* $G = (X, E)$ *be an interval graph as in Theorem 5,* σ-*bal the optimal linear ordering defined for CLAP, and LAP(G) the optimal value of the Linear Arrangement Problem on G. We denote by Strong-Tr the number of Strong Triangles. Then the following holds:*

$$BG(G, \sigma\text{-}bal) - LOP(G) \leq Strong\text{-}Tr.$$

Proof. Theorem 5 states that $BG(G, \sigma\text{-}bal) = CLOP(G) = Tr(G) + \sum_x CV(x)$. Theorem 1 shows that $LOP(G) \geq Tr(G) + \sum_x V(x)$. Theorem 7 also shows that, for any $x \in X$, $CV(x) - V(x)$ does not exceed $|E^1(x)| + |E^2(x)|/2$, with

- $E^1(x) = \{(y, z) \in E$ such that $z \subset x,\ y \subset x,\ y \cap z \neq Nil\}$.
- $E^2(x) = \{(y, z) \in E$ such that $z \subset x,\ y \cap x \neq Nil,\ Not(x \subset y)\}$.

If we consider now $\sum_{x \in X} |E^1(x)|$, then we get an evaluation of all strong triangles (x, y, z), which are such that both y and z are included in x. We denote by *Strong*[1] the number of those strong triangles. When it comes to $\sum_{x \in X} |E^2(x)|$, we see that it enumerates all strong triangles (x, y, z) such that x and y overlap, while eventually counting twice triangle (x, y, z) in the case we simultaneously have $z \subset x$ and $z \subset y$. Then, $\sum_{x \in X} |E^2(x)|/2$ does not exceed the number $Strong^2$ of strong triangles which are such that x and y overlap. We thus derive

$$BG(G, \sigma \text{ - bal}) - LOP(G) \leq \sum_x (CV(x) - V(x)) \leq \sum_x |E^1(x)| + |E^2(x)|/2$$
$$= Strong^1 + \sum_x |E^2(x)|/2 \leq Strong^1 + Strong^2 = Strong - Tr.$$

The result is thus established.   □

7 Conclusion

This paper, with theoretical focus, presents approximation results for the *Linear Arrangement* problem in the case of interval graphs. Further research could be on the extension to chordal graphs and circular graphs, and also on the design of efficient exact algorithms.

References

1. Achouri, S., Bossart, T., Munier-Kordon, A.: A polynomial algorithm for MINDSC on a subclass of series parallel graphs. RAIRO Oper. Res. **43**(2), 145–156 (2009)
2. Barahona, F., Mahjoub, A.R.: On the cut polytope. Math. Prog. **36**, 157–173 (1986)
3. Charon, I., Hudry, O.: An updated survey on the linear ordering problem for weighted or unweighted tournaments. Ann. Oper. Res. **175**, 107–158 (2010)
4. Chung, F.R.K.: On optimal linear arrangement of trees. Comp. Math./Appl. **11**, 43–60 (1984)
5. Cohen, J., Fomin, F.V., Heggernes, P., Kratsch, D., Kucherov, G.: Optimal linear arrangement of interval graphs. In: Královič, R., Urzyczyn, P. (eds.) MFCS 2006. LNCS, vol. 4162, pp. 267–279. Springer, Heidelberg (2006)
6. Corneil, D.G., Kim, H., Natarajan, S., Olarin, S., Sprague, A.P.: A simple linear time algorithm of unit interval graphs. Inf. Process. Lett. **55**, 99–104 (1995)
7. Chvatal, V., Ebenegger, C.: A note on line digraphs and the directed Max-Cut problem. Discret. Appl. Math. **29**, 165–170 (1990)
8. Even, S., Shiloach, Y.: NP-Completeness of Several Arrangement Problems, Technical Report #43, Computer Science Department, The Technion, Haifa, Israel (1975)
9. Garey, M.R., Johnson, D.S.: Computers and intractability: a guide to the theory of NP-completeness. Freeman, New York (1979) (Computer Press)
10. Grotschel, M. (ed.): The Sharpest Cut, MPS-SIAM Series on Optimization. SIAM, Philadelphia (2004)
11. Horton, S.B.: The optimal linear arrangement problem: algorithms and approximation. Ph.D. Thesis, Georgia Institute of Technology (1997)

On the Asymmetric Connected Facility Location Polytope

Markus Leitner[1], Ivana Ljubić[1], Juan-José Salazar-González[2], and Markus Sinnl[1(✉)]

[1] ISOR, University of Vienna, Vienna, Austria
{markus.leitner,ivana.ljubic,markus.sinnl}@univie.ac.at
[2] DEIOC, Universidad de La Laguna, La Laguna, Spain
jjsalaza@ull.es

Abstract. This paper is concerned with the connected facility location problem, which has been intensively studied in the literature. The underlying polytopes, however, have not been investigated. This work is devoted to the polytope associated with the asymmetric version of the problem. We first lift known facets of the related Steiner arborescence and of the facility location polytope. Then we describe other new families of facet-inducing inequalities. Finally, computational results are reported.

1 Introduction

In the last years, the *connected facility location (ConFL) problem* and variants of it have received considerable attention from the operations research community (see, e.g., [1,2] and the references therein). The problem is of practical importance, e.g., in telecommunications, to model the deployment of *fiber-to-the-curb* networks, or in the design of data management networks. In this paper we are dealing with an *asymmetric ConFL* (aConFL): Given an assignment graph with connections between a set of *customers* and a set of potential *facility locations*, and a directed graph connecting facilities with each other using a (potentially empty) set of *intermediate nodes*, the goal of aConFL is to decide which facilities to open, how to assign customers to facilities and how to connect all open facilities to a *dedicated root node* at minimum cost.

Despite the large body of work on the ConFL, to our knowledge, there are no results on the facial structure of the underlying polytopes of ConFL. Our work is a first polyhedral study on aConFL. Our motivation for studying aConFL is twofold: (1) in some practical applications traversal of an edge in two opposite directions may involve different costs, and (2) the best performing computational approaches to ConFL are based on directed reformulations. In this paper we prove that some of these inequalities used in previous computational studies are facet-defining, and derive some new families of facet-defining inequalities. The obtained theoretical results are supported by a computational study on a newly generated benchmark set of digraphs. Polyhedral results for the symmetric version of the problem and a more extensive computational study are in [3].

© Springer International Publishing Switzerland 2014
P. Fouilhoux et al. (Eds.): ISCO 2014, LNCS 8596, pp. 371–383, 2014.
DOI: 10.1007/978-3-319-09174-7_32

The remainder of this article is organized as follows. Section 2 contains our main results considering the facet-defining inequalities of the aConFL polytope. In Sect. 3 we present our computational study.

More formally, in aConFL we are given a directed graph $D = (V, A_S, A_J)$ where the node set $V = S \cup J \cup \{0\}$ is the disjoint union of *Steiner nodes* S, customer nodes J and a dedicated root node 0. Facility nodes $I \subseteq S$ can be used to open facilities in which case facility opening costs $f_i \geq 0$, $\forall i \in I$, incur. For later use, let $K := S \setminus I$ denote the intermediate nodes that cannot be used as facilities, and let $K_0 := K \cup \{0\}$ and $S_0 = S \cup \{0\}$. The arc set $A_S \subseteq \{(s,t) : s,t \in V \setminus J\}$ represents possible connections between Steiner nodes. The arc set $A_J \subseteq \{(i,j) : i \in I, j \in J\}$ represents possible assignments of customers to facilities. In the context of telecommunication, A_S represents potential fiber optic connections in the core network, and A_J represents the copper cables connecting the customers to the core network through facilities. Arcs $a \in A_S$ are associated with establishing costs $c_a \geq 0$, $\forall a \in A_S$, and arcs $(i,j) \in A_J$ are associated with assignment costs $c_{ij} \geq 0$. The aConFL problem consists of selecting a subset of I of *open facilities*, connecting them through an *arborescence* rooted at 0 (that may use other Steiner nodes) and assigning each customer to exactly one open facility at the minimum cost.

In the following, we assume that $|I| \geq 3$ and $|J| \geq 3$. We also assume that subgraph $(S \cup \{0\}, A_S)$ (also called *core graph*) is a complete digraph in which all in-going arcs to the root node are removed, and that subgraph $(I \cup J, A_J)$ (also called *assignment graph*) is complete bipartite, i.e., $A_J = \{(i,j) : i \in I, j \in J\}$. Note that any instance of the undirected ConFL can be transformed into a aConFL instance by replacing each undirected core edge by a pair of oppositely directed arcs. Additionally, if no root node is given, both aConFL and ConFL can be transformed into a rooted aConFL instance by adding an artificial root node 0 together with arcs $(0,i)$, $\forall i \in I$, and additionally ensuring that the out-degree of this artificial root node is one.

2   The aConFL Polytope

We model aConFL using node decision variables $y_s \in \{0,1\}$, $\forall s \in S$, which indicate if node s is part of the solution and facility decision variables $z_i \in \{0,1\}$, $\forall i \in I$, which indicate whether facility i is opened. Furthermore, arc decision variables $x_a \in \{0,1\}$, $\forall a \in A_S$, specify which arcs of the core graph are part of the directed arborescence, and assignment variables $a_{ij} \in \{0,1\}$, $\forall i \in I, \forall j \in J$, indicate whether facility i serves customer j. Let $A = A_S \cup A_J$ denote the union of core and assignment arcs. For a set $H \subset V$, we define $\delta^-(H) := \{(u,v) \in A : u \notin H, v \in H\}$ and $\delta^+(H) := \{(u,v) \in A : u \in H, v \notin H\}$ and for sets $H, L \subset V$, we define $(H : L) := \{(u,v) \in A : u \in H, v \in L\}$. Moreover, for any vector $\mu \in \{0,1\}^M$ over a ground set M, we write $\mu(M') = \sum_{m \in M'} \mu_m$, for any $M' \subseteq M$. The aConFL problem can now be formulated as follows:

$$\min \sum_{a \in A_S} c_a x_a + \sum_{i \in I} f_i z_i + \sum_{(i,j) \in A_J} c_{ij} a_{ij} \tag{1}$$

$$a(\delta^-(j)) = 1 \qquad\qquad\qquad \forall j \in J \tag{2}$$

$$a_{ij} \leq z_i \qquad\qquad\qquad \forall i \in I, \forall j \in J \tag{3}$$

$$z_i \leq y_i \qquad\qquad\qquad \forall i \in I \tag{4}$$

$$x(\delta^-(s)) = y_s \qquad\qquad\qquad \forall s \in S \tag{5}$$

$$x(\delta^-(H)) \geq y_s \qquad\qquad \forall H \subseteq S, \forall s \in H \tag{yCuts}$$

$$(x, y, z, a) \in \{0,1\}^{|A_S|+|S|+|I|+|A_J|} \tag{6}$$

Constraints (2) ensure that every customer is assigned to exactly one facility, while constraints (3) make sure that assignment arcs can only be used if the corresponding facility is opened. Inequalities (4) are the coupling constraints between node and facility variables and Eq. (5) link node variables to the set of arc variables corresponding to ingoing arcs. Together with the directed cutset constraints (yCuts) which ensure that there is a directed path from the root node 0 to every other node in the solution, they also ensure that the solution cannot contain cycles. Thus, the solution is a directed arborescence rooted at 0. Let $\mathcal{Q}$ denote the aConFL polytope, i.e.:

$$\mathcal{Q} = \text{conv}\{(x, y, z, a) \in \{0,1\}^{|A_S|+|S|+|I|+|A_J|} \mid (x, y, z, a) \text{ satisfies}(2)-(6)\}.$$

2.1   Dimension of the aConFL Polytope

To establish the dimension of $\mathcal{Q}$ and some of its facet-inducing inequalities, we consider the intermediate polytopes

$$\mathcal{Q}_y(S') = \{(x, y, z, a) \in \mathcal{Q} : y_s = 1, s \in S'\} \qquad \forall S' \subseteq S.$$

The projection of $\mathcal{Q}_y(S')$ into the x space is the *Steiner arborescence polytope* $\mathcal{S}(S')$ with terminal set S' which has been studied by [4] The projection of $\mathcal{Q}_y(S')$ into the (z, a) space is the *facility location polytope* $\mathcal{U}$, with facilities I and customers J, studied in (for example) [5]. Since

$$\mathcal{Q}_y(S) = \mathcal{S}(S) \times \{y_s = 1, s \in S\} \times \mathcal{U},$$

the facets of $\mathcal{S}(S)$ and the facets of $\mathcal{U}$ are also the facets of $\mathcal{Q}_y(S)$. Since the dimension of $\mathcal{S}(S)$ is $|A_S| - |S|$ and the dimension of $\mathcal{U}$ is $|A_J| + |I| - |J|$, we have the following result:

Theorem 1. $dim(\mathcal{Q}_y(S)) = |A_S| + |A_J| + |I| - |S| - |J|.$

Besides, we can show that:

Theorem 2. *For all* $S' \subseteq S$, $dim(\mathcal{Q}_y(S')) = |A_S| + |A_J| + |I| - |S'| - |J|.$

The dimension of $\mathcal{Q}$ immediately follows from Theorem 2:

Corollary 1. $dim(\mathcal{Q}) = |A_S| + |A_J| + |I| - |J|.$

The proof of Theorem 2 also reveals that the following family of valid inequalities is facet-inducing.

Theorem 3. *Inequalities $y_s \leq 1$ are facet-inducing for every $s \in S$.*

Proof. The face induced by $y_s = 1$ corresponds to $S' = \{s\}$ in the proof of Theorem 2.   □

2.2   Facets Obtained by Lifting

The proof of Theorem 2 shows that removing a node from S', $S' \subseteq S$, increases the dimension of $\mathcal{Q}_y(S')$ by one. Thus, the facet-defining inequalities given in this section can be obtained by lifting (see, e.g., [6]).

Theorem 4 provides facet-inducing inequalities obtained by lifting from the facility location polytope $\mathcal{U}$ and using the results of [7]. Let $\mathcal{H}$ be a family of injective mappings $h : I \mapsto J$, for $|I| \leq |J|$, i.e., $i \neq i' \Rightarrow h(i) \neq h(i')$.

Theorem 4. *(a) Inequalities $a_{ij} \leq z_i$ are facet-inducing for all $i \in I$ and all $j \in J$.*
(b) Inequalities $a_{ij} \geq 0$ are facet-inducing for all $i \in I$ and all $j \in J$.
(c) Inequalities $z_i \leq y_i$ are facet-inducing for all $i \in I$.
(d) Let $|I| \leq |J|$ and let $h \in \mathcal{H}$ be an injective mapping. Then inequalities $\sum_{i \in I} a_{ih(i)} + z_i \geq 2$ are facet-inducing.

The following facet-defining inequalities can be obtained by lifting from the Steiner arborescence polytope $\mathcal{S}(S)$.

Theorem 5. *(a) Inequalities $x_{st} \geq 0$ are facet-inducing for all $(s,t) \in A_S$, if $s \neq 0$ or $|S| \geq 4$.*
(b) Inequalities $x(\delta^-(H)) \geq y_s$ are facet-inducing for all $H \subseteq S$ with $|H| \geq 2$, $|H \cap I| \leq |I| - 1$ and $s \in H$.
(c) Inequalities $x(\delta^-(H)) \geq 1$ are facet-inducing for all $H \subseteq S$, $|H \cap I| = |I|$.

2.3   New Facets

For proving that some additional inequalities are facet-inducing, the following well-known result (restated appropriately for our formulation) will be used:

Lemma 1. *[6] Let $(A^=, b^=)$ be the equality set of $\mathcal{Q}$, denote its size with m, and let $\mathcal{F} = \{(x, y, z, a) \in \mathcal{Q} : \pi_x x + \pi_y y + \pi_z z + \pi_a a = \pi_0\}$ be a proper face of $\mathcal{Q}$. Then the following two statements are equivalent:*

1. $\mathcal{F}$ is a facet of $\mathcal{Q}$.
2. If $\mathcal{F} \subseteq \mathcal{G} = \{(x, y, z, a) \in \mathcal{Q} : \alpha x + \beta y + \gamma z + \delta a = \lambda_0\}$, then $(\alpha, \beta, \gamma, \delta, \lambda_0) = (s(\pi_x, \pi_y, \pi_z, \pi_a) + tA^=, s\pi_0 + tb^=)$, for some $s \in \mathbb{R}$ and $t \in \mathbb{R}^m$.

In the following proofs we will construct feasible solutions $L \in \mathcal{F}$ for the face $\mathcal{F}$ under consideration and insert them into the equality defining $\mathcal{G}$ in order to determine the coefficients of $\mathcal{G}$ and then use Lemma 1. We denote the left-hand side of $\mathcal{G}$, i.e., the evaluation of $\alpha x + \beta y + \gamma z + \delta a$ for some L by $\mathcal{L}(L)$.

First, we consider the inequalities

$$x(\delta^-(H)) + \sum_{i \notin I \cap H} a_{ij} \geq 1, \quad \forall j \in J, \forall H \subseteq S \qquad \text{(aCuts)}$$

These inequalities, already considered in [1] for ConFL, state that for each customer j and any subset H of the core nodes (excluding 0), j is either served by a facility outside of H, or there has to be an arc going into H. Note that when $|H \cap I| = |I|$, the inequalities reduce to $x(\delta^-(H)) \geq 1$.

Theorem 6. *Inequalities* (aCuts) *are facet-inducing iff* $2 \leq |H \cap I| \leq |I| - 1$.

Proof. Let $\mathcal{F} = \{(x, y, z, a) \in Q : x(\delta^-(H)) + \sum_{i \notin I \cap H} a_{ij} = 1\}$ be the proper face induced by (aCuts) for some $j \in J$ and $H \subset S, 2 \leq |H \cap I| \leq |I| - 1$. In the following, we describe feasible solutions as tuples $L_q = (V_q \cap H, V_q \setminus H, I_q, A_q \cap A_S, A_q \cap A_J)$. Thereby, $V_q \subseteq S$ is the set of core nodes of solution L_q, $I_q \subseteq I$ its set of open facilities and $A_q \subset A_S \cup A_J$ its set of arcs. For the rest of the proof, we will need the following feasible solutions from $\mathcal{F}$, where $i, i', i_1, i_2 \in I$, $s, s_1, s_2, t, t_1, t_2 \in S$, and $j' \in J, j' \neq j$. Note that in some solutions, we make use of the assumption $|H \cap I| \geq 2$.

- $L_1 = (\{i_1, i_2\}, \emptyset, \{i_1\}, \{(0, i_1), (i_1, i_2)\}, (i_1 : J))$
- $L_2 = (\{i_1, i_2\}, \emptyset, \{i_1, i_2\}, \{(0, i_1), (i_1, i_2)\}, (i_1 : J))$
- $L_3 = (\{i, s\}, \emptyset, \{i\}, \{(0, i), (i, s)\}, (i : J))$
- $L_4 = (\{i, s\}, \{t\}, \{i\}, \{(0, i), (i, s), (s, t)\}, (i : J))$
- $L_5 = (\{i\}, \emptyset, \{i\}, \{(0, i)\}, (i : J))$
- $L_6 = (\{i_1, i_2\}, \emptyset, \{i_1, i_2\}, \{(0, i_1), (i_1, i_2)\}, \{(i_2, j'), (i_1 : J \setminus \{j'\})\})$
- $L_7 = (\{i_1, i_2\}, \emptyset, \{i_1, i_2\}, \{(0, i_1), (i_1, i_2)\}, (i_1 : J))$
- $L_8 = (\{i, t\}, \{s_1\}, \{i\}, \{(0, s_1), (s_1, t), (t, i)\}, (i : J))$
- $L_9 = (\{i, t\}, \{s_2\}, \{i\}, \{(0, s_2), (s_2, t), (t, i)\}, (i : J))$
- $L_{10} = (\{i, t_1\}, \{s\}, \{i\}, \{(0, s), (s, t_1), (t_1, i)\}, (i : J))$
- $L_{11} = (\{i, t_2\}, \{s\}, \{i\}, \{(0, s), (s, t_2), (t_2, i)\}, (i : J))$
- $L_{12} = (\emptyset, \{i'\}, \{i'\}, \{(0, i')\}, (i' : J))$ (assumption: $|H \cap I| \leq |I| - 1$)

We now suppose $\mathcal{F} \subseteq \mathcal{G}$ and determine the coefficients of $\mathcal{G}$.

(a) $\gamma_i = 0, \forall i \in I$: If $i \in H$, this follows from $\mathcal{L}(L_1) = \mathcal{L}(L_2)$. Else, it is obtained from $\mathcal{L}(L_{1'}) = \mathcal{L}(L_{2'})$, where $L_{1'}$ and $L_{2'}$ are obtained from L_1 and L_2, respectively, by assuming $i_2 \in I \setminus H$.
(b) $\alpha_{st} = -\beta_t, \forall s \in H, \forall t \in S \setminus H$: Obtained from $\mathcal{L}(L_3) = \mathcal{L}(L_4)$.
(c) $\alpha_{st} = -\beta_t, \forall s, t \in H$: If $|H| \geq 3$, the relation is obtained from $\mathcal{L}(L_3) = \mathcal{L}(L_{4'})$ where $L_{4'}$ is obtained from L_4 by assuming $t \in H$. Else, $\{s, t\} \subseteq I$ by assumption and the relation is obtained from $\mathcal{L}(L_5) = \mathcal{L}(L_{5'})$ for $i = s$ and where $L_{5'}$ is obtained from L_5 by adding node $t \in H$ and arc (i, t).

(d) $\alpha_{st} = -\beta_t$, $\forall s \in S_0 \setminus H$, $\forall t \in S \setminus H$: If $s \neq 0$, this is follows from $\mathcal{L}(L_{3'}) = \mathcal{L}(L_{4'})$ where $L_{3'}$ and $L_{4'}$ are obtained from L_3 and L_4, respectively, by assuming $s \in S \setminus H$. For $s = 0$, compare an arbitrary solution with a variant of it additionally considering a new node $t \in S \setminus H$ and arc $(0, t)$.

(e) $\delta_{ij'} = \delta_{j'}^H$, $\forall i \in H \cap I$, $\forall j' \in J$: Obtained from $\mathcal{L}(L_6) = \mathcal{L}(L_7)$

(f) $\delta_{ij'} = \delta_{j'}^H$, $\forall i \in I \setminus H$, $\forall j' \in J$, $j' \neq j$: Obtained from $\mathcal{L}(L_{6'}) = \mathcal{L}(L_{7'})$, where $\mathcal{L}(L_{6'})$ and $\mathcal{L}(L_{7'})$ are obtained from $\mathcal{L}(L_6)$ and $\mathcal{L}(L_7)$, respectively, by assuming $i_2 \in I \setminus H$. Note that $j' \neq j$ must hold since, neither $\mathcal{L}(L_{6'})$ nor $\mathcal{L}(L_{7'})$ would lie on $\mathcal{F}$ otherwise.

(g) $\alpha_{s_1 t} = \alpha_{s_2 t}$, $\forall s_1, s_2 \in S_0 \setminus H$, $\forall t \in H$: Obtained from $\mathcal{L}(L_8) = \mathcal{L}(L_9)$ using the result from (d).

(h) $\alpha_{st_1} + \beta_{t_1} = \alpha_{st_2} + \beta_{t_2}$, $\forall s \in S_0 \setminus H$, $\forall t_1, t_2 \in H$: If $s \neq 0$, the result follows from $\mathcal{L}(L_{10}) = \mathcal{L}(L_{11})$ using the result from (c). For $s = 0$ consider variants of L_{10} and L_{11} obtained by contracting arc $(0, s)$.

(i) $\alpha_{st} + \beta_t = \rho$, $\forall s \in S_0 \setminus H$, $\forall t \in H$: Follows from (g) and (h).

(j) $\delta_{i'j} = \rho + \delta_j^H$, $\forall i' \in I \setminus H$: Obtained from $\mathcal{L}(L_5) = \mathcal{L}(L_{12})$ using the results of (e), (f), and (i).

Inserting the obtained coefficients in the equation defining $\mathcal{G}$, using equations (5) and inserting an arbitrary solution from $\mathcal{F}$ (yielding $\lambda_0 = \rho + \sum_{j \in J} \delta_j^I$), the equation can be simplified to

$$\rho(x(\delta^-(H)) + \sum_{i \notin I \cap H} a_{ij}) + \sum_{j \in J} \delta_j^H a(\delta^-(j)) = \rho + \sum_{j \in J} \delta_j^H$$

which is a linear combination of the equation defining $\mathcal{F}$ and Eq. (2).

To show that $|H \cap I| \geq 2$ is also a necessary condition, observe that when $|H \cap I| = 1$, inequalities (aCuts) are dominated by $x(\delta^-(H)) \geq y_i$. For $|H \cap I| = 0$, the inequality is the sum of trivial facets $x_a \geq 0$, for $a \in \delta^-(H)$ and the equation $a(\delta^-(j)) = 1$.   □

Theorem 7. *Let $h \in \mathcal{H}$ be an injective mapping and $\hat{I} \subset I$. Then, the following inequalities are valid for aConFL:*

$$z(\hat{I}) + \sum_{i \in I \setminus \hat{I}} (y_i + a_{ih(i)}) + x(K_0 : \hat{I}) \geq 2 \tag{7}$$

Proof. If $z(\hat{I}) \geq 2$ or $y(I \setminus \hat{I}) \geq 2$ the theorem holds. Since at least one facility needs to be opened, it is sufficient to consider the following two cases:

(1) $z(\hat{I}) = 0$ and $z(I \setminus \hat{I}) = 1$, i.e., there exists a unique facility $i \in I \setminus \hat{I}$ with $z_i = 1$ to which all customers are assigned. Then, validity is implied by $a_{ih(i)} = 1$.

(2) $z(\hat{I}) = 1$ and $y(I \setminus \hat{I}) = 0$. Let $i \in \hat{I}$ be the unique facility with $z_i = 1$. Since no nodes from $I \setminus \hat{I}$ are used ($y(I \setminus \hat{I}) = 0$), validity of the inequality follows since i must be connected to the root and thus $x(K_0 : \hat{I}) \geq 1$.   □

Theorem 8. *Inequalities (7) are facet-inducing iff $\hat{I} \neq \emptyset$ and $|I \setminus \hat{I}| \geq 2$.*

Proof. For some $\hat{I} \neq \emptyset$, such that $|I \setminus \hat{I}| \geq 2$, let $\mathcal{F} = \{(x, y, z, a) \in \mathcal{Q} : z(\hat{I}) + \sum_{i \in I \setminus \hat{I}} (y_i + a_{ih(i)}) + x(K_0 : \hat{I}) = 2\}$ be the proper face induced by (7).

In the following, we describe feasible solutions as tuples $L_q = (V_q \cap \hat{I}, V_q \cap (I \setminus \hat{I}), V_q \cap K, I_q, A_q \cap A_S, A_q \cap A_J)$. Thereby, $V_q \subseteq S \cup \{0\}$ is the set of core nodes of solution L_q, $I_q \subseteq I$ its set of open facilities and $A_q \subset A_S \cup A_J$ its set of arcs. For the proof, we will need the following feasible solutions from $\mathcal{F}$, where $i_1, i_2, i_3 \in I$, $s_1, s_2, t \in K$. For solutions in which exactly two facilities, say $i, i' \in I$, are open, let $\mathcal{A} = (i : J \setminus \{h(i)\}) \cup \{(i', h(i))\}$ be an assignment of customers, s.t. the sum of a-variables in (7) is zero.

- $L_1 = (\{i_1\}, \{i_2\}, \emptyset, \{i_1, i_2\}, \{(0, i_2), (i_2, i_1)\}, (i_1 : J))$
- $L_2 = (\{i_1\}, \{i_2\}, \emptyset, \{i_1\}, \{(0, i_2), (i_2, i_1)\}, (i_1 : J))$
- $L_3 = (\{i_1\}, \{i_2\}, \emptyset, \{i_1, i_2\}, \{(0, i_2), (i_2, i_1)\}, \mathcal{A})$
- $L_4 = (\{i_1\}, \emptyset, \emptyset, \{i_1\}, \{(0, i_1)\}, (i_1 : J))$
- $L_5 = (\{i_1\}, \{i_2\}, \emptyset, \{i_2\}, \{(0, i_2), (i_2, i_1)\}, (i_2 : J))$
- $L_6 = (\{i_1\}, \emptyset, \{s_1\}, \{i_1\}, \{(0, s_1), (s_1, i_1)\}, (i_1 : J))$
- $L_7 = (\{i_1\}, \emptyset, \{s_2\}, \{i_1\}, \{(0, s_2), (s_2, i_1)\}, (i_1 : J))$
- $L_8 = (\{i_2\}, \emptyset, \{s_1\}, \{i_2\}, \{(0, s_1), (s_1, i_2)\}, (i_2 : J))$
- $L_9 = (\emptyset, \{i_2\}, \emptyset, \{i_2\}, \{(0, i_2)\}, (i_2 : J))$
- $L_{10} = (\emptyset, \{i_1, i_2\}, \emptyset, \{i_1, i_2\}, \{(0, i_1), (0, i_2)\}, \mathcal{A})$
- $L_{11} = (\emptyset, \{i_1, i_2\}, \emptyset, \{i_1, i_2\}, \{(0, i_1), (i_1, i_2)\}, \mathcal{A})$
- $L_{12} = (\{i_1\}, \{i_2, i_3\}, \emptyset, \{i_2, i_3\}, \{(0, i_2), (i_2, i_3), (i_2, i_1)\}, \mathcal{A})$

We now suppose $\mathcal{F} \subseteq \mathcal{G}$ and determine the coefficients of $\mathcal{G}$.

(a) $\gamma_i = 0$, $\forall i \in I \setminus \hat{I}$: Obtained from $\mathcal{L}(L_1) = \mathcal{L}(L_2)$.

(b) $\delta_{ij} = \delta_j^I$, $\forall i \in \hat{I}$, $\forall j \in J$ and $\forall i \in I \setminus \hat{I}$, $j \in J$, $j \neq h(i)$: From $\mathcal{L}(L_1) = \mathcal{L}(L_3)$ it follows that all the coefficients δ_{ij} for a $j \in J$ (except when $j = h(i)$ for $i \in I \setminus \hat{I}$) are the same – the coefficient is denoted by δ_j^I.

(c) $\alpha_{st} = -\beta_t$, $\forall s \in S_0$, $\forall t \in K$: Obtained from $\mathcal{L}(L_s) = \mathcal{L}(L_{st})$ where L_s is an arbitrary solution lying on $\mathcal{F}$ containing $s \in S_0$, but not $t \in K$, and L_{st} is obtained from L_s by attaching arc (s, t) to it.

(d) $\alpha_{i'i} = -\beta_i$, $\forall i', i \in \hat{I}$: Follows from $\mathcal{L}(L_{i'}) = \mathcal{L}(L_{i'i})$, where $L_{i'}$ is an arbitrary solution lying on $\mathcal{F}$ containing $i' \in \hat{I}$, but not $i \in \hat{I}$, and $L_{i'i}$ is obtained from $L_{i'}$ by attaching arc (i', i) to it.

(e) $\delta_{ih(i)} = \gamma^I + \delta_{h(i)}^I$ for $i \in I \setminus \hat{I}$: Obtained from $\mathcal{L}(L_2) = \mathcal{L}(L_5)$, which gives $\delta_{i_2 h(i_2)} = \gamma_{i_1} + \delta_{h(i_2)}^I$, where we used results from step (b). This result implies that $\gamma_{i_1} = \gamma_{i'}$ for $i_1, i' \in \hat{I}$. Denote this value by γ^I.

(f) $\alpha_{si} = \alpha_i^I$, $\forall i \in \hat{I}$, $s \in K$: Obtained from $\mathcal{L}(L_6) = \mathcal{L}(L_7)$, which gives $\alpha_{s_1 i_1} = \alpha_{s_1 i_2}$, where we used results from step (c). Thus, all arcs from K to a facility $i \in \hat{I}$ have the same coefficient, denote it by α_i^I.

(g) $\alpha_{0i} = \alpha_i^I$, $\forall i \in \hat{I}$: Obtained from $\mathcal{L}(L_4) = \mathcal{L}(L_6)$.

(h) $\alpha_i^I + \beta_i = \rho$, $\forall i \in \hat{I}$: Obtained from $\mathcal{L}(L_6) = \mathcal{L}(L_8)$, which gives $\alpha_{i_1}^I + \beta_{i_1} = \alpha_{i_2}^I + \beta_{i_2}$, where we used results from steps (b), (c) and (f). Hence, this sum is a constant value for every node in $\hat{I}$ – denote it by ρ.

(i) $\alpha_{ti}+\beta_i = \rho$, $\forall i \in I\backslash\hat{I}$, $\forall t \in K_0\cup\hat{I}$: We demonstrate this result for $t = 0$, and similar solutions can be constructed for $t \in K\cup\hat{I}$. From $\mathcal{L}(L_4) = \mathcal{L}(L_9)$ and using results of (a), (e), we have: $\alpha_{0i_1}+\beta_{i_1}+\gamma^I+\delta^I_{h(i_2)} = \alpha_{0i_2}+\beta_{i_2}+\delta_{i_2h(i_2)}$. Using one more time the result of (e), we obtain: $\alpha_{0i_1}+\beta_{i_1} = \alpha_{0i_2}+\beta_{i_2}$ for all $i_1 \in \hat{I}$ and $i_2 \in I\backslash\hat{I}$. From (h), it follows that $\alpha_{0i}+\beta_i = \rho$, $\forall i \in I\backslash\hat{I}$.

(j) $\alpha_{i'i}+\beta_i = \rho$, $\forall i', i \in I\backslash\hat{I}$: From $\mathcal{L}(L_{10}) = \mathcal{L}(L_{11})$ it follows that $\alpha_{0i_2} = \alpha_{i_1i_2}$. By adding β_{i_2} to both sides and using the result from (i), the result follows.

(k) $\rho = \gamma^I$: From $\mathcal{L}(L_{12}) = \mathcal{L}(L_3)$ and using results from (b), it follows that $\gamma_{i1} = \alpha_{i2i3} + \beta_{i3}$. From (e) and (j), we have $\rho = \gamma^I$. Note that for this step we need that $|I \backslash \hat{I}| \geq 2$.

Inserting the obtained coefficients in the equation defining $\mathcal{G}$, we get

$$\rho z(\hat{I}) + \sum_{i\in\hat{I}}((\rho - \beta_i)x(K_0 : i) - \beta_i x(I : i) + \beta_i y_i) + \sum_{k\in K}(-\beta_k x(\delta^-(k)) + \beta_k y_k) +$$

$$\sum_{i\in I\backslash\hat{I}}((\rho - \beta_i)x(\delta^-(i)) + \beta_i y_i + \rho a_{ih(i)}) + \sum_{j\in J}\delta^I_j a(I : j) = \lambda_0$$

By inserting any of the used solutions into the left-hand-side of the equation, we get $\lambda_0 = 2\rho + \sum_{j\in J}\delta^I_j$. Using Eq. (5), the equation can be simplified to

$$\rho(z(\hat{I}) + x(K_0 : \hat{I}) + y(I \backslash \hat{I}) + \sum_{i\in I\backslash\hat{I}} a_{ih(i)}) + \sum_{j\in J}\delta^I_j a(\delta^-(j)) = 2\rho + \sum_{j\in J}\delta^I_j.$$

It can be seen that the equation is a linear combination of the equation defining $\mathcal{F}$ and Eq. (2). Thus, (7) are facet-inducing when $|I \backslash \hat{I}| \geq 2$, $\hat{I} \neq \emptyset$.

To see that $\hat{I} \neq \emptyset$ and $|I \backslash \hat{I}| \geq 2$ is also a necessary condition, consider the following cases:

1. For $\hat{I} = I$, the inequality reduces to $z(I)+x(K_0 : I) \geq 2$, which is dominated by $a(\delta^-(j) + x(K_0 : I) \geq 2$. The latter is a linear combination of a facet $x(\delta^-(I)) \geq 1$ and an equation of type (2).
2. For $\hat{I} = I\backslash\{i\}$, the inequality reduces to $z(I\backslash\{i\})+a_{ih(i)}+y_i+x(K_0 : I\backslash\{i\}) \geq 2$. Notice that this inequality is dominated by the inequality in which y_i is replaced by $x(K_0 : i)$. The latter reduces to $z(I \backslash \{i\})+a_{ih(i)}+x(K_0 : I) \geq 2$, which is also not facet-inducing, for the same reasons as above.
3. Finally, for $\hat{I} = \emptyset$, we obtain $\sum_{i\in I}(y_i + a_{ih(i)}) \geq 2$ which is dominated by facet-defining constraints $\sum_{i\in I}(z_i + a_{ih(i)}) \geq 2$. $\qquad\square$

For the next family of valid inequalities, we employ a direct proof to show that the inequalities are facet-inducing.

Theorem 9. *Let $h \in \mathcal{H}$ be an injective mapping, and let $\hat{I} \subset I$, and $s \in K$. Then, the following inequalities are valid for aConFL:*

$$z(\hat{I}) + \sum_{i\in I\backslash\hat{I}}(y_i + a_{ih(i)}) + y_s + x(K_0 \backslash \{s\} : \hat{I} \cup \{s\}) \geq 2 + x(s : I \backslash \hat{I}) \qquad (8)$$

Proof. We will distinguish between the following cases:

(1) $y_s = 0$: Inequality (8) corresponds to inequality (7) since $y_s = 0$ implies that $x(s : I \setminus \hat{I}) = 0$ and $x(K_0 \setminus \{s\} : \hat{I} \cup \{s\}) = x(K_0 : \hat{I})$.

(2) $y_s = 1$ and $x(s : I \setminus \hat{I}) = 0$: Since at least one facility must be opened we obtain $z(\hat{I}) + \sum_{i \in I \setminus \hat{I}}(y_i + a_{ih(i)}) \geq 1$ which trivially holds.

(3) $y_s = 1$ and $x(s : I \setminus \hat{I}) \geq 1$: Let $I' = \{i \in I \setminus \hat{I} \mid x_{si} = 1\}$ and observe that $\sum_{i \in I \setminus \hat{I}} y_i \geq \sum_{i \in I'} y_i \geq x(s : I \setminus \hat{I})$ due to (5). Further note that the path from 0 to s either contains at least one arc from the cut $(K_0 \setminus \{s\} : \hat{I} \cup \{s\})$ or at least one node $i' \in I \setminus \hat{I}$. In either case, validity of (8) follows immediately. $\qquad \square$

Theorem 10. *Inequalities* (8) *are facet-inducing if* $|I \setminus \hat{I}| \geq 2$ *and* $\hat{I} \neq \emptyset$.

Proof. Let $\mathcal{F}$ be the face induced by (8) for given $\hat{I} \subset I$, $h \in \mathcal{H}$ and $s \in K$. We show how to construct $|A_S| + |A_J| + |I| - |J|$ affinely independent solutions lying on $\mathcal{F}$, which implies that $\mathcal{F}$ is a facet. We proceed in two steps, in the first step, we construct solutions that do not contain s and in the second step, we construct solutions containing s.

(1) Let $D' = (V \setminus \{s\}, A'_S, A_J)$ be a digraph obtained by removing s from D. By Theorem 8, the corresponding inequality (7) (for the given h and $\hat{I}$) is facet-defining, and therefore we can determine $|A'_S| + |A_J| + |I| - |J|$ affinely independent solutions in the associated lower dimensional space. By setting $y_s = 0$ and $x_{0s} = x_{si} = x_{is} = 0$, for all $i \in S, i \neq s$, these solutions are extended to feasible and affinely independent solutions lying on $\mathcal{F}$. Therefore, it only remains to additionally construct $|A_S| - |A'_S| = 2|S| - 1$ affinely independent solutions lying on $\mathcal{F}$ such that $y_s = 1$. This is done in the next step.

(2) The constructed solutions will be described using 6-tuples as in the proof of Theorem 8. Moreover, we will also use the assignment $\mathcal{A}$ defined in the same proof.

(a) Fix some facility $u \in \hat{I}$ and the arc $(u, s) \in A_S$. Now, pick some facility $i' \in I \setminus \hat{I}$ and for each $i \in I \setminus \hat{I}$, $i \neq i'$ build the following feasible solutions: $L_{si} = (\{u\}, \{i, i'\}, \{s\}, \{i, i'\}, \{(0, i'), (i', u), (u, s), (s, i)\}, \mathcal{A})$. Clearly, that way we create $|I \setminus \hat{I}| - 1$ affinely independent solutions due to the arcs (s, i). One more affinely independent solution can be found by switching the roles of i and i'.

(b) Consider now solutions $L_{is} = (\emptyset, \{i, i'\}, \{s\}, \{i, i'\}, \{(0, i), (i, s), (s, i')\}, \mathcal{A})$ for $i' \neq i \in I \setminus \hat{I}$. These solutions are all affinely independent due to the arcs (i, s). Again, we can also define a solution, where i and i' switch roles so that in total we obtain $|I \setminus \hat{I}|$ more affinely independent solutions.

(c) We create now solutions L_{sk}, for each $k \in K$, $k \neq s$, by adding arc (s, k) to one fixed solution L_{si} from step (a). That way, we obtain $|K| - 1$ affinely independent solutions. Moreover, for each $k \in K$, $k \neq s$, consider solutions $L_{ks} = (\emptyset, \{i, i'\}, \{s, k\}, \{i, i'\}, \{(0, k), (k, s), (s, i), (s, i')\}, \mathcal{A})$. We additionally obtain $|K| - 1$ affinely independent solutions due to arcs (k, s).

(d) One more solution is constructed as: $L_{0s} = (\emptyset, \{i, i'\}, \{s\}, \{i, i'\}, \{(0, s), (s, i),$ $(s, i')\}, \mathcal{A})$. This solution is affinely independent from all the previous ones due to arc $(0, s)$.

(e) Take the solution from step (d) and construct solutions $L_{su'}$ for each $u' \in \hat{I}$ by adding arc (s, u') to L_{0s}. We obtain $|\hat{I}|$ affinely independent solutions this way.

(f) Consider now solutions $L_{u's} = (\{u'\}, \{i, i'\}, \{s\}, \{i, i'\}, \{(0, u'), (u', s),$ $(s, i), (s, i')\}, \mathcal{A})$ for $u' \in \hat{I}$, $u' \neq u$, where u is the facility fixed for the solutions constructed in step (a). We get $|\hat{I}| - 1$ affinely independent solutions due to arcs (u', s).

(g) Note that all $2(|I \setminus \hat{I}| + |K| - 1 + |\hat{I}|) = 2|S| - 2$ solutions constructed so far are easily seen to be affinely independent, since in every solution, a previously unused arc is involved. The last affinely independent solution is constructed as $L^* = (\{u\}, I \setminus \hat{I}, \{s\}, \{i, i'\}, \{(0, u), (u, s), (s : I \setminus \hat{I})\}, \mathcal{A})$, where u is the facility from step (a). This concludes the proof. $\qquad \square$

3   Computational Results

3.1   Instances

To our knowledge, no instance sets with asymmetric costs are available for variants of ConFL or closely related problems. Thus, we generated two sets of random instances in the following way (following procedures described in [8,9]): $|V|$ points, each corresponding to one node in $S_0 \cup J$, are randomly generated in the Euclidean plane of size 100×100. Let (u_x, u_y) and (v_x, v_y) be the coordinates of two such nodes $u, v \in S_0 \cup J$ and let $\Delta_x(uv) = v_x - u_x$ and $\Delta_y(uv) = v_y - u_y$. Then, arc costs are defined as $c_{uv} = \omega \lfloor \sqrt{\Delta_x(uv)^2 + \xi \Delta_y(uv)^2} \rfloor$. Thereby, $\omega = 1$ for core arcs, $\omega = 3$ for assignment arcs, $\xi = 1$ if $\Delta_y \leq 0$ and $\xi = 2$ if $\Delta_y > 0$. Facility opening costs f_i, $\forall i \in I$ are integers chosen uniformly at random from the interval $[30, 60]$.

The first set of instances, denoted by A, consists of 20 complete graphs with $|I| = |J| = 100$ and $|K| = 50$. The second set of instances, denoted by B, consists of 20 sparse graphs with $|I| = |J| = 150$ and $|K| = 75$. In the latter instances, an arc between u and v only exists if the Euclidean distance between them is smaller than 40% of the largest Euclidean distance between any two points in this graph.

In addition to these randomly generated asymmetric instances, we also considered the symmetric Stein+UFL instances from [1]. The instances have $|I| = |J| = 200$ or $|I| = |J| = 300$, while $|S|$ ranges between 500 and 1000. Depending on the size of $|S|$, we get two sets of instances, denoted by C and D. In these instances the core network is sparse, while the assignment graph is complete bipartite. Moreover, in these instances, the average facility opening costs are approximately 15 times higher than the average arc costs.

3.2   Separation Algorithms

It is well known that cut inequalities (yCuts) and (aCuts) can both be separated in polynomial time using a max-flow algorithm (once for each core or customer node, respectively). Since, all coefficients in the objective function are nonnegative, we also obtain a valid model for aConFL when replacing (yCuts) by the following, so-called (zCuts) inequalities:

$$x(\delta^-(H)) \geq z_i \quad \forall H \subseteq S, \forall i \in H \cap I \tag{zCuts}$$

Though (zCuts) are not facet-inducing, they performed well in practice (see [1]).

3.3   Results

The computational results have been obtained using an Intel Xeon X5500 with 2.67 Ghz and 24 GB RAM and CPLEX 12.5 as solver for the ILPs. CPLEX-cuts have been turned off and the highest branching priority was given to facility variables. Before starting the solution process, all polynomial-size constraints, plus the inequalities $x_{st} + x_{ts} \leq y_s$, $\forall s \in S$, are added to the model. We have developed a branch-and-cut approach and tested the performance of the following settings: (1) yCuts: separating (yCuts) only; (2) aCuts: separating (aCuts) only; (3) y+aCuts: separating (yCuts), and only if no (yCuts) are violated in a branch-and-bound node, (aCuts) are separated; (4) zCuts: separating (zCuts)

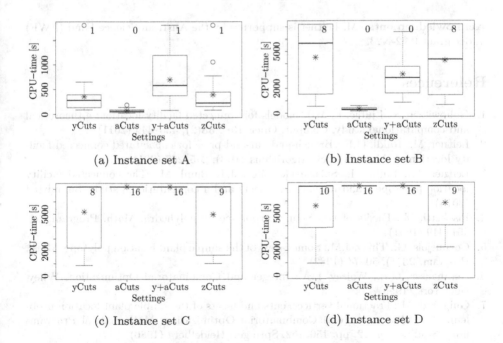

(a) Instance set A   (b) Instance set B

(c) Instance set C   (d) Instance set D

Fig. 1. Runtimes for the different settings

only. The separation routine for a node s (facility i) is called only if the corresponding LP-value on the right-hand-side is ≥ 0.5.

Figure 1a–d shows boxplots of the runtimes (in seconds) over all instances from A, B, C and D, respectively. The star in the boxplot indicates the average solution time and the number on top of each plot indicates the number of instances, which could not be solved within the given timelimit (two hours for B-D and 30 min for A).

There is a clear contrast in the performance on the instances A, B and instances C, D. For the former ones, the aCuts-setting significantly outperforms the remaining setting. On the contrary, the aCuts-setting is the worst approach for C, D. This can be explained by the different facility opening costs in the two groups: for A, B instances, opening a facility costs on average as much as establishing a link; however, it is about 15 times as expensive in groups C, D. On average, there are about twelve open facilities in optimal solutions of A, B, while only around four are open for C, D. Consequently, LP-solutions contain much less non-zero z- (and y-) variables in the latter case, and therefore, less separation calls are needed for the zCuts- and yCuts-setting. On the contrary, the numbers of separation calls for (zCuts) and (aCuts) are comparable for instances A, B. Therefore, the aCuts-setting is clearly beneficial, as it implies the strongest LP-bounds (recall that (aCuts) are facet-defining).

Comparing the performance between groups A and B, we observe that the sparsity of instances (group B) seems to deteriorate the performance of yCuts- and zCuts-settings.

Acknowledgements. M. Leitner is supported by the Austrian Science Fund (FWF) under grant I892-N23.

References

1. Gollowitzer, S., Ljubić, I.: MIP models for connected facility location: a theoretical and computational study. Comput. Oper. Res. **38**(2), 435–449 (2011)
2. Leitner, M., Raidl, G.R.: Branch-and-cut-and-price for capacitated connected facility location. J. Math. Model. Algorithms **10**(3), 245–267 (2011)
3. Leitner, M., Ljubić, I., Salazar-González, J.J., Sinnl, M.: The connected facility location polytope: valid inequalities, facets and a computational study. Submitted (2014)
4. Fischetti, M.: Facets of two Steiner arborescence polyhedra. Math. Program. **51**, 401–419 (1991)
5. Cornuejols, G., Thizy, J.M.: Some facets of the simple plant location polytope. Math. Program. **23**(1), 50–74 (1982)
6. Nemhauser, G.L., Wolsey, L.A.: Integer and Combinatorial Optimization. Wiley, New York (1999)
7. Guignard, M.: Fractional vertices, cuts and facets of the simple plant location problem. In: Padberg, M. (ed.) Combinatorial Optimization. Mathematical Programming Studies, vol. 12, pp. 150–162. Springer, Heidelberg (1980)

8. Bardossy, M.G., Raghavan, S.: Dual-based local search for the connected facility location and related problems. INFORMS J. Comput. **22**(4), 584–602 (2010)
9. Cirasella, J., Johnson, D.S., McGeoch, L.A., Zhang, W.: The asymmetric traveling salesman problem: algorithms, instance generators, and tests. In: Buchsbaum, A.L., Snoeyink, J. (eds.) ALENEX 2001. LNCS, vol. 2153, pp. 32–59. Springer, Heidelberg (2001)

Heuristic Approaches for the Robust Vehicle Routing Problem

Elyn L. Solano-Charris[1,2], Christian Prins[1], and Andréa Cynthia Santos[1](✉)

[1] ICD-LOSI, Université de Technologie de Troyes, 12, rue Marie Curie,
CS 42060, 10004 Troyes CEDEX, France
[2] Universidad de La Sabana, Campus del Puente del Común,
Km. 7, Chía, Cundinamarca, Colombia
{elyn.solano_charris,christian.prins,andrea.duhamel}@utt.fr

Abstract. In this article, the Robust Vehicle Routing Problem (RVRP) with uncertain traveling costs is studied. It covers a number of important applications in urban transportation and large scale bio-terrorism emergency. The uncertain data are defined as a bounded set of discrete scenarios associated with each arc of the transportation network. The objective is to determine a set of vehicle routes minimizing the worst total cost over all scenarios. A mixed integer linear program is proposed to model the problem. Then, we adapt some classical VRP heuristics to the RVRP, such as Clarke and Wright, randomized Clarke and Wright, Sequential Best Insertion, Parallel Best Insertion and the Pilot versions of the Best Insertion heuristics. In addition, a local search is developed to improve the obtained solutions and be integrated in a Greedy Randomized Adaptive Search Procedure (GRASP). Computational results are presented for both the mathematical formulation and the proposed heuristics.

Keywords: Vehicle routing · Robust optimization · *Min-max* objective · Heuristic · Local search · Metaheuristic

1 Introduction

The Vehicle Routing Problem (VRP) is a NP-hard problem which aims at defining routes for a fleet of vehicles, such that each vehicle starts and ends its tour at a depot node, each customer is visited once, and vehicle loads comply with vehicle capacity [1]. Introduced by Dantzig and Ramser [2], the VRP is one of the most studied problems in combinatorial optimization. One of its main assumptions is that the parameters and the data are assumed to be deterministic and known in advance [3–5]. Therefore, a perturbation on the input data could result in suboptimal or even infeasible solutions [6]. This assumption simplifies the problem but makes it less realistic since uncertainties occur in most real life contexts. Thus, a new and important trend consists in investigating extensions of the VRP with uncertain data, both in terms of theoretical and practical issues.

© Springer International Publishing Switzerland 2014
P. Fouilhoux et al. (Eds.): ISCO 2014, LNCS 8596, pp. 384–395, 2014.
DOI: 10.1007/978-3-319-09174-7_33

In the last years, VRP problems with parameters affected by uncertainties have been treated by using stochastic approaches, which models uncertainties through random variables with known probability distribution [7–10]. The robust optimization approach is an alternative to stochastic programming, designed as a mean to protect solutions against undesirable impacts due to incomplete or imprecise information on the data. It has been introduced in [11] and applied to a number of applications such as portfolio optimization [12], transportation [13], supply chain management [14] and network design problems [15].

The Robust Vehicle Routing Problem (RVRP) usually refers to uncertain data in the given instances: time windows, traveling costs, demands etc. This study considers the RVRP where each arc is weighted by an uncertain traveling cost or time. This version has important applications in urban transportation and evacuation problems such as large scale bio-terrorism emergency. The RVRP considered in this work is defined on a connected and directed graph $G = (V, A)$ with a set $V = \{0, 1, 2...n\}$ of n vertices (customers), including the depot (0), and a set $A = \{(i, j)|i, j \in V, i \neq j\}$ of arcs. Uncertain data are modeled here as a set of p discrete scenarios $S = \{1, 2, ...p\}$, where each scenario $k \in S$ specifies one cost $c_{ij}^k \in \mathbb{R}$ to each arc $(i, j) \in A$. Moreover, a demand d_i is associated with each customer $i \in V$ and a fleet of identical vehicles $F = \{1, 2, ..m\}$, located at the depot, is available. Each vehicle has capacity equal to Q. A solution is a set of vehicle routes starting and ending at the depot, visiting each customer once and respecting vehicles capacity. Its cost is the total cost of traversed arcs. We consider a *min-max* objective: the worst cost of the solution over all scenarios must be minimized.

This work brings the following contributions. We handle uncertain data as a bounded set of discrete scenarios for the costs of the arcs in a directed network (the VRP literature considers undirected graphs with symmetric costs). This situation reflects for instance transit problems in urban networks. A simple mathematical formulation is introduced for this RVRP. Then, we propose several constructive heuristics such as the Clarke and Wright (CW), Randomized CW (RCW), Parallel Best Insertion (PBI), Sequential Best Insertion (SBI), Pilot Parallel Best Insertion (PPBI) and Pilot Sequential Best Insertion (PSBI). In addition, more sophisticated strategies, such as local search and Greedy Randomized Adaptive Search Procedure (GRASP) are elaborated. To the best of our knowledge, no heuristic has been published in the literature to solve the RVRP investigated in this study.

The remaining of this work is organized as follows: a bibliographical review is introduced in Sect. 2, followed by a description of a mathematical formulation in Sect. 3. Then, the proposed heuristics are detailed in Sect. 4. Finally, the computational experiments and concluding remarks are respectively given in Sects. 5 and 6.

2   Related Works

Some works in the literature deal with the RVRP, mainly with uncertain data associated with time windows, travel times, travel costs or demands. We present in this section the main works which either apply robust optimization techniques and some entry points for research applying stochastic programming for the RVRP. The pioneer work [16] addresses the RVRP with uncertain demands and time windows. Analytical results on cluster-first route-second heuristics are given for large scale RVRPs. A survey which outlines the RVRP models with uncertainties related to demands, travel times and cost coefficients can be found in [17]. Some issues on applying stochastic programming and robust optimization are also discussed. Reference [18] provides a more extended review on the RVRP.

The RVRP with uncertain demands is probably the most investigated case. For instance, a Branch-and-Bound (B&B) algorithm is proposed by [19] which considers the *min-max* optimization criterion. The authors analyze the trade-off between robust solutions and deterministic solutions. The computational results show that the robust solution can protect from unmet demand while incurring a small additional cost over the deterministic optimal routes. Furthermore, a Particle Swarm Optimization (PSO) strategy integrating a local search is proposed in [6]. The PSO results are compared with the B&B proposed by [19], and performs well when costs are affected by small perturbations. The authors in [20] adapted the two-index and three-index VRP formulations and another one using the Miller, Tucker and Zemlin (MTZ) subtour elimination constraints. A Branch-and-Cut method is applied. The results demonstrate the computational advantages of the robust rounded capacity inequality cuts for the RVRP and the robust two-index vehicle flow formulation. Moreover, the price of robustness using different level of uncertainties is also analyzed.

More recently, the Open Vehicle Routing Problem (OVRP) with uncertain demands has been investigated by [21]. The OVRP differs from the VRP since vehicles do not return to the depot. In order to trade off the unmet demands, four heuristics strategies are considered to obtain the optimal solution when demands are disturbed, and a differential evolutionary algorithm is proposed. Instances with up to 199 customers have been tested.

Concerning the RVRP with uncertainties on time windows, a cutting-plane algorithm is embedded in a B&B method and in a column generation approach based on path inequalities and resource inequalities in [22]. Computational results are presented using the budget uncertainty polytopes and the results show that the path inequalities are almost as easy to separate as in the deterministic VRP case.

Uncertain travel times are handled via stochastic programming in [23]. The authors consider a two-stage recourse stochastic programming solved by a B&B. Two strategies "here and now" and "wait and see" are investigated. For the scenarios, the authors restrict the number of times an uncertain travel time can have the worst value. The results show that the approach obtains good solutions when the penalization over the objective function is small. As far as we know, the only work dealing with robust optimization strategies for the RVRP with uncertain travel costs is [24]. The authors considers the RVRP with uncertain travel costs modeled as interval data. An ant colony algorithm is introduced, where perturbations are performed on the objective coefficients towards the upper bounds of the interval data. This work differs from the RVRP focused here, since interval data are considered instead of discrete scenarios. Moreover, we consider in this paper a directed network with asymmetric costs.

Finally, the work [25] deals with the RVRP with uncertain data in travel times and demands. The authors also consider delays. Thus, the number of acceptable delayed segments are provided in order to determine a supported robustness. A multiflow formulation solved by Dantzig-Wolfe decomposition scheme are presented. The solutions are compared with the solutions obtained with the Monte-Carlo simulation and show that a robust solution can be improved with a small penalty in the optimal value.

3   Mathematical Formulation

A Mixed Integer Linear Programming (MILP) formulation for the *min-max* RVRP is given from (1) to (10). It makes use of binary variables x_{ij} which defines if an arc (i, j) belongs to the solution $(x_{ij} = 1)$ or not $(x_{ij} = 0)$. Variables t_i specify the vehicle load when leaving each node $i \in V$. Thus, the variable t_0 associated with the depot is considered only when vehicles leave the depot, as a consequence $t_0 = 0$.

$$\min \quad Z = \delta \qquad \text{subject to:} \tag{1}$$

$$\sum_{(i,j)\in A} c_{ij}^k x_{ij} \leq \delta \qquad\qquad \forall k \subset S \tag{2}$$

$$\sum_{i\in V} x_{ij} = 1 \qquad\qquad \forall j \in V\backslash\{0\} \tag{3}$$

$$\sum_{j\in V} x_{ij} = 1 \qquad\qquad \forall i \in V\backslash\{0\} \tag{4}$$

$$\sum_{i\in V} x_{0i} = m \tag{5}$$

$$t_j \geq t_i + d_j - Q(1 - x_{ij}) \qquad\qquad \forall(i,j) \in A,\ i,j \neq 0 \tag{6}$$

$$d_i \leq t_i \leq Q \qquad\qquad \forall i \in V \tag{7}$$

$$x_{ij} \in \{0, 1\} \qquad\qquad \forall(i,j) \in A \tag{8}$$

$$\delta \geq 0 \tag{9}$$

$$t_i \geq 0 \qquad\qquad \forall i \in V \tag{10}$$

The objective function (1) together with constraints (2) ensure that the worst total cost is minimized. Equalities (3) and (4) are the classical flow conservation constraints, which guarantee that only one vehicle arrives at each customer i and leaves it. Constraints (5) specify that m vehicles leave the depot and return to it, due to the flow conservation restrictions. Constraints (6) and (7) generalize the classical MTZ constraints for the TSP [26]. Here they are based on vehicle loads: if a vehicle visits i then j, its load increases by d_j. These constraints prevent subtours and ensure that vehicle capacity is respected. Finally, variables are defined from (8) to (10).

4   Heuristic Methods

Two Clarke and Wright-based heuristics, two insertion-based heuristics, and two pilot insertion-based heuristics are described below, as well as a local search and a GRASP. The main differences between the proposed RVRP heuristics and the similar versions for the VRP found in the literature are mainly the use of scenarios, the asymmetric arc costs, and a lexicographic approach to compare decisions in a greedy heuristic and evaluate moves in the local search.

A solution (complete or being constructed) is defined by a set of feasible routes, the total cost for each scenario k and the worst cost (maximum of these costs). We first tried implementations that do not degrade the worst cost, but the results were mitigated. Indeed, several decisions can lead to the same variation of the worst cost and cannot be distinguished. Moreover, a single move in the local search is in general not enough to decrease the worst cost: a sequence of moves is required. The lexicographic approach consists in sorting the costs of a solution (one per scenario) in non-increasing order, giving what we call the *lexicographic vector* of a solution. Then, a solution is said to be better than another solution if its vector is lexicographically smaller. This strategy is quite fruitful for instance, in the local search, a sequence of moves can improve progressively the lexicographic vector until the first component (the worst case) decreases. The price to pay is a multiplication of complexity expressions by $O(p \log p)$ to sort the costs and get the *lexicographic vector*. However, this extra cost is acceptable if the number of scenarios is relatively small compared with the number of customers. Compared to classical VRP heuristics, the algorithms are also complicated by the directed network. For instance, the cost of a sequence of customers changes when reversed, contrary to the undirected case.

4.1   Constructive Heuristics

Clarke and Wright-Based Heuristics. The CW heuristic or savings method is a well-known constructive heuristic for the VRP [27]. Its general idea consists of concatenating two routes such that the cost saving is maximized. The original CW heuristic considers symmetric costs associated with the arcs. As mentioned above, costs are asymmetric. Thus, there are more ways to concatenate two routes than in the original CW for each scenario. Furthermore, since the optimization criterion considered is a *min-max* one, edges are sorted in increasing order of savings instead of decreasing order in the CW.

A randomized version of the CW, referred as RCW is also proposed. The RCW is based on the CW, but when evaluating the merger of two routes, the resulting solution cost is increased by a random percentage in the range $[0, \theta]$. Thus, instead of selecting the best savings at each iteration, good moves, not necessarily the best, can be done. The best concatenation at each iteration is determined in $O(n^2)$, multiplied by the complexity of the sorting algorithm in $O(p \log p)$. Hence, CW runs in $O(n^3 p \log p)$, where p is the number of scenarios and n is the number of customers.

Insertion-Based Heuristics. The insertion heuristics have been proposed by [28]. The best insertion heuristics build a set of feasible routes by selecting seed customers and inserting them in one of the partial routes already created. At each iteration, the heuristic expands the current route by inserting the best unserviced customer, such that the vehicle capacity is ensured. SBI and PBI heuristics are introduced below.

SBI begins with a single route reduced to a loop on the depot. Best insertions are performed in the current route and it stops whenever all customers are attended, or if the new cannot be done because it exceeds the vehicle capacity. SBI tends to assign few customers to the last vehicle, thus routes are not balanced considering the number of customers.

PBI employs all vehicles available and fills m routes in parallel, which are initially empty. Then, at each iteration, the heuristic evaluates all feasible insertions of unrouted customers for every available routes. Since demands cannot be splitted, PBIH can fail to use m vehicles. In this case, one extra route is created and the heuristic performs similar steps as the SBIH.

The SBI and PBI mainly differ on the way the routes are built. In the SBI, a client is inserted at a time, and the routes are filled one after the other. While in the PBI heuristics, a set of routes are initially available and the customers are inserted in parallel to each route, i.e. the first customer is assigned for each route, only then, the second customers is set to the routes, etc. The resulting solutions can be found in $O(n^2 p \log p)$ by the SBI and for the PBI heuristics in $O(mn^2 p \log p)$.

Pilot Insertion-Based Heuristics. In the pilot method [29], a main heuristic calls an auxiliary heuristic (the pilot heuristic) to guide its decisions. We derived two pilot heuristics from SBI and PBI, called respectively PSBI and PPBI. It requires partial solutions, generated by the insertion heuristic, and with some customers already attended. Starting from the depot, the Pilot heuristic tests at each iteration all possible ways of extending the emerging route, adding to a route, one customer not visited at a time, following the PSBI. For both the PSBI and PPBI, a copy of the solution under construction is taken to insert the customer tested in incumbent partial solution, and the heuristic iterates by calling the pilot heuristic. The best insertion is performed. The difference between the PSBI and the PPBI is the way customers are included in the partial solution, and follows the insertions strategies previously described for the SBI and PBI. Since the pilot version of a heuristic consist in calling it as a subroutine at each iteration, the complexity of the non-pilot version is squared, making the pilot approach time-consuming.

4.2 Local Search

Relocation, Interchanges, and *2-opt* moves are applied in the local search procedure for the RVRP. Each iteration of the local search examines all ordered pairs of distinct routes (T, U) and evaluates the moves on one route if $T = U$ or otherwise on two routes $T \neq U$. The cost of the solution obtained if the move were performed is computed for each scenario. The current iteration stops as soon as a move improving the *lexicographic vector* of the current solution is detected (first improvement local search), or if no such move exists.

Interchanges exchanges two chains which may have 1 or 2 customers each. The lengths of the two swapped chains can be different.

Relocations move one or two adjacent customers to a different position.

2-opt moves intra-routes try to improve a solution by inverting a subsequence between two customers i and j (included, j must be after i). When $T \neq U$, vehicles capacity must be checked before computing the cost variations. A variant of *2-opt* is also proposed considering the asymmetric cost when a route is inverted. For this variant, the chains of nodes before i and after j are inverted.

Concerning complexity, the number of *interchanges* moves is $O(n^2)$ since they are applied to each pair of customers. There are also $O(n^2)$ *relocations*, because the n customers can be inserted in $O(n)$ different positions. There are also $O(n^2)$ *2-opt* moves. Each move traditionally evaluated in $O(1)$ is now checked in $O(p \log p)$, due to the construction of the *lexicographic vector* of the solution obtained by each move.

4.3 Greedy Randomized Adaptive Search Procedure

The GRASP is a multi-start metaheuristic proposed by [30]. It basically consists in building at each iteration, an initial solution using a randomized constructive heuristic, then in improving it by a local search. The best solution found is kept. GRASP is especially interesting as it only requires two components (a randomized heuristic and a local search) and very few parameters like the number of iterations. The GRASP for the RVRP makes use of the RCW heuristic and the local search presented in Sect. 4.2. The stopping criteria is the number of iterations which can be fine-tuned.

5 Computational Experiments

The tests were performed on a Dell Precision M6600 with a 2.2 GHz Intel Core i7-2720QM, 16 GB of RAM and Windows Professional. The proposed heuristics and metaheuristics were developed in Delphi XE, and the mathematical formulation was tested using GLPK (GNU Linear Programming Kit) under default parameters. All experiments for the mathematical formulation were carried out with a runtime limit of four hours. The goals of the experiments are to analyze the heuristics performance and the impact of using discrete scenarios for the RVRP.

Table 1. Results for the constructive and greedy heuristics

Instance name	d	Q	GLPK			Gap'(%)							
			LR	LB	UB	T(s)	Gap	SBI	PBI	PSBI	PPBI	CW	RCW
n10-m2-p10	264	150	195.5	217	*217	8.2	0.0	36.4	36.4	5.5	13.4	21.2	4.6
n10-m2-p20	264	175	250.6	284	*284	32.2	0.0	18.0	18.0	15.1	12.0	7.4	2.8
n10-m2-p30	264	200	272.1	301	*301	35.8	0.0	23.9	23.9	16.6	12.3	8.6	4.7
n15-m2-p10	346	200	320.2	346	*346	313.2	0.0	37.9	37.9	24.9	24.3	19.1	9.5
n15-m2-p20	346	230	339.3	373	*373	6,560	0.0	32.4	32.4	21.7	19.3	12.6	6.2
n15-m2-p30	346	260	368.7	398	404	-	1.5	49.0	49.0	25.1	21.4	21.4	9.5
n20-m2-p10	441	250	402.3	419	423	-	1.0	49.6	40.3	29.6	27.0	24.8	16.2
n20-m2-p20	441	300	443.3	460	470	-	2.2	48.9	37.8	26.1	24.3	16.3	9.3
n20-m2-p30	441	350	463.7	481	501	-	4.2	44.3	44.3	25.4	23.9	20.6	14.6
n10-m3-p10	264	95	227.0	255	*255	10.4	0.0	30.6	17.2	6.3	8.2	19.2	4.7
n10-m3-p20	264	105	279.5	316	*316	24.7	0.0	16.8	16.8	12.7	6.6	12.7	0.6
n10-m3-p30	264	115	302.0	337	*337	38.9	0.0	28.8	28.8	10.4	9.8	8.9	0.0
n15-m3-p10	346	120	349.1	381	*381	2,200	0.0	38.6	38.6	13.9	17.3	15.2	1.0
n15-m3-p20	346	140	365.2	399	*399	6,871	0.0	26.1	26.1	18.0	14.5	13.8	4.3
n15-m3-p30	346	160	394.4	426	433	-	1.6	36.9	36.9	22.8	16.4	11.7	5.9
n20-m3-p10	441	160	427.3	443	448	-	1.1	43.1	51.5	32.1	30.0	26.9	17.2
n20-m3-p20	441	175	466.3	481	497	-	3.3	50.3	34.1	23.3	24.9	20.6	15.0
n20-m3-p30	441	190	489.6	508	528	-	3.9	54.3	53.1	23.8	22.8	17.1	16.9
Average							1.0	37.0	34.6	19.6	18.3	16.6	8.4

For the purpose of these experiments, random instances were generated as follows. The travel cost of each arc and the demand per client is randomly chosen in $[1,50]$. The number of vehicles is either 2 or 3. The capacity of vehicles is selected to ensure a slack of $0.2Q$ to $0.8Q$ between the total demand d and fleet capacity mQ. The number of scenarios is either 10, 20 or 30. The file name format of each instance is $n\rho$-$m\beta$-$p\gamma$, where ρ, β and γ stand respectively for the numerical values of n, m and p.

Tables 1 and 2 summarize the results for the MILP, the greedy heuristics, the local search and the GRASP. The MILP results in Table 1 corresponds to the five columns LR (linear relaxation), LB (best lower bound), UB (best upper bound (UB), Gap (percentage deviation of the optimum or the best upper bound to LB), and $T(s)$ (computational time in seconds) except for the instances where the solver has attained the time limit. In this case, it is referred as "−". The last six columns indicate the percentage gap Gap' between the upper bound produced by each proposed heuristic and the lower bound achieved by GLPK. The RCW heuristic is run 50 times with $\theta = 8\%$. Due to lack of space, running times of greedy heuristics are not reported but here after they are commented. In Table 2, the results produced by the GLPK are recalled, followed by the solution values and running times for the local search and the GRASP. The local search is applied to the solution produced by the CW heuristic. The GRASP performs $ncalls = 500$ iterations, each of them calling the RCW heuristic with $\theta = 8\%$ and the local search. Optimal values are identified by asterisks "*".

Results in Tables 1 and 2 demonstrate that in spite of its simple definition, the RVRP is a very hard problem to solve. In fact, the heuristics which work

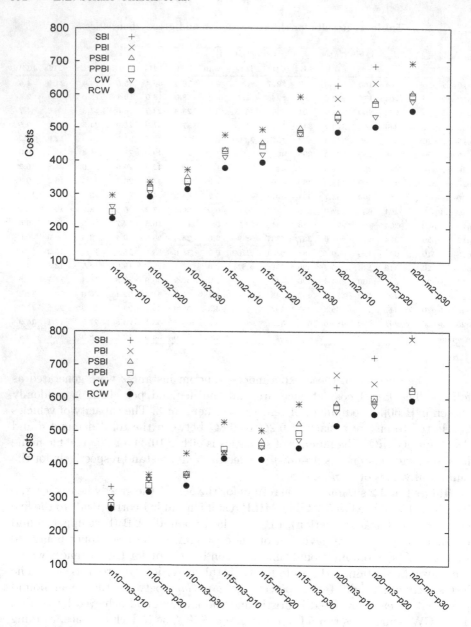

Fig. 1. Heuristic results respectively for the instances with $m = 2$ and $m = 3$.

well for the VRP are not able to find a good approximation. The two insertion heuristics are the fastest (less than 1 ms on average) but lead to very poor average gaps, 37.0 % for SBI and 34.6 % for PBI.

The pilot heuristics are able to find better solutions, with average gaps of 19.6 % for the PSBI and 18.3 % for the PPBI, and the running time reaches

Table 2. Results for the heuristic coupled with the local search and the GRASP

Instance name	GLPK					CW+ LS			GRASP		
	LR	LB	UB	T(s)	Gap	Cost	Gap'	T(s)	Cost	Gap'	T(s)
n10-m2-p10	195.5	217	*217	8.2	0.0	247	13.8	0.00	*217	0.0	0.28
n10-m2-p20	250.6	284	*284	32.2	0.0	296	4.2	0.00	*284	0.0	0.65
n10-m2-p30	272.1	301	*301	35.8	0.0	307	2.0	0.00	*301	0.0	1.11
n15-m2-p10	320.2	346	*346	313.2	0.0	385	11.3	0.00	347	0.3	1.02
n15-m2-p20	339.3	373	*373	6,560	0.0	410	9.9	0.01	376	0.8	2.39
n15-m2-p30	368.7	398	404	-	1.5	426	7.0	0.01	406	2.0	4.21
n20-m2-p10	402.3	419	423	-	1.0	458	9.3	0.01	436	4.1	2.58
n20-m2-p20	443.3	460	470	-	2.2	508	10.4	0.01	484	5.2	6.22
n20-m2-p30	463.7	481	501	-	4.2	536	11.4	0.02	513	6.7	10.48
n10-m3-p10	227.0	255	*255	10.4	0.0	279	9.4	0.00	*255	0.0	0.22
n10-m3-p20	279.5	316	*316	24.7	0.0	333	5.4	0.00	*316	0.0	0.56
n10-m3-p30	302.0	337	*337	38.9	0.0	348	3.3	0.00	*337	0.0	0.98
n15-m3-p10	349.1	381	*381	2,200.0	0.0	396	3.9	0.00	*381	0.0	0.93
n15-m3-p20	365.2	399	*399	6,871.0	0.0	411	3.0	0.00	*399	0.0	2.25
n15-m3-p30	394.4	426	433	-	1.6	476	11.7	0.01	435	2.1	3.98
n20-m3-p10	427.3	443	448	-	1.1	514	16.0	0.01	464	4.7	2.45
n20-m3-p20	466.3	481	497	-	3.3	540	12.3	0.01	511	6.2	6.01
n20-m3-p30	489.6	508	528	-	3.9	573	12.8	0.02	541	6.5	10.16
Average				9,350.4	1.0		8.7	0.01		2.1	3.14

50 ms on average. In fact, the heuristics with best performances are the CW and the RCW, as can be seen in Table 2 and in Fig. 1. Indeed, they find solutions with average gaps of 16.6 % and 8.4 %, respectively. CW is quite fast (5 ms on average) but the price to pay for RCW and its 50 iterations is an augmented average duration (around 0.3 s).

Starting from the solution returned by CW, the average gap is lowered from 16.6 % to 8.7 % by the local search, in less than 1 s. It is noticed that the best solutions for the RVRP were obtained with the GRASP procedure, which is able to retrieve 8 optima out of the 10 found by GLPK, with a small average gap of 2.1 % in 3.14 s, while GLPK achieves an averaged gap of 1.0 % in 9,350.4 s. This results show the advantages of using our local search and the GRASP to achieve small gaps and retrieve most proven optima in competitive computational time. This good performance probably comes from the combination of a constructive heuristic, a local search with moves that work well on the RVRP, and the random sampling of the local optima done by the GRASP in the solution space.

6 Conclusions

This article considers the RVRP with uncertain data associated with the costs or travel times. In this case, the variation of travel costs in the transportation network are considered. Constructive and greedy heuristics, a local search and a GRASP procedure are proposed. Such strategies have been addressed to deal with asymmetric costs and also with a set of discrete scenarios.

The experimental results show the RVRP is a hard problem in spite of its rather simple statement. Among the proposed heuristics, the CW-based heuristics outperform the others, and the local search manages to reduce their solution gaps. The GRASP is able to find even better solutions, including most proven optima, in reasonable running times.

Regarding future work, the current mathematical formulation can be strengthened and other formulations can be explored. We are currently investigating other metaheuristics and hybridizations for the RVRP. In addition, other optimization criteria can be studied such as *min-max* regret and lexicographical criterion. Finally, we are working on the design of transportation urban networks for which uncertainties are modeled as scenarios, to take into account the delays produced by traffic jams.

Acknowledgements. This research is part of the project METHODI which is funded by the Champagne-Ardenne Region.

References

1. Toth, P., Vigo, D.: Exact solution of the vehicle routing problem. In: Fleet Management and Logistics, pp. 1–31 (1998)
2. Dantzig, G.B., Ramser, J.H.: The truck dispatching problem. Manage. Sci. **6**, 80–91 (1959)
3. Achutan, N.R., Caccetta, L., Hill, S.P.: A new subtour elimination constraint for the vehicle routing problem. Eur. J. Oper. Res. **91**, 573–586 (1996)
4. Cordeau, J., Gendreau, M., Laporte, G., Potvin, J., Semet, F.: A guide to vehicle routing heuristics. J. Oper. Res. Soc. **53**, 512–522 (2002)
5. Brandão, J.: A deterministic tabu search algorithm for the fleet size and mix vehicle routing problem. Eur. J. Oper. Res. **19**, 716–728 (2009)
6. Moghaddam, B.F., Ruiz, R., Sadjadi, S.: Vehicle routing problem with uncertain demands: an advanced particle swarm algorithm. Comput. Ind. Eng. **62**, 306–317 (2012)
7. Golden, B.L., Yee, J.R.: A framework for probabilistic vehicle routing. AIIE Trans. **11**, 109–112 (1979)
8. Gendreau, M., Laporte, G., Seguin, R.: A tabu search heuristic for the vehicle routing problem with stochastic demands and customers. Oper. Res. **44**, 469–477 (1996)
9. Jula, H., Dessouky, M.M., Ioannou, P.: Truck route planning in non-stationary stochastic networks with time-windows at customer locations. IEEE Trans. Intell. Transp. Syst. **37**, 51–63 (2006)
10. Ai, J., Kachitvichyanukul, V.: A particle swarm optimization for the vehicle routing problem with simultaneous pickup and delivery. Comput. Oper. Res. **36**, 1693–1702 (2009)
11. Ben-Tal, A., Nemirovski, A.: Robust convex optimization. Math. Oper. Res. **23**, 769–805 (1998)
12. El-Ghaoui, L., Oks, M., Oustry, F.: Worst-case value-at-risk and robust portfolio optimization: a conic programming approach. Oper. Res. **51**, 543–556 (2003)

13. Erera, A.L., Morales, J.C., Savelsbergh, M.: Robust optimization for empty repositioning problems. Oper. Res. **57**, 468–483 (2009)
14. Tal, A.B., Golany, B., Nemirovski, A., Vial, J.: Supplier-retailer exible commitments contracts: a robust optimization approach. Manuf. Serv. Oper. Manage. **7**, 248–273 (2005)
15. Atamturk, A., Zhang, M.: Two-stage robust network flow and design under demand uncertainty. Oper. Res. **55**, 662–673 (2007)
16. Bertsimas, D., Simchi-Levi, D.: A new generation of vehicle routing research: robust algorithms, addressing uncertainty. Oper. Res. **44**, 286–304 (1996)
17. Ordóñez, F.: Robust vehicle routing. INFORMS TutORials Oper. Res. **7**, 153–178 (2010)
18. Solano, E., Santos, A.C., Prins, C.: An overview on solving robust vehicle routing problem. In: 14ème congrès de la Société Française de Recherche Opérationnelle et d'Aide à la Décision (ROADEF) 2 p. (2013)
19. Sungur, I., Ordóñez, F., Dessouky, M.: A robust optimization approach for the capacitated vehicle routing problem with demand uncertainty. IIE Trans. **40**, 509–523 (2008)
20. Gounaris, C., Wiesemann, W., Floudas, C.A.: The robust capacitated vehicle routing problem under demand uncertainty. Oper. Res. **61**, 677–693 (2013)
21. Erbao, C., Mingyong, L., Hongming, Y.: Open vehicle routing problem with demand uncertainty and its robust strategies. Expert Syst. Appl. **41**, 3569–3575 (2014)
22. Agra, A., Hvattum, L.M., Christiansen, M., Figuereido, R., Poss, M., Requejo, C.: The robust vehicle routing problem with time windows. Comput. Oper. Res. **40**, 856–866 (2013)
23. Han, J., Lee, C., Park, S.: A robust scenario approach for the vehicle routing problem with uncertain travel times. Transp. Sci. **63**, 1294–1306 (2013)
24. Toklu, N., Montemanni, R., Gambardella, L.M.: An ant colony system for the capacitated vehicle routing problem with uncertain travel costs, pp. 32–39 (2013)
25. Lee, C., Lee, K., Park, S.: Robust vehicle routing problem with deadlines and travel time/demand uncertainty. J. Oper. Res. Soc. **63**, 1294–1306 (2012)
26. Miller, C., Tucker, A., Zemlin, R.: Integer programming formulations and traveling salesman problems. J. ACM **7**, 326–329 (1960)
27. Clarke, G., Wright, J.: Scheduling of vehicles from a central depot to a number of delivery points. Oper. Res. **12**, 568–581 (1964)
28. Solomon, M.M.: Algorithms for the vehicle routing and scheduling problem with time window constraints. Oper. Res. **35**, 254–265 (1987)
29. Duis, C.W., Voss, S.: The pilot method: a strategy for heuristic repetition with application to the steiner problem in graphs. Networks **34**, 181–191 (1999)
30. Feo, T., Resende, M.: A probabilistic heuristic for a computationally difficult set covering problem. Oper. Res. Lett. **8**, 67–71 (1989)

A Fast Large Neighborhood Search for Disjunctively Constrained Knapsack Problems

Mhand Hifi, Sagvan Saleh, and Lei Wu[✉]

EPROAD EA 4669, Univercité de Picardie Jules Verne,
7 rue du Moulin Neuf, 80039 Amiens, France
{mhand.hifi,sagvan.saleh,lei.wu}@u-picardie.fr

Abstract. In this paper, we propose a heuristic based upon the large neighborhood search for the disjunctively constrained knapsack problem (DCKP). The proposed method combines a two-phase procedure and a large neighborhood search. First, the two-phase procedure is applied in order to provide a starting feasible solution for the large neighborhood search. The first phase serves to determine a feasible solution by successively solving two subproblems: the weighted independent set and the classical binary knapsack. The second phase tries to improve the quality of the solutions by using a descent method which applies both degrading and re-optimizing strategies. Second, a large neighborhood search is introduced in order to diversify the search space. Finally, the performance of the proposed method is computationally analyzed on a set of benchmark instances of the literature where its provided results are compared to those reached by Cplex solver and some recent algorithms. The provided results show that the method is very competitive since it is able to reach new solutions within small runtimes.

Keywords: Heuristic · Knapsack · Neighborhood · Re-optimisation

1 Introduction

In this paper we investigate the use of the large neighborhood search for solving the *Disjunctively Constrained Knapsack Problem* (DCKP). DCKP is characterized by a knapsack of fixed capacity c, a set I of n items, and a set E of incompatible couples of items, where $E \subseteq \left\{ (i,j) \in I \times I, \ i < j \right\}$. Each item $i \in I$ is represented by a nonnegative weight w_i and a profit p_i. The goal of the DCKP is to maximize the total profit of items that can be placed into the knapsack without exceeding its capacity, where all items included in the knapsack must be compatible. Formally, DCKP can be defined as follows:

$$(\mathrm{P}_{DCKP}) \max \quad \sum_{i \in I} p_i x_i$$

© Springer International Publishing Switzerland 2014
P. Fouilhoux et al. (Eds.): ISCO 2014, LNCS 8596, pp. 396–407, 2014.
DOI: 10.1007/978-3-319-09174-7_34

$$\text{s.t.} \qquad \sum_{i \in I} w_i x_i \leq c \qquad\qquad\qquad (1)$$

$$x_i + x_j \leq 1 \quad \forall\, (i,j) \in E \qquad\qquad (2)$$

$$x_i \in \{0,1\} \quad \forall\, i \in I,$$

where $x_i, \forall\, i \in I$, is equal to 1 if the i-th item is included in the knapsack (solution); 0 otherwise. Inequality (1) denotes the knapsack constraint with capacity c and inequalities (2) represent the disjunctive constraints which ensure that all items belonging to a feasible solution must be compatible. We can observe that the solution domain of a knapsack problem can be characterized by (i) the inequality (1), (ii) the integral constraints $x_i \in \{0,1\}$, $\forall i \in I$, and, (iii) those corresponding to the weighted independent set problem obtained by combining both inequalities (2) and (3). Without loss of generality, we assume that (i) all input data c, p_i, $w_i, \forall\, i \in I$, are nonnegative integers and (ii) $\sum_{i \in I} w_i > c$ for avoiding trivial solutions.

The remainder of the paper is organized as follows. Section 2 reviews some previous works on the DCKP. Section 3 discusses the two-phase procedure that provides a starting solution for P_{DCKP}. Section 4 describes the large neighborhood search-based heuristic for the DCKP. Section 5 evaluates the performance of the proposed method on a set of benchmark instances taken from the literature. Finally, Sect. 6 summarizes the contents of the paper.

2   Background

The DCKP is an NP-hard combinatorial optimization problem. It reduces to the maximum *weighted independent set* problem (Garey and Johnson [1]) when the knapsack capacity constraint is omitted and to the *classic knapsack* problem when $E = \emptyset$. It is easy to show that DCKP is a more complex extension of the multiple choice knapsack problem which arises either as a stand alone problem or as a component of more difficult combinatorial optimization problems. Its induced structure in complex problems allows the computation of upper bounds and the design of heuristic and exact methods for these complex instances. For example, DCKP was used in Dantzig-Wolfe's decomposition formulation for the two-dimensional bin packing problem (Pisinger and Sigurd [10]). It served as a local optimization subproblem for the pricing problem which consists in finding a feasible packing of a single bin verifying the smallest reduced cost. The same problem has been also used in Sadykov and Vanderbeck [12] as the pricing problem for solving the bin packing with conflicts.

Due to the complexity and hardness of the DCKP, most results on this topic are based on heuristics although exact methods have been proposed. Among papers addressing the resolution of DCKP, we found that of Yamada *et al.* [14,15] in which the problem was tackled with approximate and exact methods. The approximate heuristic generates an initial feasible solution, and improves it using a 2-opt neighborhood-search. The exact algorithm starts its search from the solution obtained by the approximate algorithm, and undertakes an implicit enumeration combined with an interval reduction technique.

Hifi and Michrafy [3] proposed three exact algorithms in which reduction strategies, an equivalent model and a dichotomous search cooperate to solve DCKP. The first algorithm reduces the size of the original problem by starting with a lower bound and successively solving relaxed DCKPs. The second algorithm combines a reduction strategy with a dichotomous search in order to accelerate the search process. The third algorithm tackles instances with a large number of disjunctive constraints using two cooperating equivalent models.

Hifi and Michrafy [4] proposed a three-step reactive local search. The first step of the algorithm starts by determining an initial solution using a greedy procedure. The second step is based on an intensification procedure which removes an item from the solution and inserts other ones. It adopts a memory list that stores swaps and/or the hashing function; thus, forbids cycling. The third step diversifies the search process by accepting to temporarily degrade the quality of the solution in hope to escape from local optima.

Pferschy and Schauer [9] presented pseudo-polynomial algorithms for special cases of the disjunctively knapsack problem which are mainly based on a graph representation: trees, graphs with bounded tree-width and chordal graphs. The authors extended their algorithms for establishing fully polynomial time approximation schemes (FPTAS).

Hifi et al. [7] investigated the use of the rounding solution procedure and an effective local branching. The method combines two procedures: (i) a rounding solution procedure and (ii) a restricted exact solution procedure. Hifi and Otmani [5] investigated the use of the scatter search for approximately solving the DCKP. The approach tried to explore some characteristics of two problems in order to tackle the DCKP: the independent set problem and the single knapsack problem. The performance of the approach was evaluated on the same instances as considered in [7] and showed that such approach was able to improve the solution quality of some instances. Hifi and Otmani [6] adapted the same approach as in [5], but by considering an equivalent model of the DCKP already proposed by Hifi and Michrafy [4]. The equivalent model was solved by applying a first level scatter search in which the model was refined by injecting some valid constraints.

Finally, Hifi [2] investigates an iterative rounding search-based algorithm. The method can be viewed as an alternate to both approaches considered in Hifi et al. [5,7] where three strategies were combined: (i) the variable-fixing technic using the rounding method applied to the linear relaxation of DCKP, (ii) the injection of successive valid constraints with bounding the objective function, and (iii) a neighbor search around solutions characterizing a series of reduced subproblems. The aforementioned steps are iterated until satisfying some stopping criteria.

3   A Two-Phase Solution Procedure

This section describes an efficient algorithm for approximately solving the DCKP by alternatively running two solution procedures. The first procedure is applied in order to determine a feasible solution and then, the second one tries to improve the solution at hand. For the rest of the paper, we assume that all items are ranked in decreasing order of their profits.

3.1 The First Phase

The first phase determines a feasible solution of the DCKP by solving two optimization problems:

- A *weighted independent set problem* (noted P_{WIS}), extracted from P_{DCKP}, is first solved in order to determine an *Independent Set* solution, noted IS.
- A *classical binary knapsack problem* (noted P_K) associated to both IS and the corresponding capacity constraint (i.e., $\sum_{i \in IS} w_i x_i \leq c$) is solved in order to provide a feasible solution for the P_{DCKP}.

Let $S_{IS} = (s_1, \ldots, s_n)$ be a feasible solution of P_{WIS}, where s_i is the binary value assigned to x_i, $\forall\, i \in I$. Let $IS \subseteq I$ be the restricted set of items denoting items of S_{IS} whose values are fixed to 1. Then, linear programs referring to both P_{WIS} and P_K may be defined as follow:

$$(P_{WIS}) \begin{cases} \max \sum_{i \in I} p_i x_i \\ \text{s.t.} \quad x_i + x_j \leq 1, \ \forall (i,j) \in E \\ \qquad x_i \in \{0,1\}, \ \ \forall\, i \in I, \end{cases} \qquad (P_K) \begin{cases} \max \sum_{i \in IS} p_i x_i \\ \text{s.t.} \ \sum_{i \in IS} w_i x_i \leq c, \\ \qquad x_i \in \{0,1\}, \ \forall\, i \in IS. \end{cases}$$

Algorithm 1. Compute a feasible solution for P_{WIS}

Require: An instance I of P_{DCKP}.
Ensure: A feasible solution (independent set) IS for P_{WIS}.
 1: **Initialization:**
       Set $IS = \emptyset$ and $I = \{1, \ldots, n\}$.
 2: **while** $I \neq \emptyset$ **do**
 3:     Let $i = \operatorname{argmax}\{p_i \mid p_i \geq p_k, \ k \in I\}$.
 4:     Set $IS = IS \cup \{i\}$.
 5:     Remove i and all items j such that $(i,j) \in E$ from I.
 6: **end while**
 7: **return** IS as a feasible solution of P_{IS}.

On the one hand, we can observe that the solution domain of P_{WIS} includes the solution domain of P_{DCKP}. On the other hand, an optimal solution of P_{WIS} is not necessary an optimal solution of P_{DCKP}. Therefore, in order to search a quick solution IS, the following procedure is applied (as described in Algorithm 1):

Set IS to an empty set (a trivial solution of P_{WIS}) and, fill iteratively the set IS with the items belonging to I. At each iteration (of Algorithm 1), the item realizing the greatest profit is selected and inserted into IS. The chosen item with its incompatible neighbors are then removed from I. Such a process is iteratively performed until no more items can be added to the current solution IS. Finally, the algorithm stops and exits with a feasible solution IS for P_{WIS}.

Algorithm 2. Compute a feasible for P_{DCKP}

Require: IS, an independent set of P_{WIS}, and I, an instance of P_{DCKP}.
Ensure: S_{DCKP}, a DCKP's feasible solution.
 1: **Initialization:**
    Let P_K be the resulting knapsack problem constructed according to items belonging to IS.
 2: **if** S_{IS} satisfies the capacity constraint (1) of P_{DCKP} **then**
 3:     Set $S_{DCKP} = S_{IS}$;
 4: **else**
 5:     Let S_{DCKP} be the resulting solution of P_K.
 6: **end if**
 7: **return** S_{DCKP} as a feasible solution of P_{DCKP}.

As mentioned above, IS may violate the capacity constraint of P_{DCKP}. Then, in order to provide a feasible solution for P_{DCKP}, the knapsack problem P_K is solved. Herein, P_K is solved using the exact solver of Martello et al. [8]. Algorithm 2 describes the main steps used for determining a feasible solution of P_{DCKP}.

3.2   The Second Phase

In order to improve the quality of the solution provided by the first phase (i.e., the feasible solution S_{DCKP} returned by Algorithm 2), a local search is performed. The used local search can be considered as a descent method that tends to improve a solution by alternatively calling two procedures: *degrading* and *re-optimizing* procedures. The *degrading procedure* serves to build a k-neighborhood of S_{DCKP} by dropping k fixed items from S_{DCKP} while the *re-optimizing procedure* tries to determine an optimal solution regarding the current neighborhood. The descent procedure is stopped when no better solution can be reached.

Algorithm 3 shows how an improved solution can be computed by using a descent method. Let S_{DCKP} be the current feasible solution obtained at the first phase and α be a constant, where $\alpha \in [0, 100]$. Then, $\alpha |I|$ represents the number of the unassigned variables of S_{DCKP}. The main loop (cf., lines 2–7) of the descent method serves to alternatively yield and explore a neighborhood issue from a local optimum. At line 3, the best solution found so far, namely $S^{\star}_{DCKP}$, is updated with the solution S_{DCKP} reached at the last iteration. Line 4 removes $\alpha |I|$ variables regarding the current solution S_{DCKP}, where the related items with the highest degree are favored. Let i be an item realizing the highest degree; that is, an item i whose variable x_i is fixed to 1 in S_{DCKP}. Then, set x_i

Algorithm 3. A descent method

Require: S_{DCKP}, a feasible solution of DCKP.
Ensure: S^*_{DCKP}, a local optimal solution of DCKP.
1: Set S^*_{DCKP} as an initial feasible solution, where all variables are fixed to 0.
2: **while** S_{DCKP} is better than S^*_{DCKP} **do**
3:    Update S^*_{DCKP} with S_{DCKP}.
4:    Set $\alpha|I|$ fixed variables of S_{DCKP} as free.
5:    Define the corresponding neighborhood of S_{DCKP}.
6:    Determine the optimal solution in the current neighborhood and update S_{DCKP}.
7: **end while**
8: **return** S^*_{DCKP}.

---

as free variable and also for its incompatible variables x_j, such that $(i,j) \in E$ and $(j,k) \notin E$, where k, $k \neq i$, corresponds to the index of the variables whose values are equal to 1 in S_{DCKP}. At line 6, S_{DCKP}, is replaced by the best solution found in the current neighborhood. Finally, the process is iterated until no better solution can be reached (in this case, Algorithm 3 exits with the best solution S^*_{DCKP}).

Algorithm 4. Remove $\beta|I|$ variables of S_{DCKP}

Require: S_{DCKP}, a starting solution of P_{DCKP}.
Ensure: An independent set IS and a reduced instance I^r of P_{DCKP}.
1: Set $counter = 0$, $I^r = \emptyset$ and IS to the set of items whose decision variable are fixed to 1 in S_{DCKP}.
2: Range IS in non decreasing order of their profit per weight.
3: **while** $counter < \beta|I|$ **do**
4:    Let r be a real number randomly generated in the interval $[0,1]$ and $i = |IS| \times r^\gamma$.
5:    Set $IS = IS \setminus \{i\}$, $I^r = I^r \cup i$ and increment $counter = counter + 1$.
6:    **for** all items j such that $(i,j) \in E$ **do**
7:       **if** item j is compatible with all items belong to IS **then**
8:          Set $I^r = I^r \cup \{j\}$ and $counter = counter + 1$.
9:       **end if**
10:    **end for**
11: **end while**
12: **return** IS and I^r.

Note that, on the one hand, the runtime of Algorithm 3 may increase when α tends to 100. Because dropping a large percentage of items involves that the reduced DCKP is closest to the original one. On the other hand, Algorithm 3 is called at each iteration of the large neighborhood search (cf., Sect. 4), a large size reduced DCKP can cause the large neighborhood search slow down. Therefore, we favor the achievement of a fast algorithm which is able to converge towards a good local optimum. This is why our choice is oriented to moderate the values of α, as shown in the experimental part (cf., Section 5).

4   A Large Neighborhood Search

Large Neighborhood Search (LNS) is a heuristic that has proven to be effective on wide range of combinatorial optimization problems. A simplest version of LNS has been presented in Shaw [13] for solving the vehicle routing problem (cf., also Pisinger and Ropke [11]). LNS is based on the concepts of *building* and *exploring* a neighborhood; that is, a neighborhood defined implicitly by a *destroy* and a *repair* procedure. Unlike the descent methods, which may stagnates in local optima, using large neighborhoods makes it possible to reach better solutions and explore a more promising search space.

For instance, the descent method discussed in Sect. 3.2 (cf., Algorithm 3) may explore some regions and stagnates in a local optimum because either *degrading* or *re-optimizing* considers a mono-criterion. In order to enlarge the chance of reaching a series of improved solutions or to escape from a series of local optima, a random destroying strategy, which depends on the value of the profit per weight of items, is applied. Algorithm 5 summarizes the main steps of LNS (noted LNSBH) which uses Algorithm 4 for determining the neighborhood of a given solution.

Algorithm 5. A large neighborhood search-based heuristic

Require: S_{DCKP}, a starting solution of P_{DCKP}.
Ensure: S^*_{DCKP}, a local optimum of P_{DCKP}.
 1: Set S^*_{DCKP} as a starting feasible solution, where all variables are assigned to 0.
 2: **while** the time limit is not performed **do**
 3:     Call Algorithm 4 in order to find IS and I^r according to S_{DCKP}.
 4:     Call Algorithm 1 with an argument I^r to complete IS.
 5:     Call Algorithm 2 with an argument IS for reaching a new solution S_{DCKP}.
 6:     Improve S_{DCKP} by applying Algorithm 3.
 7:     Update S^*_{DCKP} with the best solution.
 8: **end while**
 9: **return** S^*_{DCKP}.

---

5   Computational Results

This section evaluates the effectiveness of the proposed *Large Neighborhood Search-Based Heuristic* (LNSBH) on two groups of instances (taken from the literature [4] and generated following the schema used by Yamada *et al.* [14,15]). The first group contains twenty medium instances with 500 items, a capacity $c = 1800$ and different densities (assessed in terms of the number of disjunctive constraints). The second group contains thirty large instances, where each instance contains 1000 items, with c taken in the discrete interval $\{1800, 2000\}$ and with various densities. The proposed LNSBH was coded in C++ and run on a PC Intel Pentium Core i5-2500 with 3.3 Ghz.

The performance of LNSBH depends on certain parameters, like the percentage α of the unassigned items considered in the descent method, the percentage β of items to be removed when applying Algorithm 4 (according to the large neighborhood search), the constant γ used by Algorithm 4 and Algorithm 5's runtime limit t. In what follows, we show how the aforementioned parameters can be experimentally fixed in order to provide good performance for LNSBH.

5.1   Effect of Both Degrading and Re-optimizing Procedures

This section evaluates the effect of the descent method based upon degrading and re-optimizing procedures (as used in Algorithm 3) on the starting solution realized by Algorithm 2. We recall that the re-optimization procedure tries to solve a reduced P_{DCKP} which contains a small number of items. These problems are optimally solved using the Cplex solver (version 12.4).

Table 1 shows the variation of $Av.$ $Sol.$, the average value of solutions provided by the considered algorithm over all treated instances and $Av.$ $Time$, the average runtime needed by each algorithm for finishing the descent search. One can observe that, for the descent method (cf., Algorithm 3), the results are obtained by performing Algorithm 3 with different settings of α, where the value of α is varied in the discrete interval $\{5, 10; 15; 20\}$.

From Table 1, we can observe that the best average solution value is realized for the value $\alpha = 15\%$, but it needs an average runtime of 19.72 s. Note that LNSBH's runtime depends on the descent method's runtime. Therefore, according to the results displayed in Table 1, one can observe that the value of 5 % favors a quick resolution (0.15 s) with an interesting average solution value of 2129.98. Because our aim is to propose a fast efficient LNSBH, we then set $\alpha = 5\%$ for the rest of the paper.

5.2   Behavior of LNSBH on both Groups of Instances

Remark that, according to the results shown in Shaw [13], when γ varies over the range of the integer interval $[5, 20]$, the LNS works reasonably well. In our test, we set $\gamma = 20$ for Algorithm 4. Therefore, in order to evaluate the performance of LNSBH, we focus on both parameters β and t; that are used in Algorithm 5. The study is conducted by varying the value of β in the discrete interval $\{10, 15, 20, 25, 30\}$ and t in the interval $\{25, 50, 100, 150, 200\}$ (measured

Table 1. Effect of the descent method on the starting DCKP's solution.

	Algorithms 1–2	The descent method $\alpha =$			
		5 %	10 %	15 %	20 %
$Av.$ $Sol.$	2014.62	2129.98	2188.80	2232.36	2217.04
$Av.$ $Time$	$\approx$0.001	0.15	2.03	19.72	118.30

Table 2. The quality of the average values when varying the values of the couple (β, t).

t \ β	Variation of β				
	10%	15%	20%	25%	30%
25	2395.38	2394.48	2392.5	2391.14	2389.06
50	2397.52	2397.44	2394.74	2393.34	2391.18
100	2399.16	2398.98	2396.36	2394.2	2393.9
150	2399.28	2399	2397.82	2395.62	2394.58
200	2400.22	2399.62	2398.76	2396.8	2395.74

in seconds). Table 2 displays the average solution values realized by LNSBH using the different values of (β, t).

From Table 2, we observe what follows:

- Setting $\beta = 10\%$ provides the best average solution value. Further, the solution quality increases when the runtime limit is extended.
- All other variations of β induce smaller average values than those of $\beta = 10\%$ in 200 s.

According to the results displayed in Table 2, the objective value of the solutions determined by setting $(\beta, t) = (10\%, 100)$ and $(10\%, 200)$ are displayed in Table 3. In our computational study, ten random trials of the LNSBH are performed on the fifty literature instances and each trial is stopped respectively after 100 and 200 s.

Table 3 shows objective values reached by LNSBH and Cplex when compared to the best solutions of the literature (taken from Hifi and Otami [2, 6]). Column 1 of Table 3 displays the instance label, column 2 reports the values of the best solutions (denoted V_{Cplex}) reached by Cplex v12.4 after one hour of runtime and column 3 displays the solutions provided by the most recent algorithm of the literature, denoted V_{IRS}. Finally, column 4 (resp. 5) reports Max.Sol. (resp. Av.Sol) denoting the maximum (average) solution value obtained by LNSBH over ten trials for the first runtime limit of 100 s and columns 6 and 7 display those of LNSBH for the second runtime limit of 200 s.

From Table 3, we observe what follows:

1. First, we can observe the inferiority of the Cplex solver because it realizes an average value of 2317.88 compared to that realized by LNSBH (2390.40). In this case, Cplex matches 5 instances over 50, representing a percentage of 10% of the best solutions of the literature.
2. Second, for both runtime limits (100 and 200 s), the mean value (Mean.Sol.) of the average objective values (Av.Sol) reached by LNSBH over ten trials (2393.63 and 2395.94) are better then the mean value of the best objectives values reached by V_{IRS} (2390.40).
3. Third and last, according to the best solutions realized by LNSBH over ten trials with the first runtime limit (100 s), one can observe that LNSBH is able to

Table 3. Performance of LNSBH vs Cplex and IRS on the benchmark instances of the literature.

Instance	V_{Cplex}	V_{IRS}	LNSBH			
			$\beta = 10$ and $t = 100$		$\beta = 10$ and $t = 200$	
			Max.Sol.	Av.Sol.	Max.Sol.	Av.Sol.
1I1	2567	2567	2567	2564.2	2567	2564.6
1I2	2594	2594	2594	2594	2594	2594
1I3	2320	2320	2320	2319	2320	2319
1I4	2298	2303	2310	2310	2310	2310
1I5	2310	2310	2330	2328	2330	2329
2I1	2080	2100	2117	2116.1	2118	2117
2I2	2070	2110	2110	2110	2110	2110
2I3	2098	2128	2119	2110.2	2132	2118.1
2I4	2070	2107	2109	2106.9	2109	2108.2
2I5	2090	2103	2110	2109.7	2114	2111.2
3I1	1667	1840	1845	1788.4	1845	1814
3I2	1681	1785	1779	1759.9	1779	1769.2
3I3	1461	1742	1774	1759.3	1774	1762.9
3I4	1567	1792	1792	1792	1792	1792
3I5	1563	1772	1775	1751.6	1775	1759.2
4I1	1053	1321	1330	1330	1330	1330
4I2	1199	1378	1378	1378	1378	1378
4I3	1212	1374	1374	1374	1374	1374
4I4	1066	1353	1353	1352.7	1353	1353
4I5	1229	1354	1354	1336.4	1354	1336.4
5I1	2680	2690	2690	2684	2690	2686
5I2	2690	2690	2690	2683.9	2690	2685.9
5I3	2670	2689	2680	2675.7	2690	2679.7
5I4	2680	2690	2698	2683.2	2698	2689.2
5I5	2660	2680	2670	2668	2670	2669.9
6I1	2820	2840	2850	2850	2850	2850
6I2	2800	2820	2830	2823.9	2830	2827.7
6I3	2790	2820	2830	2819.9	2830	2821.9
6I4	2790	2800	2820	2817	2822	2820.2
6I5	2800	2810	2830	2823.7	2830	2825.6
7I1	2700	2750	2780	2771.9	2780	2773
7I2	2720	2750	2770	2769	2770	2770
7I3	2718	2747	2760	2759	2760	2760

Table 3. *(Continued).*

Instance	V_{Cplex}	V_{IRS}	LNSBH			
			$\beta = 10$ and $t = 100$		$\beta = 10$ and $t = 200$	
			Max.Sol.	Av.Sol.	Max.Sol.	Av.Sol.
7I4	2728	2773	2800	2791	2800	2793
7I5	2730	2757	2770	2763	2770	2765
8I1	2638	2720	2720	2719.1	2720	2719.1
8I2	2659	2709	2720	2719	2720	2720
8I3	2664	2730	2740	2733	2740	2734
8I4	2620	2710	2710	2708.7	2719	2709.9
8I5	2644	2710	2710	2709	2710	2710
9I1	2589	2650	2676	2670.9	2677	2671.3
9I2	2580	2640	2665	2661.5	2665	2663
9I3	2580	2635	2670	2665.8	2670	2668.6
9I4	2540	2630	2660	2659.8	2660	2659.9
9I5	2594	2630	2669	2663.5	2670	2664.9
10I1	2500	2610	2620	2616.7	2620	2619.7
10I2	2549	2642	2630	2627.5	2630	2629.9
10I3	2527	2618	2620	2617.1	2627	2620.5
10I4	2509	2621	2620	2617	2620	2618.6
10I5	2530	2606	2620	2619.3	2625	2620.5
Mean.Sol.	2317.88	2390.40	2399.16	2393.63	2400.22	2395.94

improve the best solutions of the literature on 30 cases, matches 14 instances and fails in 6 occasions. On the other hand, by extending the runtime limit to 200 s, the best solutions reached by LNSBH over ten trials become more interesting. Indeed, in this case, LNSBH realizes 33 new solutions, matches 13 solutions and fails in 4 occasions.

6   Conclusion

In this paper, we proposed a fast large neighborhood search-based heuristic for solving the disjunctively constrained knapsack problem. The proposed method combines a two-phase procedure and a large neighborhood search. First, the two-phase procedure is applied in order to reach a starting feasible solution. This solution is obtained by combining the resolution of two complementary problems: the weighted independent set and the classical binary knapsack. Second, a descent method, based upon degrading and re-optimization strategies, is applied in order to improve the solution provided by the first phase. Once a local optimal solution is reached, the large neighborhood search is used in order to diversify the search space. Finally, the computational results show that the proposed algorithm is very competitive when compared to both Cplex solver and one of the most recent algorithm of the literature.

References

1. Garey, M.R., Johnson, D.S.: Computers and Intractability: A Guide to the Theory of NP-completeness. W.H. Freeman and Company, San Francisco (1979)
2. Hifi, M.: An iterative rounding search-based algorithm for the disjunctively constrained knapsack problem. Eng. Optim. doi:10.1080/0305215X.2013.819096 (Published online: 19 Sep 2013)
3. Hifi, M., Michrafy, M.: Reduction strategies and exact algorithms for the disjunctively constrained knapsack problem. Comput. Oper. Res. **34**, 2657–2673 (2007)
4. Hifi, M., Michrafy, M.: A reactive local search algorithm for the disjunctively constrained knapsack problem. J. Oper. Res. Soc. **57**, 718–726 (2006)
5. Hifi, M., Otmani, N.: An algorithm for the disjunctively constrained knapsack problem. Int. J. Oper. Res. **13**, 22–43 (2012)
6. Hifi, M., Otmani, N.: An algorithm for the disjunctively constrained knapsack problem. In: IEEE - International Conference on Communications, Computing and Control Applications, pp. 1–6 (2011)
7. Hifi, M., Negre, S., Ould Ahmed Mounir, M.: Local branching-based algorithm for the disjunctively constrained knapsack problem. In: IEEE Proceedings of the International Conference on Computers and Industrial Engineering, pp. 279–284 (2009)
8. Martello, S., Pisinger, D., Toth, P.: Dynamic programming and strong bounds for the 0-1 knapsack problem. Manage. Sci. **45**, 414–424 (1999)
9. Pferschy, U., Schauer, J.: The knapsack problem with conflict graphs. J. Graph Algorithms Appl. **13**, 233–249 (2009)
10. Pisinger, D., Sigurd, M.: Using decomposition techniques and constraint programming for solving the two-dimensional bin-packing problem. INFORMS J. Comput. **19**, 36–51 (2007)
11. Pisinger, D., Ropke, S.: Large neighborhood search. In: Gendreau, M., Potvin, J.-Y. (eds.) Handbook of Metaheuristics. International Series in Operations Research & Management Science, vol. 146, pp. 399–419. Springer, New York (2010)
12. Sadykov, R., Vanderbeck, F.: Bin packing with conflicts: a generic branch-and-price algorithm. INFORMS J. Comput. **25**(2), 244–255 (2013)
13. Shaw, P.: Using constraint programming and local search methods to solve vehicle routing problems. In: Maher, M.J., Puget, J.-F. (eds.) CP 1998. LNCS, vol. 1520, pp. 417–431. Springer, Heidelberg (1998)
14. Yamada, T., Kataoka, S., Watanabe, K.: Heuristic and exact algorithms for the disjunctively constrained knapsack problem. Inf. Process. Soc. Jap. J. **43**, 2864–2870 (2002)
15. Yamada, T., Kataoka, S.: Heuristic and exact algorithms for the disjunctively constrained knapsack problem. In: EURO 2001, Rotterdam, The Netherlands, pp. 9–11 (2001)

Approximating the k-Set Packing Problem
by Local Improvements

Martin Fürer and Huiwen Yu[✉]

Department of Computer Science and Engineering,
The Pennsylvania State University, University Park, PA, USA
{furer,hwyu}@cse.psu.edu

Abstract. We study algorithms based on local improvements for the k-Set Packing problem. The well-known local improvement algorithm by Hurkens and Schrijver [14] has been improved by Sviridenko and Ward [15] from $\frac{k}{2} + \epsilon$ to $\frac{k+2}{3}$, and by Cygan [7] to $\frac{k+1}{3} + \epsilon$ for any $\epsilon > 0$. In this paper, we achieve the approximation ratio $\frac{k+1}{3} + \epsilon$ for the k-Set Packing problem using a simple polynomial-time algorithm based on the method by Sviridenko and Ward [15]. With the same approximation guarantee, our algorithm runs in time singly exponential in $\frac{1}{\epsilon^2}$, while the running time of Cygan's algorithm [7] is doubly exponential in $\frac{1}{\epsilon}$. On the other hand, we construct an instance with locality gap $\frac{k+1}{3}$ for any algorithm using local improvements of size $O(n^{1/5})$, where n is the total number of sets. Thus, our approximation guarantee is optimal with respect to results achievable by algorithms based on local improvements.

Keywords: k-set packing · Tail change · Local improvement · Color coding

1 Introduction

Given a universe of elements U and a collection $\mathcal{S}$ of subsets with size at most k of U, the k-Set Packing problem asks to find a maximum number of disjoint sets from $\mathcal{S}$. The most prominent approach for the k-Set Packing problem is based on local improvements. In each round, the algorithm selects p sets from the current packing and replaces them with $p + 1$ sets such that the new solution is still a valid packing. It is well-known that for any $\epsilon > 0$, there exists a constant p, such that the local improvement algorithm has an approximation ratio $\frac{k}{2} + \epsilon$ [14]. In quasi-polynomial time, the result has been improved to $\frac{k+2}{3}$ [11] and later to $\frac{k+1}{3} + \epsilon$ for any $\epsilon > 0$ [8] using local improvements of size $O(\log n)$, here n is the size of $\mathcal{S}$. In [8], the algorithm looks for any local improvement of size $O(\log n)$, while in [11], only sets which intersect with at most 2 sets in the current solution are considered and the algorithm looks for improvements of a binocular shape.

Research supported in part by NSF Grant CCF-0964655 and CCF-1320814.

© Springer International Publishing Switzerland 2014
P. Fouilhoux et al. (Eds.): ISCO 2014, LNCS 8596, pp. 408–420, 2014.
DOI: 10.1007/978-3-319-09174-7_35

One can obtain a polynomial-time algorithm which looks for local improvements of logarithmic size using the color coding technique [7,15]. The algorithm in [15] looks for local improvements similar to [11] and has an approximation ratio $\frac{k+2}{3}$. In [7], local improvements of bounded pathwidth are considered and an approximation ratio $\frac{k+1}{3} + \epsilon$, for any $\epsilon > 0$ is achieved.

In this paper, we obtain an approximation ratio $\frac{k+1}{3} + \epsilon$ for the k-Set Packing problem, for any $\epsilon > 0$. On the other hand, we improve the lower bound given in [15] by constructing an instance that any algorithm using local improvements of size $O(n^{1/5})$ has a performance ratio at least $\frac{k+1}{3}$. Thus, our result is optimal with respect to the performance guarantee achievable by algorithms using local improvements. Our algorithm extends the types of local improvements considered in [11,15] by first looking for a series of set replacements which swap some sets in the current packing $\mathcal{A}$ with a same number of disjoint sets $\mathcal{T}$ which are not in $\mathcal{A}$. We then look for local improvements which can be decomposed into cycles and paths, from sets in $\mathcal{S} \setminus (\mathcal{A} \cup \mathcal{T})$ which intersect with at most 2 sets in $\mathcal{A}$. We also use the color-coding technique [2,9] to ensure a polynomial time complexity when the local improvement has logarithmic size. Our algorithm is more efficient as it runs in time singly exponential in $\frac{1}{\epsilon^2}$, while the running time of Cygan's algorithm [7] is doubly exponential in $\frac{1}{\epsilon}$. We believe that our approach makes an important step towards a practical algorithm for the k-Set Packing problem.

Related works. The Set Packing problem has been studied for decades. Hastad has shown that the general Set Packing problem cannot be approximated within $N^{1-\epsilon}$ unless $NP \subseteq ZPP$ [12]. Here N is the size of the universe U. The bounded Set Packing problem assumes an upper bound of the size of the sets. In the unweighted case, i.e. the k-Set Packing problem, besides algorithms based on local improvements [7,8,11,14,15], Chan and Lau have shown that the standard linear programming algorithm has an integrality gap $k - 1 + 1/k$ [5]. They have also constructed a polynomial-sized semi-definite program with integrality gap $\frac{k+1}{2}$, but no rounding strategy is provided. The problem is also known to have a lower bound $\Omega(\frac{k}{\log k})$ [13]. In the weighted case, Chandra and Halldórsson have given a nontrivial approximation ratio $\frac{2(k+1)}{3}$ [6]. The result was improved to $\frac{k+1}{2} + \epsilon$ by Berman [3], which remains the best so far.

The paper is organized as follows. In Sect. 2, we review previous local search algorithms and define useful tools for analysis. In Sect. 3, we introduce the new local improvement and analyze its performance guarantee. In Sect. 4, we give an efficient implementation of our algorithm and a tight example of algorithms based on local improvements. Omitted proofs and analysis can be found in the full version [10].

2   Preliminaries

2.1   Local Improvements

Let $\mathcal{S}$ be a collection of subsets of size at most k of the universe U and the size of $\mathcal{S}$ is n. Let $\mathcal{A}$ be the collection of disjoint sets chosen by the algorithm. In this paper, we are interested in the unweighted k-set packing problem. We assume without loss of generality that every set is of uniform size k. Otherwise, we could add distinct elements to any set until it is of size k. In the following context, we use calligraphic letters to represent collections of k-sets and capital letters to represent sets of vertices which correspond to k-sets.

The most widely used algorithm for the k-Set Packing problem is local search. The algorithm starts by picking an arbitrary maximal packing. If there exists a collection of $p + 1$ sets $\mathcal{P}$ which are not in $\mathcal{A}$ and a collection of p sets $\mathcal{Q}$ in $\mathcal{A}$, such that $(\mathcal{A} \setminus \mathcal{Q}) \cup \mathcal{P}$ is a valid packing, the algorithm will replace $\mathcal{Q}$ with $\mathcal{P}$. We call it a $(p+1)$-*improvement*.

With p being a constant which depends on ϵ, it is well-known that this local search algorithm achieves an approximation ratio $\frac{k}{2} + \epsilon$, for any $\epsilon > 0$ [14]. Halldórsson [11] and later Cygan et al. [8] show that when p is $O(\log n)$, the approximation ratio can be improved at a cost of quasi-polynomial time complexity. Based on the methods of [11], Sviridenko and Ward [15] have obtained a polynomial-time algorithm using the color coding technique [2]. We summarize their algorithm as follows. Let $\mathcal{A}$ be the packing chosen by the algorithm and $\mathcal{C} = \mathcal{S} \setminus \mathcal{A}$. Construct an *auxiliary multi-graph* G_A as follows. The vertices in G_A represent sets in $\mathcal{A}$. For any set in $\mathcal{C}$ which intersects with exactly two sets $s_1, s_2 \in \mathcal{A}$, add an edge between s_1 and s_2. For any set in $\mathcal{C}$ which intersects with only one set $s \in \mathcal{A}$, add a self-loop on s. The algorithm searches for local improvements which can be viewed as *binoculars* in G_A. They call them *canonical improvements* [15]. A binocular can be decomposed into paths and cycles. The color coding technique [2] and the dynamic programming algorithm are employed to efficiently locate paths and cycles of logarithmic size. This algorithm has an approximation ratio at most $\frac{k+2}{3}$.

Cygan [7] has shown that an approximation ratio $\frac{k+1}{3} + \epsilon$ can be obtained in polynomial time by restricting the local improvements from anything of size $O(\log n)$ [8] to local improvements of bounded pathwidth. Namely, let $G(A, C)$ be the *bipartite conflict graph* where $\mathcal{A}$ and $\mathcal{C} = \mathcal{S} \setminus \mathcal{A}$ represent one part of vertices respectively. For any $u \in A, v \in C$, if the corresponding sets are not disjoint, we put an edge between u and v. For any disjoint collection $P \subseteq C$, if the subgraph induced by P and the neighbors of P, $N(P)$ in A have bounded pathwidth, a set replacement of P with $N(P)$ is called a local improvement of bounded pathwidth. The color coding technique is also employed for efficiently locating such an improvement.

Theorem 1 [7]. *For any $\epsilon > 0$, there exists a local search algorithm which runs in time $2^{O(kr)} n^{O(pw)}$ with an approximation ratio $\frac{k+1}{3} + \epsilon$ of the k-Set Packing problem. Here $r = 2(k+1)^{\frac{1}{\epsilon}} \log n$ is the upper bound of the size of a local improvement, $pw = 2(k+1)^{\frac{1}{\epsilon}}$ is the upper bound of pathwidth.*

2.2   Partitioning the Bipartite Conflict Graph

Consider a bipartite conflict graph $G(A, B)$ where one part of the vertices A representing sets $\mathcal{A}$ chosen by the algorithm and the other part B representing an arbitrary disjoint collection of sets $\mathcal{B}$. We assume without loss of generality that $\mathcal{B} \cap \mathcal{A} = \emptyset$. The collection $\mathcal{B}$ can be thought of an optimal solution. It is only used for analysis.

Given $\epsilon > 0$, let $c_k = k - 1$, $b = |B|$, we further partition $G(A, B)$ iteratively as follows. Let B_1^1 be the set of vertices in B with degree 1 to A. Denote the neighbors of B_1^1 in A by A_1^1. If $|B_1^1| < \epsilon b$, stop the partitioning. Otherwise, we consider B_1^2 which is the set of vertices whose degree drops to 1 if we remove A_1^1. Denote the neighbors of B_1^2 in $A \setminus A_1^1$ by A_1^2. If $|B_1^1 \cup B_1^2| < c_k \epsilon b$, stop the partitioning. In general for any $j \geq 2$, let B_1^j be the set of vertices with their degree dropping to 1 if the vertices in $\cup_{l=1}^{j-1} A_1^l$ are removed, and let A_1^j be the neighbors of B_1^j which are not in $\cup_{l=1}^{j-1} A_1^l$. If $|\cup_{l=1}^{j} B_1^l| < c_k^{j-1} \epsilon b$, we stop. Otherwise continue the partitioning. Let i be the smallest integer such that $|\cup_{l=1}^{i} B_1^l| < c_k^{i-1} \epsilon b$. This integer i exists as $c_k^{i-2} \epsilon b \leq |\cup_{l=1}^{i-1} B_1^l| \leq b$, we have $i \leq 2 + \log_{c_k} \frac{1}{\epsilon}$. Let $B_1^{\leq j}$ ($A_1^{\leq j}$) be the set union $\cup_{l=1}^{j} B_1^l$ ($\cup_{l=1}^{j} A_1^l$), for $j \geq 1$.

3 Canonical Improvements with Tail Changes

In this section, we present a local search algorithm based on [15], and show that it achieves an approximation ratio $\frac{k+1}{3} + \epsilon$ for any $\epsilon > 0$.

3.1 The New Local Improvement

In this section, we introduce a new type of local improvements. Let $\mathcal{A}$ be a packing chosen by the algorithm, and let $\mathcal{C} = \mathcal{S} \setminus \mathcal{A}$. We create the bipartite conflict graph $G(A, C)$ as in Sect. 2.1. Recall that only those sets in $\mathcal{C}$ which intersect with at most 2 sets in $\mathcal{A}$ are considered in [15]. Our approach tries to include sets of higher degree in a local improvement by swapping p sets in $\mathcal{A}$ with p sets in $\mathcal{C}$. In this way, if the degree of a vertex in $\mathcal{C}$ drops to 2, it could be included in a local improvement.

Definition 1 (Tail change). *Consider any vertex $v \in C$ of degree at least 3, we call a swapping of p sets U in A with p disjoint sets V in C a **tail change** associated with an edge (v, u) of v if the following three requirements are satisfied: (1) $v \notin V$. (2) u is the unique neighbor of v in U. (3) The neighbors of V in A are exactly U. The **size** of this tail change is defined to be p.*

We denote a tail change associated with e which swaps U with V by $T_e(U, V)$. We say that two tail changes $T_e(U, V), T_{e'}(U', V')$ of vertices v, v' respectively are *consistent*, if either $v \neq v'$ and $(\{v\} \cup V) \cap (\{v'\} \cup V') = \emptyset$, or $v = v'$, $e \neq e'$ and $V \cap V' = \emptyset$. Moreover we require that the degrees of v, v' after the tail changes remain at least 2. Therefore, for any vertex $v \in C$ of degree $d \geq 3$, we could perform at most $d - 2$ tail changes for v.

We are now ready to introduce the new local search algorithm. We first consider an algorithm that runs in quasi-polynomial time. Given parameter $\epsilon > 0$, in each iteration, the algorithm starts by performing local improvements of constant size $O(\frac{k}{\epsilon})$. If no such local improvement is present, the algorithm starts looking for improvements of size $O(\log n)$. Construct the bipartite conflict graph $G(A, C)$. For any set I of at most $\frac{4}{\epsilon} \log n$ vertices in C, let $I_3 \subseteq I$ be the set of vertices of degree at least 3 in $G(A, C)$. The algorithm checks if there exists a collection of consistent tail changes each of size at most $\frac{2(k-1)}{\epsilon}$ for I_3, which together replace $U \subseteq A$ with $V \subseteq C$, such that $V \cap I = \emptyset$, and after the replacement the degree of every vertex in I_3 drops to 2. If so, the algorithm goes on checking in the auxiliary multi-graph G_A where edges are constructed from vertices in I assuming the swapping of U with V is performed, whether there is a subgraph which is one of the following **six types**: (1) two disjoint cycles connecting by a path, (2) two cycles intersecting at a single point, (3) two cycles with a common arc, (those three types are binoculars), (4) a path, (5) a cycle, (6) a path and a cycle intersecting at a single point. Let U' be the vertices in this subgraph, and V' be the edges. The algorithm checks if a replacement of $U \cup U'$ with $V \cup V'$ is an improvement. We call this new local improvement the **canonical improvement with tail changes**, and this quasi-polynomial time algorithm, **Algorithm LI** (LI stands for local improvement). We will explain the parameter settings in Algorithm LI in the next section.

Before showing how to efficiently locate a canonical improvement with tail changes, we first show that the approximation ratio of Algorithm LI is $\frac{k+1}{3} + \epsilon$.

3.2   Analysis

Given a packing A chosen by Algorithm LI and for an arbitrary packing B, consider the bipartite conflict graph $G(A, B)$ defined in Sect. 2.2. The notations in this section are taken from Sect. 2.2. First, we remark that since we make all $O(\frac{k}{\epsilon})$-improvements at the beginning of Algorithm LI, for any set $V \subseteq B$ of size $O(\frac{k}{\epsilon})$, there are at least $|V|$ neighbors of V in A. In $G(A, B)$, we make every vertex a in A full degree k by adding self-loops of a which we call *null edges*. We define a *surplus edge* which is either a null edge, or an edge incident to some vertex in B which is of degree at least 3. We first show that there exists a one-to-one matching from almost all vertices in B_1^1 to surplus edges with the condition that after excluding the matched surplus edges of any vertex in B, the degree of this vertex remains at least 2. We define such a matching in the following matching process.

The matching process. Pick an arbitrary order of vertices in B_1^1. Mark all edges and vertices as unmatched. Try to match every vertex with a surplus edge in this order one by one. For any vertex $v_1 \in B_1^1$, starting from v_1, go to its neighbor $u_1 \in A_1^1$. If u_1 has an unmatched null edge, match v_1 to it, mark this null edge as matched and stop. Otherwise, if u_1 has a neighbor v in B, such that the degree of v is at least 3, (u_1, v) is unmatched and v is unmatched, match v_1 to (u_1, v) and mark this edge as matched. If the degree of v drops to 2 by excluding all matched edges of v, mark v as matched. If u_1 does not have a neighbor satisfying the requirement, try every neighbor v_2 (except v_1) of u_1 and continue the process from v_2. In general, suppose we are at a vertex $v_j \in B_1^j$

and it has a neighbor $u_j \in A_1^j$. We try to match v_1 with a null edge of u_j, or a surplus edge of an unmatched neighbor of u_j. If no matching edge is found, continue by trying every neighbor of u_j in $B_1^{j_1}$ for $j_1 > j$, until either v_1 is matched, or $j > 2 + \log_{c_k} \frac{1}{\epsilon}$. In the latter case, we mark v_1 as unmatched.

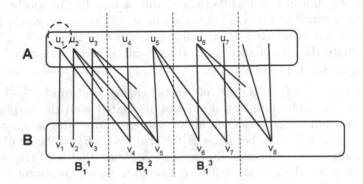

Fig. 1. The bipartite conflict graph. $k = 3$.

We give an example of the matching process illustrated in Fig. 1. We match v_1 to the null edge (dotted line) of its neighbor u_1. v_2 is matched to a surplus edge (u_2, v_5) of v_5. After that, the degree of v_5 drops to 2 by excluding the edge (u_2, v_5) and v_5 is marked as matched. For v_3, we go on to u_3, v_5, u_5, v_6, u_6, then v_8 with a surplus edge (u_6, v_8). We match v_3 to this edge.

Lemma 1. *For any $\epsilon > 0$, there exists a set of surplus edges E_1, such that except for at most $\epsilon|B|$ vertices, B_1^1 can be matched to E_1 one-to-one. Moreover, every endpoint of E_1 in B has degree at least 2 after excluding E_1.*

Proof. It is sufficient to prove that at most $\epsilon|B|$ vertices in B_1^1 are unmatched. Let v be an unmatched vertex in B_1^1. The neighbor of v in A, u has no null edges and thus has degree k, and none of the neighbors of u have an unmatched surplus edge. The matching process runs in exactly $i - 1 = 1 + \log_{c_k} \frac{1}{\epsilon}$ iterations for v and has tried $k - 1 + (k-1)^2 + \cdots + (k-1)^{i-1} = \frac{(k-1)^i - (k-1)}{k-2}$ vertices in $B_1^{\leq i}$. Notice that for two different vertices $v, v' \in B_1^1$ which are marked unmatched, the set of vertices the matching process has tried in $B_1^{\leq i}$ must be disjoint. Otherwise, we can either find a matching for one of them, or there exists a local improvement of size $O(\frac{k}{\epsilon})$. Suppose there are n_{um} unmatched vertices in B_1^1. Recall that $|B_1^{\leq i}| \leq c_k^{i-1}\epsilon|B|$. Therefore, $n_{um}(1 + \frac{(k-1)^i - (k-1)}{k-2}) \leq |B_1^{\leq i}| \leq c_k^{i-1}\epsilon|B|$. We have $n_{um} \leq \epsilon|B|$, where $c_k = k - 1$. □

Consider any surplus edge $e = (u, v)$ matching to $w \in B_1^1$, we can obtain a tail change $T_e(U, V)$ associated with e by viewing the matching process reversely. Assume $u \in A_1^i$, for $i < 2 + \log_{c_k} \frac{1}{\epsilon}$. Let $U_i = \{u\}$ and V_i be the neighbor of U_i in B_1^i. In general, let V_j be the set of neighbors of U_j in B_1^j, we define U_{j-1} to be the neighbors of V_j excluding U_j. This decomposition ends when $j = 1$ and $V_1 \subseteq B_1^1$. As w is matched to e, we know that $w \in V_1$. Let $U = \cup_{j=1}^i U_j, V = \cup_{j=1}^i V_j$, then a swapping of U with V is a tail change associated with edge e. First, we have $|V_j| = |U_j|$ for $1 \leq j \leq i$, otherwise there exists a $O(\frac{k}{\epsilon})$-improvement. Hence $|U| = |V|$. Secondly, the set of neighbors of V is U by the construction. And u is the only neighbor of v in U, otherwise w will be matched to another surplus edge. As an example in Fig. 1, $U = (u_6, u_5, u_3, u_2), V = (v_6, v_5, v_3, v_2)$ is a tail change associated with edge (u_6, v_8) which is matched to v_3.

We estimate the size of such a tail change. Since every vertex in U has at most k neighbors, there are at most $\sum_{j=1}^i (k-1)^{j-1} = \frac{(k-1)^i - 1}{k-2}$ vertices in V. Let $i = 2 + \log_{k-1} \frac{1}{\epsilon}$. Then the size of this tail change is at most $\frac{2(k-1)}{\epsilon}$.

Let $\mathcal{T}_S$ be the collection of tail changes associated with the surplus edges which are not null edges as defined above. Assume those tail changes together replace $U \subseteq A$ with $V \subseteq B$. Let $B_L = B_1^1 \cap V$. Let $B_N \subseteq B_1^1$ be the set of vertices which are matched to null edges. By Lemma 1, we know that $|B_1^1 \setminus (B_L \cup B_N)| \leq \epsilon|B|$. Moreover, we show in the following Corollary that a consistent collection of tail changes with the same property can be extracted from $\mathcal{T}_S$.

Corollary 1. *There exists a subcollection $\mathcal{T}_c$ of consistent tail changes from $\mathcal{T}_S$, which together replace a set $U_c \subseteq A$ with $V_c \subseteq B$, such that $V_c \cap B_1^1 = B_L$.*

Proof. We consider the tail changes associated with the surplus edges one by one. $\mathcal{T}_c$ is initialized to be empty. If the tail change $T_{e_i}(U_i, V_i)$ is consistent with every tail change in $\mathcal{T}_c$, we include it in $\mathcal{T}_c$. If there exists any tail change $T_{e_j}(U_j, V_j)$ such that $V_i \cap V_j \neq \emptyset$, assume $e_j = (u_j, v_j)$ is matched with the vertex w_j and $e_i = (u_i, v_i)$ with w_i, we know that at the time the matching for w_i tries the edges of v_j, v_j has been marked as matched. Hence, $V_j \subseteq V_i$. We discard $T_{e_j}(U_j, V_j)$ and include $T_{e_j}(U_j, V_j)$ in $\mathcal{T}_c$. $\qquad \square$

Theorem 2. *For any $\epsilon > 0$, Algorithm LI has an approximation ratio $\frac{k+1}{3} + \epsilon$.*

Before proving the theorem, we state the following result from [15] which is derived from a lemma in [4]. The lemma in [4] states that when the density of a graph is greater than a constant $c > 1$, there exists a subgraph of size $O(\log n)$ with more edges than vertices. If the underlying graph is the auxiliary multigraph G_A defined in Sect. 2.1 and this condition holds, we know from [15] that there exists a binocular of size $O(\log n)$.

Lemma 2 [15]. *For any integer $s \geq 1$ and any undirected multigraph $G = (V, E)$ with $|E| \geq \frac{s+1}{s}|V|$, there exists a binocular of size at most $4s \log n - 1$.*

Proof (Theorem 2). For a given ϵ, let $\epsilon' = \frac{2k+5}{3}\epsilon > 3\epsilon$. Let A be the packing returned by Algorithm LI with parameter ϵ and for any other packing B, we show that $(\frac{k+1}{3} + \epsilon)|A| \geq |B|$. In the following, we use the corresponding small letter of a capital letter (which represents a set) to represent the size of this set.

In Corollary 1, the collection of consistent tail changes $\mathcal{T}_c$ together replace a set of a_t vertices A_t in A with $b_t = a_t$ vertices B_t in B. We exclude theses vertices from the original bipartite conflict graph $G(A, B)$. Denote the remaining graph by $G(A', B')$. We add null edges to vertices in A' until every vertex in A' has degree k. There are ka' edges counting from A'.

Let $B'_N = B_N \cap B'$ and $b^1_n = |B'_N|$. We can also think of that there is a null edge at each vertex in B'_N when we count the number of edges from B'. By Lemma 1, there are at most $\epsilon' b$ unmatched vertices in B^1_1. We further partition the vertices in B' as follows. Let B^2_3 ($B^{2'}_3$) be the set of vertices in B' whose degree drops to 2 after performing the tail changes in $\mathcal{T}_c$, and for any vertex $v \in B^2_3$ ($v' \in B^{2'}_3$), there is at least one (no) tail change in $\mathcal{T}_c$ associated with v. Let B^2_2 be the set of vertices in B' with degree 2 in $G(A', B')$ and no neighbors in A_t. Let B^1_2 (B^1_3) be the set of vertices in B' whose degree drops from 2 (at least 3) in $G(A, B)$ to 1 in $G(A', B')$. Let B^3_3 be the set of vertices in B' with degree at least 3 in $G(A', B')$. Moreover, there is no vertex in $G(A', B')$ of degree 0, otherwise, there exists a local improvement.

By Lemma 1, the number of edges in $G(A', B')$ is at least $2b^1_n + \epsilon' b + b^1_2 + 2b^2_2 + 2b^2_3 + 3b^3_3 + 2b^{2'}_3 + b^1_3$. Therefore,

$$k(a - a_t) \geq 2b^1_n + \epsilon' b + b^1_2 + 2b^2_2 + 2b^2_3 + 3b^3_3 + 2b^{2'}_3 + b^1_3. \tag{1}$$

Next, we show that $b^1_n + b^1_2 + b^2_2 + b^2_3 \leq (1 + \epsilon')(a - a_t)$. Suppose the set of neighbors of $B'_N, B^2_2, B^1_2, B^2_3$ in A' is A_2. Construct an auxiliary multi-graph G_{A_2} as in Sect. 2.1, where the vertices are A_2, every vertex in B'_N, B^1_2 creates a self-loop, and every vertex in B^2_2, B^2_3 creates an edge. Assume G_{A_2} has at least $(1 + \epsilon')|A_2|$ edges, implied by Lemma 2, there exists a binocular of size at most $\frac{4}{\epsilon'} \log |A_2| - 1$ in G_{A_2}.

Let G_A be the auxiliary multi-graph with vertices being A, every degree-1 vertex in B creates a self-loop, every degree-2 vertex in B creates an edge, and every vertex v with degree dropping to 2 by performing some consistent tail changes of this vertex in $\mathcal{T}_c$ creates an edge between the two neighbors u_1, u_2 of v, where $(u_1, v), (u_2, v)$ are not associated with any tail change in $\mathcal{T}_c$. (Notice that contrary to the auxiliary multi-graph considered in [15], here some vertices in B might simultaneously create an edge in G_A and involve in tail changes.) We have the following claim for sufficiently large n ($n > (\frac{k}{\epsilon})^{O(\epsilon)}$).

Claim 1. *If there is a binocular of size $p \leq \frac{4}{\epsilon'} \log |A_2| - 1$ in G_{A_2}, there exists a canonical improvement with tail changes in G_A of size at most $\frac{12}{\epsilon'} \log n$.*

Implied by the claim, we know that there exists a canonical improvement with tail changes in G_A of size $\frac{12}{\epsilon'} \log n < \frac{4}{\epsilon} \log n$, which can be found by Algorithm LI. Therefore

$$(1 + \epsilon')(a - a_t) \geq (1 + \epsilon')|A_2| \geq b^1_n + b^1_2 + b^2_2 + b^2_3. \tag{2}$$

Combining (1) and (2), we have

$$
\begin{aligned}
(k + 1 + \epsilon')(a - a_t) &\geq 3b^1_n + \epsilon' b + 2b^1_2 + 3b^2_2 + 3b^2_3 + 3b^3_3 + 2b^{2'}_3 + b^1_3 \\
&= 3(b - b_t - \epsilon' b) - b^1_2 - b^1_3 - 2b^{2'}_3 + \epsilon' b. \tag{3}
\end{aligned}
$$

Hence, $(3 - 2\epsilon')b \leq (k + 1 + \epsilon')a - (k - 2 + \epsilon')a_t + b_2^1 + b_3^{2'} + 2b_3^1$.

Since every vertex in A_t can have at most $k - 2$ edges to B', we have $b_2^1 + b_3^{2'} + 2b_3^1 \leq (k - 2)a_t$. Therefore, $(3 - 2\epsilon')b \leq (k + 1 + \epsilon')a$. As $\epsilon' = \frac{2k+5}{3}\epsilon$, we have $b \leq (\frac{k+1}{3} + \epsilon)a$. $\qquad\square$

The proof of Claim 1 helps understand why we consider three more types of local improvements in addition to binoculars and helps explain the algorithm design in the next section.

Proof (Claim 1). Consider any binocular I in G_{A_2}. If there is no edge in I which is from B_2^1, we have a corresponding improvement I' in G_A by performing tail changes for any edge from B_3^2 in I. Otherwise, we assume that there is one self-loop in I from $v \in B_2^1$. By definition, one neighbor u_1 of v lies in A_t and the other neighbor u_2 in A'. Suppose u_1 belongs to a tail change in $\mathcal{T}_c$ which is associated with $w \in B_3^2$. If $w \in I$, we associate w with tail changes in G_A. In G_A, we remove the self-loop on u_2 and add edge (u_1, u_2). In this way, we have a path together with the other cycle in I which form an improvement in G_A, assuming the other cycle in I is not a self-loop from B_2^1. If the other cycle in I is also a self-loop from $v' \in B_2^1$, let u_1' be a neighbor of v' in A_t and u_2' be the other neighbor of v' in A'. If u_1' belongs to the tail change associated with $w' \in B_3^2$ and $w' \in I$, the path between u_2, u_2' in I together with the edges $(u_1, u_2), (u_1', u_2')$ form an improvement. If $u_1 = u_1'$, we have an improvement in G_A as a cycle. Other cases can be analyzed similarly. $\qquad\square$

4   The Algorithm and Main Results

In this section, we give an efficient implementation of Algorithm LI using the color coding technique [2] and dynamic programming. Let U be the universe of elements and K be a collection of kt colors, where $t = \frac{4}{\epsilon} \log n \cdot \frac{2(k-1)}{\epsilon} \cdot (k - 2) \leq \frac{4}{\epsilon} \log n \cdot \frac{2k^2}{\epsilon}$. We assign every element in U one color from K uniformly at random. If two k-sets contain $2k$ distinct colors, they are recognized as disjoint. Applying color coding is crucial to obtain a polynomial-time algorithm for finding a logarithmic-sized local improvement.

4.1   Efficiently Finding Canonical Improvements with Tail Changes

In this section, we show how to efficiently find canonical improvements with tail changes using the color coding technique. Let $C(S)$ be the set of distinct colors contained in sets in S. We say a collection of sets is colorful if every set contains k distinct colors and every two sets contain different colors.

Tail changes. We say a tail change $T_e(U, V)$ of a vertex v is *colorful* if V is colorful, and the colors in $C(V)$ are distinct from $C(v)$. A surplus edge can be associated with many tail changes. Let $\mathcal{T}_v(e)$ be all colorful tail changes of size at most $\frac{2(k-1)}{\epsilon}$ which are associated with an edge e of v. We enumerate all subsets of $\mathcal{S} \setminus \mathcal{A}$ of size at most $\frac{2(k-1)}{\epsilon}$ and check if they are colorful, and if they are tail changes associated with e. The time to complete the search for all vertices is at most $n^{O(k/\epsilon)}$.

The next step is to find all colorful groups of tail changes associated with v such that after performing one group of tail changes, the degree of v drops to 2. Notice that the tail changes in a colorful group are consistent. For every two edges e_i, e_j of v, we can compute a collection of colorful groups of tail changes which associate with all edges of v except e_i, e_j by comparing all possible combinations of tail changes from $E(v) \setminus \{e_i, e_j\}$. There are at most $(n^{O(k/\epsilon)})^{k-2}$ combinations. For every group of colorful tail changes which together replace V with U, we explicitly keep the information of which vertices are in U, V and the colors of V. It takes at most $n^{O(k^2/\epsilon)}$ space. To summarize, the time of finding colorful groups of tail changes for every vertex of degree at least 3 is $n^{O(k^2/\epsilon)}$.

Canonical improvements with tail changes. After finding all colorful tail changes for every vertex of degree at least 3, we construct the auxiliary multi-graph G_A. For vertices a_1, a_2 in G_A, we put an edge $e(a_1, a_2)$ between a_1 and a_2 if first, there is a set $b \in C = S \setminus A$ intersecting with only a_1, a_2, or secondly, there is a set $b \in C$ of degree $d_b \geq 3$ intersecting with a_1, a_2, and for other edges of b, there exists at least one group of $d_b - 2$ colorful tail changes. In the first case, we assign the colors of b to $e(a_1, a_2)$. In the second case, we add as many as $n^{O(k^2/\epsilon)}$ edges between a_1 and a_2, and assign to each edge the colors of b together with the colors of the corresponding group of $d_b - 2$ tail changes. The number of edges between two vertices in G_A is at most $n \cdot n^{O(k^2/\epsilon)}$. The number of colors assigned to each edge is at most $\frac{2k^3}{\epsilon}$ (Notice that the number of colors on an edge is at most $k(1 + \frac{(k-1)^2/\epsilon-1}{k-2})(k-2)$. This is at most $\frac{2k^3}{\epsilon}$ for $\epsilon < k+5$, which is usually the case.) Moreover, we add a self-loop for a vertex a in G_A if there exists a set $b \in C$ such that b intersects only with set a and assign the colors of b to this self-loop.

We use the dynamic programming algorithm to find all colorful paths and cycles of length $p = \frac{4}{\epsilon} \log n$ in G_A. A path/cycle is colorful if all the edges contain distinct colors. If we did not consider improvements containing at most one cycle, we could use a similar algorithm as in [15]. In our case, when extending a path by an edge e by dynamic programming, we would like to keep the information of the vertices replaced by the tail changes of e in this path. This would take quasi-polynomial time when backtracking the computation table. By Claim 1, it is sufficient to check for every path with endpoints u, v, if there is an edge of this path containing a tail change $T_e(U, V)$, such that $u \in U$ or $v \in U$. We sketch the algorithm as follows. For a given set of colors C, let $\mathcal{P}(u, v, j, C, q_u, q_v)$ be an indicator function of whether there exists a path of length j from vertex u to v with the union of the colors of these edges equal C. $q_u(q_v)$ are indicator variables of whether there is a tail change $T_e(U, V)$ of some edge in the path such that $u \in U(v \in U)$. The computation table can be initialized as $\mathcal{P}(u, u, 0, \emptyset, 0, 0) = 1$ and $\mathcal{P}(u, v, 0, \emptyset, 0, 0) = 1$, for every $u, v \in A$. In general, for a fixed set of colors C and integer $j \geq 1$, $\mathcal{P}(u, v, j, C, q_u, q_v) = 1$ if there exists a neighbor w of v, such that $\mathcal{P}(u, w, j - 1, C', q_u, q_w) = 1$, $C' \cup C((w, v)) = C$ and $C' \cap C((w, v)) = \emptyset$.

If $C((w, v)) > k$ (i.e., there are tail changes), we check every edge between w and v which satisfies the previous conditions. If there exists an edge associated with a tail change $T_e(U, V)$ such that $u, v \in U$, we mark $q_u = 1, q_v = 1$. Otherwise, if there exists an edge associated with a tail change $T_e(U, V)$ such that $u \in U$, we mark $q_u = 1$. To find colorful cycles, we query the result of $\mathcal{P}(u, u, j, C, q_u, q_u)$ for $j \geq 1$. Recall that we use kt many colors, where $t \leq p \cdot \frac{2k^2}{\epsilon}$. The running time of finding all colorful paths and cycles is $O(n^3 k p 2^{kt})$, which is $n^{O(k^3/\epsilon^2)}$.

The final step is to find canonical improvements with tail changes by combining colorful paths and cycles to form one of the six types defined in Sect. 3.1 by enumerating all possibilities. The running time of this step is $n^{O(k^3/\epsilon^2)}$. We omit the details of the time complexity analysis. In conclusion, the total running time of finding colorful tail changes, colorful paths/cycles, and canonical improvements with tail changes is $n^{O(k^3/\epsilon^2)}$. We call this color coding based algorithm **Algorithm CITC** (canonical improvement with tail changes).

4.2 Main Results

In this section, we present our main results. We first present a randomized local improvement algorithm. The probability that Algorithm CITC succeeds in finding a canonical improvement with tail changes if one exists can be calculated as follows. The number of sets involved in a canonical improvement with tail changes is at most $\frac{2k^2}{\epsilon} \cdot \frac{4 \log n}{\epsilon}$. The probability that an improvement with i sets having all ki elements of distinct color is

$$\frac{\binom{kt}{ki}(ki)!}{(kt)^{ki}} = \frac{(kt)!}{(kt - ki)!(kt)^{ki}} \geq \frac{(kt)!}{(kt)^{kt}} > e^{-kt} \geq n^{-8k^3/\epsilon^2}. \tag{4}$$

Let $N = n^{8k^3/\epsilon^2} \ln n$. We run Algorithm CITC $2N$ times and each time with a fresh random coloring. From (4), we know that the probability that at least one call of CITC succeeds in finding an improvement is at least $1 - (1 - n^{-8k^3/\epsilon^2})^{2N} \geq 1 - exp(n^{-8k^3/\epsilon^2} \cdot 2n^{-8k^3/\epsilon^2} \ln n) = 1 - n^{-2}$.

Since there are at most n local improvements for the problem, the probability that all attempts succeed is at least $(1 - n^{-2})^n \geq 1 - n^{-1} \longrightarrow 1$ as $n \longrightarrow \infty$. Hence this randomized algorithm has an approximation ratio $\frac{k+1}{3} + \epsilon$ with high probability. We call this algorithm **Algorithm RLI** (R for randomized). The running time of the algorithm is $2N \cdot n^{O(k^3/\epsilon^2)}$, which is $n^{O(k^3/\epsilon^2)}$.

We can obtain a deterministic implementation of Algorithm RLI, which always succeeds in finding a canonical improvement with tail changes if one exists. We call this deterministic algorithm **Algorithm DLI** (D for deterministic). The general approach is given by Alon et al. [1]. The idea is to find a collection of colorings $\mathcal{K}$, such that for every improvement there exists a coloring $K \in \mathcal{K}$ that assigns distinct colors to the sets involved in this improvement. Then Algorithm CITC can be implemented on every coloring until an improvement is found. The collection of colorings satisfying this requirement can be constructed using perfect hash functions as in [1]. In our case, we need $n^{O(k^3/\epsilon^2)}$

hash functions, therefore a total $n^{O(k^3/\epsilon^2)}$ runs of dynamic programming to find an improvement.

Theorem 3. *For any $\epsilon > 0$, Algorithm DLI achieves an approximation ratio $\frac{k+1}{3} + \epsilon$ of the k-Set Packing problem in time $n^{O(k^3/\epsilon^2)}$.*

Theorem 4. *For any $t \le \left(\frac{3e^3 n}{k}\right)^{1/5}$, there exist two disjoint collections of k-sets $\mathcal{A}$ and $\mathcal{B}$ with $|\mathcal{A}| = 3n$ and $|\mathcal{B}| = (k+1)n$, such that any collection of t sets in $\mathcal{B}$ intersect with at least t sets in $\mathcal{A}$.*

Theorem 4 gives a tight example of Algorithm DLI. The construction is as follows. We index the sets in $\mathcal{A}$ from 1 to $3n$. Every set in $\mathcal{B}$ induced on $\mathcal{A}$ is a 2-set or a 3-set. There are $b_2 = 3n$ 2-sets and $b_3 = (k-2)n$ 3-sets in $\mathcal{B}$. We index the 2-sets in $\mathcal{B}$ from 1 to $3n$. The i-th 2-set intersects with the $(i-1)$-th and the i-th set in $\mathcal{A}$ (the 0-th set is the n-th set in $\mathcal{A}$). The 3-sets are constructed by partitioning the elements not covered by 2-sets in U into groups of three uniformly at random. The proof of Theorem 4 is omitted. This lower bound matches the performance guarantee of Algorithm DLI. This indicates that this is possibly the best result that can be achieved by a local improvement algorithm for the k-Set Packing problem.

References

1. Alon, N., Naor, M.: Derandomization, witnesses for boolean matrix multiplication and construction of perfect hash functions. Algorithmica **16**(4/5), 434–449 (1996)
2. Alon, N., Yuster, R., Zwick, U.: Color-coding. J. ACM **42**(4), 844–856 (1995)
3. Berman, P.: A d/2 approximation for maximum weight independent set in d-claw free graphs. Nord. J. Comput. **7**(3), 178–184 (2000)
4. Berman, P., Fürer, M.: Approximating maximum independent set in bounded degree graphs. In: SODA, pp. 365–371 (1994)
5. Chan, Y., Lau, L.: On linear and semidefinite programming relaxations for hypergraph matching. In: SODA, pp. 1500–1511 (2010)
6. Chandra, B., Halldórsson, M.: Greedy local improvement and weighted set packing approximation. In: SODA, pp. 169–176 (1999)
7. Cygan, M.: Improved approximation for 3-dimensional matching via bounded pathwidth local search. In: FOCS, pp. 509–518 (2013)
8. Cygan, M., Grandoni, F., Mastrolilli, M.: How to sell hyperedges: the hypermatching assignment problem. In: SODA, pp. 342–351 (2013)
9. Fellows, M.R., Knauer, C., Nishimura, N., Ragde, P., Rosamond, F., Stege, U., Thilikos, D.M., Whitesides, S.: Faster fixed-parameter tractable algorithms for matching and packing problems. Algorithmica **52**(2), 167–176 (2008)
10. Fürer, M., Yu, H.: Approximating the k-set packing problem by local improvements. http://arxiv.org/abs/1307.2262
11. Halldórsson, M.: Approximating discrete collections via local improvements. In: SODA, pp. 160–169 (1995)
12. Håstad, J.: Clique is hard to approximate within $n^{1-\epsilon}$. In: FOCS, pp. 627–636 (1996)

13. Hazan, E., Safra, S., Schwartz, O.: On the complexity of approximating k-set packing. Comput. Complex. **15**(1), 20–39 (2006)
14. Hurkens, C.A., Shrijver, J.: On the size of systems of sets every t of which have an SDR, with an application to the worst-case ratio of heuristics for packing problems. SIAM J. Discrete Math. **2**(1), 68–72 (1989)
15. Sviridenko, M., Ward, J.: Large neighborhood local search for the maximum set packing problem. In: Fomin, F.V., Freivalds, R., Kwiatkowska, M., Peleg, D. (eds.) ICALP 2013, Part I. LNCS, vol. 7965, pp. 792–803. Springer, Heidelberg (2013)

Multi-Objective Cuckoo Search
with Leader Selection Strategies

Kamel Zeltni[1,2]([✉]) and Souham Meshoul[1]

[1] College of NTIC, Computer Science and its Applications Department,
Constantine 2 University, Constantine, Algeria
[2] MISC Laboratory, Constantine, Algeria
kamel.zeltni@gmail.com, smeshoul@umc.edu.dz

Abstract. Cuckoo Search has been recently added to the pool of nature inspired metaheuristics. Its promising results in solving single objective optimization motivate its use in multiobjetive context. In this paper we describe a Pareto based multiobjective Cuckoo search algorithm. Like swarm based metaheuristics, the basic algorithm needs to specify the best solutions in order to update the population. As the best solution is not unique in multiobjective optimization, this requires the use of a selection strategy. For this purpose, we propose in this paper investigation of five leader selection strategies namely random selection, sigma method, crowding distance method, hybrid selection method and MaxiMin method. Performance of the proposed algorithm has been assessed using benchmark problems from the field of numerical optimization. Impact of selection strategies on both convergence and diversity of obtained fronts has been studied empirically. Experimental results show in one hand the great ability of the proposed algorithm to deal with multiobjective optimization and in other hand no strategy has been shown to be the best in all test problems from both convergence and diversity points of view. However they may impact significantly the performance of the algorithm in some cases.

Keywords: Cuckoo search · Multiobjective optimization · Convergence · Diversity · Leader selection · Pareto dominance

1 Introduction

In many areas of science and business and even in our daily life, selecting the best element with regard to some objectives from a set of available alternatives is a common problem. This later is essentially an optimization task. During the last decades optimization became an essential tool in different domains ranging from engineering, telecommunications to bioinformatics and business to name just few. Optimization problems can be of different types with regard to some classification criteria. A good encyclopedia on optimization can be found in [5]. In most cases, it is rare for any real world decision problem to require optimization

© Springer International Publishing Switzerland 2014
P. Fouilhoux et al. (Eds.): ISCO 2014, LNCS 8596, pp. 421–432, 2014.
DOI: 10.1007/978-3-319-09174-7_36

of only one objective. Usually decision problems are complex and involve many tradeoffs because many objectives that are often conflicting need to be taken into account. Such problems are known as Multi-Objective Problems (MOPs). Basically, a MOP consists in finding solutions to problems in a way to optimize a set of objectives while satisfying a set of constraints. In the literature, multiobjective optimization has been performed in different ways [6]. Methods for multiobjective optimization fall into four main classes depending on the way objectives are handled during the optimization process. These classes include aggregation methods, lexicographic ordering based methods, Pareto based methods and non-pareto based mothods. Pareto based methods are particularly interesting because they offer a suitable way namely the dominance concept to deal with the conflicting relationship between objectives that makes amelioration of one objective causes deterioration of another. According to dominance relation, a solution is said to dominate another one if it is as good as it in all objectives and if it is strictly better than it in at least one objective. Therefore, solving a MOP using Pareto dominance concept consists in identifying the set of non dominated solutions that represent the possible tradeoffs between objectives. With this regard, nature inspired metaheuristics and especially population based ones offer an attractive way to solve MOPs as they are able to find a set of solutions in a single run. Genetic Algorithm (GA) [9] and Particle Swarm Optimization methods (PSO) [11] have been largely investigated in the field of multiobjective optimization. Many Evolutionary Algorithms have been proposed like NSGAII [7], SPEA2 [24] and PAES [12]. Also there have been diverse Multiobjective algorithms based on PSO called MOPSO [14]. However, recent metaheuristics like Cuckoo Search (CS) need to be studied in the context of multiobjective optimization.

CS is a population based metaheuristic recently developed by Xin-She Yang et al. in 2009 [21,22]. It is based on the cuckoo parasitism, by laying their eggs in the nests of other birds. CS is based on two fundamental concepts namely the reproductive behavior of cuckoos and levy flight rather than simple random walk. Recent studies have shown the competitiveness of CS compared to GA and PSO in single-objective optimization. Therefore, its extension to MOPs is a natural progression. To the best of our knowledge, only two papers have been published [18,20] in which authors proposed an application of a multiobjective CS to design optimizationan and a real-world manufacturing process respectively. According to the dynamics of the basic CS as proposed in [21,22], best solutions are used to update the current generation in order to compute the new generation. Translating this fact to multiobjective context imposes the use of a leader selection strategy. In both papers [18,20] it is not mentioned how this task has been performed. In this paper, we propose investigating the impact of leader selection strategies to study the extent to which they may improve the performance of a CS algorithm in multiobjective optimization. Two performance measures have been used for this purpose namely convergence and diversity of obtained Pareto fronts.

The remainder of the paper is organized as follows. In Sect. 2, the basic concepts related to multiobjective optimization are given. Section 3 provides a description of the basic single objective CS algorithm. Section 4 is devoted to the proposed multiobjective CS algorithm with leader selection strategies. In Sect. 5, experimental results using benchmark functions are given and discussed. Finally, conclusions and perspectives are drawn.

2 Multiobjective Optimization: Basic Concepts and Related Work

2.1 Basic Concept

A MOP is defined by a decision space $S \subset \mathbb{R}^n$, a set of objectives F and a set of constraints. It aims at determining the decision vector $X^* \in S$ that optimizes objectives in F. As it is not possible to optimize all objectives simultaneously, several decision vectors may be solution candidates and form the so called Pareto optimal set the image of which in the objective space is known as the Pareto front. Mathematically a MOP can be formulated as:

$$Minimize\ (Maximize)\ \overrightarrow{F}(\overrightarrow{X}) = (f_1(\overrightarrow{X}), \ldots, f_m(\overrightarrow{X}))$$
$$Subject\ to\ (g_1(\overrightarrow{X}), \ldots, g_k(\overrightarrow{X})) \leqslant 0\ (inequality\ constraints)$$
$$and\ (h_1(\overrightarrow{X}), \ldots, h_k(\overrightarrow{X})) = 0\ (equality\ constraints)$$
$$where\ \overrightarrow{X} = (x_1, \ldots, x_n) \in S$$

where $\overrightarrow{X}$ is the decision vector and $\overrightarrow{F}$ the objective vector.

Comparing solutions in multiobjective context is based on dominance relation. A vector $\overrightarrow{U}$ in the decision space is said to dominate a vector $\overrightarrow{V}$ denoted by $\overrightarrow{U} \succ \overrightarrow{V}$ if and only if (when considering a minimization task)

$$\forall i \in [1, n], f_i(\overrightarrow{U}) \leqslant f_i(\overrightarrow{V})\ and\ \exists j\ f_j(\overrightarrow{U}) < f_j(\overrightarrow{V})$$

A solution $\overrightarrow{U}$ is said to be Pareto optimal if and only if it is not dominated by any other solution $\overrightarrow{V}$ in S

2.2 Related Work

Like evolutionary algorithms, much interest has been devoted to swarm based metaheuristics to solve multiobjective optimization problems during the last few years. This class of metaheuristics is also inspired by nature, but instead by evolution, their development has been motivated by the flocking and swarm behavior of birds like particle swarm optimization (PSO) and insects like artificial bee colony optimization (ABC) [19]. These approaches operate on a population of candidate solutions that are updated within an iterative process. Generally, such updates are done by taking into account the self performance of each member of the swarm as well as the global performance within the swarm. In another way, search is guided by two mechanisms related to the influence of self performance and the influence of global performance. Therefore, the use of these metaheuristics for multiobjective optimization raises an issue about the selection of the global guide or leader as it is not unique in this case. To deal with this issue several techniques and strategies have been proposed in the literature with the aim to obtain a good approximation of the optimal Pareto set. The simplest way to choose a global leader is random selection. Mostaghim and Teich [15] proposed a leader selection method for a MOPSO based on a density measure called Sigma method. Its principle is to select as a leader for a particular particle the member of the external archive (i.e. the set of nondominated solutions) that is close to the particle in terms of sigma values computed for the particle and each member of the external archive. Crowding distance has been also used as a leader selection strategy. Originally, it has been proposed by Deb et al. for NSGAII [7] to promote diversity of the obtained Pareto fronts. The mechanism of crowding distance computation was incorporated into a PSO algorithm for global best selection in [16]. Since, this strategy is largely used in context of MOP for selecting leaders. Its principle is to favor selection of members that are located in less crowded regions. Baling was the first to propose the use of the Maximin fitness for Multiobjective Evolutionary optimization [2] for archiving a set of optimal solutions. Later it has been applied to select leader in a maximinPSO algorithm [13]. More recently a hybrid method that combines Sigma and Crowding distance has been proposed in [1]. It consists in determining the k neighbors of each particle in terms of sigma values then to select the leader among these k candidates using crowding distance. Other strategies were recently reported like density estimation [10] and Step by step Rejection SR [4].

3   Basic Cuckoo Search Algorithm

Cuckoos are species with amazing reproduction strategy. They do not build their own nests, but they engage the obligate brood parasitism by laying their eggs in the nest of other birds, though some cuckoos species may remove others' eggs to increase a survival probability of their own eggs [3]. Sometimes, the egg of cuckoo in the nest is discovered by host bird, in this case the host bird will either throw away the alien eggs or simply abandon the nest and build a new nest elsewhere. To develop the standard Cuckoo Search algorithm by Yang and Deb [21, 22] the authors supposed the following three idealized rules:

– Each cuckoo lays one egg at a time, and dumps it in a randomly chosen nest.
– The best nests with high quality of eggs (solutions) will carry over to the next generations.
– The number of available host nests is fixed, and a host can discover an alien egg with probability $p_\alpha \in [0, 1]$. In this case, the host bird can either throw the egg away or abandon the nest to build a completely new nest in a new location. This can be approximated by a fraction p_α of the n nests being replaced by new nests (with new random solutions at new locations).

Knowing that a nest represent a potential solution, the three previous rules have been inplemented as follows: The first rule can be applied using randomization process, where the new solution x^{t+1} is generated randomly by using Lévy flight equation (1).

$$x_i^{t+1} = x_i^t + \alpha \oplus levy\,(\beta) \tag{1}$$

Lévy flights essentially provide a random walk while their random steps are drawn from a Lévy distribution for large steps which has an infinite variance with an infinite mean.

The Second rule corresponds to elitism where the best solutions are passed onto the next generation. The third rule can be considered as a mutation operation where the fraction p_α in solutions dimension is replaced by new generated one according to a probability.

4   Proposed Multiobjective Cuckoo Search with Leader Selection

Motivated by the success and promising results of single objective CS algorithm, we propose in this section its use for multiobjective optimization. For this purpose two issues need to be addressed. The first one is related to the way to keep and manage non-dominated solutions that are found during the search process and the second one is related to the strategy to use in order to select leaders to update solutions. Like other multiobjective evolutionary algorithms, using an archive to serve as a repository of nondominated solutions is adopted in our work to solve the first issue. As for the second one, we propose investigating several leader selection strategies namely random selection, sigma method, crowding distance method, a hybrid selection strategy and MaxiMin method. The proposed multiobjective CS algorithm with leader selection is outlined below in Algorithm 2.

Algorithm 1. Select_Best(X, A, Sigma)

Input : X: current solution
$\quad\quad\quad$ A: current archive
Output: *leader*

$n = |A|$

$X.\sigma=$ *Sigma*($X.objective\_vector$)
For each non dominated solution j in A
$\quad\quad\quad$ A(j).σ = Sigma(A(j).objective_vector)
$\quad\quad\quad\quad$ *A(j).distance=calcdist*($X.\sigma, A(j).\sigma$)
endFor
Sort distance according to A.distance in ascending order
leader $= A(1)$

Algorithm 2. CSMO

Step 1: Initialization

- *1.1.* t=0
- *1.2.* P_t=create a population of N candidate solutions
- *1.3.* **For** each solution X_i in P_t $objectives\_vector_i^t = evaluate(X_i^{(t)})$ **endFor**
- *1.4.* A_t= Create initial global archive

Step 2: Iteration search process

- *2.1.* $P_{t+1} = \{\}$
- *2.2.* **For** each solution $X_i^{(t)}(i = 1..N)$ in P_t // get new solution according to equation 1
  - 2.2.a. $Best_i = Select\_Best(X_i, A_i, SelectStrategy)$
  - 2.2.b. Step=Compute_step according to lévy flight distribution
  - 2.2.c. Step_size= $f * step * (X_i^{(t)} - Best_i)$
  - 2.2.d. $X_i^{(t+1)} = X_i^{(t)} + Step\_size * randn(size(X_i^{(t)}))$
  - 2.2.e. $objective\_vector_i^t = evaluate(X_i^{(t)})$
  - 2.2.f. Update solution X_i
  If $X_i^{(t+1)}$ dominates $X_i^{(t)}$ in the objective space **Then**
  $\qquad P_{t+1} = P_{t+1} \cup \{X_i^{(t+1)}\}$
  else
  $\qquad P_{t+1} = P_{t+1} \cup \{X_i^{(t)}\}$
  endIf
- *2.3.* A_t =Updade global archive
  endFor
- *2.4.* **For** each solution X_i in P_{t+1}
  - 2.4.a. new_i= apply mutation on fraction P_α of solution dimention;
  - 2.4.b. Update solution X_i
  If new_i dominates X_i in the objective space **then**
  $\qquad X_i = new_i$
  endIf
  endFor
- *2.5.* A_t =Updade global archive
- *2.6.* t=t+1
- *2.7.* Check stopping criterion

  If stopping criterion not satisfied
  $\qquad$ go to **step 2**
  else
  $\qquad$ stop search and output A_{t-1}
  endIf

As shown in the Algorithm 2, the process starts with an initialization phase **step (1)** during which a population of N candidate solutions are generated randomly according to the domain of each problem dimension (*1.2*). A solution corresponds to a nest in the biological metaphor. Then these solutions are evaluated in order to get their vector of objectives' values (*1.3*). Using dominance relation, non-dominated solutions within the initial population are determined to create the initial archive (1.4). Then the search process undergoes an iterative phase where the population is updated according to a specific dynamics **Step 2**. At each iteration, each solution in the current population is modified as follows (2.2). First, a guide or leader is selected from the current archive (2.2.a). This is done using a leader selection strategy. The investigated strategies are outlined in Algorithms 3, 1 and 4 . Then a step size is calculated based on a factor f, a step drawn from Levy flights distribution and the gap between the current solution and its selected leader (2.2.c). The factor f is related to the typical step size of flights. Whenever the current solution is equal to its leader it remains unchanged.

The new solution is compared to the old one based on dominance relation. The nondominated one among them is kept in the new population (2.2.f). Once all solutions are processed in this manner, the global archive is updated based on the current nondominated solutions and the new population (2.3). After these steps, the new population undergoes a mutation operation (2.4). A mutated solution is allowed to join the new population only if it dominates the original one. In this case, this later is removed from the current population and the global archive is updated again (2.5). At this stage (2.7), if the stopping criterion is satisfied the current global archive is given as the output of the algorithm otherwise the process undergoes a new iteration.

Algorithm 3. Select_Best(X, A, Crowding_Distance)

Input : X: current solution
       A: current archive
Output: *leader*

$n = |A|$ // Archive size

For each non dominated solution i in A
    $A(i).distance = 0$
endFor
For each objective j
    $A = sort(A.objective\_vector(j))$
    $A(1).distance = \infty$
    $A(n).distance = \infty$
    For $i = 2$ to $(n-1)$
        $A(i).distance = A(i).distance + \frac{[A(i+1).objective\_vector(j) - A(i-1).objective\_vector(j)]}{[A(n).objective\_vector(j) - A(1).objective\_vector(j)]}$
    endFor
endFor
sort A according to A.distance in descending order
$leader = A(1)$

Algorithm 4. Select_Best(X, A, Maximin)

Input : X: current solution
         A: current archive
Output: *leader*

For each member i in A
    For each member j in A j≠ i
        For each objective m
            $temp(j) = A(i).objective\_vector(m) - A(j).objective\_vector(m)$
        endFor
        $Minimum(i) = min(temp)$
    endFor
    $maximin = max(Minimum)$
endFor
$leader = max(Maximin)$

---

5   Experimental Result

The performance of the proposed algorithm have been assessed using two metrics that measure convergence and diversity of obtained Pareto fronts [7]. Convergence metric shows the extent to which an obtained Pareto front is close to the optimal front. It is described by the following equation:

$$\Upsilon = \frac{\sum_{i=1}^{n} \min_{j=1...m} (eucldist(F_i, H_j))}{n} \tag{2}$$

where H is known set of pareto-optimal solutions with m solutions, F is the set of non-dominated solutions found by the algorithm with n solutions and eucldist is the euclidean distance between two given solutions.

Diversity metric gives an indication about the uniformity of distribution of solutions along the Pareto front. It is given by the following equation

$$\Delta = \frac{d_f + d_l + \sum_{i=1}^{N-1} |d_i - \overline{d}|}{d_f + d_t + (N-1)\overline{d}} \tag{3}$$

where, d_f and d_l are the euclidean distance between the extreme solution in the Pareto-optimal set, d_i is the ith solution in the obtained non-dominated set and $\overline{d}$ is the avrage of all distances. Smaller values of these two metrics indicate better convergence and diversity.

In our experiments, five benchmark problems related to numerical optimization have been considered. ZDT1 3 from the ZDT serie proposed by Ziztler et al. [23], Fonseca and Flemings study (FON) [8] and Schaffers study (SCH) [17], All problems have two objective functions without constraints.

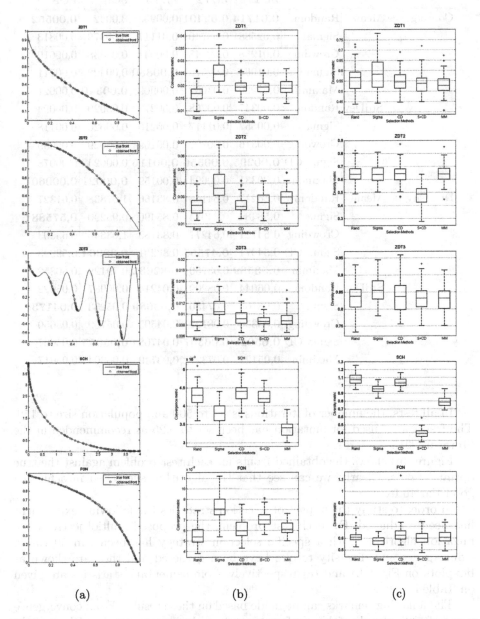

Fig. 1. Experimental result: (a) Obtained fronts, (b) Convergence results, (c) Diversity results.

Table 1. Complementary statistics of convergence and diversity metrics

			ZDT1	ZDT2	ZDT3	SCH	FON
Convergence	Mean	Random	**0,01714**	**0,02401**	0,00953	0,00429	**0,00552**
		Sigma	0,02588	0,03740	0,01117	0,00379	0,00813
		Crowding	0,01938	0,02448	0,00941	0,00438	0,00610
		Sigma CD	0,01964	0,02576	**0,00880**	0,00422	0,00711
		Maximin	0,01929	0,03032	0,00909	**0,00340**	0,00621
	StdDev	Random	0,00347	0,00758	0,00272	0,00029	0,00094
		Sigma	0,00557	0,01117	0,00210	0,00026	0,00118
		Crowding	0,00316	0,00707	0,00169	0,00029	0,00099
		Sigma CD	**0,00295**	**0,00556**	**0,00116**	0,00024	0,00078
		Maximin	0,00331	0,00694	0,00159	**0,00021**	**0,00060**
Diversity	Mean	Random	0,56217	0,64163	0,83166	1,07838	0,61327
		Sigma	0,58484	0,64014	0,85490	0,95290	**0,57588**
		Crowding	0,54721	0,64271	0,83785	1,03345	0,63503
		Sigma CD	0,54071	**0,47555**	0,83201	**0,38964**	0,59993
		Maximin	**0,53539**	0,63750	**0,82632**	0,79129	0,60852
	StdDev	Random	0,06044	0,08209	0,04214	0,06987	0,05022
		Sigma	0,08165	0,07450	**0,03689**	0,05253	**0,04173**
		Crowding	0,05350	0,08122	0,04595	0,05709	0,08650
		Sigma CD	**0,05019**	**0,06307**	0,04708	**0,05055**	0,05667
		Maximin	0,05122	0,07327	0,03749	0,06304	0,05617

In all tests, the number of iterations is set to 500 and population size to 150. The fraction p_α used for mutation has been set to 0.25 as recommended in the literature [21].

Figure 1(a) shows the obtained fronts for each test problem against the true optimal fronts. Clearly, we can see that the algorithm succeeded in achieving good Pareto fronts.

In order to study the impact of selection strategies the following experiment has been conducted. For each test problem, the proposed multiobjective optimization algorithm with a specific selection strategy has been run 50 times. Convergence and diversity results have been gathered and they are shown as boxplots on Fig. 1(b) and (c) respectively. Complementary statistics are given on Table 1.

The following remarks can be made based on these results. From convergence point of view, the algorithm performs poorly with sigma method on ZDT1 problem while the other strategies achieves competitive results. For ZDT2 all strategies are competitive with a slight advantage for crowding distance method in terms of median value and stability. Once again sigma method gave worst results in this case although it was able to find the best convergence value in one of the runs. With ZDT3, results reveal that sigma method is the worst both in terms of median and best value. The others strategies have nearly similar convergence with little advantage to the hybrid method in term of stability. Concerning FON and SCH the results are almost similar or competitive. Sigma method achieves the worst results and MaxiMin has been found to be slightly better in SCH case.

From diversity point of view, the hybrid method is significantly better than the other strategies on ZDT2 and SCH test problems. For example, the worst value achieved for ZDT2 is in the neighborhood of the median of the other strategies. MaxiMin and hybrid method seem to be relatively better in terms of median and best value on ZDT3 problem. For ZDT1, better results have been obtained with MaxiMin. in other cases, results are competitive. In the light of these results, we can say that compared to a random selection, the other strategies may impact significantly the performance of the algorithm. No strategy has been shown to be the best in all test problems from both convergence and diversity points of view.

6   Conclusion and Perspectives

In this paper a multiobjective Cuckoo Search algorithm has been proposed. Updating candidate solutions during the search process requires selection of a leader. To handle this issue properly, leader selection strategies have been incorporated in the proposed algorithm. Experimental results using benchmark problems have shown the ability of the proposed algorithm to solve MOPs. Although no strategy has been found to be the best in all cases, it is worth mentioning that they may impact significantly the performance of the algorithm in terms of convergence and diversity.

References

1. AlBaity, H., Meshoul, S., Kaban, A.: On extending quantum behaved particle swarm optimization to multiobjective context. In: proceedings of the 2012 IEEE Congress on Evolutionary Computation (CEC), pp. 1–8 (2012)
2. Balling, R.J.: The maximin fitness function for multiobjective evolutionary optimization. In: Parmee, I.C., Hajela, P. (eds.) Optimization in Industry, pp. 135–147. Springer, London (2002)
3. Bonabeau, E., Dorigo, M., Theraulaz, G.: Swarm Intelligence: From Natural to Artificial Systems. Oxford University Press, New York (1999)
4. Cheng, S., Chen, M.-Y., Hu, G.: An approach for diversity and convergence improvement of multi-objective particle swarm optimization. In: Yin, Z., Pan, L., Fang, X. (eds.) Proceedings of The Eighth International Conference on Bio-Inspired Computing: Theories and Applications (BIC-TA), 2013. AISC, vol. 212, pp. 495–503. Springer, Heidelberg (2013)
5. Floudas, C.A., Pardalos, P.M.: Encyclopedia of Optimization: With 247 Tables. Springer, New York (2009)
6. Deb, K.: Multi-Objective Optimization Using Evolutionary Algorithms. Wiley, Chichester (2001)

7. Deb, K., Pratap, A., Agarwal, S., Meyarivan, T.A.M.T.: A fast and elitist multiobjective genetic algorithm: NSGA-II. IEEE Trans. Evol. Comput. **6**, 182–197 (2002)
8. Fonseca, C.M., Fleming, P.J.: Multiobjective optimization and multiple constraint handling with evolutionary algorithms. i. a unified formulation. IEEE Trans. Syst. Man Cybern. Part A Syst. Humans **28**, 26–37 (1998)
9. Goldberg, D.E.: Genetic Algorithms in Search Optimization and Machine Learning. Addison-Wesley, Reading (1988)
10. Hu, W., Yen, G.G.: Density estimation for selecting leaders and maintaining archive in MOPSO. In: Proceedings of the 2013 IEEE Congress on Evolutionary Computation (CEC), pp. 181–188 (2013)
11. Kennedy, J., Eberhart, R.: Particle swarm optimization. In: Proceedings of the 1995 IEEE International Conference on Neural Networks, pp. 1942–1948 (1995)
12. Knowles, J., Corne, D.: The Pareto archived evolution strategy: a new baseline algorithm for Pareto multiobjective optimisation. In: Proceedings of the 1999 Congress on Evolutionary Computation, CEC 99 (1999)
13. Li, X.: Better spread and convergence: particle swarm multiobjective optimization using the maximin fitness function. In: Deb, K., Tari, Z. (eds.) GECCO 2004. LNCS, vol. 3102, pp. 117–128. Springer, Heidelberg (2004)
14. Mohankrishna, S., Maheshwari, D., Satyanarayana, P., Satapathy, S.C.: A comprehensive study of particle swarm based multi-objective optimization. In: Satapathy, S.C., Avadhani, P.S., Abraham, A. (eds.) Proceedings of the InConINDIA 2012. AISC, vol. 132, pp. 689–701. Springer, Heidelberg (2012)
15. Mostaghim, S., Teich, J.: Strategies for finding good local guides in multi-objective particle swarm optimization (MOPSO). In: Proceedings of the 2003 IEEE Swarm Intelligence Symposium, SIS'03, pp. 26–33 (2003)
16. Raquel, C.R., Naval, P.C. Jr.: An effective use of crowding distance in multiobjective particle swarm optimization. In: Proceedings of the Genetic and Evolutionary Computation Conference (GECCO), pp. 257–264 (2005)
17. Schaffer, J.D.: Multiple objective optimization with vector evaluated genetic algorithms. In: Proceedings of the 1st International Conference on Genetic Algorithms, pp. 93–100. L. Erlbaum Associates Inc, (1985)
18. Syberfeldt, A.: Multi-objective optimization of a real-world manufacturing process using cuckoo search. In: Yang, X.-S. (ed.) Cuckoo Search and Firefly Algorithm. SCI, vol. 516, pp. 179–193. Springer, Heidelberg (2014)
19. Talbi, E.-G.: Metaheuristics: From Design to Implementation. Wiley, Chichester (2009)
20. Yang, X.-S., Deb, S.: Multiobjective cuckoo search for design optimization. Comput. Oper. Res. **40**, 1616–1624 (2006–2013)
21. Yang, X.-S., Deb, S. Cuckoo search via lévy flights. In: Proceedings of the World Congress on Nature and Biologically Inspired Computing, NaBIC 2009, December 2009, India, pp. 210–214. IEEE Publications, USA (2009)
22. Yang, X.-S., Deb, S.: Engineering optimisation by cuckoo search. Int. J. Math. Model. Numer. Optim. **1**, 330–343 (2010)
23. Zitzler, E., Deb, K., Thiele, L.: Comparison of multiobjective evolutionary algorithms: empirical results. In: Proceeding of the Evolutionary Computation, vol. 8, pp. 173–195, (summer 2000)
24. Zitzler, E., Laumanns, M., Thiele, L.: SPEA2: improving the strength Pareto evolutionary algorithm. Eidgenssische Technische Hochschule Zrich (ETH), Institut für Technische Informatik und Kommunikationsnetze (TIK) (2001)

Vulnerability Assessment of Spatial Networks: Models and Solutions

Eduardo Álvarez-Miranda[1,2](✉), Alfredo Candia-Véjar[2], Emilio Carrizosa[3], and Francisco Pérez-Galarce[2]

[1] DEI, Università di Bologna, Bologna, Italy
e.alvarez@unibo.it
[2] DMGI, Universidad de Talca, Talca, Chile
{ealvarez,acandia,franperez}@utalca.cl
[3] Faculdad de Matemáticas, Universidad de Sevilla, Seville, Spain
ecarrizosa@sevilla.es

Abstract. In this paper we present a collection of combinatorial optimization problems that allows to assess the vulnerability of spatial networks in the presence of disruptions. The proposed measures of vulnerability along with the model of failure are suitable in many applications where the consideration of failures in the transportation system is crucial. By means of computational results, we show how the proposed methodology allows us to find useful information regarding the capacity of a network to *resist* disruptions and under which circumstances the network collapses.

1 Introduction

Shortest path problems correspond to an old and very known class of problems in combinatorial optimization. A variant of one of these basic problem consists on analyzing the effects of removing arcs from a network. In [14] the problem of removing k arcs that cause the greatest decrease in the maximum flow from a source to a sink in a planar network is studied. This problem is a special case of a broad class of network optimization problems known as *interdiction* problems. Applied to the shortest s, t-path problem, the interdiction problem can be defined in the following way. Given a graph $G = (V, E)$ with a non-negative length function on its arcs $l : E \to \mathbb{R}$ and two terminals $s, t \in V$, the goal is to destroy all (or the *best*) paths from s to t in G by *optimally* eliminating as many arcs of A as possible (usually respecting a so-called *interdiction budget*). Interdiction problems are often used to measure the robustness of solutions of network optimization problems. In [9] several versions of these problems are studied; they consider the case of *total limited interdiction* when a fixed number of k arcs can be removed, and *node-wise limited interdiction* (for each node $v \in V$ a fixed number $k(v)$ of out-going arcs can be removed). For a complete survey on early interdiction problems with different underlying network properties the reader is referred to [2]. For a more general discussion regarding network vulnerability approaches we suggest to see [10].

© Springer International Publishing Switzerland 2014
P. Fouilhoux et al. (Eds.): ISCO 2014, LNCS 8596, pp. 433–444, 2014.
DOI: 10.1007/978-3-319-09174-7_37

Based on a well-known network interdiction model we formulate a framework of combinatorial optimization problems whose solutions can be used for assessing the vulnerability of spatial networks in the case of disruptions. We design a flexible model of network disruption based on the geometric characteristics of spatial networks. This model incorporates the nature of the disruptions present in different situations such as military planning [5,8], terrorist attacks [12] or emergency control of infectious disease spreading [1]. The proposed problems, along with the model of disruption, span several realizations of network interdiction providing a useful tool to characterize network vulnerability. Our aim is to propose a methodology that uses network optimization problems to characterize the robustness of a network in the presence of multiple failures.

In Sect. 2 we present the optimization framework for vulnerability assessment; in Sect. 3 we report computational results on realistic instances; these results show the versatility of the proposed models to characterize the robustness of the network infrastructure. Finally, in Sect. 4 we draw final conclusions and propose paths for future work.

2   Vulnerability Measures as Optimization Problems

Notation. Let $G = (V, E)$ be a spatial network such that $|V| = n$ and $|E| = m$. Let $s, t \in V$ be a source and a target node respectively; $l_e, \forall e : \{i, j\} \in E$, be the cost of edge e (distance between i and j); and ℓ be the cost of the shortest s, t-path on G with edge costs given by $l_e, \forall e \in E$.

Let $\mathcal{X} \subset \mathbb{R}^2$ be an arbitrary sub-region of $\mathbb{R}^2$. An element $x \in \mathcal{X}$ is a *point* in $\mathcal{X}$; for a given point x and a given edge e, let $d(x, e)$ be the minimum distance between x and the line segment defined by e (recall that $e : \{i, j\}$ links node i with node j, whose positions are given). For a given $R \in \mathbb{R}^{>0}$ and a given $x \in \mathcal{X}$, let $E_x = \{e \in E \mid d(e, x) > R\}$ and $\bar{E}_x = \{e \in E \mid d(e, x) \le R\}$. In other words, E_x is the set of edges that are not *reached* by the disk of radius R centered at x (the *disruption disk* $\rho(x, R)$), and $\bar{E}_x$ is the set of disrupted or interdiced edges. We will refer to $G_x = (V, E_x)$ as the *operating network* with respect to $\rho(x, R)$. Note that G_x might be disconnected.

The model of failure represented by $\rho(x, R)$ embodies a characteristic of disruption produced by many different sources: instead of having isolated failures, we have a set of failures all of them circumscribed within a delimited area. This naturally occurs in the application contexts that we have already mentioned.

2.1   The Max-Cost Single-Failure Shortest Path Problem

Let us assume that $\mathcal{X}$ is a finite set of points x in $\mathbb{R}^2$ and that R can take values in $\mathcal{R}$ which is a finite subset of $\mathbb{R}^{>0}$. Given a radius $R \in \mathcal{R}$ and a discrete set $\mathcal{X}$, we are interested in knowing what is the *maximum* length Ω of a *shortest* s, t-path across all possible locations $x \in \mathcal{X}$ of the disruption disk $\rho(x, R)$.

Knowing Ω is threefold: (i) It tells us how severe a disruption can be by comparing the value of Ω with respect to ℓ; in other words, the increase of the

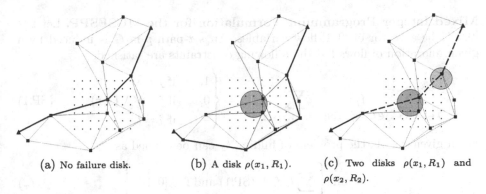

(a) No failure disk.　　　(b) A disk $\rho(x_1, R_1)$.　　　(c) Two disks $\rho(x_1, R_1)$ and $\rho(x_2, R_2)$.

Fig. 1. Example of a network $G = (V, E)$, nodes s and t, a region $\mathcal{X}$ and different interdiction cases. (a) No failure disk, (b) A disk $\rho(x_1, R_1)$, (c) Two disks $\rho(x_1, R_1)$ and $\rho(x_2, R_2)$.

transportation time between s and t induced by a failure located in the *worst* location $x^* = \arg_{x \in \mathcal{X}} \{\Omega\}$. (ii) From the tactical point of view, *preventive* actions can be taken in order to reduce the chances that a failure can be produced at x^* or the edges $\bar{E}_{x^*}$ can be reinforced to increase their reliability. And (iii) we can know whether the network is *so* vulnerable that s and t might be disconnected, which can be verify if $\Omega = \infty$.

The problem of calculating Ω will be called the Max-Cost Single-Failure Shortest Path Problem (MCSFSPP). Therefore, the MCSFSPP is an optimization problem whose objective function value is a vulnerability measure of the network on which it is solved. Intuitively, the MCSFSPP can be solved as follows. For a given $x \in \mathcal{X}$, let ℓ_x be the cost of the shortest s, t-path on G_x with edge costs l_e^x defined as $l_e^x = l_e$ if $e \in E_x$ and $l_e^x = M$ if $e \in \bar{E}_x$, with $M = O(m \max_{e \in E} l_e)$; therefore, $\Omega = \max_{x \in \mathcal{X}} \ell_x$. If $\Omega > M$ then there is at least one x for which s and t cannot be connected.

In Fig. 1(a) it is shown a network $G = (V, E)$ where s and t correspond to the nodes represented with triangles and $\mathcal{X}$ is represented by a grid of 8×7 points in the background of part of G; an optimal s, t-path is shown with bold edges. In Fig. 1(b) we show the case where a disruption disk $\rho(x_1, R_1)$ interdicts the network such that an alternative (an more expensive) s, t-path has to be established ($\Omega < M$). And in Fig. 1(c) a more complex situation is shown; here two disruption disks, $\rho(x_1, R_1)$ and $\rho(x_2, R_2)$, are simultaneously interdicting the network. In the latter case all possible s, t-paths (one of them is shown in bold dashed lines) have at least one interdicted edge, i.e., $\Omega > M$.

The MCSFSPP is closely related with the network interdiction problems studied in [3–5,7,8,11]. In the following, we will use this basic definition to construct generalizations addressing different, but complementary, measures of vulnerability under different models of failure.

Mixed Integer Programming Formulation for the MCSFSPP. Let $\mathbf{f} \in [0,1]^m$ be a vector of $[0,1]$-flow variables. An s,t-path p in G is induced by a given allocation of flows $\mathbf{f}$ if the following constraints are satisfied:

$$\sum_{k \in V | e:\{j,k\} \in E} f_{j,k} - \sum_{i \in V | e:\{i,j\} \in E} f_{i,j} = \begin{cases} 1, & \text{if } j = s \\ 0, & \text{if } j \in V \setminus \{s,t\} \\ -1, & \text{if } j = t. \end{cases} \quad \text{(SP.1)}$$

For a given $x \in \mathcal{X}$, the problem of finding ℓ_x can be defined as

$$\ell_x = \min \left\{ \sum_{e \in E} l_e^x f_e \mid \text{(SP.1) and } \mathbf{f} \in [0,1]^m \right\}. \quad (\ell_x)$$

Let $\mathbf{y} \in \{0,1\}^{|\mathcal{X}|}$ be a vector of binary variables such that $y_x = 1$ if the failure disc is centered at x and $y_x = 0$ otherwise. Now, let $\mathbf{z} \in \{0,1\}^m$ be a set of binary variables such that $z_e = 1$ if edge e is *operative* and $z_e = 0$ otherwise for any given $x \in \mathcal{X}$. Variables $\mathbf{y}$ and $\mathbf{z}$ are related as follows

$$y_x + z_e \leq 1, \ \forall e \in E \mid d(e,x) \leq R, \ \forall x \in \mathcal{X} \quad \text{(YZ.1)}$$

$$\sum_{x \in \mathcal{X} | d(e,x) > R} y_x - z_e \leq 0, \ \forall e \in E. \quad \text{(YZ.2)}$$

Constraints (YZ.1) and (YZ.2) state that, for any $x \in \mathcal{X}$, an edge e has to be operative ($z_e = 1$) if is not reached by the disruption disk $\rho(x, R)$. Since a single disruption disk affects the network, we have that

$$\sum_{x \in \mathcal{X}} y_x = 1. \quad \text{(YZ.3)}$$

Using (YZ.1) and (YZ.2), for a given $x \in \mathcal{X}$ the edge costs l_e^x can be written as $l_e^x = l_e z_e + (1 - z_e)M, \ \forall e \in E$. Hence, the MCSFSPP is as follows

$$\Omega = \max_{x \in \mathcal{X}} \left\{ \ell_x \mid \text{(YZ.1), (YZ.2), (YZ.3) and } (\mathbf{y}, \mathbf{z}) \in \{0,1\}^{m+|\mathcal{X}|} \right\}. \quad (\Omega)$$

Problem (Ω), as it is, is non-linear. To linearize it, we will convert the max min objective into a pure max one; to do so, let us consider the dual of (ℓ_x), which is given by

$$\ell_x = \max \{ \gamma_t - \gamma_s \mid \gamma_j - \gamma_i \leq l_{ij} z_{ij} + (1 - z_{ij})M, \ \forall e : \{i,j\} \in E \text{ and } \gamma \in \mathbb{R}^n \}. \quad (\lambda)$$

Embedding (λ) into (Ω), we get the next Mixed Integer Programming Formulation (MILP) formulation for the MCSFSPP:

$$\Omega = \max \quad \gamma_t - \gamma_s \quad \text{(MCSF.1)}$$
$$\text{s.t} \quad \text{(YZ.1), (YZ.2), and (YZ.3)} \quad \text{(MCSF.2)}$$
$$\gamma_j - \gamma_i \leq l_{ij} z_{ij} + (1 - z_{ij})M, \ \forall e : \{i,j\} \in E \quad \text{(MCSF.3)}$$
$$(\mathbf{y}, \mathbf{z}) \in \{0,1\}^{m+|\mathcal{X}|} \text{ and } \gamma \in \mathbb{R}^n. \quad \text{(MCSF.4)}$$

Note that in our approach we assume that $\rho(x, R)$ can be located in any point $x \in \mathcal{X}$ without any stochastic characterization. That is, any point $x \in \mathcal{X}$ is likely to "host" the center of the failure.

In the proposed setting we assume that if an edge e is disrupted by at least one failure disk $\rho(x, R)$, then it becomes inoperative. However, one can easily extend this to a more general case by defining a coefficient $d_e \geq 0 \ \forall e \in E$ representing the *delay* on edge e in case of interdiction (in our setting $d_e = M$ $\forall e \in E$). The MCSFSPP can be redefined by replacing constraint (MCSF.3) with

$$\gamma_j - \gamma_i \leq l_{ij} + (1 - z_{ij})d_{ij}, \ \forall e : \{i, j\} \in E. \tag{MCSF.3b}$$

The Shortest-Path Network Interdiction problem presented in [8] is very similar to the definition of the MCSFSPP using (MCSF.3b) instead of (MCSF.3). In that problem, edges can be interdicted without any geometrical pattern among them; instead, they consider interdiction costs so that any feasible disruption of the network should not cost more than a given interdiction *budget*. Later we formally define these concepts and adapt them to our setting.

2.2 The Multiple Failures Case

As described above, in the MCSFSPP only a single failure $\rho(x, R)$ occurs. However, there are applications in which this characteristic does not hold and, instead, multiple failures occur simultaneously. More precisely, we now have that k failure disks $\rho(x_1, R), \ldots, \rho(x_k, R)$ of radius R are located in $\mathcal{X}$, resulting in an operative network $G_{\mathbf{x}^k} = (V, E_{\mathbf{x}^k})$ where $E_{\mathbf{x}^k} = \{e \in E \mid \min_{x \in \{x_1, \ldots, x_k\}} d(e, x) \leq R\}$. Under these conditions, finding the maximum cost, across all possible $\{x_1, \ldots, x_k\} \in \mathcal{X}^k$, of the shortest s, t-path on $G_{\mathbf{x}^k}$ can be done by modifying MCSFSPP as follows. Instead of (YZ.3), we have

$$\sum_{x \in \mathcal{X}} y_x = k. \tag{YZ.3k}$$

Besides, constraint (YZ.2) should be now adapted in order to impose that $z_e = 1$ if none of the k failure disks reaches e; the new constraint is

$$\sum_{x \in \mathcal{X} \mid d(e, x) > R} y_x - z_e \leq 1 - \sum_{x \in \mathcal{X}} y_x, \ \forall e \in E, \tag{YZ.2k}$$

clearly if $k = 1$, then (YZ.2k) corresponds to (YZ.2). Therefore, the Max-Cost Multiple-Failure Shortest Path Problem (MCMFSPP) can be formulated as

$$\Omega^k = \max\{\gamma_s - \gamma_t \mid (YZ.1), \ (YZ.3k), \ (YZ.2k), \ (MCSF.3) \text{ and } (MCSF.4)\} \tag{MCMF}$$

Note that in formulation (MCMF) it is assumed that $R \in \mathcal{R}$ is known in advance.

Maximal Disruption for an interdiction budget. Similar as in [3,7,8], let us consider that associated with each point $x \in \mathcal{X}$ there is a disruption *cost* $c_x > 0$. Assume that the interdictors have a *budget* B of interdiction *resources*, so that they can disrupt the network using several disks $\rho(x, R)$ as long as the total cost does not exceed B. Formally, the interdiction-budget constraint is given by

$$\sum_{x \in \mathcal{X}} c_x y_x \leq B; \tag{IB}$$

so the Budget Constrained MCMFSPP is formulated as

$$\Omega^k = \max\{\gamma_s - \gamma_t \mid (\text{YZ.1}), (\text{YZ.2k}), (\text{IB}), (\text{MCSF.3}) \text{ and } (\text{MCSF.4})\} \tag{B}$$

By solving (B) we can know *how* vulnerable the network is if the interdictors are able to optimally use their resources to disrupt it. Models as the one presented in [7,8] are particular cases of (B) in which $\mathcal{X}$ coincides with the midpoint of every edge $e \in E$ and $R = \epsilon$ (ϵ being infinitesimally small).

Minimum Simultaneity for Complete Vulnerability: Critical k. One might be interested in knowing the minimum number of failures (the *critical k* or k^c) that should occur simultaneously in order to have at least one set $\rho(x_1, R), \ldots, \rho(x_k, R)$ that damages the network so that s and t cannot be connected anymore or the shortest length between them is greater than a threshold Θ.

The value k^c and the corresponding collection $\{x_1, \ldots, x_{k^c}\}$ will enable a decision maker to perform more general preventive actions to endure the network not in a single but a in several areas. In many practical contexts, the possibility of multiple and synchronized failures might be the rule, so knowing k^c might play a strategical role. Clearly, for a given R, the larger k^c is the more robust the network is. Mathematically, one can formulate the search for k^c as

$$k^c = \min\{k \mid (\text{YZ.1}), (\text{YZ.2K}), (\text{YZ.3K}), (\text{MCSF.3}), (\text{MCSF.4}), \tag{k^c}$$
$$\gamma_s - \gamma_t \geq \Theta \text{ and } k \in \mathbb{Z}_{\geq 0}\}$$

If $\Theta = M$, then (k^c) aims at finding the minimum k such that allocating k disks produces a disconnection between s and t. A similar model is presented in [7] in the context of interdiction in stochastic networks.

If instead of k^c one is interested in knowing the minimum *cost* needed to produce a damage represented by Θ, model (k^c) can be easily modified by replacing the objective function of (k^c) with $C^c = \min \sum_{x \in \mathcal{X}} c_x y_x$.

3   Computational Results

3.1   Instance Benchmark and Solver Setting

Instance Benchmark. For our experiments we consider three sets of instances: ND, US and Bangladesh.

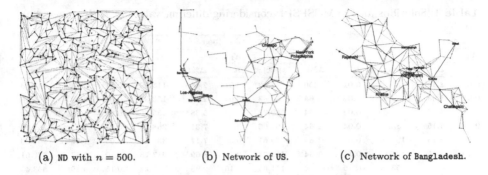

(a) ND with $n = 500$.   (b) Network of US.   (c) Network of Bangladesh.

Fig. 2. Representation of the instances used for computations.

In the first set, the instances are generated as follows: (i) n points are randomly located in a unit Euclidean square; (ii) a minimum spanning tree connecting all points is calculated; (iii) $\beta \times n$ additional edges are added to the network such that an edge is added if l_{ij} (euclidean distance) satisfies $l_{ij} \leq \alpha/\sqrt{n}$ and the planarity of the network is still preserved; (iv) the set $\mathcal{X}$ is created by randomly located K points within the are area defined by points (x_1, y_1), (x_2, y_1), (x_1, y_2) and (x_2, y_2).

For experiments we have considered $n \in \{500, 1000\}$, $\beta = 1.5$, $\alpha = 1.6$, $(x_1, x_2, y_1, y_2) = (0.3, 0.7, 0.0, 1.0)$ ($\mathcal{X}_1$) and $(x_1, x_2, y_1, y_2) = (0.1, 0.9, 0.1, 0.9)$ ($\mathcal{X}_2$), and $K = 100$.

In Fig. 2(a) it is shown an example of an instance with 500 nodes and $\mathcal{X}$ contained in $(0.3, 0.0)$, $(0.7, 0.0)$, $(0.3, 1.0)$ and $(0.7, 1.0)$.

In the case of groups US and Bangladesh we consider the geographical coordinates of the most populated cities in each case (see [13]) to define the set V. Then, we used an approximation of their highway and interurban road system with the information available in [6] to approximate the set of edges E. The set $\mathcal{X}$ is created by randomly located K points within the are area defined by points (x_1, y_1), (x_2, y_1), (x_1, y_2) and (x_2, y_2). In Fig. 2(b) and (c) we show the networks used to generate the instances US and Bangladesh respectively. In the case of US, the area $\mathcal{X}$ is given by placing 100 points in the so-called *south* area. With this we intend to represent possible cases of failure produced by hurricanes and other natural disasters. For the Bangladesh instances, we have created $\mathcal{X}$ by placing 100 points in squared area in the very center that covers around the 15 % of the total area.

In the case of instances ND, nodes s and t are selected as those with the longest euclidean distance. In the case of instances US we have used $s \in \{$NY:New York, CH:Chicago$\}$ and $t \in \{$LA:Los Ángeles, HS:Houston$\}$; likewise, in the case of instances Bangladesh we have used $s = $ Rajshahi and $t = $ Silhat.

Solver Setting. Models (MCSF.1)–(MCSF.4), (MCMF) and (k^c) were solved using CPLEX 12.5 (all CPLEX parameters were set to their default values). The experiments were performed on a Intel Core i7-3610QM machine with 8 GB RAM.

Table 1. Solutions for the MCSFSPP considering different values of R (Instances ND)

$n = 500$						$n = 1000$					
$\mathcal{X}_1$			$\mathcal{X}_2$			$\mathcal{X}_1$			$\mathcal{X}_2$		
R	$\Delta\%\Omega$	t[sec]	R	$\Delta\%\Omega$	t[sec]	R	$\Delta\%\Omega$	t[sec]	R	$\Delta\%\Omega$	t[sec]
0.01	2.17	38.92	0.01	0.00	32.93	0.01	2.48	115.39	0.01	0.14	144.66
0.02	3.93	46.46	0.02	1.64	36.83	0.02	2.93	153.58	0.02	0.69	215.16
0.03	3.93	65.63	0.03	1.64	49.73	0.03	5.38	235.95	0.03	1.52	240.88
0.04	5.15	80.79	0.04	1.64	63.06	0.04	7.31	258.85	0.04	1.52	265.00
0.05	5.15	103.01	0.05	1.64	79.67	0.05	7.17	395.46	0.05	1.52	259.80
0.10	5.15	97.83	0.10	1.64	112.76	0.10	8.69	917.78	0.10	3.87	373.61
0.15	-	53.70	0.15	10.64	111.65	0.15	9.93	587.27	0.15	6.19	843.45

3.2   Vulnerability Assessment of Spatial Networks: Solutions

From the operative perspective, the value of R corresponds to the *intensity* of a disruption. If we consider the MCSFSPP or the MCMFSPP we would expect that a *vulnerable* network is such that Ω increases quickly (up to M) when R increases marginally. On the other hand, a *reliable* network is such that the cost of the shortest s,t-path does not change *too much* even if R increases considerably.

In Table 1 we report solutions for the MCSFSPP for instances of group ND considering different values of n, different compositions of set $\mathcal{X}$ and different values of R (columns 1, 4, 7 and 10). In columns $\Delta\%\Omega$ is reported the relative increase of Ω, for a given $\mathcal{X}$ and a given R, with respect to cost of the shortest s,t-path without any failure. In this column, "-" means that all paths have been disrupted. In columns t[sec] are reported the running times in seconds needed to reach optimality. One can observe from this table that when the area where the failure can occur, $\mathcal{X}$, is such that covers a *stripe* on the network (as $\mathcal{X}_1$) then it is more vulnerable (see the values $\Delta\%\Omega$ for different R) than a network in which the failure area, although larger, still leaves *corridors* where s,t-paths can be constructed, as for $\mathcal{X}_2$. In a warfare context, if we were to be the enemies, this analysis would suggest us that is better to concentrate our resources in a narrower area potentially spanning a complete stripe of the network than in a larger area (which might be more expensive) that does not properly covers the network. On the other hand, who wants to protect the network should concentrate the efforts in protecting at least one corridor connecting s and t.

In Tables 2 and 3, results for (MCMF) and (k^c), respectively, are reported. The analysis is similar as for Table 1. From Table 2 we can see that the increase of $\Delta\%\Omega$ (due to a larger k), is greater for $\mathcal{X}_1$ than for $\mathcal{X}_2$. Along the same lines, we see from Table 3 that the minimum resources needed to disconnect s and t (see columns k^c) are greater for $\mathcal{X}_2$ than for $\mathcal{X}_1$. In Table 3, when results for a given R are not reported (e.g., $R = 0.01$ for $n = 500$ and $\mathcal{X}_1$) is because not even $|\mathcal{X}|$ failure disks are enough to make the s,t connectivity collapse. This applies for all the remaining Tables.

From the algorithmic point of view, we can notice in Tables 1, 2 and 3 that the search for an alternative path in a disrupted network is not *for free*. In all

Table 2. Solutions for the MCMFSPP considering different values of R and k (Instances ND)

500								1000							
$\mathcal{X}_1$				$\mathcal{X}_2$				$\mathcal{X}_1$				$\mathcal{X}_2$			
R	k	$\Delta\%\Omega$	t[sec]	R	k	$\Delta\%\Omega$	t[sec]	R	k	$\Delta\%\Omega$	t[sec]	R	k	$\Delta\%\Omega$	t[sec]
0.01	1	2.17	24.15	0.01	1	0.00	22.51	0.01	1	2.48	88.16	0.01	1	0.01	88.64
	2	3.93	25.07		2	0.00	21.96		2	3.03	94.80		2	0.09	96.35
	3	4.74	25.02		3	0.00	22.11		3	3.03	93.90		3	1.63	93.57
	4	5.56	24.87		4	0.00	22.21		4	3.03	95.08		4	1.63	95.08
	5	6.79	24.52		5	0.00	22.14		5	3.03	94.15		5	1.63	94.4
0.1	1	5.15	59.94	0.1	1	1.64	69.61	0.1	1	8.68	889.61	0.1	1	3.87	222.36
	2	-	26.43		2	13.82	142.93		2	17.89	3448.93		2	8.16	1012.06
	3	-	98.05		3	13.92	297.95		3	21.24	75319.9		3	12.45	5434.01
	4	-	526.47		4	-	63.01		4	28.14	26123.3		4	21.99	13130.3
	5	-	147.39		5	-	149.54		5	-	16576.5		5	-	742.63

Table 3. Solutions of (k^c) considering different values of R (Instances ND)

n	$\mathcal{X}$	R	k^c	t[sec]	n	$\mathcal{X}$	R	k^c	t[sec]
500	$\mathcal{X}_1$	0.10	2	170.76	1000	$\mathcal{X}_1$	0.10	5	628.76
		0.15	1	43.09			0.15	2	805.32
	$\mathcal{X}_2$	0.03	10	39.05		$\mathcal{X}_2$	0.02	19	219.38
		0.04	9	109.70			0.03	13	621.60
		0.05	7	161.06			0.04	11	933.85
		0.10	4	274.20			0.05	8	1759.32
		0.15	3	162.47			0.10	5	1277.24
							0.15	3	1214.14

cases we see an increase of the algorithmic effort (time) needed to find such a path (if exists). This is due to the high combinatorial nature of the problem when more edges are subject to be interdicted (when R increases and/or when k is either greater than 1 or when it is a variable).

In the case of USA Instances, we report in Table 4 results of the MCSFSPP considering different pairs of s and t and different values of R. In this case, we can see that different combinations of s and t yield to different levels of vulnerability in the system. For instance, the network is considerably more vulnerable when it is intended to host a path from Chicago to Los Ángeles than when the path should be established from Chicago to Houston. This is due to the fact that, in our instance, the system of roads connecting the north of the Midwest with the south of the West Coast is composed by relatively few elements. Hence, a single disruption disk (that is optimally placed) is enough to interrupt the communication between the cities. In this case the values of $\Delta\%\Omega$ are

Table 4. Solutions for the MCSFSPP considering different values of R (Instances USA)

{NY, LA}			{CH, LA}			{NY, HS}			{CH, HS}		
R	$\Delta\%\Omega$	t[sec]	R	$\Delta\%\Omega$	t[sec]	R	$\Delta\%\Omega$	t[sec]	R	$\Delta\%\Omega$	t[sec]
0.01	9.00	17.22	0.01	0.00	17.43	0.01	13.00	17.25	0.01	0.00	17.22
0.02	10.00	21.42	0.02	20.00	25.05	0.02	13.00	26.83	0.02	2.00	20.12
0.03	10.00	24.77	0.03	20.00	33.17	0.03	15.00	22.25	0.03	2.00	35.27
0.04	10.00	23.57	0.04	20.00	31.29	0.04	18.00	23.07	0.04	2.00	28.31
0.05	10.00	25.30	0.05	20.00	35.51	0.05	19.00	25.04	0.05	8.00	36.15
0.10	30.00	34.16	0.10	-	33.67	0.10	45.00	42.31	0.10	-	30.67
0.15	-	40.22	0.15	-	28.41	0.15	-	29.69	0.15	-	28.17
0.20	-	28.00	0.20	-	48.14	0.20	-	27.16	0.20	-	37.13

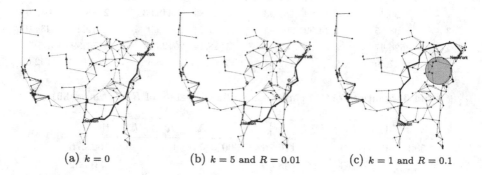

(a) $k = 0$ (b) $k = 5$ and $R = 0.01$ (c) $k = 1$ and $R = 0.1$

Fig. 3. Solutions for the MCMFSPP for different k and R (Instances USA), (a)$k = 0$, (b)$k = 5$ and $R = 0.01$, (c)$k = 1$ and $R = 0.1$

particularly important from the tactic point of view; if it is up to the decision maker to decide where to establish both the source and the target of the transportation system, then it might preferable to have New York - Houston than, for instance, Chicago - Los Ángeles. However, this analysis is valid only when a single failure occurs. For an approximate equivalence to real distances in kilometers, R should be multiply by 1700.

In Fig. 3(a) we show the solution of the shortest path problem between New York and Houston when there is no disruption. In Fig. 3(b) is shown the solution of the MCMFSPP when 5 disruption disks with $R = 0.01$ are optimally located. In Fig. 3(b) is shown the solution of the MCMFSPP with $k = 1$ and $R = 0.10$. These figures show how different the optimal s, t-paths can be when the network is disrupted by failures of different magnitude.

Finally, in Table 5 we report results for the Instances Bangladesh. From the solutions of the MCSFSPP (reported in columns 1–3) we can see that the relatively dense road system of this country is able to *resist* (small values of $\Delta\%\Omega$), reasonably well the optimal location of a single failure disk up to $R = 0.05$. For greater values, the network can be dramatically damaged. This later observation is reinforced by the results reported in columns 4–6 in the same table: a critical k can be found only if $R \geq 0.05$. When looking at the results

Table 5. Solutions for MCSFSPP, MCSFMPP and k^c, s =Rajshahi and t =Silhat (Instances **Bangladesh**)

							$R = 0.01$			$R = 0.1$	
R	$\Delta\%\Omega$	t[sec]	R	k^c	t[sec]	k	$\Delta\%\Omega$	t[sec]	k	$\Delta\%\Omega$	t[sec]
0.01	3.76	8.72	0.05	3	10.06	1	3.76	7.69	1	24.17	8.11
0.02	2.93	7.44	0.10	2	28.52	2	5.48	8.75	2	-	38.28
0.03	3.76	8.99	0.15	1	24.01	3	5.48	8.42	3	-	26.75
0.04	3.76	9.53				4	5.48	8.41	4	-	28.32
0.05	4.78	19.03				5	5.48	7.64	5	-	10.55
0.10	24.17	36.97									
0.15	-	11.19									

of the MCSFMPP (columns 7–9 for $R = 0.01$ and 10–12 for $R = 0.1$) we can see that the network resists *well* ($\Delta\%\Omega \approx 5\,\%$) several failures with $R = 0.01$; however, if $R = 0.1$ then the network collapses even if $k = 2$.

4   Conclusions and Future Work

We have presented a collection of combinatorial optimization problems that in combination allow to measure the vulnerability of a network. Vulnerability is represented by the relative increase of the cost of a s, t-shortest path when part of the network is disrupted. By analyzing the solutions of these problems for different instances, we have highlighted how different aspects of both the failure and the network yield to different levels of vulnerability.

Two main paths of future work can be identified. First, we should consider the case in which $\mathcal{X}$ is not given by a discrete set of points, but rather as continuous area. Second, at the light of the large computational effort needed to solve some of the instance considered here, we think it is important to design and implement more sophisticated algorithmic techniques such as decomposition approach in order to be able to consider larger and more complex instances.

Acknowledgements. This research was supported by Fondecyt Project #1121095, CONICYT, Ministerio de Educación, Chile. Eduardo Álvarez-Miranda thanks the Institute of Advanced Studies of the Università di Bologna from where he is a PhD Fellow.

References

1. Assimakopoulos, N.: A network interdiction model for hospital infection control. Comput. Biol. Med. **17**(6), 413–422 (1987)
2. Church, R., Scaparra, M., Middleton, R.: Identifying critical infrastructure: the median and covering facility interdiction problems. Ann. Assoc. Am. Geogr. **94**(3), 491–502 (2004)

3. Cormican, K., Morton, D., Wood, R.: Stochastic network interdiction. Oper. Res. **46**(2), 184–197 (1998)
4. Fulkerson, D., Harding, G.: Maximizing the minimum source-sink path subject to a budget constraint. Math. Program. **13**(1), 116–118 (1977)
5. Golden, B.: A problem in network interdiction. Nav. Res. Logist. Q. **25**(4), 711–713 (1978)
6. Google. Google Maps (2013)
7. Hemmecke, R., Schültz, R., Woodruff, D.: Interdicting stochastic networks with binary interdiction effort. In: Woodruff, D. (ed.) Network Interdiction and Stochastic Integer Programming. Operations Research/Computer Science Interfaces Series, vol. 22, pp. 69–84. Springer, New York (2003)
8. Israeli, E., Wood, R.: Shortest-path network interdiction. Networks **40**(2), 97–111 (2002)
9. Khachiyan, L., Boros, E., Borys, K., Elbassioni, K., Gurvich, V., Rudolf, G., Zhao, J.: On short paths interdiction problems: total and node-wise limited interdiction. Theor. Comput. Syst. **43**(2), 204–233 (2008)
10. Murray, A.: An overview of network vulnerability modeling approaches. GeoJournal **78**(2), 209–221 (2013)
11. Phillips, C.: The network inhibition problem. In: Proceedings of the Twenty-Fifth Annual ACM Symposium on Theory of Computing, STOC'93, pp. 776–785. ACM (1993)
12. Salmeron, J., Wood, K., Baldick, R.: Worst-case interdiction analysis of large-scale electric power grids. IEEE Trans. Power Syst. **24**(1), 96–104 (2009)
13. United Nations Statistics Division. UNSD Statistical Databases (2013)
14. Wollmer, R.: Removing arcs from a network. Oper. Res. **12**(6), 934–940 (1964)

Author Index

Álvarez-Miranda, Eduardo   433
Amaldi, Edoardo   1
Argiroffo, Gabriela   13

Baffier, Jean-François   26
Baïou, Mourad   38
Barahona, Francisco   38
Bartolini, Sandro   50
Baumann, Frank   62
Bender, Marco   75
Bianchi, Silvia   13
Bogue, Eduardo T.   87
Bonomo, Flavia   100
Brčić, Mario   112
Buchheim, Christoph   62, 125
Bulteau, Laurent   137

Candia-Véjar, Alfredo   433
Carrieri, Anna Paola   137
Carrizosa, Emilio   433
Casini, Iacopo   50
Coniglio, Stefano   1

Darties, Benoit   161
da Cunha, Alexandre Salles   149
de Campos, Cassio Polpo   173
de Souza, Cid C.   87
de Souza, Fernanda Sumika Hojo   149
Della Croce, Federico   186, 198
Detti, Paolo   50
Disser, Yann   208
Dondi, Riccardo   137

Escalante, Mariana   221

Fernandes, Cristina G.   230
Ferreira, Carlos E.   230
Fortz, Bernard   242
Franco, Álvaro J.P.   230
Freire, Alexandre, S.   87
Freivalds, Kārlis   255
Fürer, Martin   408
Furini, Fabio   267

Garraffa, Michele   186
Giroudeau, Rodolphe   161
Glagoļevs, Jans   255
Gouveia, Luís   242

Hifi, Mhand   396
Hinrichsen, Erica G.   325
Hiraishi, Hidefumi   26
Hoeksma, Ruben   280

Ilyina, Anna   62
Imai, Hiroshi   26
Ishii, Toshimasa   292

Kacem, Imed   305
Kalpić, Damir   112
Katić, Marija   112
Kellerer, Hans   305
Kidd, Martin Philip   267
König, Jean-Claude   161
Koulamas, Christos   198
Kristianto, Yohanes   315

Leitner, Markus   371
Leoni, Valeria Alejandra   325
Ljubić, Ivana   371

Manthey, Bodo   280
Mattia, Sara   336
Méndez Díaz, Isabel   347
Meshoul, Souham   421
Mihalák, Matúš   208
Moniz, Martim   242
Montanari, Sandro   208

Nasini, Graciela   221, 347

Ono, Hirotaka   292

Pérez-Galarce, Francisco   433
Persiani, Carlo Alfredo   267
Prins, Christian   384

Quilliot, Alain   359

Rebaine, Djamal   359
Rossi, Fabrizio   336

Salassa, Fabio   186
Salazar-González, Juan-José   371
Saleh, Sagvan   396
Santos, Andréa Cynthia   384
Schaudt, Oliver   100
Schouery, Rafael C.S.   230
Seifaddini, Maryam   305
Servilio, Mara   336
Severín, Daniel   347
Simonin, Gilles   161
Sinnl, Markus   371
Smriglio, Stefano   336
Solano-Charris, Elyn L.   384
Stamoulis, Georgios   173
Stein, Maya   100
Suppakitpaisarn, Vorapong   26

Taccari, Leonardo   1
T'kindt, Vincent   198
Toth, Paolo   267
Trieu, Long   125

Uetz, Marc   280
Uno, Yushi   292

Valencia-Pabon, Mario   100

Wagler, Annegret   13
Westphal, Stephan   75
Weyland, Dennis   173
Widmayer, Peter   208
Wu, Lei   396

Xavier, Eduardo C.   87

Yu, Huiwen   408

Zeltni, Kamel   421

Printed in the United States
by Bookmasters

Printed in the United States
By Bookmasters